D0320870

SECOND EDITION

Combustion Engineering

SECOND EDITION

Combustion Engineering

Kenneth W. Ragland
Kenneth M. Bryden

CRC Press
Taylor & Francis Group
Boca Raton   London   New York

CRC Press is an imprint of the
Taylor & Francis Group, an **informa** business

CRC Press
Taylor & Francis Group
6000 Broken Sound Parkway NW, Suite 300
Boca Raton, FL 33487-2742

© 2011 by Taylor & Francis Group, LLC
CRC Press is an imprint of Taylor & Francis Group, an Informa business

No claim to original U.S. Government works

Version Date: 20110829

International Standard Book Number: 978-1-4200-9250-9 (Hardback)

Library of Congress Cataloging-in-Publication Data

Combustion engineering / editors, Kenneth W. Ragland, Kenneth M. Bryden. -- 2nd ed.
    p. cm.
  Includes bibliographical references and index.
  ISBN 978-1-4200-9250-9 (hardcover : alk. paper)
  1. Combustion engineering. I. Ragland, Kenneth W. II. Bryden, Kenneth M.

TJ254.5.B67 2011
621.402′3--dc23   2011017384

Visit the Taylor & Francis Web site at
http://www.taylorandfrancis.com

and the CRC Press Web site at
http://www.crcpress.com

To our wives Nancy and Kristy

Contents

PART II   Combustion of Gaseous and Vaporized Fuels

PART III   *Combustion of Liquid Fuels*

PART IV   Combustion of Solid Fuels

Preface to Second Edition

Since the first edition of *Combustion Engineering* was published in 1998, worldwide consumption of fossil and biomass fuels has increased by 20%. Additional population growth and unmet human needs continue to drive the demand for more energy. Issues of global climate change and environmental sustainability have added more urgency to the quest to develop renewable energy. The need for improved combustion systems for heat, power, and transportation using fossil fuels and biomass is more urgent than ever.

The goal of the book remains the same: to provide a balance of combustion fundamentals and combustion engineering applications for seniors and first-year graduate engineering students that builds on the students' knowledge of chemistry, thermodynamics, and fluid mechanics. The breadth of scope also remains the same: combustion of gaseous, liquid, and solid fuels in applications needed by all communities on planet Earth.

We have endeavored to improve the clarity and the accuracy of the book. We have updated the applications and the references. More information has been added on efficiency improvements, emission reductions, and biomass utilization. Analysis using conservation of mass and energy for one-dimensional, reacting homogenous and two-phase systems in differential form has been retained, while the discussion of computational fluid dynamics of reacting systems has been reduced because the authors believe this subject is best done in a separate advanced course.

Sadly the co-author of the first edition, Professor Gary Borman, has passed away. The current authors wish to honor his memory.

Kenneth W. Ragland

Kenneth M. Bryden

Preface to First Edition

Currently, 90% of the energy used for transportation, power production and heating is produced by combustion of liquid, solid and gaseous fuels. Although serious shortages of crude oil are certain to occur during the early part of the next century, it is unlikely that this percentage will change significantly for many years. Thus the study of combustion is of continuing importance, especially if we are to conserve our sources of energy and reduce air pollution in a world of increasing population and increasing energy needs.

The engineer intending to study combustion will find that current sources of information fall into two categories—literature on the scientific aspects of combustion, and literature on the design and performance of specific technologies such as engines, turbines and furnaces. Although a considerable number of books on combustion are available, they tend to emphasize the scientific aspects with little reference to specific engineering applications. Books on applications, however, typically focus on one technology and contain only one chapter on combustion as applied to that technology.

The authors feel that a need exists for a book that bridges the gap between the scientific monograph and the specific technology texts. This book attempts to meet this need by presenting a broad coverage of combustion technology with enough accompanying combustion theory to allow a rudimentary understanding of the phenomena. The level of the text is such that a senior in mechanical or chemical engineering should have little difficulty in comprehension. Mathematical treatment has been kept to a minimum, although modeling concepts have been emphasized and state-of-the-art of theoretical approaches have been explained in simple terms so as to indicate the current possibility of theoretical solutions to specific design questions.

This book is divided into four sections, covering: I—basic concepts, II—gaseous fuel combustion, III—liquid fuel combustion, and IV—solid fuel combustion. Basic concepts pertaining to fuels, thermodynamics, chemical kinetics, flames, detonations, sprays, and particle combustion are presented. The basic concepts are applied to gas fired furnaces, gasoline engines, oil fired furnaces, gas turbine combustors, diesel engines, fixed bed combustors, suspension burners and fluidized bed combustors. The application chapters describe how the combustion system works, evaluate the thermodynamic and fluid dynamic states of the system, provide a physical description and simplified model of how the combustion proceeds in the system, and discuss the emissions from the system. In several chapters computational fluid dynamic modeling (CFD) is discussed to indicate the rich possibilities of this approach for combustion systems. Because the mathematical aspects of CFD are beyond the scope of this book, the theory is presented only in a heuristic fashion.

Ample text material and homework problems are provided for a three-credit course covering 45 periods of 50 minutes each over a 15 week semester. Our approach in teaching the course has been to give equal time to each of the four sections because we feel that at the senior/MS level breadth is important. However, some instructors may prefer to emphasize one section over another. Although the material has been prepared for use as a classroom text, it is hoped that the practicing engineer will also find the material useful and the treatment amenable to self-study.

The authors wish to thank the many students and faculty who have provided valuable input along the way as this text has evolved. We especially thank Dave Foster, Glen Myers, Phil Myers, Rolf Reitz, Chris Rutland, Mark Bryden and Danny Aerts at the University of Wisconsin-Madison, Dave Hofelt at the University of Minnesota, Eric Van den Bulck at the University of Leuven in Belgium, Duane Abata at Michigan Technological University, and Bill Brown at Caterpillar Corp. Sally Radeke faithfully did much of the word processing during the extended period of manuscript preparation.

Gary L. Borman

Kenneth W. Ragland

Acknowledgments

In addition to the many faculty and students who were acknowledged in the first edition, we especially wish to thank the following people for generously making available new concepts and test results for this edition: Professor Jeff Naber and Yeliana Yeliana at Michigan Technological University; Professors Dave Foster, Jaal Ghandhi, Rolf Reitz, and Chris Rutland at the University of Wisconsin–Madison; Dr. Robert Cheng at the U.S. DOE Lawrence Berkeley National Laboratory; and David Ostlie at Energy Performance Systems Inc. Finally, we wish to thank Dr. Kris Bryden at Iowa State University for managing the editing and review of the final manuscript, as well as our editors at Taylor & Francis Group, Jonathan Plant and Glen Butler.

Authors

Dr. Kenneth Ragland is an emeritus professor of mechanical engineering at the University of Wisconsin–Madison. Throughout his career, he taught courses in thermodynamics, fluid dynamics, combustion, and air pollution control. His early research was on solid fuel ram jet combustion, and gaseous and heterogeneous detonations. His research at UW–Madison focused on solid fuel combustion of coal and biomass as single particles, combustion in shallow and deep fixed beds, fluidized bed combustion, and combustion emissions. He served as chair of the Department of Mechanical Engineering from July 1995 until his retirement in July 1999. In retirement, his research has focused on the development of systems for planting, harvesting, and combusting biomass crops for energy. Currently, he is vice president of Energy Performance Systems Inc.

Dr. Kenneth "Mark" Bryden joined the faculty of the Mechanical Engineering Department at Iowa State University in 1998 after receiving his doctoral degree in mechanical engineering from the University of Wisconsin–Madison. Prior to his studies at the University of Wisconsin–Madison, he worked for 14 years in a wide range of engineering positions at Westinghouse Electric Corporation. This included 8 years in power plant operations and 6 years in power plant engineering, of which more than 10 years were spent in engineering management. Mark has an active research and teaching program in the areas of energy, combustion, and appropriate technology. He is particularly interested in biomass combustion and small cookstoves for the developing world. He is president of Engineers for Technical and Humanitarian Opportunities for Service (ETHOS) and is the program director for the Simulation, Modeling and Decision Science Program at the U.S. Department of Energy's Ames Laboratory. He teaches classes in combustion, sustainability, energy systems, and design for the developing world. He is the recipient of numerous teaching and research awards, including three R&D 100 awards within the past 5 years.

Nomenclature and Abbreviations

a	speed of sound, m/s
A	area, m^2
$\bar{A}$	mean surface area per unit volume, m^{-1}
A_R	surface area of the reaction zone in contact with the tube walls in a spray flow detonation, m
A_S	frontal area of the leading shock wave, m
A_v	area per unit volume, m^{-1}
[A]	molar concentration of species A, kgmol/m^3
B	fraction of air that flows through the dense phase in a fluidized bed; mass transfer driving force
CR	compression ratio
c	specific heat, kJ/kg·K
C_H	heat transfer coefficient for spray detonations
C_D	drag coefficient
c_p	specific heat at constant pressure, kJ/kg·K
c_v	specific heat at constant volume, kJ/kg·K
d	diameter, m
$\bar{d}$	surface mean diameter, m
$\bar{d}_1$	mean diameter, m
$\bar{d}_2$	area mean diameter, m
$\bar{d}_3$	volume mean diameter, m
$\bar{d}_{32}$	Sauter mean diameter, m
D_{AB}	binary diffusion coefficient, m^2/s
E	activation energy in the Arrhenius form of a reaction rate, kJ/kg
EA	excess air
ER	emission reduction
f	fuel-air mass ratio; friction factor
f_s	stoichiometric fuel-air mass ratio
F	equivalence ratio $= f/f_s$; force, N
g	Gibbs free energy per unit mass, kJ/kg
G	API gravity; Gibbs free energy, kJ
H	enthalpy, kJ; heat of reaction, kJ/kg
h	specific enthalpy, kJ/kg
$\tilde{h}$	convective heat transfer coefficient, W/m^2·K
$\tilde{h}^*$	convective heat transfer coefficient corrected for mass transfer, W/m^2·K
$\tilde{h}_D$	mass transfer coefficient, cm/s

h_{fg}	latent heat of vaporization, kJ/kg
HR	heat rate, Btu/kWh
HR''	heat input rate per unit area, kW/m^2
Δh°	enthalpy of formation, kJ/kg
h_{sorp}	heat of sorption, kJ/kg
I	combustion intensity, kW/m^3
$\tilde{k}$	thermal conductivity, W/m·K
k	reaction rate, units vary
k_0	pre-exponential factor in the Arrhenius form of a reaction rate
k_a	attrition rate constant
k_i	kinetic rate constant, units vary
K_p	thermodynamic equilibrium constant
L	length, m
L_f	penetration with cross flow, m
L_I	integral scale of turbulence, m
L_K	Kolmogorov scale of turbulence, m
m	mass, kg
$\dot{m}$	mass flow rate, kg/s
$\dot{m}''$	mass flux, kg/m^2·s
M	molecular weight, kg/kgmol
MC	Moisture content
n	molar concentration, kgmol/m^3
$\tilde{n}_{ji}$	number of j atoms in species i
n'	droplet number concentration, drops/cm^3
N	moles, mol
ΔN_i	fractional number of drops in size range i
$\dot{N}$	molar flow rate, kgmol/s
$\dot{N}''$	molar flux, kgmol/m^2·s
p	pressure, kPa
q	heat transfer rate, W
q_{chem}	rate of heat release from combustion, W
q_{loss}	extraneous heat loss, W
q''	heat transfer rate per unit area, W/m^2
q'''	heat transfer rate per unit volume, W/m^3
Q	heat, MJ
Q_{12}	total heat input for process from state 1 to state 2, kJ
Q_p	heat transfer for reaction at constant pressure, kJ
Q_v	heat transfer for reaction at constant volume, kJ
r	reaction rate (rate of production or destruction of a chemical species per unit area or volume), g/cm^2·s or g/cm^3·s; radius, cm
r_{bed}	fraction of heat extracted from a fluidized bed
$\hat{R}$	universal gas constant, kJ/kgmol·K

R	specific gas constant $= \hat{R}/M$, kJ/kg·K
$R\bullet$	radical species
S	entropy, kJ/K
s	entropy per unit mass, kJ/kg·K
sg	specific gravity (density at 20°C relative to water)
t	time, s
T	temperature, K
u	specific internal energy, kJ/kg
v	specific volume, m³/kg
V	volume, m³
$\dot{V}$	volume flow rate, m³/s
$\underline{V}$	velocity, m/s
$\underline{V}'$	velocity of turbulent fluctuations, m/s
$\underline{V}'_{rms}$	root mean square of turbulent velocity fluctuations, m/s
$\tilde{\underline{V}}$	diffusion velocity, m/s
$\check{\underline{V}}$	velocity relative to shock or detonation wave, m/s
$\underline{V}_L$	laminar flame speed or laminar burning velocity, m/s
$\underline{V}_T$	turbulent flame speed or turbulent burning velocity, m/s
$\underline{V}_D$	detonation velocity, m/s
W	work, kJ
$\dot{W}$	power, W
$\dot{W}_b$	brake power, W
x	distance, m
x_i	mole fraction of species i
X	fraction of fuel burned in bed
y_i	mass fraction of species i
Z	correction for effect of mass transfer on heat transfer
α	thermal diffusivity $= \tilde{k}/\rho c_p$, m²/s
β	droplet burning rate constant, m²/s
γ	ratio of specific heats
δ	laminar flame thickness, m
ε	emissivity; porosity; void fraction
η	efficiency
η_t	thermal efficiency
λ	Lagrange multipliers (element potentials); detonation cell size, m
μ	absolute viscosity, N·s/m²
υ	kinematic viscosity, m²/s
ρ	density, kg/m³
σ	surface tension, N/m
τ	torque, N·m
τ_{chem}	characteristic chemical reaction time, s

τ_{flow}	characteristic flow time, s
ϕ	screening factor used in solid fuel combustion
Φ	fraction of the bed volume occupied by tubes
ω	rotation rate, rad/s

Subscripts

af	adiabatic flame
air	air
ash	ash
as-recd	as-received
attr	attrition
aux	auxiliaries
B	bubble
b	bound water; droplet breakup; background; burned; boiler
bed	bed (in a fluidized bed)
blower	blower
bulk	bulk
C	carbon
chem	chemical
char	char
CO	carbon monoxide
CO_2	carbon dioxide
coal	coal
comb	combustion
d	diameter
daf	dry ash-free basis
drop	droplet
dry	dry fuel basis
dry	drying (time)
D	dense phase; detonation
eff	effective
f	fuel
flame	flame
fsp	fiber saturation point
g	gas
H_2	hydrogen
I	number of species in the system; turbulent intensity; interstitial
i	species *i*; counter
ig	*ignition*
in	in; inlet
init	initial
j	atoms of element *j*

jet	jet
k	counter
ℓ	liquid
L	laminar
LM	log mean
loss	loss
m	mean
mf	minimum fluidization
n	surface area n
N_2	nitrogen
out	out; outlet
O_2	oxygen
0	reference condition
orf	orifice
p	particle; pintle
pm	porous media
prod	products of combustion
pump	pump
pyr	pyrolysis
react	reactants
s	sensible; solid; superficial
(s)	stoichiometric
slip	slip
smooth	smooth
surf	surface
T	terminal
total	total
tube	tube
u	unburned
v	vaporization; reactants in a well stirred reactor
v	volatiles
vapor	water vapor
void	void
w	water
wall	wall
wood	wood
+	forward reaction
−	back reaction

Overbars

^	quantity per mole
-	average value
˘	with respect to shock or detonation front

Dimensionless Numbers

Bi   Biot number $= \dfrac{\tilde{h}L}{\tilde{k}}$

Da   Damkohler number $= \dfrac{L_l}{\delta} \cdot \dfrac{V_L}{V'_{rms}}$

Ka   Karlovitz number $= \dfrac{\delta V'_{rms}}{L_K V_L}$

Ma   Mach number $= \dfrac{V}{a}$

Nu   Nusselt number $= \dfrac{\tilde{h}L}{\tilde{k}_g}$

Oh   Ohnesorge number $= \dfrac{\mu_\ell}{\sqrt{\rho_\ell \sigma d_j}}$

Re   Reynolds number $= \dfrac{VL}{\upsilon}$

Sc   Schmidt number $= \dfrac{\nu}{D_{AB}}$

Sh   Sherwood number $= \dfrac{\tilde{h}_D d}{D_{AB}}$

We   Weber number $= \dfrac{\rho V^2 L}{\sigma}$

Abbreviations

AMD	area mean diameter
ASTM	ASTM International
ATDC	after top dead center
BDC	bottom dead center
BMEP	brake mean effective pressure
BSFC	brake specific fuel consumption
BTDC	before top dead center
CA°	crank angle degrees
CN	cetane number
CNF	cumulative number fraction
CVF	cumulative volume fraction
EGR	exhaust gas recirculation
EPA	U.S. Environmental Protection Agency
FSR	flame speed ratio

HC	hydrocarbons
HHV	higher heating value
IMEP	indicated mean effective pressure
IVC	intake valve closing
LHV	lower heating value
LNG	liquefied natural gas
LPG	liquefied petroleum gas
MBT	maximum brake torque
NO_x	nitric oxide plus nitrogen dioxide
PAH	polycyclic aromatic hydrocarbons
PAN	peroxyacetyl nitrate
RDF	refuse derived fuel
SAE	SAE International
SI	spark ignition
SMD	Sauter mean diameter
SOC	start of combustion
SOI	start of injection
SO_x	sulfur dioxide plus sulfur trioxide
TDC	top dead center
VMD	volume mean diameter
WI	Wobbe Index

1 Introduction to Combustion Engineering

This chapter introduces the broad scope of phenomena that make up the subject of combustion. Combustion impacts many aspects of our lives, especially sustainability, global climate change, and the utilization of energy. For engineers, the continuing challenge is to design safe, efficient, and non-polluting combustion systems for many different types of fuels in such a way as to protect the environment and enable sustainable lifestyles.

Improving the design of combustion systems requires an understanding of combustion from both a scientific and an engineering standpoint. Understanding the details of combustion is challenging and requires utilizing chemistry, mathematics, thermodynamics, heat transfer, and fluid mechanics. For example, a detailed understanding of even the simplest turbulent flame requires knowledge of turbulent reacting flows, which is at the frontier of current science. However, the engineer cannot wait for such an understanding to evolve, but must use a combination of science, experimentation, and experience to find practical design solutions.

1.1 THE NATURE OF COMBUSTION

Combustion is such a commonly observed phenomenon that it hardly seems necessary to define the term. From a scientific viewpoint, combustion stems from chemical reaction kinetics. The term combustion is saved for those reactions that take place very rapidly, with a large conversion of chemical energy to sensible energy. Such a definition is not precise because the point at which a reaction is characterized as combustion is somewhat arbitrary. It is easy to see that an automobile rusting is not combustion even though the oxidation reaction might be much faster than we desire. Wood burning in a fireplace, however, is clearly combustion.

There are several ways to increase the reaction rate and achieve effective combustion. One way is to increase the surface area, which can dramatically increase the reaction rate. For example, powdered metals can burn rapidly, but a bar of iron will only rust slowly. Pulverized coal combustors and liquid spray combustors (such as those used in boilers, diesel engines, and gas turbines) use small particles or droplets for quick combustion. Another common method of increasing the reaction rate is to increase the temperature. As the temperature is increased, the rate of a chemical exothermic reaction increases exponentially, such that heating the reactants to a sufficiently high temperature causes combustion. Because fuel reactions are exothermic, heating can cause a runaway condition under certain conditions. As the reactants are heated, thermal energy is released, and if this energy is released faster than it can be transported away by heat transfer, the temperature of the system will rise, causing the reaction to speed up. The accelerating reaction will cause an even greater rate of energy release that can cause an explosion.

If the temperature of a combustible mixture is raised uniformly, for example by adiabatic compression, the reaction may take place homogeneously throughout the volume. However, this is not typical. The combustion most commonly observed involves a *flame*, which is a thin region of rapid exothermic chemical reaction. For example, as shown in Figure 1.1, a Bunsen burner and a candle each exhibit a thin region in which the fuel and oxygen react, hence giving off heat and light.

Combustion can be classified by whether the mixture is homogeneous or heterogeneous, depending on if the oxidizer (typically air) and the fuel are premixed, or meet only at the point of reaction; whether the fluid flow conditions of the reaction are laminar or turbulent; and whether the fuel is gaseous, liquid, or solid.

In the case of the Bunsen burner, the reactants, consisting of a gaseous fuel such as methane and air, are premixed before ignition. When ignited, a fuel and air mixture flows into a thin cone-shaped reaction zone or *flame front* that produces combustion products. The conversion of a chemical to sensible energy takes place in the thin flame zone and causes the flame temperature to be high. Additionally, heat transfer and mixing with the surrounding air cause the temperature to rapidly decrease as the products move away from the flame. The Bunsen burner is an example of premixed combustion that is a *premixed flame*. A candle flame is different from a Bunsen burner flame in that the fuel is not premixed with the oxidizer. The solid candle wax is heated and melted by the flame, whereupon the melted wax is drawn into the wick and vaporized. The vaporized candle wax then mixes with air, which is drawn into the flame by the buoyant motion of the upward flowing products. This type of flame is called a *diffusion flame*.

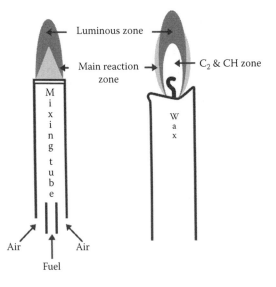

FIGURE 1.1 Diagram of premixed combustion of a Bunsen burner and diffusion combustion of a candle.

Premixed as well as diffusion flames occur in laminar and also turbulent flows. Turbulence speeds up the rate at which reactants are mixed and increases the surface area of the flame zone. As a result, turbulence greatly increases the *flame speed*. Furthermore, the flame can be stationary, as in a gas turbine burner, or propagating, as in a spark ignition engine combustion chamber.

The nature of the combustion also depends on whether the fuel is gaseous, liquid, or solid. Gaseous fuels such as natural gas are easy to feed and mix, and are relatively clean burning. Liquid fuels are typically broken into small droplets by being sprayed through a nozzle at high pressures. When heated, liquid fuels vaporize and then burn as a gaseous diffusion flame. Many solid fuels are pulverized or ground before being fed into a burner or combustion chamber. Larger-sized solid fuels combust in a fuel bed with air flowing through the bed. When heated, solid fuels (such as wood, switchgrass, or coal) release gaseous volatiles with the remainder being solid *char*. The char is mostly porous carbon and ash, and burns out as a surface reaction, while the volatiles burn as a diffusion flame. Vaporization of liquid fuel sprays and devolatilization of solid fuels occur much more slowly than gas phase chemical reactions and hence become important aspects of the combustion process. Char burnout, in turn, occurs more slowly than devolatilization.

Understanding each of the factors noted above is critical in the engineering and design of practical combustion systems, and each of these factors presents challenges and opportunities. Combustion systems must limit harmful pollutant emissions to the atmosphere. In addition, combustion systems need to take into account the carbon and other emissions responsible for global climate change as well as to address the need for a future based on sustainable energy.

1.2 COMBUSTION EMISSIONS

Until the 1800s, most cities were small, and the use of energy was limited. As a result there was little concern about emissions from combustion. With the developments of steam engines in the 1700s and then of automobiles and electricity in the late 1800s, energy consumption began to rise, and with it, combustion-based air pollution. As cities and energy consumption grew, particulate and sulfur dioxide emissions became a concern. The term "smog" (smoke plus fog) originated in the early 1900s in Great Britain, where burning of high sulfur coal plus natural fog produced deadly sulfuric acid aerosols. After World War II, with the continued growth of cities and extended urban areas with associated vehicle traffic, a new type of smog appeared—photochemical smog. The necessary ingredients for this type of smog are hydrocarbons, nitrogen oxides, air, and strong sunlight. Photochemical and chemical reactions in the atmosphere produce ozone and convert nitric oxide (a relatively harmless gas) to nitrogen dioxide and photochemical aerosols that irritate the eyes and lungs. Faced with increasing public concern, federal emission controls for automobiles began with the 1972 models in the United States. Federal regulation of emissions from stationary industrial sources and power plants also began in the United States at that time.

Today, nearly all combustion systems must satisfy governmentally imposed emission standards for combustion products, such as carbon monoxide, hydrocarbons, nitrogen oxides, sulfur dioxide, and particulate emissions. The emissions standards are set at sufficiently low levels to keep the ambient air clean enough to protect human health and the natural environment. Low emissions are achieved by a combination of fuel selection and preparation, combustion system design, and treatment of the products of combustion. There are challenging engineering tradeoffs between low emissions, high efficiency, and low cost. Until recently, carbon dioxide emissions have not been considered harmful to human health or the environment, but this is beginning to change. In December 2009, the United States Environmental Protection Agency ruled that current and projected concentrations of carbon dioxide in the atmosphere threaten the public health and welfare of current and future generations. The anticipated regulation of carbon dioxide emissions has broad repercussions for combustion engineers and the energy industry.

1.3 GLOBAL CLIMATE CHANGE

Global climate change has become a widespread concern. Carbon dioxide levels in the global atmosphere are increasing, and carbon dioxide emissions from combustion are a major contributor to the greenhouse effect. Carbon dioxide traps long-wave radiation reflected from the surface of the earth, decreasing global heat loss. Prior to the industrial revolution, the carbon dioxide content of the atmosphere was fairly stable at 280 parts per million (ppm). By 1900, the carbon dioxide level had reached 300 ppm. Beginning in 1958, direct measurements of atmospheric carbon dioxide concentrations were made at the Mauna Loa Observatory in Hawaii. These measurements have been plotted in Figure 1.2. In 1958 the carbon dioxide concentration was 315 ppm; by 1980 it was 337 ppm; by 1990 it was 355 ppm; and by 2009 it was 385 ppm. If current trends continue, the carbon dioxide concentration could reach

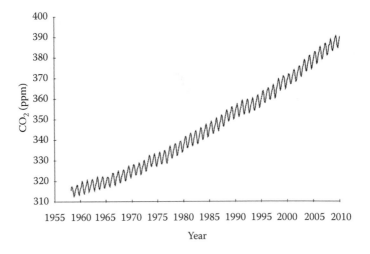

FIGURE 1.2 Atmospheric carbon dioxide concentrations at the Mauna Loa Observatory in Hawaii from 1958 to 2009. (Data from Tans, P., NOAA/ESRL, www.esrl.noaa.gov/gmd/ccgg/trends (accessed April 4, 2010).)

500–550 ppm by 2050. In addition, the current world reserves of fossil fuels that are profitable to extract under current policies are more than enough to take the world beyond 750 ppm if carbon capture of emissions is not implemented (Kirby 2009).

Carbon dioxide has an atmospheric residence time of several hundred years, and the buildup of carbon dioxide in the atmosphere is a balance between sources and sinks. The major source is combustion of fossil fuel. Worldwide emissions of carbon dioxide from fossil fuels were 28,962 million metric tons per year in 2007 up from 15,640 million metric tons per year in 1973. Currently, coal and oil produce roughly 80% of the worldwide anthropogenic (human caused) carbon dioxide emissions, while the remaining 20% comes from natural gas, as shown in Table 1.1.

The climate is changing. For example, in September 2008 the north polar ice cap was only half the size it was only 50 years ago. The sea level has been rising since direct measurements began in 1870, but the rate of rise is now five times faster than it was 135 years ago. To address concerns of global climate change, The Intergovernmental Panel on Climate Change (IPCC) was established in 1988 to evaluate the risk of global climate change. To date, the IPCC has issued four assessment reports. In the Fourth Assessment Report completed in early 2007, the IPCC concluded that the evidence for "warming of the climate system is unequivocal" and that "most of the increase in global temperatures since the mid-20th century is most likely due to the observed increase in anthropogenic (human) greenhouse gas concentrations."

The worldwide pressure for growth in fuel consumption is tremendous. As a result, carbon dioxide emissions are expected to grow due to continuing population growth and strong economic growth in developing countries including Brazil, India, and China. For example, China is building new coal-fired power plants at a rapid pace, and in India automobile sales are growing rapidly as income rises.

To encourage a decline in carbon emissions, many countries have either legislated or are considering legislating some form of carbon credit trading or carbon tax on large sources of carbon dioxide such as electrical power plants. There are many opportunities to improve the combustion technologies that we use, and even small increases in efficiency or emissions performance can result in significant savings. For example, today approximately one-third of the world's population uses an open fire for cooking (Jetter and Kariher 2009). Replacing one open wood cooking fire with an improved wood cookstove can reduce carbon dioxide emissions by approximately 2–3 metric tons per year. At today's carbon market prices, this would result in an annual payment

TABLE 1.1

World Carbon Dioxide Emissions from Fossil Fuels in 2007

Fuel	CO_2 Emissions (%)
Coal and peat	42.2
Oil	37.6
Natural gas	19.8
Other non-renewables	0.4

Source: International Energy Agency, *Key World Energy Statistics*, IEA, Paris, 2009a.

of \$8–\$25 per metric ton of carbon dioxide. And this is for a stove that costs less than \$50. On a much larger scale, capture of carbon dioxide emissions from coal-fired electric power plants and sequestration in geological formations is under intensive investigation. Clearly climate change issues and the advent of carbon credits provide a number of new opportunities and challenges for combustion engineers.

1.4 SUSTAINABILITY

Like many things, sustainability does not have a universally accepted definition. The definition of sustainability most often cited and discussed is "meeting the needs of the present without compromising the ability of future generations to meet their own needs." This was the definition given by the World Commission on Environment and Development (the Brundtland Commission, named after its chair, G. H. Brundtland 1987). The WCED was convened in 1983 by the United Nations to address the issues of growth and environment. The United States Environmental Protection Agency and a number of other organizations use this as their definition.

Sustainability is likely to be one of the most significant issues of our time. To pursue sustainability means that our lifestyles will change and our society will change. The extent of this impact will be determined in large part by how our energy resources are handled. The United States National Academy of Engineering has written in the *Grand Challenges for Engineering:* "Foremost among the challenges are those that must be met to ensure the future itself. The Earth is a planet of finite resources, and its growing population currently consumes them at a rate that cannot be sustained. Widely reported warnings have emphasized the need to develop new sources of energy, at the same time as preventing or reversing the degradation of the environment." And while non-combustion-based energy sources (e.g., solar, wind, geothermal, tidal, and nuclear) and conservation are critical to reaching a sustainable energy portfolio, combustion-based energy sources (e.g., oil, coal, gas, and biomass) will be important energy sources for the foreseeable future.

The combustion engineer's role in contributing to sustainability is to improve the efficiency, mitigate harmful emissions, and reduce the carbon footprint of our energy conversion devices, all the while maintaining cost effectiveness. The energy needs for cooking, heating, transportation, and electric power generation are global, and today in many parts of the world these needs are not being satisfied. For example, 1.5 billion of the world's 6.5 billion people live entirely without electricity (Legros et al. 2009). The grand challenge is to meet this need sustainably.

1.5 WORLD ENERGY PRODUCTION

In 2007, worldwide production of energy was 12,029 Mtoe (million metric tons of oil equivalent). This total worldwide energy production included crude oil, coal, natural gas, and biomass, plus nuclear, hydroelectric, geothermal, wind, and solar energy. Table 1.2 shows production and energy values for each of these energy sources. In 1973, world energy production was 6138 Mtoe. The International Energy Agency estimates that by the year 2030 world energy production will increase to 17,000 Mtoe.

TABLE 1.2

World Energy Production in 2007

Source	Production (Mtoe)[a]	Energy (%)
Crude oil	4090	34.0
Coal and peat	3188	26.5
Natural gas	2514	20.9
Biomass	1179	9.8
Nuclear	710	5.9
Hydroelectric	265	2.2
Geothermal, wind, solar	84	0.7
Total	12,029	100.0

Source: International Energy Agency, *Key World Energy Statistics*, IEA, Paris, 2009a.
[a] Million metric tons of crude oil equivalent on an energy content basis (1 Mtoe = 4.187×10^4 TJ).

The extent of global fossil fuel reserves is a subject of debate. Natural gas and crude oil reserves are more limited than coal reserves. World natural gas and oil production rates are expected to begin to decline by 2050, whereas coal supplies will last much longer. Estimates of the time to consume the fossil fuel reserves of the world are very difficult to predict because new exploration and new technologies expand known reserves, while increasing consumption driven by rising population and unmet human needs increases the depletion rate.

Biomass has an important role to play in reducing fossil fuel usage and transitioning to a more sustainable energy usage. When grown on a sustainable basis, biomass used for energy can be nearly carbon neutral because the carbon dioxide released in combustion is taken up by growing new plants, whereas combustion of fossil fuels releases carbon from plants grown millions of years ago. Most countries have biomass resources available or could develop them.

1.6 STRUCTURE OF THE BOOK

This book provides a practical introduction to the science and engineering of combustion required for the design of the clean and efficient combustion devices needed to meet today's challenges. These challenges include declines in traditional fuel stocks, the development of new renewable fuels, and global climate change. This book addresses both the fundamentals and applications of combustion. There are eight chapters that address the fundamentals of combustion including fuels, thermodynamics, chemical kinetics, flames, detonations, sprays, and solid fuel combustion mechanisms. There are also eight chapters that apply these fundamentals to combustion devices including internal combustion engines, gas turbines, biomass cookstoves, furnaces, and boilers. Combustion of gaseous fuel systems is considered first, followed by liquid fuel systems, and then solid fuel systems. Before starting the study of combustion engineering, students might find it interesting to read the brief historical perspective on combustion technology provided in Appendix D at the end of this book.

REFERENCES

Brundtland, G. H., *Report of the World Commission on Environment and Development: Our Common Future*, United Nations document A/42/427, 1987.

International Energy Agency, *Key World Energy Statistics*, IEA, Paris, 2009a.

——,*World Energy Outlook 2009*, OECD Publishing, Paris, 2009b.

Jetter, J. J. and Kariher, P., "Solid-Fuel Household Cook Stoves: Characterization of Performance and Emissions," *Biomass Bioenergy* 33(2):294–305, 2009.

Kirby, A., *Climate in Peril: A Popular Guide to the Latest IPCC Reports*, UNEP/Earthprint, Stevenage, Hertfordshire, UK, 2009 and UNEP/GRID-Arendal, Arendal, Norway, 2009.

Legros, G., Havet, I., Bruce, N., and Bonjour, S., *The Energy Access Situation in Developing Countries: A Review Focusing on Least Developed Countries and Sub-Saharan Africa*, United Nations Development Program, New York, 2009.

National Academy of Engineering, *Grand Challenges for Engineering*, Washington, DC, 2008.

Pachauri, R. K., and Reisinger, A., eds., *Climate Change 2007: Synthesis Report.* IPCC, Geneva, Switzerland, 2007.

Tans, P., NOAA/ESRL, www.esrl.noaa.gov/gmd/ccgg/trends (accessed April 4, 2010).

United States Environmental Protection Agency, "Endangerment and Cause or Contribute Findings for Greenhouse Gases under Section 202(a) of the Clean Air Act," Federal Register 74(239):66496–66546, December 15, 2009.

Part I

Basic Concepts

The physical and chemical properties of the various fuels in use today provide the basic information needed to begin the study of combustion. The thermodynamics of combustion provides a basic framework for every combustion engineers' toolbox. Although combustion engineering involves many aspects other than chemical kinetics, kinetics is what drives combustion. In addition to fuel properties, thermodynamics, and chemical kinetics, as the book unfolds basic concepts regarding flames, detonations, sprays, and solid particle combustion will be presented.

2 Fuels

This chapter discusses the types of fuels commonly used in combustion systems and summarizes their properties. As discussed in Chapter 1, energy is a critical part of our lifestyle, and it is likely that combustion systems will be the primary source of this energy for the foreseeable future. Because of this, the fuels used in energy systems need to be abundant, affordable, easily transported, clean, and preferably renewable. In this text, we focus on commercially important fuels that, when heated, undergo a chemical reaction with an oxidizer (typically oxygen in the air) to liberate heat.

Fuels can be classified in several ways. From a combustion device perspective, the critical aspects of a fuel are how it is transported within the combustion device and how it is combusted. Based on this, combustion engineering classifies fuels primarily as gaseous, liquid, or solid, and in this book we will follow this convention. After an initial discussion of combustion fundamentals, we address gaseous, liquid, and solid fuels in separate sections, in which fossil fuels and biofuels are discussed.

In contrast to the engineering classification of fuels, in society fuels are generally differentiated according to their source and their impact on the environment. Thus, in society, fuels are divided into fossil fuels and biofuels, renewable fuels and non-renewable fuels, or carbon neutral and non-carbon neutral fuels. The classification of renewable and non-renewable fuels is often fuzzy. Fossil fuels are non-renewable, whereas biomass fuels are generally thought of as renewable. But questions arise about the rate of use and the rate of resupply. For example, in much of the world, wood is being harvested at a rate that cannot be sustained. Trees can be a renewable source of fuel; however, when they are harvested at a greater rate than they can be replaced, they are no longer a renewable fuel. Instead, they are being "mined" just like coal. Peat is another fuel which some propose is a renewable fuel. However, peat beds grow at the rate of only 3 cm per 100 years, and the rate of consumption is far greater than the rate of replacement. Fossil fuels primarily consist of natural gas, crude oil-derived fuels, and coal. Biomass fuels consist primarily of wood and wood waste, agricultural crops and crop residues, native grasses, the organic fraction of municipal and industrial wastes, and bioderived fuels such as alcohol, biodiesel, and fuel pellets.

2.1 GASEOUS FUELS

Gaseous fuels are those fuels that are transported within a combustion device and used in gaseous form. The primary gaseous fuels are natural gas and liquefied petroleum gas (LPG). Thermal gasification of solid fuels such as coal and biomass is used to create a gaseous fuel, called syngas. Hydrogen is a gaseous fuel of increasing interest because it can be combusted without releasing greenhouses gases.

Natural gas is found compressed in porous rock and shale formations sealed in rock strata below the ground. Natural gas frequently exists near or above oil deposits. Natural gas is a mixture of hydrocarbons and small quantities of various non-hydrocarbons existing in the gaseous phase or in solution with crude oil. Raw natural gas contains methane and lesser amounts of ethane, propane, butane, and pentane. Sulfur and organic nitrogen are typically negligible in natural gas. Carbon dioxide, nitrogen, and helium are sometimes present; however, the amounts of these non-combustible gases are generally low. Prior to distribution and use, natural gas is processed to remove the non-combustible gases and higher molecular weight hydrocarbons. Dry pressurized natural gas is transmitted long distances in pipelines. At some wells natural gas is liquefied (LNG) by cooling it to $-164°C$, and it is then transported to selected ports around the world in supertankers. The LNG is then transferred to tanks and subjected to heat and pressure to return it to a gaseous state for transport by pipeline.

Methane is created in landfills by methanogenic bacteria. Municipal solid waste and agricultural waste contain significant portions of organic materials that produce methane when dumped, compacted, and covered in landfills and digesters. Methanogenic bacteria decompose these organic materials to produce primarily carbon dioxide and methane. The carbon dioxide created typically leaches out because it is soluble in water. Methane is less soluble in water and lighter than air and can be collected in imbedded tubes. Because methane is a potent greenhouse gas, its release to the environment should be limited. Instead of being released, methane should be collected and utilized as a fuel whenever possible.

LPG comes in a variety of mixes. LPG produced at natural gas processing plants consists of ethane, propane, and butane. LPG produced from crude oil at refineries also includes refinery gases such as ethylene, propylene, and butylene. LPG is stored in tanks under pressure as a liquid and is a gas at atmospheric pressure. At 38°C, the maximum vapor pressure is 208 kPa for commercial LPG.

Syngas (synthetic gas, historically called producer gas or town gas) is created by passing a less than stoichiometric amount of air or oxygen through a hot bed of coal or biomass particles. This is a process called gasification and is accomplished in a gasifier, where input air is restricted so that the main output is hydrogen and carbon monoxide along with nitrogen and carbon dioxide. The gasifier output may be used directly without cooling, or depending on the application, cooled and cleaned to remove the tars, soot, and mineral matter from the gasified fuel. Air-blown syngas is a low heating value gas because of the nitrogen. Using substoichiometric oxygen can produce syngas with a higher heat content, but the process is more expensive than using air.

Hydrogen is manufactured by reforming natural gas, partial oxidation of liquid hydrocarbons, or extraction from syngas. For example, natural gas can be converted to H_2, CO, and CO_2 by a reaction with steam over a catalyst at 800°C–900°C. A further shift of CO and H_2O to H_2 and CO_2 is then carried out, and the gas is cooled and scrubbed to remove CO_2. Hydrogen can also be generated from the electrolysis of water and hence could be used for energy storage from wind and solar electric generation facilities. Some people envision hydrogen as a fuel for fuel cells that might replace gasoline engines.

2.1.1 Characterization of Gaseous Fuels

Important characteristics of gaseous fuels include volumetric analysis, density, heating value, and autoignition temperature. The volumetric analyses of natural gas, LPG, and producer gas from coal and wood gasification are shown in Table 2.1. The volumetric analysis varies depending on how the fuel is processed and the source of the fuel. Fuel density is a function of the volumetric analysis and molecular weights of the component gases and can be determined based on a specific volumetric analysis (this is discussed in Chapter 3).

The *heating value* is the heat release per unit mass when the fuel, initially at 25°C, reacts completely with oxygen, and the products are returned to 25°C. The heating value is reported as the *higher heating value* (HHV) when the water in the combustion products is condensed or as the *lower heating value* (LHV) when the water in the combustion products is not condensed. The LHV is obtained from the HHV by subtracting the heat of vaporization of water in the products,

$$\text{LHV} = \text{HHV} - \left(\frac{m_{H_2O}}{m_f} \right) h_{fg} \tag{2.1}$$

where m_{H_2O} is the mass of water in the products, m_f is the mass of fuel, and h_{fg} is the latent heat of vaporization of water at 25°C, which is 2440 kJ/kg water. The mass of water includes the moisture in the fuel as well as water formed from hydrogen in the fuel. The heating value of a gaseous fuel may be obtained experimentally in a flow calorimeter and can also be calculated from thermodynamics if the composition is known. The heating values of various gaseous fuels are given in Table 2.2.

Example 2.1

Given that the HHV of methane is 36.4 MJ/m³ at 25°C and 101.3 kPa, calculate the LHV of methane. Assume the density of methane is 0.654 kg/m³ at 25°C and 101.3 kPa.

TABLE 2.1
Typical Volumetric Analysis of Some Gaseous Fuels

Species	Natural Gas	LPG	Coal Producer Gas	Wood Producer Gas
CO	–	–	20%–30%	18%–25%
H_2	–	–	8%–20%	13%–15%
CH_4	80%–95%	–	0.5%–3%	1%–5%
C_2H_6	<6%	–	Trace	Trace
$>C_2H_6$ [a]	<4%	100%	Trace	Trace
CO_2	<5%	–	3%–9%	5%–10%
N_2	<5%	–	50%–56%	45%–54%
H_2O	–	–	–	5%–15%

[a] Contains hydrocarbons heavier than C_2H_6

TABLE 2.2

Typical Heating Value of Some Gaseous Fuels

Fuel	HHV		LHV	
	(MJ/m³)ᵃ	MJ/kg	(MJ/m³)ᵃ	MJ/kg
Hydrogen (H_2)	11.7	142.2	9.9	121.2
Carbon monoxide (CO)	11.6	10.1	11.6	10.1
Methane (CH_4)	36.4	55.5	32.8	50.0
Ethane (C_2H_6)	63.8	51.9	58.4	47.8
Propane (C_3H_8)	90.8	50.4	83.6	46.4
Butane (C_4H_{10})	117	49.5	108	45.8
Ethylene (C_2H_4)	57.7	50.3	54.1	47.2
Acetylene (C_2H_2)	53.2	49.9	51.4	48.2
Propylene (C_3H_6)	84.2	48.9	78.8	45.8
Natural gas (typical)	38.3	53.5	34.6	48.3
Coal producer gas (typical)	5.2	5.3	4.3	4.4
Wood producer gas (typical)	4.8	5.1	4.0	4.2

ᵃ At 1 atm, 25°C

Solution

Apply Equation 2.1 but note that for this example, the heating values are on a volumetric basis. For complete combustion, 1 mol of methane (CH_4) yields 2 mol of water (H_2O)

$$CH_4 + 2O_2 \rightarrow CO_2 + 2\,H_2O$$

so that

$$\frac{m_{H_2O}}{m_{CH_4}} = \frac{N_{H_2O}}{N_{CH_4}} \cdot \frac{M_{H_2O}}{M_{CH_4}} = \left(\frac{2\ kgm\ ol_{H_2O}}{1\ kgm\ ol_{CH_4}} \right) \left(\frac{18\ kg_{H_2O}}{kgm\ ol_{H_2O}} \cdot \frac{kgm\ ol_{CH_4}}{16\ kg_{CH_4}} \right) = \frac{2.25\ kg_{H_2O}}{kg_{CH_4}}$$

From this we can write

$$LHV = \frac{36.4\ MJ}{m^3_{CH_4}} - \frac{2.25\ kg_{H_2O}}{kg_{CH_4}} \cdot \frac{2.440\ MJ}{kg_{H_2O}} \cdot \frac{0.654\ kg_{CH_4}}{m^3_{CH_4}} = \frac{32.8\ MJ}{m^3_{CH_4}}$$

Autoignition temperature is the lowest temperature at which ignition occurs spontaneously in a standard container with atmospheric air in the absence of a spark or flame and without regard to the ignition delay time. The autoignition temperature of alkanes (hydrocarbons of the form C_nH_{2n+2}) decreases with increasing molecular weight (Table 2.3). As shown, carbon monoxide has a high autoignition temperature, and hydrogen has a low autoignition temperature. In general, autoignition temperature is an indication of the relative difficulty of combusting a fuel. Autoignition temperature varies with the geometry of the hot surface and other factors such as pressure.

TABLE 2.3

Flash Point and Autoignition Temperatures of Pure Fuels in Air at 1 atm

Fuel	Flash Point (°C)	Autoignition (°C)
Methane	−188	537
Ethane	−135	472
Propane	−104	470
n-Butane	−60	365
n-Octane	10	206
Isooctane	−12	418
n-Cetane	135	205
Methanol	11	385
Ethanol	12	365
Acetylene	Gas	305
Carbon monoxide	Gas	609
Hydrogen	Gas	400

Source: From Bartok, W. and Sarofim, A. F. eds., *Fossil Fuel Combustion: A Source Book,* ©1991, Wiley, by permission of John Wiley and Sons, Inc.

2.2 LIQUID FUELS

Liquid fuels are those fuels that are transported within the combustion device as a liquid. Because liquid fuels cannot be used directly, they are generally vaporized and then combusted. The vaporization process may occur as a part of the combustion process, such as in a diesel engine, or it may occur upstream in a vaporizer, such as in a liquid fuel cookstove. Because of this need to vaporize and then combust the fuel, clean combustion of liquid fuels is more complicated and challenging than clean combustion of gaseous fuels.

Currently, liquid fuels are derived primarily from crude oil. Increasingly, liquid fuels are derived from biomass, oil shale, tar sands, and coal. Crude oil is a mixture of naturally occurring liquid hydrocarbons with small amounts of sulfur, nitrogen, oxygen, trace metals, and minerals. Crude oil is generally found trapped in certain rock formations that were originally part of the ocean floor. Organic marine matter on the ocean bottom was encased in rock layers at elevated pressure and temperature, and over millions of years gradually formed crude oil.

The *ultimate analysis* (elemental chemical composition) of crude oil does not vary greatly around the world. Crude oil is made up of roughly 84% carbon, up to 3% sulfur, and up to 0.5% nitrogen, and 0.5% oxygen with the remainder being primarily hydrogen. Crude oil is sometimes burned directly; however, because of its wide range of densities, viscosities, and impurities, crude oil is generally refined. The refining processes of fractional distillation, cracking, reforming, and impurity removal are used to produce many products including gasoline, diesel fuels, gas turbine fuels, and fuel oils.

Figure 2.1 shows typical products from crude oil refineries with the light, more volatile components at the top. Refinery gas consists mainly of hydrogen, methane, ethane, and olefins. Naptha is a mixture of volatile liquid hydrocarbons that are used as feedstock for other products including gasoline. Petroleum coke is a carbonaceous solid. Some adjustments of the product amounts can be made at the refinery. For example, a particularly cold winter may require more heating fuel, typically resulting in production of less gasoline. Before considering the properties and types of liquid fuels, the molecular structure of various fuel hydrocarbons is reviewed.

2.2.1 MOLECULAR STRUCTURE

Chemically, crude oil consists primarily of alkanes, cycloalkanes, and aromatics. Petroleum fuels also contain alkenes (olefins), which are formed during the cracking part of the refining process.

Alkanes are saturated hydrocarbons and are composed of carbon and hydrogen atoms connected by single bonds (C–C or C–H bonds). The general formula of alkanes is C_nH_{2n+2}. All the carbon bonds are shared with hydrogen atoms except for a minimum number of required carbon–carbon bonds. For simplicity the C–H bonds are not shown here; only the C–C bonds are shown. Carbon bonds require four hydrogen bonds or carbon bonds. Alkanes (sometimes called paraffins) having a straight chain structure are referred to as normal and are designated by the prefix n. The first four alkanes are gaseous at standard pressure and temperature,

Methane	CH_4
Ethane	$CH_3 \!-\! CH_3$
n-Propane	$CH_3 \!-\! CH_2 \!-\! CH_3$
n-Butane	$CH_3 \!-\! CH_2 \!-\! CH_2 \!-\! CH_3$

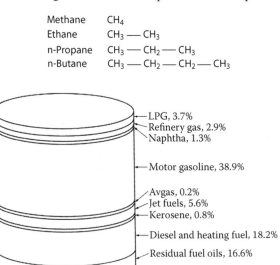

FIGURE 2.1   Typical crude oil refinery products.

Alkanes containing more carbon atoms are designated by the following prefixes: 5-pent, 6-hex, 7-hept, 8-oct, 9-non, 10-dec, 11-undec, 12-dodec, etc. Alkanes are not necessarily straight chained. If a side carbon chain or isomer exists, the name of the longest continuous chain of carbon atoms is taken as the base name. For example,

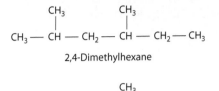

2,4-Dimethylhexane

$$CH_3 - CH_2 - \underset{\underset{CH_3}{|}}{CH} - CH_3$$

2-Methylbutane (isopentane)

Cycloalkanes are alkanes that have one or more carbon ring structures in their molecular structure. Cycloalkanes (sometimes called naphthenes) have the formula C_nH_{2n}, and the nomenclature follows the number of carbon atoms. For example,

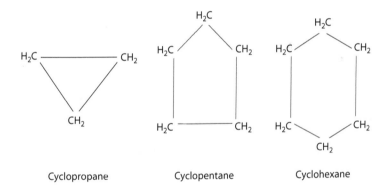

Cyclopropane   Cyclopentane   Cyclohexane

Alcohols have an OH group substituted for one of the hydrogen atoms in one of the paraffin series. For example,

$$CH_3 - OH \qquad CH_3 - CH_2 - OH$$

Methanol   Ethanol

Alkenes (olefins) also have the formula C_nH_{2n}, but two neighboring carbon atoms share a pair of electrons forming a double bond. The location of the double bond is indicated by a prefix. For example,

Ethylene $CH_2 = CH_2$
Propylene $CH_3 - CH = CH_2$
1-Butene $CH_3 - CH_2 - CH = CH_2$
2-Butene $CH_3 - CH = CH - CH_3$

Diolefins have two double bonds and have the general formula C_nH_{2n-2}. Their names end with the letters "diene", for example, hexadiene.

$$CH_3 - CH = CH - CH_2 - CH = CH_2$$

1,4-Hexadiene

Double bonded hydrocarbons may also be arranged in a ring structure. A basic building block is benzene, C_6H_6, which has three double bonds and is shown in two notations below. This class of compounds is called aromatics. Other aromatics can be formed by adding to the basic benzene ring by displacing hydrogen, for example, toluene (methylbenzene).

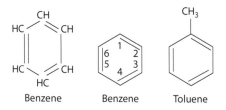

Benzene   Benzene   Toluene

Two or more rings may be united to form many compounds such as naphthalene, $C_{10}H_8$, or anthracene, $C_{14}H_{10}$. These compounds, known as polycyclic aromatic hydrocarbons (PAH), can be formed in the refining processes and can also be formed during incomplete combustion. For example, benzo[a]pyrene is a known carcinogen that is sometimes formed in rich fuel flames.

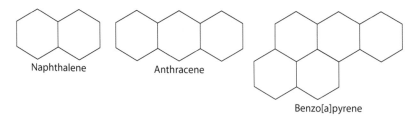

Naphthalene   Anthracene

Benzo[a]pyrene

For detailed information for hydrocarbon notation see Chapter 3 of the *CRC Handbook of Chemistry and Physics* (Lide and Haynes 2009).

2.2.2 CHARACTERIZATION OF LIQUID FUELS

Important properties of liquid fuels include heating value, specific gravity, viscosity, flash point, distillation curve, sulfur content, vanadium and lead content, octane number (for gasoline), cetane number (for diesel fuel), and smoke point (for gas turbine fuel). Fuel property data is given in this section and in Appendix A. The typical properties of fuel for internal combustion engines and gas turbines, and for fuel oils, are given in Tables 2.4 through 2.7.

The HHV for liquid fuels is determined by combustion with pressurized oxygen in a bomb calorimeter. This device is a stainless steel container that is surrounded by a large water bath. The large bath ensures that the final product temperature will be only very slightly above the initial reactant temperature of 25°C. The combustion is carried out with excess oxygen to ensure complete combustion.

Specific gravity is the density of the fuel divided by the density of water at the same temperature. In some cases, the American Petroleum Institute (API) gravity is used. The relation between API gravity (*G*), and specific gravity (sg) at 16°C is

$$G = \frac{141.5}{sg} - 131.5 \tag{2.2}$$

Viscosity is a measure of a liquid's resistance to flow; the higher the viscosity, the more resistant the liquid is to flowing. For liquid fuels, viscosity indicates the ease at which it can be pumped and atomized. The viscosity of liquids decreases with increasing temperature. There are several standard tests for viscosity. Sometimes pour point is used as a simple indicator of viscosity. The pour point is an indication

TABLE 2.4
Typical Properties of Automotive Fuels

Property	Automotive Gasoline	No. 2 Diesel Fuel	Ethanol	B100 Biodiesel
Chemical formula	C_4 to C_{12}	C_8 to C_{25}	C_2H_5OH	C_{12} to C_{22}
Molecular weight	100–105	~200	32	~292
Specific gravity at 16°C	0.72–0.78	0.85	0.794	0.88
Kinematic viscosity at 20°C (m²/s)	0.8×10^{-6}	2.5×10^{-6}	1.4×10^{-6}	–
Boiling point range (°C)	30–225	210–235	78	182–338
Reid vapor pressure (kPa)	48–69	<2	148	<0.3
Flash point (°C)	–43	60–80	13	100–170
Autoignition temp (°C)	257	~315	423	–
Octane No. (Research)	88–98	–	109	–
Octane No. (Motor)	80–88	–	90	–
Cetane No.	<15	40–55	–	48–65
Stoichiometric air-fuel ratio by weight	14.7	14.7	9.0	13.8
Carbon content (wt %)	85–88	87	52.2	77
Hydrogen content (wt %)	12–15	13	13.1	12
Oxygen content (wt %)	2.7–3.5	0	34.7	11
Heat of vaporization (kJ/kg)	380	375	920	–
LHV (MJ/kg)	43.5	45	28	42

TABLE 2.5
Typical Characteristics of Gasoline (United States, Summer 2005)

Characteristics	Conventional	Reformulated
Reid vapor pressure (kPa)	57	48
Sulfur content (wt ppm)	102[a]	69[a]
Benzene content (vol %)	1.19	0.66
Aromatics content (vol %)	27.8	20.9
Olefins content (vol %)	11.8	11.9

[a] Starting in 2008 sulfur <30 ppm.

of the lowest temperature at which a fuel oil can be stored and still be capable of flowing under very low forces in a standard apparatus.

Volatility is measured by Reid vapor pressure, which is the equilibrium pressure exerted by vapor over liquid at 37.8°C. For example, conventional gasoline has Reid vapor pressures of 48–69 kPa, and ethanol has a Reid vapor pressure of 148 kPa. Fuel volatility affects engine startup and transient performance as well as evaporative emissions during filling.

Flash point (Table 2.3) is an indication of the maximum temperature at which a liquid fuel can be stored and handled without serious fire hazard. Flash point is the minimum temperature at which fuel will rapidly catch fire when exposed to an open flame located above a mixture. An example of a flash point of interest is the ignitability of the mixture above the liquid fuel in a partially full fuel tank. Gasoline, which has a flash point of −43°C, is typically so volatile that the mixture above the liquid fuel is too rich to burn. No. 2 diesel fuel (flash point of 60°C–80°C) is so non-volatile that the mixture above the liquid fuel is too lean to burn.

TABLE 2.6
Typical Aviation Turbine Fuel Properties

Property	Units	Jet A	Jet B
Naphthalenes	% vol max	3	3
Aromatics	% vol max	20	20
Specific gravity	°API	37–51	45–57
LHV	MJ/kg, min	42.8	42.8
Viscosity	cST at −4°F, max	8	–
Freezing point	°C, max	−40	−50
Existent gum	mg/100 mL, max	7	7
Total sulfur	wt %, max	0.3	0.3
Flash point	°C, min	38	–

TABLE 2.7
Typical Properties of Fuel Oils

Fuel Grade No.	1	2	4	5	6
Property	Kerosene	Distillate	Very Light Residual	Light Residual	Residual
Color	Clear	Amber	Black	Black	Black
Specific gravity at 16°C	0.825	0.865	0.928	0.953	0.986
Kinematic viscosity at 38°C (m²/s)	1.6×10^{-6}	2.6×10^{-6}	15×10^{-6}	50×10^{-6}	360×10^{-6}
Pour point (°C)	<−17	<−18	−23	−1	19
Flash point (°C)	38	38	55	55	66
Autoignition temp. (°C)	230	260	263	–	408
Carbon (wt %)	86.5	86.4	86.1	85.5	85.7
Carbon residue (wt %)	Trace	Trace	2.5	5.0	12.0
Hydrogen (wt %)	13.2	12.7	11.9	11.7	10.5
Oxygen (wt %)	0.01	0.04	0.27	0.3	0.38–0.64
Ash (wt %)	–	<0.01	0.02	0.03	0.04
HHV (MJ/kg)	46.2	45.4	43.8	43.2	42.4

The *autoignition temperature* is the lowest temperature required to initiate self-sustained combustion in a standard container in atmospheric air in the absence of a spark or flame. For example, the autoignition temperature of gasoline is 257°C. Flash point and autoignition temperatures for selected gaseous and liquid fuels are given in Tables 2.3, 2.4, 2.6, and 2.7. In general, autoignition temperatures are an indication of the relative difficulty of combusting a fuel. Because the autoignition temperature varies with the geometry of the hot surface and various factors such as pressure, other tests—octane number and cetane number for instance—are used for engine fuels.

The *octane number* indicates the tendency of gasoline to knock (onset of auto-ignition) when the compression ratio in a spark ignition engine is raised. The octane number of a fuel is measured by comparing the performance of the fuel with the performance of mixtures of isooctane and n-heptane in a standardized spark ignition engine. Isooctane is arbitrarily set at 100, and n-heptane, which is more prone to knock, is arbitrarily set at zero. The octane number is the percentage of isooctane in the isooctane-heptane mixture that most closely matches the performance of the test fuel. Two octane test methods are used for automobile gasoline: one provides the research octane number, and the other provides the motor octane number. The research method is run with 52°C inlet air at 600 rpm with a spark advance of 13 crank angle degrees (CA°) before the top dead center of the piston. The research octane number gives a higher octane rating than the motor octane number, which is run with 149°C inlet air at 900 rpm with 19–26 CA° spark advance.

The *cetane number* (CN) ranks fuels according to their ignition delay when undergoing compression ignition. Because cetane (n-hexadecane) is one of the fastest igniting hydrocarbons in fuel, it is assigned a cetane number of 100. Isocetane (heptamethylnonane) ignites slowly and is arbitrarily assigned a cetane number of 15. A diesel fuel is compared to mixtures of the reference fuels in a standardized diesel engine and rated by the mixture that most closely matches the ignition delay of the test fuel. Ignition delay in an engine is the time between the start of injection and the onset of combustion. The cetane number of the reference mixture is defined by

$$CN = \% \text{ n-cetane} + 0.15 \left(\% \text{ heptane} \right) \tag{2.3}$$

In the cetane number test, the injection is fixed at 13° before the top dead center of the piston, and the compression ratio is changed until combustion of the test fuel starts at the top dead center. The standard mixture is found by determining which mixture gives the same ignition delay at these fixed conditions of injection timing and compression ratio. The tests are run at 900 rpm with 100°C water temperature and 65°C inlet air. Because the test engine is a prechamber design, the cetane number is at best only a relative scale when applied to open chamber engines. The test is questionable for low cetane fuels (CN < 35).

2.2.3 LIQUID FUEL TYPES

The primary uses of liquid fuels are gasoline and diesel for transportation, gas turbine fuels for industrial turbines and jet turbines, and fuel oil for household heating, industrial process heating, and small power applications.

Automotive gasoline is a carefully selected blend of alkanes (paraffins), alkenes (olefins), cycloalkanes (naphthenes), and aromatics, which varies slightly at different refineries and is blended slightly differently depending on geographic region and season of the year. Gasoline must be volatile enough to vaporize readily in the engine but not so volatile as to cause excessive evaporative emissions during handling. Gasoline is blended to control the octane number and to assist in controlling emissions of volatile organic compounds, nitrogen oxides, carbon monoxide, and particulates. Ethanol is added to improve the octane number. Various other antiknock compounds such as methyl tertiary butyl ether (MTBE) were added in the past to improve the octane number, but these created ground water problems and other environmental hazards and are no longer used. Gasoline also has additives to control deposits and inhibit corrosion.

The United States Clean Air Act Amendments of 1990 require the use of reformulated gasoline in regions that fail to meet federal ambient air standards for ozone. These regions are primarily the northeast, Detroit, Atlanta, Chicago-Milwaukee, St Louis, Dallas-Fort Worth, Houston, Denver, and large parts of California. Reformulated gasoline limits volatile organic compounds and sulfur (Table 2.5) so as to reduce evaporative emissions as well as benzene, aldehydes, and nitrogen oxides in the exhaust. Sulfur in gasoline was reduced to improve the effectiveness of the exhaust catalyst in controlling nitrogen oxide emissions. The oxygen content of reformulated gasoline is no longer regulated as of 2006.

Significant blending of ethanol with gasoline began in 1978 when ethanol received a federal highway tax exemption, and increased further in the 1990s when ethanol began to be used as a fuel oxygenator. The heating value of ethanol is lower than gasoline because of the attached oxygen (Table 2.4). Currently, all vehicles can use up to 10% ethanol. Blends between 10% and 85% require flex-fuel vehicles. Ethanol is made from sugars extracted from corn, sugarcane, and sugar beets for example, and probably will soon be made on a large-scale basis from grasses such as switchgrass and miscanthus, and from woody crops. In the future, biorefineries will probably make biofuels that are similar to gasoline. Biofuels are renewable and do not contribute directly to global warming, but some fossil fuels are required to grow, harvest, and refine the biomass. In addition, biofuels give rise to other environmental concerns, including water consumption and fertilizer runoff.

Diesel fuel is a mixture of C_{10} to C_{15} hydrocarbons with a higher boiling point range than gasoline. Grade 1-D is a light distillate fuel for applications requiring higher volatility for rapidly fluctuating loads and speeds, such as in light trucks and busses. Grade 2-D is a middle distillate fuel for high-speed engines. Sulfur levels less than 15 ppm are now required for highway use for Grade 2-D. Grade 4-D is a heavy distillate fuel used in low-speed industrial and marine diesels. Comparison of gasoline and No. 2 diesel (Table 2.4) shows the higher density of diesel fuel, which gives it more heating value on a volume basis. Note too the lower volatility (Reid vapor pressure) and higher viscosity of diesel fuel. The autoignition temperatures also partially reveal the reason for the large difference in cetane numbers bearing in mind that the cetane number reflects the ease of compression ignition. Biodiesel fuel can be made from oil seed crops such as rapeseed (canola) or soybeans. Biodiesel consists of methyl esters of C_{12} to C_{22} fatty acids (designated as FAME, fatty acid methyl ester), and is non-toxic and biodegradable.

Gas turbine fuels are not limited by antiknock or ignition delay requirements and have a wide range of boiling points. Jet A aircraft fuel is similar to kerosene and 1-D diesel fuel. Jet B fuel has a lower boiling point range than Jet A fuel. Turbine fuels limit the amounts of trace metals such as vanadium and lead, which tend to form deposits on turbine blades. Soot formation is controlled by limiting the aromatic content. Similar considerations are used for industrial gas turbine fuel oils. Figure 2.2 shows typical temperature versus percent fuel evaporated curves for gasoline, Jet A, Jet B, and No. 2 diesel. Some aviation turbine fuel properties are given in Table 2.6.

Fuel oil for heating covers a wide range of petroleum products, which have been divided into six grades. Table 2.7 shows various properties of these grades with the omission of Grade No. 3, which is rarely used. Grade No. 1 fuel oil is kerosene. Grade No. 2 fuel oil is domestic fuel oil, which boils between 218°C and 310°C. The heavier fuel oil grades are specified by viscosity and are used in industrial and utility heat and power applications. The heavier grades contain significant amounts of ash. Preheating the oil is required for No. 6 fuel oil and may be required for Nos. 4 and 5 fuel oils, depending on the climate. No. 6 fuel oil is a heavy residual fuel consisting of the remains after all distillation processes are completed; it has high viscosity and tends to have relatively high amounts of asphaltenes, sulfur, vanadium, and sodium. Residual fuel oils are burned directly in some large boilers, and after some treatment

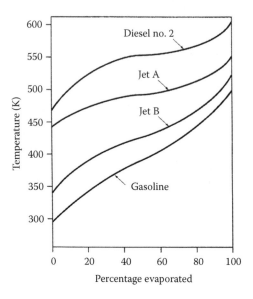

FIGURE 2.2 Distillation curves for various fuels using ASTM D86. (*Annual Book of ASTM Standards*, Vol. 5.01, 2009.)

they can be used in heavy-duty industrial gas turbines. Sulfur is limited to 0.5% for Nos. 1 and 2 fuel oils, but can be as high as 4% for No. 6 fuel oil.

2.3 SOLID FUELS

Solid fuels include wood and other forms of biomass, peat, lignite, coal, and refuse-derived fuels. Solid fuels are more challenging to transport, handle, and combust than gaseous or liquid fuels. Solid fuels come in a variety of sizes and shapes and for most uses they need to be processed to reduce their size and moisture content. In addition to carbon and hydrogen constituents, solid fuels contain significant amounts of oxygen, water, and mineral matter, as well as nitrogen and sulfur. The oxygen is chemically bound in the fuel and varies by weight from 45% for wood to 2% for anthracite coal on a dry, ash-free basis (Table 2.8).

Moisture content is a significant issue in biomass and low-ranked coals. Within the solid fuel matrix, water exists as water vapor, liquid water chemically bound to the fuel (adsorbed or bound water), and free liquid water within the pores of the solid fuel. Bound water is held by physical adsorption and exhibits a small heat of sorption. This heat of sorption is a function of moisture content and is in addition to the heat of vaporization. Green wood typically consists of 50% water on an as-received basis. After being air-dried for 1 year, the moisture content of wood typically falls to 15%–20% on an as-received basis. Lignite coals contain 20%–40% moisture on an as-received basis, most of which is free water, whereas bituminous coals contain about 5% moisture on an as-received basis that is bound water. Fuel moisture significantly influences the rate of combustion and the overall efficiency of the combustion system.

TABLE 2.8
Typical Percent Oxygen, Water, and Ash (wt) in Solid Fuels

Fuel	Oxygen (Dry, Ash-free)	Moisture (Ash-free)	Ash (Dry)
Wood	45%	15%–50%	0.1%–1.0%
Peat	35%	90%	1%–10%
Lignite coal	25%	30%	>5%
Bituminous coal	5%	5%	>5%
Anthracite coal	2%	4%	>5%
Refuse-derived fuel	40%	24%	10%–15%

The inorganic residue remaining after the fuel is completely burned is called ash. Depending on how it was handled and transported, wood usually has less than 1% ash, while coal has 3%–10% or more of ash. Typically, the ash begins to soften at 1200°C and becomes fluid at 1300°C, although this varies significantly between fuels and even between coals of the same rank or biofuels from similar crops. Ash characteristics play an important role in system design in order to minimize slagging, fouling, erosion, and corrosion. Representative inorganic elements in wood and coal ash are given in Table 2.9.

The composition of solid fuels is reported on an as-received basis, on a dry basis, or on a dry, ash-free basis. Specifically

$$m_{\text{as-recd}} = m_{\text{f}} = m_{\text{daf}} + m_{\text{w}} + m_{\text{ash}} \tag{2.4}$$

TABLE 2.9
Representative Mineral Elements and Chlorine in Pine and Bituminous Coals

Element	Pine[a]	Illinois coal
Ca	760 ppm	>5000 ppm
Na	28	200–5000
K	39	200–5000
Mg	110	200–5000
Mn	97	6–210
Fe	10	>5000
P	40	10–340
Si	–	>5000
Al	6	>5000
Cl	48	200–1000

[a] Average values for whole wood, mixed pine species.

$$m_{\text{dry basis}} = m_{\text{daf}} + m_{\text{ash}} \tag{2.5}$$

$$m_{\text{dry ash-free basis}} = m_{\text{daf}} \tag{2.6}$$

where m_{daf} is the dry, ash-free mass of the fuel, m_{w} is the mass of water in the fuel, and m_{ash} is the mass of ash in the fuel. The moisture content (MC), HHV, LHV, and most other properties may be reported on an as-received basis, a dry basis, or a dry, ash-free basis. Moisture content on an as-received basis is

$$MC_{\text{as-recd}} = \frac{m_{\text{w}}}{m_{\text{as-recd}}} = \frac{m_{\text{w}}}{m_{\text{daf}} + m_{\text{w}} + m_{\text{ash}}} \tag{2.7}$$

On a dry basis moisture content is

$$MC_{\text{dry}} = \frac{m_{\text{w}}}{m_{\text{dry}}} = \frac{m_{\text{w}}}{m_{\text{daf}} + m_{\text{ash}}} \tag{2.8}$$

and on a dry, ash-free basis moisture content is

$$MC_{\text{daf}} = \frac{m_{\text{w}}}{m_{\text{daf}}}. \tag{2.9}$$

As can be seen, the basis of the calculation must be specified.

Example 2.2

Wood chips with 40% moisture (as-received) are supplied to a biomass power plant. What is the moisture content on a dry basis?

Solution
From Equation 2.7,

$$M\,C_{\text{as-recd}} = \frac{m_{\text{w}}}{m_{\text{daf}} + m_{\text{w}} + m_{\text{ash}}}$$

rearranging and simplifying

$$m_{\text{w}} = \frac{M\,C_{\text{as-recd}}}{1 - M\,C_{\text{as-recd}}} \left(m_{\text{daf}} + m_{\text{ash}} \right)$$

$$m_{\text{w}} = \frac{0.4}{0.6} \left(m_{\text{daf}} + m_{\text{ash}} \right)$$

Substituting for m_{w} in Equation 2.8 yields

$$M\,C_{\text{dry}} = 0.667 = 66.7\%$$

2.3.1 BIOMASS

Biomass refers to a range of organic materials recently produced from plants and waste from animals that feed on plants. Biomass includes crop residues, forest and wood processing residues, dedicated energy crops including grasses and trees, livestock and poultry wastes, municipal solid waste (excluding plastics and non-organic components), and food processing waste. Biomass is a cellulosic material that can be broadly classified as either woody or herbaceous (non-woody), and as either residues or dedicated energy crops.

Dedicated *woody energy crops* are grown on tree farms and harvested in two to seven years depending on the tree type and location. Many types of trees can be farmed. In the United States, cottonwoods, hybrid poplars, and willows have been found to have high growth rates in northern climates, whereas eucalyptus and lobolly pines are fast-growing tree crops in subtropical and tropical climates. The best varieties on well-managed sites currently yield 5–17 dry-kg/ha/yr with demonstrated genetic potential to yield up to 30 dry-kg/ha/yr. There are 17 million hectares (42 million acres) of tree farms in the United States, much of it currently dedicated to paper and cardboard. *Herbaceous energy crops* are harvested annually, and some examples include corn, sugar beets, switchgrass, and miscanthus. Currently in the United States there are 12 million hectares (30 million acres) dedicated to corn for ethanol as a replacement for gasoline. High yield herbaceous crops have a similar or higher yield compared to short rotation woody crops but require more fertilizer.

The as-received moisture content of biomass varies widely from approximately 5% for well-dried material to greater than 50% for green wood. Within wood, water is first taken up as bound water until all available adsorption sites are occupied, which is the fiber saturation point, and then becomes free water within the pores of the wood. At the fiber saturation point, the heat of sorption is near zero. The adsorbed water is held with increasing energy as the wood moisture content decreases. This can be expressed as

$$h_{sorp} = 0.4 \, h_{fg} \left(1 - \frac{MC_b}{MC_{fsp}} \right)^2 \tag{2.10}$$

where MC_b and MC_{fsp} are the moisture content of the bound water and the moisture content at the fiber saturation point, respectively. The fiber saturation point of wood is approximately 30% moisture on a dry basis.

Dry biomass consists of cellulose, hemicellulose, lignin, resins (extractives), and ash forming minerals. Cellulose ($C_6H_{10}O_5$) is a condensed polymer of glucose ($C_6H_{12}O_6$). Hemicellulose consists of various sugars other than glucose that encase the cellulose fibers and represent 20%–35% of the dry weight for wood. Lignin ($C_{40}H_{44}O_6$) is a non-sugar polymer that gives strength to the fiber, accounting for 15%–30% of the dry weight. Extractives include oils, resins, gums, fats, and waxes. Inorganic constituents that make up the ash are mainly calcium, potassium, magnesium, manganese, and sodium oxides and lesser amounts of other oxides such as silica, iron, and aluminum. The mineral matter is dispersed throughout the cells in molecular form.

Major elements in wood are carbon, hydrogen, oxygen, nitrogen, and sulfur. The carbon content of wood is about 50%, slightly higher in softwoods and slightly lower in hardwoods, due to differences in lignin and extractive content. Hydrogen content is about 6% in all species. Oxygen content is 40%–44%, nitrogen content is 0.1%–0.2%, and sulfur content is less than 0.1%. The HHV of wood species varies less than 15%, with softwoods at 20–22 MJ/kg and hardwoods at 19–21 MJ/kg. The elemental analysis of switchgrass is similar to that of wood: carbon content is 46%–48%, hydrogen content is about 6%, and oxygen content is 40%–42% by weight. Ash content of switchgrass is generally in the range of 6%. Nitrogen content is a function of the fertilization schedule but is in the range of 0.6%. The HHV of switchgrass is approximately 18 MJ/kg. As shown in Table 2.10, woods tend to have the highest carbon content and the lowest nitrogen and ash contents of the biofuels. Straw and grass tend to have lower levels of carbon and higher nitrogen and ash content.

Several researchers have proposed empirical formulas that link the HHV and the ultimate analysis of biofuels. Reed presented the following generalized equation correlating the dry basis HHV with the elemental composition on a weight basis.

$$HHV_{dry} = 0.341C + 1.322H - 0.12(O+N) - 0.153A + 0.0686 \ (MJ/kg) \quad (2.11)$$

TABLE 2.10
Ultimate Analysis (wt %) and HHV for Selected Solid Biofuels (Dry Basis)

Biomass	C	H	O	N	S	Ash	HHV (MJ/kg)
Kelp, giant brown, Monterey	26.6	3.7	20.2	2.6	1.1	45.8	10.3
Mango wood	46.2	6.1	44.4	0.3	0.0	3.0	19.2
Maple	50.6	6	41.7	0.3	0.0	1.4	19.9
Oak	49.9	5.9	41.8	0.3	0.0	2.1	19.4
Pine	51.4	6.2	42.1	0.1	0.1	0.1	20.3
Pine, bark	52.3	5.8	38.8	0.2	0.0	2.9	20.4
Poplar, hybrid	50.2	6.1	40.4	0.6	0.0	2.7	19.0
Rice hulls	38.5	5.7	39.8	0.5	0.0	15.5	15.3
Rice straw	39.2	5.1	35.8	0.6	0.1	19.2	15.2
Sudan grass	45.0	5.5	39.6	1.2	0.0	8.7	17.4
Switchgrass, Dakota Leaf, MN	47.4	5.8	42.4	0.7	0.1	3.6	18.6

Source: Bain, R. L., Amos, W. P., Downing, M., and Perlack, R. L., *Biopower Technical Assessment: State of the Industry and the Technology*, NREL Report No. TP-510-33123, National Renewable Energy Laboratory, Department of Energy Laboratory, Golden, CO, 2003.

where C, H, O, N, and A are the weight percents of carbon, hydrogen, oxygen, nitrogen, and ash, respectively. This equation has been used with biomass, charcoals, pyrolysis oils, and tars; and the average absolute error was within 1.6%, 1.6%, 2.5%, and 2.5%, respectively.

Charcoal is made by heating wood in the absence of air to produce wood char, which is mostly solid carbon. The most prevalent use of charcoal is for cooking and heating in small stoves in developing countries. Because the energy density by weight of charcoal is high relative to the energy density of wood, it can be more easily transported to market than wood. Many users of small cookstoves prefer charcoal to wood or other biofuels because charcoal combustion does not produce as much smoke as wood, and a charcoal stove requires less tending. However, charcoal stoves can produce high levels of carbon monoxide and present significant risk when used indoors. In many commercial applications, charcoal is a relatively clean burning fuel. In some countries "charcoal" briquettes are made commercially from a mixture of pulverized coal, clay, and a binder such as starch. The heating value and composition of charcoal varies depending on the material used to make the charcoal and the processing method. Carbon content in charcoal is generally 65%–95% by weight. The remainder constituents of charcoal are primarily oxygen and ash. The HHV of char is generally 27–31 MJ/kg.

Biomass fuels are often sold by volume rather than weight. For example, in the United States, wood fuel is generally sold by the cord. In the developing world, charcoal is often sold by the sack. The bulk density of a fuel is the mass of the fuel particles divided by the volume occupied by the particles. The bulk density is a function of particle shape and size, how the fuel was stacked (randomly or in a fixed pattern), and the density of the fuel. The bulk density of fuels varies widely but will generally be about 40% of the density of the biomass fuel. For example, the bulk density of dry softwood chips is roughly 150 kg/m^3 and the dry density of softwood is in the range of 400 kg/m^3.

2.3.2 Peat

Peat is formed from decaying woody plants, reeds, sedges, and mosses in watery bogs and marshlands and is usually formed in northern climates. Peat forms in wet environments in which air is largely excluded. In the presence of bacterial action, chemical decomposition proceeds by a process called humification. Peat is a mixture of carbon, hydrogen, and oxygen in a ratio similar to cellulose and lignin. Some of the plant material in bogs decomposes into humic acid, bitumens, and other compounds rather than peat.

Since the rate of formation of a peat bed is about 3 cm per 100 years, peat is generally not a renewable resource, but it has been used as a fuel in some northern regions. Peat is usually brown in color and fibrous in character. Because freshly harvested peat typically contains 80%–90% water, it must be dried before being used as a fuel. Air drying peat reduces the water content to 25%–50%. Peat contains 1%–10% mineral matter on a dry basis.

2.3.3 COAL

Coal is a heterogeneous mineral consisting principally of carbon, hydrogen, and oxygen, with lesser amounts of sulfur and nitrogen. Other constituents are the ash forming inorganic compounds distributed throughout the coal. Coal originated through the accumulation of wood and other biomass that was covered, compacted, and transformed over a period of hundreds of thousands of years. Most bituminous coal seams were deposited in wetlands that were regularly flooded with nutrient-containing water that supported abundant peat-forming vegetation. The lower levels of the wetlands were anaerobic and acidic, which promoted structural changes and biochemical decompositions of the plant remnants. This microbial and chemical alteration of the cellulose, lignin, and other plant substances, and later the increasing depth of burial, resulted in a decrease in the percentage of moisture and a gradual increase in the percentage of carbon. This change from peat through the stages of lignite, bituminous coal, and ultimately to anthracite coal (the process called coalification) is characterized physically by decreasing porosity and increasing gelification and vitrification. Chemically, there is a decrease in volatile matter content as well as an increase in the percentage of carbon, a gradual decrease in the percentage of oxygen, and, as the anthracite stage is approached, a decrease in the percentage of hydrogen.

The changes involved in the coalification process are known as advances in the rank of coal. Rank expresses the progressive metamorphism of coal from lignite (low rank) to anthracite (high rank). Rank is based on the heating value for low rank coals and on the percentage of fixed carbon for higher rank coals (Table 2.11). As shown in Figure 2.3, the heating value and percentage of fixed carbon increase as the rank moves from lignite to low volatile bituminous coal, and the volatile matter decreases. The percentage of oxygen, which is contained in the volatile matter, also decreases.

Lignite is a brownish-black coal of low rank. Lignite is also referred to as brown coal. Chemically, lignite is similar to peat in that it contains a large percentage of water and volatiles. Mechanically, lignite is easily fractured and is not spongy like peat. Subbituminous coal is dull black, shows little woody material, and often appears banded. This type of coal usually fractures along the banded planes. In addition, the moisture content of subbituminous coal is reduced. Bituminous coal is dark black and is often banded. The moisture content is low and the volatile content ranges from high to medium. Bituminous coal is more resistant to disintegration in air than lignite and subbituminous coals. Anthracite coal is hard, brittle, and has a bright luster. It has almost no volatiles or moisture and is not banded. The heating value of anthracite is slightly lower than that of bituminous coal because of its reduced hydrogen content.

The chemical composition of coal is complex and varies widely from location to location. Coal is composed of a vast array of organic compounds. Benzenoid ring units play an important role in the coal structure. Hydrogen, oxygen, nitrogen, and sulfur are attached to the carbon skeleton. Inorganic minerals form the ash that remains when coal is burned. Nitrogen in coal is organic and varies up to a few percent by weight. Sulfur in coal consists of organic and inorganic forms. Organic

TABLE 2.11

Classification of Different Types of Coal by Rank (Dry, Ash-free Basis)

Rank	Fixed Carbon (%)	HHV (MJ/kg)
Meta-anthracite	>98	
Anthracite	92–98	
Semianthracite	86–92	
Low volatile bituminous	78–86	
Medium volatile bituminous	69–78	
High volatile A bituminous		>32.5
High volatile B bituminous		30.2–32.5
High volatile C bituminous		26.7–30.2
Subbutiminous A		24.4–26.7
Subbituminous B		22.1–24.4
Subbituminous C		19.3–22.1
Lignite A		14.6–19.3
Lignite B		<14.6

Source: *Annual Book of ASTM Standards*, Volume 5.06: Gaseous Fuels, Coal and Coke. ASTM International, West Conshohocken, PA, 2009.

sulfur, which is bound into the coal, varies widely from a small fraction of a percent to 8%. Inorganic sulfur is predominantly found as iron pyrite, FeS_2, and varies from zero up to a few percent. Pyritic sulfur may be removed by coal cleaning methods, whereas organic sulfur is distributed throughout and requires chemical degradation to release the sulfur.

Mineral matter in coal consists of kaolinite, detrital clay, pyrite, and calcite, and hence includes oxides of silicon, aluminum, iron, and calcium. Lesser but significant

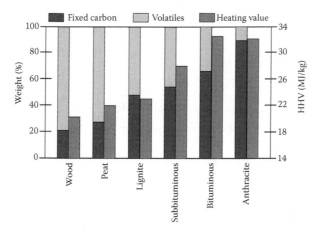

FIGURE 2.3 Typical fixed carbon, volatile matter, and HHV in solid fuels.

amounts of magnesium, sodium, potassium, manganese, and phosphorus are found. Many trace elements are found at lower levels. Mineral matter in coal varies widely and is present in molecular form as bands between layers of coal and in some instances is added from the overburden during mining.

Some coals melt and become plastic when heated and give off tars, liquors, and gases, leaving a residue called *coke*. Coke is a strong, porous residue, consisting of carbon and mineral ash that is formed when the volatile constituents of bituminous coal are driven off by heat in the absence of or in a limited supply of air. Coals that do not melt also give off tars, liquors, and gases when heated and leave a residue of friable *char* instead of coke.

Coal may be classified according to rank and grade. The grade of coal, which is independent of rank, depends on ash content, ash fusion temperature, sulfur content, and the presence of other deleterious constituents. Coal grade is used more qualitatively than rank. The grade of coal can be improved by coal cleaning methods to remove some ash and pyrite sulfur. Mechanical cleaning processes include crushing, washing, dewatering, and drying. Coarse and medium sizes are cleaned by gravity separation, while the finer sizes are cleaned by froth floatation. Coal typically has a specific gravity of 1.1–1.3. When tiny particles are added to water to bring the specific gravity to about 1.5, coal particles float while free mineral impurities tend to sink. In froth floatation, the small coal particles are buoyed up to the top of a controlled surface froth, while the heavier impurities sink to the bottom. Mechanical cleaning can reduce the ash content by 10%–70% and the sulfur content by up to 35%. Ultrafine grinding (micronization) of coal to an average particle size of 10 μm followed by washing has produced some coals with less than 1% ash.

2.3.4 Refuse-Derived Fuels

Refuse solid fuel includes municipal solid waste (MSW) and agricultural waste. MSW includes waste from residential, commercial, institutional, and industrial sources. MSW includes packaging, newspapers, waste paper, bottles and cans, boxes, wood pallets, food scraps, grass clippings, clothing, furniture, vehicle tires, and durable goods. Often these waste materials are disposed of in landfills without recycling or energy recovery. In some communities, an effort is being made to recycle some of the waste materials and to recover energy. In the United States in 2007, over 254 million tons of MSW was generated, of which 32 million tons was combusted for energy recovery (Table 2.12).

Refuse can be burned directly (referred to as mass burn) in specially designed boilers, but to reduce harmful emissions, the refuse is processed to separate the combustibles from the non-combustibles using shredding, magnetic separation, screening, and air classification. Processing facilitates recovery of metals and glass as well as controlling the fuel size. The processed refuse is called refuse-derived fuel (RDF). Non-combustibles in RDF typically vary from 10% to 15%. Because of various undesirable compounds in refuse and RDF, composting and recovery of the resulting gas for energy conversion rather than direct combustion is often considered preferable because composting does not create hazardous air

TABLE 2.12

Municipal Solid Waste Generation and Recovery in the U.S. in 2007 (in millions of tons)

Generation	254
Recovery for recycling	63
Recovery for composting	22
Combustion with energy recovery[a]	32
Discards to landfill[b]	137

Source: United States Environmental Protection Agency, *Municipal Solid Waste in the United States: 2007 Facts and Figures*, Office of Solid Waste, EPA530-R-08-010, 2008a.

[a] Includes combustion of MSW in mass burn and RDF form with energy recovery.

[b] Discards include combustion without energy recovery.

emissions. When combusting MSW in any form, effective emission control equipment is required.

2.3.5 CHARACTERIZATION OF SOLID FUELS

Standard testing and analysis of solid fuels is prescribed by ASTM standards. The proximate analysis, ultimate analysis, heating value, grindability, free swelling index, and ash fusion temperature are discussed here.

The *proximate analysis* in ASTM D3172 (*Annual Book of ASTM Standards*, Vol. 5.06 2009) determines the moisture, volatile combustible matter, fixed carbon, and ash in a coal sample. Although the determination is done quite accurately, the name proximate indicates the empirical nature of the method; a change in procedure can change the results. A sample of coal is crushed and dried in an oven at 105°C–110°C to constant weight to determine residual moisture. The sample is then heated in a covered crucible (to prevent oxidation) at 900°C to constant weight. This weight loss is referred to as volatile matter. The remaining sample is then placed in the oven at 750°C with the cover removed so that the sample is combusted. The weight loss on completion of combustion is termed fixed carbon or char. The remaining residue is defined as ash. The components of a proximate analysis are rather arbitrary. There is no sharp distinction between free water and water chemically bound to the fuel. The split between volatile matter and fixed carbon depends on the rate of heating as well as the final temperature. Moreover, a small amount of ash can be volatilized during char determination. Nevertheless the proximate analysis provides a useful comparison between fuels.

The *ultimate analysis* in ASTM D3176 (*Annual Book of ASTM Standards*, Vol. 5.06 2009) provides the major elemental composition of the coal, usually reported on a dry, ash-free basis. Burning a fuel sample in oxygen in a closed system and quantitatively measuring the combustion products is used to determine the carbon and hydrogen content of the fuel. The carbon includes organic carbon as well as carbon

TABLE 2.13

Representative Proximate Analysis, Ultimate Analysis, and Heating Value of Solid Fuels (Dry, Ash-free)

Fuel Type	Wood	Peat	Lignite	Bituminous Coal	Refuse-Derived Fuel
Proximate analysis, wt %					
Volatile matter	81	65	55	40	85
Fixed carbon	19	35	45	60	15
Ultimate analysis, wt %					
Hydrogen	6	6	5	5	7
Carbon	50	55	68	78	52
Sulfur	0.1	0.4	1	2	0.3
Nitrogen	0.1	0.6	1	2	0.7
Oxygen	44	38	25	13	40
HHV					
(Btu/lb$_m$)	8700	9500	10,700	14,000	9700
(MJ/kg)	20.2	22.1	24.9	32.5	22.5

from mineral carbonates. The hydrogen includes organic hydrogen as well as any hydrogen from the moisture of the dried sample and mineral hydrates. The extraneous carbon and hydrogen are usually negligible. Nitrogen and sulfur are determined chemically. Oxygen is determined by the difference between 100 and the sum of the percentages of carbon, hydrogen, nitrogen, and sulfur. Sometimes chlorine is included in the ultimate analysis.

Heating value is determined in a bomb calorimeter in ASTM D5865 (*Annual Book of ASTM Standards*, Vol. 5.06 2009). A small sample of coal is placed in the calorimeter, which is pressurized with excess oxygen. A spark ignites the sample, and the temperature rise in a surrounding water jacket is measured. Although the bomb calorimeter gives the heating value at a constant volume, the difference between the constant volume and constant pressure heating values is essentially negligible. A HHV means that the water of combustion has condensed, whereas for a LHV the moisture has not condensed.

The analysis procedures for biomass and RDF are very similiar to those of coal. The analysis procedures for proximate analysis, ultimate analysis, and heating value of wood are given in ASTM E870 (*Annual Book of ASTM Standards*, Vol. 11.06 2009). The analysis procedure for determining the heating value of RDF is given in ASTM E711 (*Annual Book of ASTM Standards*, Vol. 11.04 2009). The proximate analysis, ultimate analysis, and HHV of representative samples of wood, peat, coal, and refuse-derived fuel are given in Table 2.13.

Example 2.3

A solid fuel contains 6% hydrogen, 30% moisture, and 10% ash (wt %, as-received) and has a HHV of 11.6 MJ/kg (as-received). What is the dry, ash-free LHV?

Solution

To find the LHV on a dry, ash-free basis, we first determine the HHV on a dry, ash-free basis.

$$HHV_{daf} = HHV_{as\text{-}recd} \frac{m_{as\text{-}recd}}{m_{daf}}$$

Assume a 1 kg basis,

$$HHV_{daf} = \frac{11.6 \, MJ}{kg_{as\text{-}recd}} \cdot \frac{1 \, kg_{as\text{-}recd}}{(1 - 0.3 - 0.1) \, kg_{daf}} = \frac{19.3 \, MJ}{kg_{daf}}$$

Dry, ash-free fuel contains

$$\frac{0.06}{1 - 0.3 - 0.1} = 10\% \text{ hydrogen}$$

and thus,

$$\frac{m_{H_2O}}{m_{daf}} = \frac{0.1 \, kg_{H_2}}{kg_{daf}} \cdot \frac{kgm \, o_{H_2}}{2 kg_{H_2}} \cdot \frac{1 \, kgm \, o_{H_2O}}{kgm \, o_{H_2}} \cdot \frac{18 \, kg_{H_2O}}{kgm \, o_{H_2O}} = \frac{0.90 \, kg_{H_2O}}{kg_{daf}}$$

Since the latent heat of vaporization of water is 2.44 MJ/kg at 25°C, and using Equation 2.1, the dry, ash-free LHV is

$$LHV_{daf} = \frac{19.3 \, MJ}{kg_{daf}} - \frac{0.9 \, kg_{H_2O}}{kg_{daf}} \cdot \frac{2.44 \, MJ}{kg_{H_2O}} = \frac{17.1 \, MJ}{kg_{daf}}$$

The *Hardgrove grindability* test in ASTM D409 (*Annual Book of ASTM Standards*, Vol. 5.06 2009) is used to determine the relative ease of pulverization of coals. The coal is ground in a stationary grinding bowl that holds eight steel balls each 25 mm in diameter. The balls are driven by an upper grinding ring that rotates at 20 rpm by a spindle that exerts a vertical force of 284.4 N. A 50 g sample is ground for 60 revolutions. The amount of coal passing a No. 200 sieve (75 μm opening) relative to a standard sample is the Hardgrove grindability index.

The *free swelling index* in ASTM D720 (*Annual Book of ASTM Standards*, Vol. 5.06 2009) is an indication of the caking characteristics of coals when burned as a fuel. A 1 g sample of ground coal is placed in a small crucible of a specific size. The crucible is covered and placed in an oven at 820°C for 2.5 min. The coke button is then removed from the crucible, and the increase in projected cross-sectional area is noted; a free swelling index of 2 means that the projected area has doubled.

When ash is heated to a softened state, it has a tendency to foul boiler tubes and surfaces. The *ash fusion temperature* in ASTM D1857 (*Annual Book of ASTM Standards*, Vol. 5.06 2009) is determined by heating a ground fuel sample at 850°C

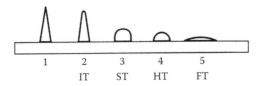

FIGURE 2.4 Determination of ash fusion temperature points: (1) before heating, (2) initial deformation temperature, (3) softening temperature (height = width), (4) hemispherical temperature (height = width/2), (5) fluid temperature. (From ASTM D1857 in the *Annual Book of ASTM Standards*, Vol. 5.06. ASTM International, West Conshohocken, PA, 2009.)

in air and then in oxygen to ensure complete oxidation of the fuel. The powdered ash is then mixed with a solution of dextrin to form a stiff paste, which is pressed into small cone-shaped molds. The ash is removed from the mold and placed in a furnace in which the temperature is slowly rising. The shape of the cone is observed at the five positions indicated in Figure 2.4, and the temperatures reported are at each position. The deformation temperature comes up at the first rounding of the apex of the cone, and the softening temperature occurs when the cone has fused into a spherical lump in which the height is equal to the width at the base. The hemispherical temperature is where the height is one-half the width of the base, and the fluid temperature is where the fused mass has spread out in a nearly flat layer with a maximum height of 1.6 mm. Ash fusion temperatures are reported for oxidizing atmospheres (air) and reducing atmospheres (60% CO and 40% CO_2). Ash fusion temperatures for a representative bituminous coal ash under oxidizing and reducing conditions are given in Table 2.14. The temperature differential between initial deformation and fluid temperature gives an indication of the type of deposit on furnace tube surfaces. A small temperature difference indicates that the surface slag will be thin and tenacious. A larger temperature difference indicates the slag deposit may build up in thicker layers.

Ash fusion temperatures depend on the composition of the ash. Acidic oxide constituents, such as SiO_2, Al_2O_3, and TiO_2, tend to produce higher melting temperatures, whereas basic oxides, such as Fe_2O_3, CaO, MgO, Na_2O, and K_2O, tend to produce lower melting temperatures. The combustion engineer should note that the ash fusion temperatures are typically below the flame temperature and above the steam and wall temperatures, thus resulting in fouling and slagging. Herein lies one of the challenges of utilizing solid fuels.

TABLE 2.14

Typical Ash Fusion Temperatures for a Representative Bituminous Coal

Deformation	Reducing (°C)	Oxidizing (°C)
Initial deformation temperature	1100	1125
Softening (height = width) temperature	1205	1230
Hemispherical (height = 1/2 width) temperature	1230	1255
Fluid temperature	1330	1365

2.4 PROBLEMS

1. In the United States, electricity is sold using the SI units of kilowatt (kW) or megawatt (MW), and fuel for power generation stations is generally purchased using English units, e.g., British thermal unit (Btu), gallon (gal), and ton. An industrial power plant has an average annual load of 100 MW (electrical). If the overall thermal efficiency is 33% (based on HHV), what is the annual cost of fuel if the plant is fired with (a) natural gas, (b) No. 2 fuel oil, and (c) bituminous coal? Use the data from Tables 2.2, 2.7, and 2.13. Assume the cost of natural gas is $5/million Btu, the cost of No. 2 fuel oil is $3/gal, and the cost of bituminous coal is $60/ton (1 ton = 2000 lb_m).

2. In the United States, energy for household heating is generally sold using English units, e.g., therm, gal, and cord. A house in Wisconsin uses 1200 therms of thermal energy during the heating season. Calculate the cost of fuel if the furnace uses (a) natural gas with an efficiency of 70%; (b) No. 2 fuel oil, efficiency 65%; (c) kerosene, efficiency 99.9% (unvented); and (d) wood with 15% moisture with an efficiency of 50%. Use the data in Tables 2.2, 2.7, and 2.13. The efficiencies are based on the HHV. Assume the cost of natural gas is $8/MBtu, the cost of No. 2 fuel oil is $3/gal, the cost of kerosene is $3.50/gal, and the cost of wood is $100/cord. Assume the bulk density of cord wood is 30 lb_m/ft^3.

3. Using Table 2.4, calculate the percent difference in the LHV for ethanol when introduced as a liquid at 16°C and when introduced as a vapor at 16°C. Assume that the vapor is an ideal gas. Repeat the analysis for gasoline and compare the results to ethanol. Assume the molecular weight of gasoline is 100.

4. Find the HHV on a volume basis and weight basis for a mixture of 50% methane and 50% hydrogen by volume. Use the data given in Table 2.2.

5. Natural gas can be simulated in the laboratory by a mixture of 83.4% (vol) methane, 8.6% propane, and 8.0% nitrogen. What is the HHV (MJ/m^3) at 1 atm and 25°C for this mixture, and how does it compare to the HHV given in Table 2.2 for natural gas?

6. Indicate the molecular structure of (a) n-hexane, (b) 3,4 diethylhexane, (c) 1-2,3,3 trimethylbutene, (d) methylnaphthalene, and (e) heptamethylnonane.

7. Kerosene has an API gravity of 42.5. What is the specific gravity of kerosene at 16°C? How does this compare with the value in Table 2.7? What is the energy density (MJ/L) of kerosene at 16°C? Use the data given in Table 2.7.

8. If the price of gasoline, diesel fuel, and ethanol were the same per gallon (after taxes), which would be the best buy for automobile use from a consumer standpoint? Why? Use the data given in Table 2.4.

9. As shown in Table 2.10, maple wood has a HHV (dry basis) of 19.9 MJ/kg. Calculate and tabulate the ash-free HHV and LHV for maple as a function of moisture content from 0% to 60% (as-received).

10. Pelletized peanut shells have been proposed as a solid fuel for household cookstoves in rural villages in developing countries. The ultimate analysis (wt %, dry) of the peanut shells is

C	H	O	N	S	Ash
46.8	5.5	40.1	1.6	0.1	5.9

Determine the as-received HHV and LHV of the peanut shells. If the as-received moisture content of the pellets is 10%, and 10% clay (dry basis) is added as a binder, what will be the as-received LHV of the pelletized fuel? If the cookstove efficiency is 18%, what weight of pellets (with binder) is required to heat 2 L of water from 20°C to 100°C?

11. A bituminous coal contains 10% ash and 5% moisture on an as-received basis, and a refuse-derived fuel contains 13% ash and 20% moisture on an as-received basis. Compare the as-received HHV and LHV of this bituminous coal and refuse-derived fuel using the data of Table 2.13.

12. A coal–water slurry contains 70% dry powdered bituminous coal and 30% water by weight. If the coal has a dry HHV of 30 MJ/kg and contains 5% (wt) hydrogen, find the HHV and LHV of the slurry. Assume the coal is ash-free.

13. A proposed alternative diesel fuel contains 50% finely powdered bituminous coal and 50% ethanol by weight. The coal contains 4% moisture, has been cleaned of ash, and has an as-received HHV of 25 MJ/kg. Assume the coal has 5% hydrogen and 40% volatiles as-received. Find the as-received LHV of this fuel. Estimate the as-received mass fraction (%) of volatiles and the mass fraction (%) of fixed carbon in this fuel.

14. 2,5-Dimethylfuran (DMF), which can be made from cellulosic biomass, has been suggested as a possible liquid transportation fuel. DMF has 40% greater energy density by mass than ethanol based on HHV. DMF is C_6H_8O. Compare the LHV of dimethylfuran and ethanol on a weight and volume basis. Use the data in Table 2.4 for ethanol. The specific gravity of DMF is 0.9.

15. Heating water from 20°C to 100°C using an open fire for cooking and washing is a daily task in the developing world. Approximately one-third of the world's population uses an open fire with a heat transfer efficiency of approximately 5%. Consider the difference between cooking with wood and cooking with braided grass.

    a. Assume the wood has 25% moisture and 0.8% ash on an as-received basis. The bulk density of stacked wood is 45% of the density of the wood. Assume that the dry, ash-free composition of the wood is 51.2%, 5.8%, 42.4%, 0.6%, and 0% for C, H, O, N, and S, respectively. The dry density of the wood is 640 kg/m³. The dry, ash-free HHV of the wood is 19.7 MJ/kg. What is the as-received LHV of the wood (MJ/kg)? What is the as-received density of the wood (kg/m³)? What is the as-received bulk density of the wood (kg/m³)? What is the bulk volume (L) and weight (kg) of wood needed to heat 5 L of water?

    b. Some cultures use braided grasses for cooking. Assume the grass has 2% moisture and 10% ash on an as-received basis. Assume that the dry, ash-free composition of the grass is 49.4%, 5.9%, 43.4%, 1.3%, and 0.0% for C, H, O, N, and S; respectively. The bulk density of braided grass is 20% of the density of the grass. The dry density of the grass is 250 kg/m³. The dry, ash-free HHV of the grass is 17.4 MJ/kg. What is the as-received LHV of the grass (kJ/kg)? What is the as-received density of the grass (kg/m³)? What is the as-received bulk density of the grass (kg/m³)? What is the bulk volume (L) and weight (kg) of grass needed to heat 5 L of water?

    c. Report your results in a table comparing wood and grass. What are your observations?

REFERENCES

Annual Book of ASTM Standards, Vol. 4.10: Wood, Vols. 5.01–5.03: Petroleum Products and Lubricants, Vol. 5.06: Gaseous Fuels; Coal and Coke, Vol. 11.04: Waste Management, Vol. 11.06: Biological Effects and Environmental Fate; Biotechnology, ASTM International, West Conshohocken, PA, 2009.

API Research Project No. 44, National Bureau of Standards, Washington, DC, December 1952.

Bain, R. L., Amos, W. P., Downing, M., and Perlack, R. L., *Biopower Technical Assessment: State of the Industry and the Technology*, NREL Report No. TP-510-33123, National Renewable Energy Laboratory, Department of Energy Laboratory, Golden, CO, 2003.

Bartok, W. and Sarofim, A. F., *Fossil Fuel Combustion: A Source Book*, Wiley, New York, 1991.

Bennethum, J. E. and Winsor, R. E., "Toward Improved Diesel Fuel," SAE paper 912325, 1991.

Koehl, W. J., Painter, L. J., Reuter, R. M., Benson, J. D., Burns, V. B., Gorse, R. A., and Hochhauser, A. M., "Effects of Gasoline Sulfur Level on Mass Exhaust Emissions," SAE paper no. 912323, 1991.

Brown, R. C., *Biorenewable Resources: Engineering New Products from Agriculture*, Iowa State Press, Ames, IA, 2003.

CFR (Code of Federal Regulations) Part 86 Subpart A, para. 86.113–82, p. 482, "Protection of the Environment," July 1, 1985.

Food and Agriculture Organization of the United Nations, *State of the World's Forests, 2009*, Rome, 2009.

Francis, W. and Peters, M. L., *Fuels and Fuel Technology: A Summarized Manual*, 2nd ed., Pergamon Press, Oxford, 1980.

Hancock, E. G., ed., *Technology of Gasoline,* Critical Reports on Applied Chemistry, 10, Blackwell Scientific, London, 1985.

Hochhauser, A. M., Benson, J. D., Burns, V. R., Gorse, R. A., Koehl, W. J., Painter, L. J., Reuter, R. M., and Rutherford, J. A., "Fuel Composition Effects on Automotive Fuel Economy— Auto/Oil Air Quality Improvement Research Program," SAE paper 930138, 1993.

International Energy Agency, *Key World Energy Statistics 2009*, IEA, Paris, 2009.

International Energy Agency Bioenergy, *Potential Contribution of Bioenergy to the World's Future Energy Demand*, IEA Bioenergy: ExCo: 2007:02, 2007.

Leppard, W. R., "The Chemical Origin of Fuel Octane Sensitivity," SAE paper 902137, 1990.

Lide, D. R. and Haynes, W. M., eds., *CRC Handbook of Chemistry and Physics*, 90th ed., CRC Press, Boca Raton, FL, 2009.

Lieuwen, T., Yang, V., and Yetter, R., eds., *Synthesis Gas Combustion: Fundamentals and Applications*, CRC Press, Boca Raton, FL, 2010.

Owen, K., Coley, T. R., and Weaver, C. S., *Automotive Fuels Reference Book*, 3rd ed., SAE International, Warrendale, PA, 2005.

Perlack, R. D., Wright, L. L., Turhollow, A. F., Graham, R. L., Stokes, B. J., and Erbach, D. C., *Biomass as Feedstock for a Bioenergy and Bioproducts Industry: The Technical Feasibility of a Billion-Ton Annual Supply*, USDA/GO-102005-2135 and ORNL/TM-2005/66, US Department of Agriculture and US Department of Energy, Washington, DC, 2005.

Green, D. W. and Perry, R. H., *Perry's Chemical Engineers Handbook*, 8th ed., McGraw-Hill Professional, New York, 2007.

Ragland, K. W., Aerts, D. J., and Baker, A. J., "Properties of Wood for Combustion Analysis," *Bioresour. Technol* 37(2):161–168, 1991.

Reed, R. J., *North American Combustion Handbook: A Basic Reference on the Art and Science of Industrial Heating with Gaseous and Liquid Fuels*, Vol. 1, North American Mfg. Co., 3rd ed., 1985.

Singer, J. G., ed., *Combustion Fossil Power: A Reference Book on Fuel Burning and Steam Generation*, 4th ed., Combustion Engineering Power Systems Group, Windsor, CT, 1993.

Springer, K. J., "Energy, Efficiency and the Environment: Three Big Es of Transportation," *J. Eng. Gas Turbines Power* 114(3):445–458, 1992.

United States Department of Energy, *Thermal Systems for Conversion of Municipal Solid Waste*, Argonne National Laboratory, ANL/CNSV-TM-120, 6 vols., 1983.

United States Environmental Protection Agency, *Municipal Solid Waste in the United States: 2007 Facts and Figures*, Office of Solid Waste, EPA530-R-08-010, 2008a.

———, *Fuel Trends Report: Gasoline 1995–2005*, Office of Transportation and Air Quality, EPA420-R-08-002, 2008b.

3 Thermodynamics of Combustion

The engineering and design of combustion devices begins with the application of the laws of thermodynamics. Thermodynamics provides the laws of conservation of mass, species, and energy; defines efficiency; and deals with equilibrium states and chemical composition of the reactants and products. In this chapter we review the first law of thermodynamics for closed and open systems, the properties of mixtures, and combustion stoichiometry. In addition we examine the principles of equilibrium, which state how the chemical composition can be determined for a system of known atomic or molecular composition if two independent thermodynamic properties are known. Although systems undergoing chemical reaction are generally far from chemical equilibrium, in many cases of interest a small enough control volume can be selected so that pressure and temperature are uniform within the control volume. The chemical composition at a given instant in time is dictated by the thermodynamic properties, chemical reaction rates, and fluid dynamics of the system. In this chapter we assume that chemical reactions are in equilibrium (i.e., not rate controlled) and that there are no gradients and hence no fluid dynamics.

3.1 REVIEW OF FIRST LAW CONCEPTS

For a *closed thermodynamic system* (i.e., a quantity of matter of fixed mass) the first law of thermodynamics in rate form states that

$$
\left\{ \begin{array}{c} \text{the rate of change} \\ \text{of energy} \\ \text{of a system} \end{array} \right\} = \left\{ \begin{array}{c} \text{rate of work} \\ \text{done} \\ \text{on the system} \end{array} \right\} + \left\{ \begin{array}{c} \text{the rate at which} \\ \text{heat is transferred} \\ \text{to the system} \end{array} \right\}
$$

If we neglect potential and kinetic energy within the system, the energy in the system consists of the internal energy. In this case the first law of thermodynamics may be written as

$$
\frac{d(mu)}{dt} = \dot{W} + q \tag{3.1}
$$

Where m is the mass of the system, u is the internal energy per unit mass, $\dot{W}$ is the rate at which work is done on the system (i.e., the shaft power), and q is the rate at which heat is transferred to the system. The internal energy of the system includes (1) the thermal (or sensible) energy due to translation, rotation, and vibration of the molecules, and (2) the chemical energy due to chemical bonds between atoms in the molecules. Shaft power generally includes the mechanical power, electrical power, and power due to moving boundaries. It does not include flow work in open systems. This text follows the sign convention in which heat transfer into the system and work done on the system are positive. Conversely heat transfer out of the system and work done by the system are negative.

If the work done on or by a system of volume V is restricted to the mechanical work due to a uniform system pressure pushing against a moving boundary of area A which is moving outward with a velocity dx/dt, then the power is

$$\dot{W} = -pA\frac{dx}{dt} = -p\frac{dV}{dt} \tag{3.2}$$

Substituting, the first law for a closed system becomes

$$\frac{d(mu)}{dt} = -p\frac{dV}{dt} + q \tag{3.3}$$

This form of the first law of thermodynamics emphasizes that time is the independent variable and can be numerically integrated for complex systems, such as internal combustion engines.

Integrating Equation 3.3 with respect to time, the closed system energy balance becomes

$$m(u_2 - u_1) = W_{12} + Q_{12} \tag{3.4}$$

where

$$W_{12} = -\int_{V_1}^{V_2} p\,dV$$

and

$$Q_{12} = \int_{t_1}^{t_2} q\,dt$$

For a *uniform system with constant pressure*, Equation 3.3 simplifies to

$$\frac{d(mu + pV)}{dt} = q \tag{3.5}$$

or

$$\frac{d(mh)}{dt} = q \tag{3.6}$$

where h is the enthalpy per unit mass. Integration of Equation 3.6 yields

$$m(h_2 - h_1) = Q_{12} \tag{3.7}$$

Assuming that the chemical composition is constant, the chemical energy does not change and only the sensible enthalpy need be considered. For an ideal gas the change in internal energy and enthalpy are given by

$$u_2 - u_1 = \int_{T_1}^{T_2} c_v \, dT \tag{3.8}$$

$$h_2 - h_1 = \int_{T_1}^{T_2} c_p \, dT \tag{3.9}$$

where $c_p = c_v + R$, c_p is the constant pressure specific heat, c_v is the constant volume specific heat, and R is the specific gas constant for the fluid.

Open thermodynamic systems have mass flow across the boundaries of the control volume. This mass flow convects energy (e.g., internal energy, kinetic energy, and potential energy) in and out of the system. In addition, flow work is done on the system to move the fluid across the boundary. Because of this, additional terms must be added to Equation 3.1 to apply conservation of energy to an open system. Neglecting potential energy, conservation of energy for an open system is

$$\frac{d(mu)}{dt} = \sum_n \dot{m}_n \left(u_n + \frac{V_n^2}{2} + p_n v_n \right) + \dot{W} + q \tag{3.10}$$

where $V_n^2/2$ is the kinetic energy and pv is the flow work. Noting that, $h = u + pv$, the open system energy equation can be simplified to

$$\frac{d(mu)}{dt} = \sum_n \dot{m}_n \left(h_n + \frac{V_n^2}{2} \right) + \dot{W} + q \tag{3.11}$$

The first term in this equation represents the time rate of change of energy in the system. In this equation $\dot{m}_n$ is the mass flow rate of the fluid through the boundary surface area n, and h_n is the enthalpy of the fluid as it crosses boundary surface area n. The flow rates are given positive values for flow into the system and negative values for flow out of the system. The enthalpy of the fluid includes the sensible enthalpy and chemical energy of the mixture of gases at surface area n. $V_n^2/2$ is the kinetic energy of the flow crossing boundary surface area n, q is the rate of heat transfer to the system, and $\dot{W}$ is the shaft power added to the system. For steady flow with one stream flowing in and one stream flowing out of the control volume, the open system energy equation (Equation 3.11) can be simplified to

$$\dot{m}\left(h_{out} - h_{in} + \frac{V_{out}^2}{2} - \frac{V_{in}^2}{2}\right) = q + \dot{W} \tag{3.12}$$

3.2   PROPERTIES OF MIXTURES

As a starting place for thinking about mixtures of gases, remember that the ideal gas law states

$$pV = mRT \tag{3.13}$$

For mixtures of gases the system mass is obtained from the sum of the mass of each species:

$$m = \sum_i m_i \tag{3.14}$$

Since each species occupies the entire volume, the density of the mixture is the sum of the species' densities:

$$\rho = \sum_i \rho_i \tag{3.15}$$

The mass fraction, y_i, is the mass of species i divided by total mass:

$$y_i = \frac{m_i}{m} = \frac{\rho_i}{\rho} \tag{3.16}$$

and by definition

$$\sum_i y_i = 1 \tag{3.17}$$

Similarly, the mole fraction is the moles of species i divided by the total moles:

$$x_i = \frac{N_i}{N} = \frac{n_i}{n} \tag{3.18}$$

where N_i is the number of moles of species i and N is the total number of moles in the mixture. In the same way n_i is the molar concentration of species i, and n is the molar concentration of all species in the mixture. By definition

$$\sum_i x_i = 1 \tag{3.19}$$

In chemical kinetics the molar concentration of a species is given by the chemical symbol of the species with square brackets around it; for example, [CO], which is in units of mol_{CO}/cm^3.

The specific gas constant, R, for the mixture is

$$R = \frac{\hat{R}}{M} \tag{3.20}$$

where $\hat{R}$ is the universal gas constant (for values and units see the inside cover), and the molecular weight, M, is

$$M = \frac{m}{N} \tag{3.21}$$

From this we see that molecular weight is expressed in units of mass per moles (e.g., kg/kgmol). For a mixture

$$M = \sum_i x_i M_i \tag{3.22}$$

To find the relation between mole fraction and mass fraction we write

$$x_i = \frac{N_i}{N} = \frac{m_i}{M_i} \frac{M}{m} = \frac{M}{M_i} \frac{m_i}{m} \tag{3.23}$$

Thus

$$x_i = \frac{M}{M_i} y_i \tag{3.24}$$

The mixture internal energy, u, and enthalpy, h, per unit mass are given by

$$u = \sum_i y_i u_i \tag{3.25}$$

and

$$h = \sum_i y_i h_i \tag{3.26}$$

Similarly the mixture internal energy, $\hat{u}$, and the enthalpy, $\hat{h}$, per mole of mixture are

$$\hat{u} = \sum_i x_i \hat{u}_i \tag{3.27}$$

and

$$\hat{h} = \sum_i x_i \hat{h}_i \tag{3.28}$$

In the same way c_p, $\hat{c}_p$, c_v, and $\hat{c}_v$ can be found.

The pressure, p, of a mixture of ideal gases is equal to the sum of partial pressures that the component gases would exert if each existed alone in the mixture volume at the mixture temperature:

$$\sum_i p_i = \sum_i x_i p = p \tag{3.29}$$

The volume, V, of a mixture of ideal gases is equal to the sum of the partial volumes that the component gases would occupy if each existed alone at the pressure and temperature of the mixture:

$$\sum_i V_i = \sum_i x_i V = V \tag{3.30}$$

Thus the ideal gas law for a mixture of gases can be written as

$$pV = mRT \tag{3.31}$$

where R is the gas constant for the mixture as defined by Equations 3.20 and 3.22.

Example 3.1

A gaseous mixture at 1000 K contains 25% CO, 10% CO_2, 15% H_2, 4% CH_4, and 46% N_2 by volume. Find (a) the molecular weight, M, and (b) the constant volume specific heat on a mass basis of the mixture, c_v, in SI units.

Solution

Part (a)

We remember that the volume fraction and the mole fraction are the same for an ideal gas. That is, from Equation 3.30

$$x_i = \frac{V_i}{V}$$

Molecular weight can then be found from Equation 3.22:

$$M = \sum_i x_i M_i$$

Species	x_i	M_i	$x_i M_i$
CO	0.25	28	7.0
CO_2	0.10	44	4.4
H_2	0.15	2	3.0
CH_4	0.04	16	0.6
N_2	0.46	28	12.9
sum	1.00		25.2

Thus for the mixture

$$M = 25.2 \, \text{kg/kgmol}$$

Part (b)

Now the challenge is that we need the specific heat for a mixture at a constant volume on a mass basis, but we do not know the mass fractions for the mixture. There are two approaches available: (1) convert the molar gas fractions to a mass fractions basis (Equation 3.24) and then sum the constant volume specific heat on a mass basis for each component or (2) find the molar basis constant volume specific heat for the mixture and then divide by the molecular weight to obtain specific heat on a mass basis.

In this example we use the second method and use an analogous form of Equation 3.27 to find the constant volume specific heat. That is,

$$\hat{c}_v = \sum_i x_i \hat{c}_{v,i} \quad \text{kJ/(kgmol·K)}$$

Values of $\hat{c}_p$ are found in Appendix C, and $\hat{c}_v$ may be found by using $\hat{c}_v = \hat{c}_p - \hat{R}$. From the inside front cover $\hat{R} = 8.314$ kJ/(kgmol·K). Computing these using a spread sheet,

Species	x_i	$\hat{c}_{p,i}$ (kJ/kgmol·K)	$\hat{c}_{v,i}$ (kJ/kgmol·K)	$x_i\hat{c}_{v,i}$ (kJ/kgmol·K)
CO	0.25	33.18	24.866	6.216
CO_2	0.10	54.31	45.996	4.600
H_2	0.15	30.20	21.886	3.283
CH_4	0.04	71.80	63.486	2.539
N_2	0.46	32.70	24.386	11.218
sum	1.00			27.856

Dividing by M yields the constant volume specific heat of the mixture on a mass basis:

$$c_v = \frac{\hat{c}_v}{M} = \frac{27.856 \text{ kJ}}{\text{kgmol·K}} \cdot \frac{\text{kgmol}}{25.2 \text{ kg}} = \frac{1.105 \text{ kJ}}{(\text{kg·K})}$$

3.3　COMBUSTION STOICHIOMETRY

When molecules undergo chemical reaction, the reactant atoms are rearranged to form new combinations. For example, when hydrogen and oxygen react to form water we write

$$H_2 + \frac{1}{2}O_2 \rightarrow H_2O$$

That is, two atoms of hydrogen (H) and one atom of oxygen (O) form one molecule of water (H_2O) since the number of atoms of H and O must be conserved. In the same way we can say two moles of hydrogen (H) and one mole of oxygen (O) form one mole of water (H_2O) since the number of moles of H and O must be conserved.

Such reaction equations represent initial and final results and do not indicate the actual path of the reaction, which may involve many intermediate steps and intermediate species. This overall or global approach is similar to thermodynamic system analysis where only end states and not path mechanisms are considered.

The relative masses of the molecules are obtained by multiplying the number of moles of each species in the molecule by their respective molecular weights. For the hydrogen-oxygen reaction above,

$$\left(1 \text{ kgmol}_{H_2}\right)\left(\frac{2 \text{ kg}_{H_2}}{\text{kgmol}_{H_2}}\right) + \left(\frac{1}{2} \text{ kgmol}_{O_2}\right)\left(\frac{32 \text{ kg}_{O_2}}{\text{kgmol}_{O_2}}\right) \rightarrow \left(1 \text{ kgmol}_{H_2O}\right)\left(\frac{18 \text{ kg}_{H_2O}}{\text{kgmol}_{H_2O}}\right)$$

simplifying

$$2\,kg_{H_2} + 16\,kg_{O_2} \rightarrow 18\,kg_{H_2O}$$

Thus as expected, we see that mass is conserved. The mass of the reactants equals the mass of the products, although the moles of the reactants do not equal the moles of the products. In the same way the volume may change even if ideal gases react at constant temperature and pressure. For example, consider that

$$1 \text{ volume } H_2 + \frac{1}{2} \text{ volume } O_2 \rightarrow 1 \text{ volume } H_2O$$

In most combustion devices fuel is reacted with air rather than oxygen. Air is composed of 21% O_2, 79% N_2, and small amounts of argon, carbon dioxide and hydrogen. In combustion calculations it is conventional to approximate dry air as a mixture of 79% (vol) N_2 and 21% (vol) O_2 or 3.764 moles of N_2 per mole O_2. Using $M_{O_2} = 32.00$ and $M_{N_2} = 28.01$ yields $M_{air} = 28.85$; however, the molecular weight of pure air is 28.96 because of the small amounts of other gases. The most straightforward way to correct for this difference is to use an apparent molecular weight of N_2 of 28.16. This value is obtained by substitution into Equation 3.22,

$$M_{air} = \sum_i x_i M_i = x_{N_2} M_{N_2} + x_{O_2} M_{O_2}$$

rearranging and simplifying,

$$M_{N_2} = \frac{\left(M_{air} - x_{O_2} M_{O_2}\right)}{x_{N_2}} = \frac{\left(28.96 - 0.21 \times 32\right)}{0.79} = 28.16$$

In this text we assume dry air for all calculations and use a molecular weight of 29.0. In practice for certain applications the effect of water vapor in the air may need to be considered. For example, at 80°F saturated water vapor in air occupies 6.47% by volume, and hence this air contains only 19.6% O_2.

Stoichiometric air is the amount of air required to burn a fuel completely to products with no dissociation. Stoichiometric calculations are done by performing an atom balance for each of the elements in the mixture. For example, consider the combustion of methane (CH_4) with air. This can be written as

$$1\left(CH_4\right) + a\left(O_2 + 3.76\,N_2\right) \rightarrow b\left(CO_2\right) + c\left(H_2O\right) + 3.76a\left(N_2\right)$$

Performing an element balance for C yields

$$1 = b$$

In the same way an element balance for H yields

$$4 = 2c$$

simplifying to

$$c = 2$$

An element balance for O balance yields

$$2a = 2b + c.$$

Substituting for b and c,

$$a = 2.$$

Thus the stoichiometric equation for the combustion of CH_4 is

$$CH_4 + 2(O_2 + 3.76 \ N_2) \rightarrow CO_2 + 2 \ H_2O + 7.52 \ N_2$$

Example 3.2

For a stoichiometric hydrogen-air reaction at 1 atm pressure, find (a) the fuel-air mass ratio, f, (b) the mass of fuel per mass of reactants, and (c) the partial pressure of water vapor in the products.

Solution
Part (a)
React 1 mole of H_2 with enough air to form the complete products H_2O and N_2,

$$1(H_2) + a(O_2 + 3.76 \ N_2) \rightarrow b(H_2O) + 3.76a(N_2)$$

Performing a hydrogen atom balance,

$$1(H_2) + a(O_2 + 3.76 \ N_2) \rightarrow b(H_2O) + 3.76a(N_2)$$
$$1(2) = b(2)$$
$$b = 1$$

Performing an oxygen atom balance,

$$1(H_2) + a(O_2 + 3.76 \ N_2) \rightarrow b(H_2O) + 3.76a(N_2)$$
$$a(2) = b$$
$$a = \frac{1}{2}$$

Hence the stoichiometric equation for the combustion for H_2 is

$$H_2 + \frac{1}{2}\left(O_2 + 3.76\,N_2\right) \rightarrow H_2O + 1.88\,N_2$$

That is, 1 kgmol of H_2 reacts with

$$\frac{1}{2}(1 + 3.76) = 2.38 \text{ kgmol}_{air}$$

On a mass basis

$$m_{H_2} = \frac{1 \text{ kgmol}_{H_2}}{1} \cdot \frac{2 \text{ kg}_{H_2}}{\text{kgmol}_{H_2}} = 2 \text{ kg}_{H_2}$$

reacts with 69.02 kg of air:

$$m_{air} = \frac{2.38 \text{ kgmol}_{air}}{1} \cdot \frac{29.0 \text{ kg}_{air}}{\text{kgmol}_{air}} = 69.02 \text{ kg}_{air}$$

Therefore

$$f = \frac{m_{H_2}}{m_{air}} = \frac{2}{69.02} = 0.029$$

Part (b)

The mass of H_2 per unit mass of reactant mixture is

$$\frac{m_f}{\left(m_{air} + m_f\right)}$$

From the definition of the fuel-air ratio,

$$m_f = m_{air}\, f$$

The mass of fuel to the total mass of the reactants is

$$\frac{m_f}{\left(m_{air} + m_f\right)} = \frac{m_{air}\, f}{\left(m_{air} + m_{air}\, f\right)} = \frac{f}{\left(1 + f\right)} = \frac{0.029}{1.029} = 0.0282$$

Part (c)

The partial pressure of the water vapor in the products is obtained from the mole fraction of water in the products;

$$x_{H_2O} = \frac{\text{mol}_{H_2O}}{\text{mol}_{prod}} = \frac{1}{2.88} = 0.347$$

From Equation 3.29,

$$p_{H_2O} = x_{H_2O} p = 0.347p$$

Therefore,

$$p_{H_2O} = 0.347 \text{ atm } \left(\text{for } T > 73°C \right)$$

Example 3.3

Consider the stoichiometric combustion of dry pine at 1 atm pressure. Assume that the ultimate analysis of pine is 51% C, 7% H, 42% O, <0.1% N, and <0.1% S (wt %). Find (a) the fuel-air mass ratio, f, (b) the composition of the products on a mass basis, and (c) the partial pressure of water vapor in the products.

Solution

Part (a)

Using 100 kg of fuel as a basis for calculation,

	m (kg)	M (kg/kgmol)	N (kgmol)	Normalized moles
C	51	12	4.25	1.00
H	7	1	7.00	1.65
O	42	16	2.63	0.62

The stoichiometric reaction is

$$CH_{1.65}O_{0.62} + 1.10(O_2 + 3.76 N_2) \rightarrow CO_2 + 0.82 H_2O + 4.14 N_2$$

The mass of fuel is

$$m_f = 12 + 1.65 + 0.62(16.0) = 23.6 \text{ kg}_{fuel}$$

The mass of air is

$$m_{air} = \frac{1.1(4.76) \text{ kgmol}_{air}}{1} \cdot \frac{29.0 \text{ kg}_{air}}{\text{kgmol}_{air}} = 151.8 \text{ kg}_{air}$$

The fuel-air ratio is

$$f = \frac{m_f}{m_{air}} = \frac{23.6}{151.6} = 0.155$$

Part (b)

Referring back to the stoichiometric balance in Part a,

	N (kgmol)	M (kg/kgmol)	m (kg)	x_i
CO_2	1.00	44.0	44.0	0.251
H_2O	0.82	18.0	14.8	0.084
N_2	4.14	28.16	116.6	0.665
sum			175.4	1.000

As noted earlier we use an apparent molecular weight of 28.16 for N_2 to account for the small amounts of argon, carbon dioxide, and hydrogen in air. Based on this, the mass of products equals the mass of reactants.

Part (c)

The partial pressure of the water vapor in the products is obtained from the mole fraction of water in the products;

$$x_{H_2O} = \frac{mol_{H_2O}}{mol_{prod}} = \frac{0.82}{(1+0.82+4.14)} = 0.138$$

From Equation 3.29,

$$p_{H_2O} = x_{H_2O}\, p = 0.138\, p$$

Therefore,

$$p_{H_2O} = 0.138\ \text{atm}\ (\text{for } T > 52°C)$$

For a fuel containing carbon, hydrogen, and oxygen that is burnt to completion with a stoichiometric amount of air, atom balances on C, H, O, and N atoms yield the following general expression:

$$C_\alpha H_\beta O_\gamma + \left(\alpha + \frac{\beta}{4} - \frac{\gamma}{2}\right)(O_2 + 3.76\ N_2)$$

$$\rightarrow \alpha\ CO_2 + \frac{\beta}{2}\ H_2O + 3.76\left(\alpha + \frac{\beta}{4} - \frac{\gamma}{2}\right) N_2 \tag{3.32}$$

where α, β, and γ are the number of carbon, hydrogen, and oxygen atoms in a molecule of fuel. Alternatively, α, β, and γ are the mole fractions of the carbon, hydrogen, and oxygen from the ultimate analysis of the fuel. The moles of stoichiometric air, $n_{air(s)}$, per mole of fuel, is

$$\frac{n_{air(s)}}{n_f} = 4.76\left(\alpha + \frac{\beta}{4} - \frac{\gamma}{2}\right) \tag{3.33}$$

The stoichiometric fuel-air ratio is

$$f_s = \frac{m_f}{m_{air(s)}} = \frac{n_f M_f}{n_{air(s)} M_{air}} = \frac{1(M_f)}{4.76(\alpha + \beta/4 - \gamma/2)(M_{air})} \tag{3.34}$$

substituting $M_{air} = 29.0$ and simplifying yields

$$f_s = \frac{M_f}{138.0(\alpha + \beta/4 - \gamma/2)} \tag{3.35}$$

The *percent excess air* is

$$\% \text{ excess air} = 100 \frac{\left(m_{air} - m_{air(s)}\right)}{m_{air(s)}} \tag{3.36}$$

By dividing the numerator and denominator by M_{air}, this can be rewritten as

$$\% \text{ excess air} = 100 \frac{\left(n_{air} - n_{air(s)}\right)}{n_{air(s)}} \tag{3.37}$$

or

$$\% \text{ excess air} = 100 \frac{\left(n_{O_2} - n_{O_2(s)}\right)}{n_{O_2(s)}} \tag{3.38}$$

Percent theoretical air is the amount of air actually used divided by the stoichiometric air,

$$\% \text{ thoretical air} = 100 \frac{m_{air}}{m_{air(s)}} = 100 \frac{n_{air}}{n_{air(s)}} \tag{3.39}$$

Hence,

$$\% \text{ excess air} = \% \text{ theoretical air} - 100 \tag{3.40}$$

For example, 110% theoretical air is a fuel-lean mixture (lean combustion) with 10% excess air; 85% theoretical air is a fuel-rich mixture (rich combustion) that is 15% deficient in air.

Sometimes the equivalence ratio is used instead of excess air to describe a combustible mixture. The *equivalence ratio*, F, is the actual fuel-air mass ratio, f, divided by the stoichiometric fuel-air mass ratio, f_s. That is,

$$F = \frac{f}{f_s} \tag{3.41}$$

Excess air is directly related to the equivalence ratio. Using Equation 3.38 it follows that

$$\% \text{ excess air} = \frac{100(1-F)}{F} \tag{3.42}$$

Figure 3.1 shows percent excess air as a function of the equivalence ratio. For lean mixtures excess air tends to infinity, and this is why the use of the equivalence ratio is preferred for internal combustion engines, which often run lean.

In practice excess air is often determined by measuring the composition of the products. If combustion is complete (e.g., the products are CO_2, H_2O, O_2, and N_2), then we can write

$$\% \text{ excess air} = \frac{m_{air} - m_{air(s)}}{m_{air(s)}} = \left(\frac{M_{air}}{M_{air}}\right) \frac{m_{air} - m_{air(s)}}{m_{air(s)}} = \frac{n_{air} - n_{air(s)}}{n_{air(s)}} \tag{3.43}$$

Continuing,

$$\frac{n_{air} - n_{air(s)}}{n_{air(s)}} = \frac{n_{O_2} - n_{O_2(s)}}{n_{O_2(s)}}$$

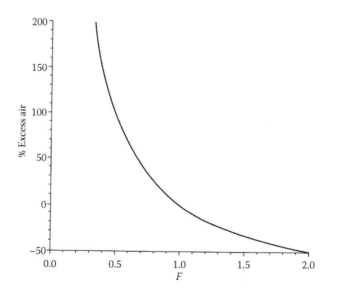

FIGURE 3.1   Excess air versus equivalence ratio.

Relating the O_2 in the reactants and products,

$$\left(n_{O_2} - n_{O_2(s)} \right)\Big|_{react} = n_{O_2}\Big|_{prod}$$

We can rearrange and substitute to obtain

$$\frac{n_{O_2} - n_{O_2(s)}}{n_{O_2(s)}}\Bigg|_{react} = \frac{n_{O_2}\big|_{prod}}{n_{O_2}\big|_{react} - n_{O_2(s)}\big|_{prod}} \qquad (3.44)$$

Equation 3.44 can be expressed in terms of product measurements by remembering that

$$n_{N_2}\Big|_{react} = 3.76\; n_{O_2}\Big|_{react}$$

and that

$$n_{N_2}\Big|_{prod} = n_{N_2}\Big|_{react}$$

Since gaseous measurements are normally done in terms of volume (mole) fraction, n's are replaced by x's by dividing the numerator and denominator by the total moles in the products. This yields

$$\% \text{ excess air} = 100 \left(\frac{x_{O_2}\big|_{prod}}{\dfrac{x_{N_2}\big|_{prod}}{3.76} - x_{O_2}\big|_{prod}} \right) \qquad (3.45)$$

If the products contain species such as CO, H_2, and fuel fragments from incomplete combustion, it is still possible to determine the excess air by measuring the composition of the products. For example, exhaust gas emissions from engines can be used to compute the overall fuel-air ratio (Spindt 1965).

Example 3.4

The dry exhaust products from a burner that uses natural gas and air contain 5% oxygen and 9% carbon dioxide. Assuming complete combustion, find the excess air used for this burner. Note that the term "dry exhaust products" is used in this example. This means that the water vapor in the products is condensed before the measurements are made and hence the products are dry. This is a common measurement method.

Solution

From the measurements, $x_{O_2} = 0.05$ and $x_{CO_2} = 0.09$. Noting that the mole fraction of exhaust products must sum to 1 (e.g., $x_{O_2} + x_{CO_2} + x_{N_2} = 1$), $x_{N_2} = 0.86$. From Equation 3.45,

$$\% \text{ excess air} = 100 \left(\frac{0.05}{\dfrac{0.86}{3.76} - 0.05} \right) = 28\%$$

Typical values of the stoichiometric air-fuel ratio, f_s, and the volume percent of CO_2 in the dry products for stoichiometric combustion are shown in Table 3.1 for various fuels. Concentrations in the dry products are reported because the water is condensed out of the products before gas analysis to protect the gas monitoring instruments.

Example 3.5

Bituminous coal is burned to completion with 50% excess air. Find (a) the fuel-air ratio, f, and (b) the volumetric analysis of the dry products. The as-received ultimate analysis of the coal is 70% carbon, 5% hydrogen, 15% oxygen, 5% moisture, and 5% ash on a weight basis.

Solution

Part (a)

The starting place for this problem is

$$\text{Coal} + (1.5 \times \text{Stoichiometric Air}) \rightarrow \text{Products}$$

TABLE 3.1

Stoichiometric Complete Combustion of Several Fuels in Air

Fuel	$m_{air(s)}/m_f$	f_s	CO_2 (Vol % in Dry Products)
Methane	17.2	0.0581	11.7
Gasoline	14.7	0.0680	14.9
Methanol	6.5	0.154	15.1
Ethanol	9.0	0.111	15.1
No. 1 fuel oil	14.8	0.0676	15.1
No. 6 fuel oil	13.8	0.0725	15.9
Bituminous coal[a]	10.0	0.100	18.2
Wood[a]	5.9	0.169	20.5

[a] Dry basis.

Remembering that

$$N = \frac{m}{M}$$

For 100 kg of coal,

Species	m_i (kg)	M_i (kg/kgmol)	N_i (kgmol)
C	70	12	5.833
H	5	1	5.000
O	15	16	0.937
H_2O	5	18	0.278

The balance equation for 50% excess air for 100 kg of coal is

$$(5.833\,C + 5\,H + 0.937\,O + 0.278\,H_2O) + 1.5a\,(O_2 + 3.76\,N_2)$$
$$\rightarrow 5.833\,CO_2 + (2.5 + 0.278)\,H_2O + 0.5a\,O_2 + 1.5(3.76)a\,N_2$$

Performing an oxygen atom balance,

$$(0.937(1) + 0.278(1)) + 1.5a(2) = 5.833(2) + (2.5 + 0.278)(1) + 0.5a(2)$$

Solving for a,

$$a = 6.614$$

The combustion reaction can then be written as

$$(C_{5.833}H_5O_{0.937} + 0.278\,H_2O) + 9.921(O_2 + 3.76\,N_2)$$
$$\rightarrow 5.833\,CO_2 + 2.778\,H_2O + 3.307\,O_2 + 32.303\,N_2$$

where $C_{5.833}H_5O_{0.937}$ is a pseudo molecule with the correct C, H, O ratios for 100 kg of coal with 5% moisture. This can be simplified to $CH_{0.857}O_{0.161}$ by dividing by 5.833, but then it would no longer represent 100 kg of coal. Hence the mass of air required to burn 100 kg of this coal with 50% excess air is

$$m_{air} = \frac{29\ kg_{air}}{kmol_{air}} \cdot \frac{(9.921)(1 + 3.76)\ kmol_{air}}{1} = 1369\ kg_{air}$$

And the fuel-air mass ratio is

$$f = \frac{100\ kg_{coal}}{1369\ kg_{air}} = 0.0730$$

Part (b)

The volumetric analysis of the gaseous combustion products is

Species (N_i)	N (kgmol)	x_i	x_i (dry)
CO_2	5.833	0.1185	0.1256
H_2O	2.778	0.0564	–
O_2	3.307	0.0672	0.0712
N_2	37.303	0.7579	0.8032
Total	49.221	1.0000	1.0000

3.4 CHEMICAL ENERGY

Determining the energy released by a given fuel in a particular combustion situation is a critical aspect of combustion engineering. This section introduces the tools needed by the combustion engineer to make this determination. This includes the concepts of heat of reaction at constant pressure and constant volume, higher and lower heating value, heat of formation, and absolute enthalpy.

3.4.1 HEAT OF REACTION

Heat of reaction is the chemical energy released when a fuel reacts with air to form products. The starting point for determining the heat of reaction is to specify the chemical species and their states in the reactants and products. This is done by writing a balanced reaction with the phase of each species noted. The heat of vaporization of liquid fuels and the heat of pyrolysis of solid fuels are typically small compared to the chemical energy released by combustion. However, the effect of water condensation can be important. For example, there is a substantial increase in furnace efficiency when water is condensed in the heat exchanger of a home gas-fired furnace.

For very lean hydrocarbon-air mixtures where the temperature is low, the products may be assumed to be complete (usually CO_2, H_2O, O_2, and N_2). However, with high product temperatures and for rich mixtures, it is generally necessary to include other species and assume chemical equilibrium to determine the species mole fractions. If the products are not in equilibrium, then a chemical kinetic analysis or direct measurement is required to determine the end state. For example, the amount of solid carbon produced by very rich mixture combustion is typically not predicted correctly by equilibrium. The gaseous composition of rich mixture products is also somewhat uncertain, even if equilibrium prevails, because the identity of the unburned hydrocarbon species is not easily determined. A solution to this problem is to complete the oxidation in a separate reactor and measure the additional energy release.

To understand the heat of reaction, consider the reaction of a fuel and air mixture of mass m (the total mass of the system). For constant volume combustion with heat transfer there is no work and the first law of thermodynamics (Equation 3.4) is

$$Q_v = m\left[\left(u_2 - u_1\right)_s + \left(u_2 - u_1\right)_{chem}\right] \qquad (3.46)$$

where the subscripts s and *chem* refer to sensible and chemical energy. Note that in Equation 3.46, the values of u at State 1 and State 2 are obtained by summing over all I species, $i = 1, 2, \ldots, I$. Thus

$$[u_1]_s = [u_{T_1}]_s = \sum_{i=1}^{I} y_i [u_{i,T_1}]_s = \sum_{i=1}^{I} y_i \int_{T_0}^{T_1} (c_{v,i})_{react} dT = \int_{T_0}^{T_1} (c_v)_{react} dT$$

where T_0 is the reference temperature. The quantity $(c_v)_{react}$ is the specific heat of the reactant mixture. Similarly

$$[u_2]_s = \int_{T_0}^{T_2} (c_v)_{prod} dT$$

where T_0 is the reference temperature, and the quantity $(c_v)_{prod}$ is the specific heat of the product mixture.

If the heat transfer is just large enough to bring the product's temperature back to the reactant temperature, and if this temperature is taken as the reference temperature, T_0, for the sensible energy, then $(u_2 - u_1)_s = 0$ by definition, and Q_v is the chemical energy released by the reaction. For constant volume combustion if the water in the products is not condensed, then the lower heating value can be expressed as

$$\text{LHV} = \frac{-Q_v}{m} \left(\frac{1+f}{f} \right) \tag{3.47}$$

If the water in the products is condensed, then the higher heating value,

$$\text{HHV} = \frac{-Q_v}{m} \left(\frac{1+f}{f} \right) \tag{3.48}$$

is obtained for constant volume combustion.

If the reaction takes place at constant pressure and the total heat transfer is Q_p, then the energy equation (Equation 3.12) becomes

$$m \left[(h_2 - h_1)_s + (h_2 - h_1)_{chem} \right] = Q_p \tag{3.49}$$

Again, if $T_1 = T_2 = T_0$, then Q_p is the chemical energy released. For the constant pressure case if the number of moles of gaseous products, N_{prod}, is larger than the number of moles of gaseous reactants, N_{react}, then some of the chemical energy is expended as work as the expanding gases push aside the ambient pressure. Thus for reactants and products which are ideal gases,

$$Q_p - Q_v = \Delta(pV) = (N_{prod} - N_{react}) \hat{R} T_0 = \Delta N \hat{R} T \tag{3.50}$$

Consider the generalized combustion reaction

$$C_\alpha H_\beta O_\gamma + \left(\alpha + \frac{\beta}{4} - \frac{\gamma}{2}\right)(O_2 + 3.76\, N_2)$$

$$\to \alpha\, CO_2 + \frac{\beta}{2} H_2O + 3.76\left(\alpha + \frac{\beta}{4} - \frac{\gamma}{2}\right) N_2$$

(3.51)

Assuming that the fuel is in the gas phase and the water remains water vapor, then

$$\Delta N = \frac{\beta}{4} + \frac{\gamma}{2} - 1$$

For methane (CH_4) $\beta = 4$ and $\gamma = 0$. Therefore, the number of moles of reactants equals the number of moles of reactants ($\Delta N = 0$) for methane. Because both Q_p and Q_v are typically negative, $\Delta N > 0$ implies $|Q_v| > |Q_p|$ as anticipated. For most cases of interest the difference, $Q_p - Q_v$, is only a few kilocalories and is thus often neglected.

The heat of reaction may be calculated for reactions taking place at temperatures other than T_0 and for cases where the initial and final temperature are not equal by use of the heat of reaction data taken at T_0. Consider the reaction at constant pressure with reactant temperature T_1 and product temperature T_2. Assume, for example, $T_2 > T_1 > T_0$. To use the $Q_p(T_0)$ value, first imagine the reactants are cooled from T_1 to T_0, then the reaction takes place at T_0, and finally the products are heated from T_0 to T_2.

$$Q_p = m\int_{T_1}^{T_0} c_{p,\text{react}}\, dT + Q_{p,T_0} + m\int_{T_0}^{T_2} c_{p,\text{prod}}\, dT$$

or

$$Q_p - Q_{p,T_0} = m\left(h_{s2} - h_{s1}\right)$$

(3.52)

Remember that Q_{p,T_0} is negative for an exothermic reaction.

Example 3.6

The higher heating value of gaseous methane and air at 25°C is 55.5 MJ/kg. Find the heat of reaction at constant pressure of a stoichiometric mixture of methane and air if the reactants and products are at 500 K.

Solution

The reaction is

$$CH_4 + 2\left(O_2 + 3.76\, N_2\right) \to CO_2 + 2\, H_2O + 7.52\, N_2$$

Determining the molecular weight of the reactant mixture,

Reactants	N_i (moles)	x_i	M_i (kg/kgmol)	$M_i x_i$ (kg/kgmol)
CH_4	1	0.095	16	1.5
O_2	2	0.190	32	6.1
N_2	7.52	0.715	28	20.0
	10.52	1.000		27.6

$$N_{react} = 1 + 2(4.76) = 10.52$$

$$M_{react} = \sum_i x_i M_i = 27.6 \text{ kg/kgmol}$$

Analysis of the reactants using sensible enthalpies from Appendix C yields

Reactants	x_i	$\hat{h}_i$ (MJ/kgmol)	$\hat{h}_i x_i$ (kg/kgmol)
CH_4	0.095	8.20	0.779
O_2	0.190	6.09	1.157
N_2	0.715	5.91	4.226
	1.000		6.162

$$\hat{h}_{s,react} = \sum_i x_i h_{si} = 6.16 \text{ MJ/kgmol}$$

$$h_{s,react} = \frac{6.16 \text{ MJ}}{\text{kgmol}} \cdot \frac{\text{kgmol}}{27.6 \text{ kg}} = \frac{223 \text{ kJ}}{\text{kg}}$$

The fuel-air mass ratio is

Reactants	x_i	M_i	M_i/M	y_i
CH_4	0.095	16	0.580	0.055
O_2	0.190	32	1.159	0.220
N_2	0.715	28	1.014	0.725
	1.000			1.000

$$f = \frac{y_f}{y_{air}} = \frac{0.055}{0.220 + 0.725} = 0.0582$$

In the same way analysis of the products yields

Products	x_j	$\hat{h}_{sj}$ (MJ/kgmol)	M_j
CO_2	0.095	8.31	44
H_2O	0.190	6.92	18
N_2	0.715	5.91	28

$$N_{prod} = 10.52$$
$$\hat{h}_{s,prod} = 6.33 \ \text{MJ/kgmol}$$
$$M_{prod} = 2.76 \ \text{kg/kgmol}$$
$$h_{s,prod} = 229 \ \text{kJ/kg}$$

Because the water vapor does not condense at 500 K, the lower heating value is used;

$$\text{LHV} = \frac{55{,}000 \ \text{kJ}}{\text{kg}_{H_2O}} - \frac{2394 \ \text{kJ}}{\text{kg}_{H_2O}} \cdot \frac{2 \ \text{kmol}_{H_2O}}{\text{kmol}_{CH_4}} \cdot \frac{18 \ \text{kg}_{H_2O}}{\text{kgmol}_{H_2O}} \cdot \frac{\text{kgmol}_{CH_4}}{16 \ \text{kg}_{CH_4}}$$

$$= \frac{50{,}113 \ \text{KJ}}{\text{kg}_{CH4}}$$

Remembering that

$$\frac{m_f}{m_{react}} = \frac{m_f}{m_{air} + m_f} = \frac{f}{1+f}$$

we can write

$$\frac{Q_{p,T_0}}{m} = \frac{f}{1+f}(\text{LHV}) = \frac{0.0582}{1+0.0582}(50{,}113)\text{kJ/kg}_{react}$$

$$= -2747 \ \text{kJ/kg}_{react}$$

From Equation 3.52 the heat of reaction per unit mass of reactants at 500 K is

$$Q_p = -2747 + (229 - 223) = -2741 \ \text{kJ/kg}_{react}$$

The negative sign indicates that the heat flows out of the system.

The heat of reaction of fuels combusting in air (or oxygen) with the starting and ending points at 25°C and 1 atm gives the fuel heating value. These heating values are then tabulated for common fuels. Following this same practice for all possible reactions would, however, lead to an enormous amount of tabulated data. The solution to this problem comes by noting that we may simply add selected reactions and their heats of reaction to obtain any given reaction and its heat of reaction. The most basic set of reactions are thus for the formation of compounds from their elements. Once the data for this basic set is tabulated, any reaction can be constructed from it.

3.4.2 HEAT OF FORMATION AND ABSOLUTE ENTHALPY

The *heat of formation* of a particular species is the heat of reaction per mole of product formed isothermally from elements in their standard states. The standard state is chosen as the most stable form of the element at 1 atm and 25°C. For carbon the most stable form is solid graphite. For oxygen and nitrogen the standard state is gaseous O_2 and N_2. The heat of formation of elements in their standard states is assigned a value of zero. Heats of formation are given for alkane fuels in Appendix A.1. Heats of formation for other species are given in Appendix C and the JANAF tables (Chase 1998). The heat of formation is denoted as $\Delta h°$ in this text. For fuels that are complex mixtures, such as gasoline or coal for example, the heat of formation is generally not reported. However, the heat of formation can be calculated from the heating value and the hydrogen and carbon content of the fuel. This is typically necessary when using computer programs that use absolute enthalpies.

The absolute enthalpy of a substance is the sensible enthalpy relative to the reference temperature, T_0, plus the heat of formation at the reference temperature.

$$\hat{h} = \int_{T_0}^{T} \hat{c}_p \, dT + \Delta\hat{h}° \tag{3.53}$$

Sensible enthalpies and heats of formation of selected species are tabulated in Appendix C for a reference temperature of 25°C. The absolute enthalpy of elements is equal to the sensible energy and therefore is always positive above the reference temperature. However, for compounds, because the heat of formation is typically a large negative number, the absolute enthalpy is usually negative up to a relatively high temperature. For example, $\Delta\hat{h}°$ of water is –241.83 MJ/kgmol; from this we can see that the absolute enthalpy is negative up to 5000 K.

Example 3.7

In a flow calorimeter 24 mg/s of graphite particulate reacts completely with oxygen initially at 25°C to form carbon dioxide at 1 atm and 25°C. The rate of heat absorbed by the calorimeter water is 787.0 W. Find the heat of formation of CO_2.

Solution

The reaction is

$$C_{(s)} + O_{2(g)} \rightarrow CO_{2(g)}$$

where (s) and (g) refer to solid and gas phases respectively. Noting that one mole of C yields one mole of CO_2,

$$\dot{N}_{CO_2} = \frac{0.024 \text{ g}_C}{s} \cdot \frac{1 \text{ gmol}_C}{12 \text{ g}_C} \cdot \frac{1 \text{ gmol}_{CO_2}}{1 \text{ gmol}_C} = \frac{0.002 \text{ gmol}_{CO_2}}{s}$$

is formed. The energy balance (Equation 3.12 in molar form) for this problem is

$$\left[\dot{N}\hat{h} \right]_{react} = \left[\dot{N}\hat{h} \right]_{prod} + q$$

and because the sensible energies are all zero at the reference temperature of 25°C and because the C and O_2 are in their standard states, their heats of formation are zero:

$$0 = \left[\dot{N} \ \hat{h}^{\circ} \right]_{CO_2} + q$$

or

$$\hat{h}^{\circ}_{CO_2} = \frac{-787.0 \text{ W}}{1} \cdot \frac{J}{W \cdot s} \cdot \frac{s}{0.002 \text{ gmol}_{CO_2}} = \frac{-393.5 \text{ MJ}}{kgmol}$$

This is in agreement with the value given in Appendix C for CO_2.

Example 3.8

The higher heating value of a dry ash-free bituminous coal is 29,050 kJ/kg. The coal contains 70% (wt) carbon and 5% (wt) hydrogen on a dry, ash-free basis. Find the enthalpy of formation on a mass basis (kJ/kg) for this coal.

Solution

Consider coal plus air at T_0 reacting to produce products at T_0. An energy balance yields

$$\left[m \ h^{\circ} \right]_{coal} - \left[m \ h^{\circ} \right]_{CO_2} - \left[m \ h^{\circ} \right]_{H_2O(\ell)} = \left[m(HHV) \right]_{coal}$$

Dividing by the mass of coal,

$$h^{\circ}_{coal} = HHV + \left(\frac{m_{CO_2}}{m_{coal}} \right) h^{\circ}_{CO_2} + \left(\frac{m_{H_2O}}{m_{coal}} \right) h^{\circ}_{H_2O(\ell)}$$

Because

$$\frac{m_{CO_2}}{m_{coal}} = \frac{0.7\ kg_C}{1\ kg_{coal}} \cdot \frac{kgmol_C}{12\ kg_C} \cdot \frac{1\ kgmol_{CO_2}}{1\ kgmol_C} \cdot \frac{44\ kg_{CO_2}}{kgmol_{CO_2}} = \frac{2.57\ kg_{CO_2}}{kg_{coal}}$$

and

$$\frac{m_{H_2O}}{m_{coal}} = \frac{0.05\ kg_{H_2}}{1\ kg_{coal}} \cdot \frac{kgmol_{H_2}}{2\ kg_{H_2}} \cdot \frac{1\ kgmol_{H_2O}}{1\ kgmol_{H_2}} \cdot \frac{18\ kg_{H_2O}}{kgmol_{H_2O}} = \frac{0.45\ kg_{H_2O}}{kg_{coal}}$$

then

$$h^\circ_{coal} = \frac{29,050\ kJ}{kg_{coal}} + \frac{2.57\ kg_{CO_2}}{kg_{coal}} \left(\frac{-393,520\ kJ}{kgmol_{CO_2}} \right) \frac{kgmol_{CO_2}}{44\ kg_{CO_2}}$$

$$+ \frac{0.45\ kg_{H_2O}}{kg_{coal}} \left(\frac{-285,750\ kJ}{kgmol_{H_2O}} \right) \frac{kgmol_{H_2O}}{18\ kg_{H_2O}}$$

Solving

$$h^\circ_{coal} = -1079\ kJ/kg$$

3.5   CHEMICAL EQUILIBRIUM

To obtain thermodynamic equilibrium, complete equilibrium between the molecular internal degrees of freedom, complete chemical equilibrium, and complete spatial equilibrium is required. Before discussing chemical equilibrium, internal and spatial equilibrium will be briefly discussed.

Internal molecular energies are the ways that molecules store energy. The major forms of energy for polyatomic molecules are translational, vibrational, rotational, electronic level excitation, and nuclear spin. For most engineering combustion applications, it is safe to assume equilibrium among the internal degrees of freedom. One case where it is not safe to assume equilibrium is in shock waves. The times for relaxation of the various internal degrees of energy are typically translation 10^{-13} s, rotation 10^{-8} s, and vibration 10^{-4} s. Therefore, the molecular internal energy will briefly overshoot the equilibrium value, and one cannot strictly assume internal equilibrium until approximately 0.1 ms have elapsed.

Homogeneity is assumed for a purely thermodynamic system. This means that the system is described by a set of single valued properties. If the system contains gradients, it can be divided into a number of subsystems such that each subsystem has negligible gradients. In a system with species gradients there will be mass transfer, and in the general case, the equations of reacting fluid mechanics are required to completely describe the system.

Chemical equilibrium is achieved for constant temperature and pressure systems when the rate of change of concentration goes to zero for all species. In a complex reaction some species may come to equilibrium rapidly due to fast reaction rates or a very small change in concentration, while others approach equilibrium more slowly. For example, consider the flow of high temperature hydrogen in a nozzle. Suppose that at the stagnation temperature, all the hydrogen is dissociated to H atoms. If the expansion is slow, the reaction $2H \leftrightarrow H_2$ will follow the dropping temperature rapidly enough to give a series of equilibrium values (called *shifting equilibrium*). If the expansion is very rapid, hardly any reaction can occur, and the result will be a constant concentration of H atoms (*frozen equilibrium*). Between these two extremes the concentration will be determined by the rate of reaction and is said to be *kinetically limited*. This case is the subject of Chapter 4 and is important because chemical equilibrium does not exist in a flame zone. The temperature gradients are very steep, and many short-lived species are found in the flame zone. In the post flame zone many of the combustion products are in chemical equilibrium or possibly shifting equilibrium. An example of practical interest is nitric oxide production. The major combustion product species may follow an equilibrium path while the NO reacts too slowly to stay in equilibrium as the temperature is lowered by heat transfer or expansion.

3.5.1 CHEMICAL EQUILIBRIUM CRITERION

When the products have reached chemical equilibrium, the problem is to determine the composition of the products at a known pressure and temperature, and a given reactant composition. Thermodynamics alone cannot determine what species may be in the product mixture. However, given an assumed set of constituents, thermodynamics can determine the proportions of each species that exist in the equilibrium mixture. Once the composition is determined, the thermodynamic properties of the mixture, such as internal energy, u; enthalpy, h; and entropy, s, may be calculated.

For a system in chemical equilibrium, the pressure and temperature do not change. This may be specified by stating that the Gibbs free energy of the system $(G = H - TS)$ does not change:

$$(dG)_{T,p} = 0 \qquad (3.54)$$

where

$$G = \sum_{i=1}^{I} N_i \hat{g}_i \qquad (3.55)$$

and where

$$\hat{g}_i = \hat{h}_i - T\hat{s}_i$$

Because

$$\hat{s}_i = \hat{s}_i^\circ - \hat{R} \ln\left(\frac{p_i}{p_0}\right) \qquad (3.56)$$

and

$$\hat{s}_i^{\circ} = \int_{T_0}^{T} \frac{\hat{c}_{pi}}{T} dT \tag{3.57}$$

the Gibbs free energy for species i can be written as

$$\hat{g}_i = \hat{h}_i - T\hat{s}_i^{\circ} + \hat{R}T \ln\left(\frac{p_i}{p_0}\right) \tag{3.58}$$

Introducing $\hat{g}_i^{\circ} = \hat{h}_i - T\hat{s}_i^{\circ}$ and noting that $x_i = p_i/p$, then

$$\hat{g}_i = \hat{g}_i^{\circ} + \hat{R}T \ln(x_i) + \hat{R}T \ln\left(\frac{p}{p_0}\right) \tag{3.59}$$

Substituting Equation 3.55 into Equation 3.54, the equilibrium criterion becomes

$$d\left(\sum_{i=1}^{I} N_i \hat{g}_i\right) = 0 \tag{3.60}$$

Equation 3.60 is subject to atom balance constraints that hold the number of C, H, O, etc. atoms constant;

$$A_j = \sum_{i=1}^{I} \tilde{n}_{ji} N_i \tag{3.61}$$

where i refers to the species,
            j refers to the atoms,
            I is the total number of species in the system,
            $\tilde{n}_{ji}$ is the number of j atoms in species i, and
            A_j is the moles of j atoms in the system.

One approach to solving chemical equilibrium problems is to minimize G (Equation 3.60) with constraints (Equation 3.61). However, for reactions involving many species, there are many degrees of freedom and so more practical approaches are needed.

A more computationally tractable approach is to use *element potentials* to minimize G. This is the approach which was utilized in the well known StanJan software package developed by the late Professor Reynolds. Copies of StanJan can still be found on the web. To start expand Equation 3.60 and note that $d\hat{g}_j = 0$ at constant p and T. Then the equilibrium criterion becomes

$$\sum_{i=1}^{I} \hat{g}_i dN_i = 0 \tag{3.62}$$

For a mixture containing I species and J atom types, it can be shown that minimizing G using J Lagrange multipliers, λ_j, (which in this context are called *element potentials*) results in the following equation set that must be satisfied (Powell and Sarner 1959):

$$x_i = \frac{\exp\left(-\dfrac{\hat{g}_i^{\circ}}{RT} + \sum_{j=1}^{J} \lambda_j \tilde{n}_{ji}\right)}{\dfrac{p}{p_0}} \tag{3.63}$$

and

$$\sum_{i=1}^{I} x_i = 1$$

where x_i is the mole fraction of species i ($i = 1, \ldots, I$) and j is the atom type. In addition, the constraints of Equation 3.61 must also hold.

Another approach is the *method of equilibrium constants*. This is the more traditional approach. Note that for a given reaction $dN_j = a_j d\varepsilon$, where a_j is the stoichiometric coefficient and ε represents the progress of the reaction. For example, for the reaction

$$aA + bB \rightarrow cC + dD$$

it follows that

$$d(N_A) = -a \, d\varepsilon$$
$$d(N_B) = -b \, d\varepsilon$$
$$d(N_C) = c \, d\varepsilon$$
$$d(N_D) = d \, d\varepsilon$$

Substituting, Equation 3.62 becomes

$$\left(a\hat{g}_A + b\hat{g}_B - c\hat{g}_C - d\hat{g}_D\right) d\varepsilon = 0$$

Using the definition of $\hat{g}$ and dividing by $\hat{R}T$ yields

$$\frac{\left(a\hat{g}_A^\circ + b\hat{g}_B^\circ - c\hat{g}_C^\circ - d\hat{g}_D^\circ\right)}{\hat{R}T} = \ln\left(\frac{p_C^c p_D^d}{p_A^a p_B^b}\right) + \ln\left(p_0\right)^{a+b-c-d} \tag{3.64}$$

If the left-hand side of Equation 3.64 is defined as $\ln(K_p)$ of the reaction and if p_0 is assumed to be 1 atm, then Equation 3.64 for chemical equilibrium becomes

$$K_p = \frac{p_C^c p_D^d}{p_A^a p_B^b} = \frac{x_C^c x_D^d}{x_A^a x_B^b} \cdot p^{c+d-a-b} \tag{3.65}$$

Pressure p must be in atmospheres, and K_p is evaluated from thermodynamic data (Appendix C) as follows

$$\ln K_p = a\frac{\hat{g}_A^\circ}{\hat{R}T} + b\frac{\hat{g}_B^\circ}{\hat{R}T} - c\frac{\hat{g}_C^\circ}{\hat{R}T} - d\frac{\hat{g}_D^\circ}{\hat{R}T} \tag{3.66}$$

When solving for equilibrium products using Equation 3.66, the reactions to be considered are identified, and the equilibrium constants are evaluated at the specified temperature. Then the atom balance constraints are specified for the system, and an equilibrium equation is written for each of the specified reactions using the form of Equation 3.65. This set of equations is solved simultaneously to obtain the species mole fractions and other thermodynamic properties of the system. For combustion processes the important gas phase equilibrium reactions include

$$H_2O \leftrightarrow H_2 + \frac{1}{2}O_2 \tag{i}$$

$$CO_2 \leftrightarrow CO + \frac{1}{2}O_2 \tag{ii}$$

$$CO + H_2O \leftrightarrow CO_2 + H_2 \tag{iii}$$

$$H_2 + O_2 \leftrightarrow 2\,OH \tag{iv}$$

$$O_2 \leftrightarrow 2\,O \tag{v}$$

$$N_2 \leftrightarrow 2\,N \tag{vi}$$

$$H_2 \leftrightarrow 2\,H \tag{vii}$$

$$O_2 + N_2 \leftrightarrow 2\,NO \tag{viii}$$

Reactions i, ii, v, vi, and vii are dissociation reactions. Reaction iii is the water-gas shift reaction. Reaction iv accounts for equilibrium OH formation, which is an important species in chemical kinetics reactions. Reaction viii accounts for equilibrium nitric oxide, an important air pollutant.

For a solid-gas equilibrium reaction, such as carbon-oxygen, carbon-water vapor, or carbon-carbon dioxide, the equilibrium constant is determined using the Gibbs free energies of each constituent, and Equation 3.65 is used in the same way as with gas–gas reactions. However, it should be noted that a solid has zero partial pressure.

The thermodynamic equilibrium products from kerosene-air combustion are shown in Figures 3.2 and 3.3. The major products of lean combustion are H_2O, CO_2, O_2, and N_2, and for rich combustion the major products are H_2O, CO_2, CO, H_2, and N_2. At stoichiometric conditions at flame temperature, O_2, CO, and H_2 are present. However, for the assumption of complete combustion, i.e., no dissociation, these three species are zero. Minor species of equilibrium combustion products at flame

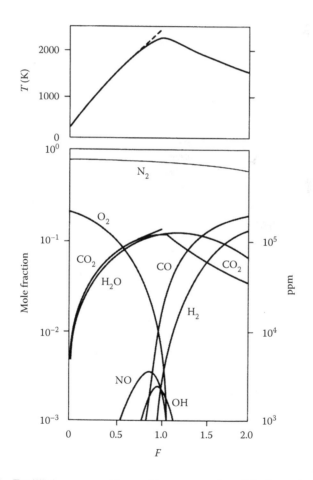

FIGURE 3.2 Equilibrium composition and temperature for adiabatic combustion of kerosene, $CH_{1.8}$, as a function of the equivalence ratio. (From Flagan, R. C. and Seinfeld, J. H., *Fundamentals of Air Pollution Engineering*, Prentice Hall, Englewood Cliffs, NJ, 1988.)

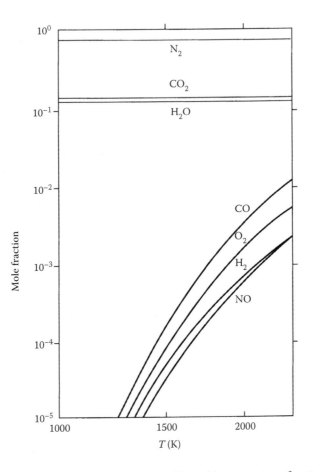

FIGURE 3.3 Variation of equilibrium composition with temperature for stoichiometric combustion of kerosene, $CH_{1.8}$. (From Flagan, R. C. and Seinfeld, J. H., *Fundamentals of Air Pollution Engineering*, Prentice Hall, Englewood Cliffs, NJ, 1988.)

temperature include O, H, OH, and NO. CO is a minor species in lean products, while O_2 is a minor species in rich products.

Several examples of simple thermodynamic equilibrium calculations are given in the example problems that follow this section. The equilibrium constant method (Examples 3.9, 3.11, and 3.12) is conceptually and computationally straightforward because only one equilibrium reaction is involved. In Chapter 4 the equilibrium constant will be related to the ratio of the forward to backward kinetic reaction rate constants. Example 3.10 demonstrates the method of minimizing G.

Example 3.9

Equal moles of H_2 and O_2 react to produce H_2O, H_2, and O_2 at 2500 K and 1 atm. Find the percent by volume of H_2, O_2, and H_2O due to the equilibrium reaction

$$H_2O \leftrightarrow H_2 + \frac{1}{2}O_2$$

using the method of equilibrium constants.

Solution

It is given that equal moles of H_2 and O_2 react to form an equilibrium mixture of H_2, O_2, and H_2O. Written as molar fractions this reaction is

$$\frac{1}{2}H_2 + \frac{1}{2}O_2 \rightarrow x_{H_2}H_2 + x_{H_2O}H_2O + x_{O_2}O_2$$

A hydrogen atom balance gives

$$\frac{1}{2}(2) = x_{H_2}(2) + x_{H_2O}(2)$$

simplifying to

$$1 = 2x_{H_2} + 2x_{H_2O} \tag{a}$$

An oxygen atom balance gives

$$\frac{1}{2}(2) = x_{H_2O} + x_{O_2}(2)$$

simplifying to

$$1 = x_{H_2O} + 2x_{O_2} \tag{b}$$

Dividing Equation a by Equation b, the H to O atom ratio in the equilibrium mixture is

$$1 = \frac{2x_{H_2} + 2x_{H_2O}}{x_{H_2O} + 2x_{O_2}} \tag{c}$$

The sum of the mole fractions of the equilibrium mixture is 1:

$$1 = x_{H_2} + x_{H_2O} + x_{O_2} \tag{d}$$

Remembering that the equilibrium reaction is

$$H_2O \leftrightarrow H_2 + \frac{1}{2}O_2$$

from Equation 3.65 we can write

$$K_p = \frac{x_{H_2} x_{O_2}^{0.5}}{x_{H_2O}} \cdot p^{1+0.5-1} = \frac{x_{H_2} \sqrt{x_{O_2}}}{x_{H_2O}} \cdot \sqrt{p} \tag{e}$$

and from Equation 3.66

$$\ln K_p = \frac{\hat{g}_{H_2O}^o}{\hat{R}T} - \frac{\hat{g}_{H_2}^o}{\hat{R}T} - \frac{1}{2} \frac{\hat{g}_{O_2}^o}{\hat{R}T}$$

Using the data from Appendix C,

$$\ln K_p = -40.103 - (-20.198) - \frac{1}{2}(-29.570) = -5.120$$

and

$$K_p = 0.005976$$

Substituting for K_p and using $p = 1$ atm in Equation e,

$$0.00598 = \frac{x_{H_2} \sqrt{x_{O_2}}}{x_{H_2O}} \tag{f}$$

Solving Equations c, d, and f simultaneously yields

$$x_{H_2O} = 0.658, x_{O_2} = 0.335, x_{H_2} = 0.007$$

Note that there is no OH, H, or O in this example because only the equilibrium reaction for H_2O to H_2 and O_2 (reaction i given above) was specified. To account for these additional species, equilibrium reactions iv, v, and vii given above with associated equilibrium constants and equilibrium equations would need to be included in the analysis. The solution to the expanded problem is

$$x_{H_2O} = 0.6264, x_{O_2} = 0.3168, x_{H_2} = 0.00664$$

$$x_H = 0.00204, x_{OH} = 0.03994, x_O = 0.00809$$

Example 3.10

Repeat Example 3.9 using the method of minimizing Gibbs free energy.

Solution
The equilibrium mixture contains H_2, O_2, and H_2O, thus

$$G = N_{H_2} \hat{g}_{H_2} + N_{O_2} \hat{g}_{O_2} + N_{H_2O} \hat{g}_{H_2O}$$

Dividing both sides by $\hat{R}T$,

$$\frac{G}{\hat{R}T} = N_{H_2} \frac{\hat{g}_{H_2}}{\hat{R}T} + N_{O_2} \frac{\hat{g}_{O_2}}{\hat{R}T} + N_{H_2O} \frac{\hat{g}_{H_2O}}{\hat{R}T}$$

Remembering Equation 3.58,

$$\hat{g}_j = \hat{g}_j^\circ + \hat{R}T \ln(x_j) + \hat{R}T \ln\left(\frac{p}{p_0}\right)$$

At 2500 K and 1 atm this becomes

$$\frac{G}{\hat{R}T} = N_{H_2}\left[-20.198 + \ln\left(\frac{N_{H_2}}{N}\right)\right] + N_{O_2}\left[-29.570 + \ln\left(\frac{N_{O_2}}{N}\right)\right]$$

$$+ N_{H_2O}\left[-40.103 + \ln\left(\frac{N_{H_2O}}{N}\right)\right]$$

(a)

The number of moles of elements in the equilibrium mixture is

$$N = N_{H_2} + N_{O_2} + N_{H_2O}$$ (b)

Arbitrarily assuming that there is 1 mole of H atoms and 1 mole of O atoms in the system maintains the specified O/H ratio of 1. The atom balance constraints (Equation 3.61) for H and O, respectively, are

$$2\,N_{H_2} + 2\,N_{H_2O} = 1$$

$$2\,N_{O_2} + N_{H_2O} = 1$$

(c)

Now minimize Equation a subject to conservation of elements (Equation b) and the atom balance constraints (Equation c) by varying one of the molar concentrations such as N_{O_2}. Using an equation solver yields

$$\left.\frac{G}{\hat{R}T}\right|_{minimum} = -27.926 \text{ kgmol}$$

and

$$N = 0.756 \text{ kgmol}$$

and

$$N_{H_2} = 0.005 \text{ kgmol}; \quad N_{O_2} = 0.253 \text{ kgmol}; \quad N_{H_2O} = 0.495 \text{ kgmol}$$

Note that the selection of 1 mole of atomic hydrogen is arbitrary here provided that the H/O atom ratio meets the specified ratio of 1. The mole fractions are

$$x_{H_2O} = 0.658, x_{O_2} = 0.335, x_{H_2} = 0.007$$

which is the same result as in Example 3.9.

Example 3.11

Charcoal (wood char) is an easily made, high energy fuel that is lightweight and easy to transport. Assume that charcoal is composed of only C and consider the reaction of carbon with stoichiometric air to produce CO_2, CO, and O_2 at 2200 K and 2 atm. What is the volume fraction of CO when the products are in equilibrium at 2200 K due to the dissociation of CO_2?

Solution
The stoichiometric reaction is

$$C + (O_2 + 3.76\,N_2) \rightarrow CO_2 + 3.76\,N_2$$

Considering the disassociation CO_2 into CO and O_2, the reaction becomes

$$C + (O_2 + 3.76\,N_2) \rightarrow a\,CO_2 + b\,CO + c\,O_2 + 3.76\,N_2$$

Note that only disassociation of CO_2 is considered, and it is assumed that no solid carbon remains and the N_2 and O_2 do not dissociate. Using the method of equilibrium constants, we obtain molar fractions of the dissociated products. Rewriting the reaction in terms of molar fractions of the dissociated products yields

$$d[C + (O_2 + 3.76\,N_2)] \rightarrow x_1\,CO_2 + x_2\,CO + x_3\,O_2 + x_4\,N_2 \qquad \text{(a)}$$

A carbon atom balance gives

$$d = x_1 + x_2 \qquad \text{(b)}$$

An oxygen atom balance gives

$$2d = 2x_1 + x_2 + 2x_3 \qquad \text{(c)}$$

A nitrogen balance gives

$$3.76d = x_4 \qquad \text{(d)}$$

The sum of the product mole fractions is 1:

$$x_1 + x_2 + x_3 + x_4 = 1 \tag{e}$$

The equilibrium constant K_p can be determined from Equation 3.65,

$$K_p = \frac{x_C^c x_D^d}{x_A^a x_B^b} \cdot p^{c+d-a-b}$$

For the dissociation reaction

$$CO_2 \leftrightarrow CO + \frac{1}{2}O_2$$

$$a = 1, b = 0, c = 1, \text{ and } d = 1/2$$

and

$$K_p = \frac{x_2 \sqrt{x_3}}{x_1} \sqrt{p} \tag{f}$$

where p is in atmospheres. From Equation 3.66

$$\ln K_p = \frac{\hat{g}_1^o}{\hat{R}T} - \frac{\hat{g}_2^o}{\hat{R}T} - \frac{1}{2}\frac{\hat{g}_3^o}{\hat{R}T}$$

Using data from Appendix C

$$\ln K_p = -53.737 - (-34.062) - \frac{1}{2}(-29.062) = -5.127$$

and

$$K_p = 0.005934$$

Substituting K_p and $p = 2$ into Equation f and rearranging,

$$x_1 = 238.3 x_2 \sqrt{x_3} \tag{g}$$

Solving Equations b, c, d, e, and g simultaneously yields

$$x_1 = 0.1978, x_2 = 0.0111, x_3 = 0.0056, x_4 = 0.7855, d = 0.2089$$

And CO is 1.11% by mole (volume) fraction.

Example 3.12

Solid carbon, C(s), reacts with steam at 1000 K and 1 atm to produce carbon monoxide and hydrogen. Find the equilibrium composition if the initial C/O atom mole ratio is 1/1 and the initial C/H atom mole ratio is 1/1.

Solution

The reaction is

$$C_{(s)} + H_2O \leftrightarrow CO + H_2$$

The products in the equilibrium mixture are $C_{(s)}$, CO, H_2O, and H_2. This is a case of heterogeneous equilibrium—in heterogeneous equilibrium there are chemical reactions that involve more than one phase. In this case a solid phase with $C_{(s)}$ and a gas phase comprised of H_2O, CO, and H_2.

In the case of heterogeneous equilibrium, the equilibrium composition of the gas phase is not dependent on how much of the solid phase (or liquid phase) exists as long as some of the solid phase exists. To see this consider the vaporization of water

$$H_2O_{(\ell)} \leftrightarrow H_2O_{(g)}$$

The vapor pressure is a function of temperature but not the quantity of water.

Based on this, the approach to finding the equilibrium composition of a heterogeneous mixture is similar to finding the equilibrium mixture of a homogeneous mixture. In the first step, determine the equilibrium constant(s), K_p, for the reaction(s). Then using Equation 3.65 and the appropriate conservation equations for the gas phase components, find the equilibrium gas phase mixture. In the second step, the overall mole fraction for the combined solid and gas phases can be determined from conservation of atoms. The gas phase mixture computed in the first step remains constant.

Using the data from Appendix C, the equilibrium constant for this reaction is

$$\ln K_p = -\frac{\hat{g}_{CO}^{\,\circ}}{\hat{R}T} - \frac{\hat{g}_{H_2}^{\,\circ}}{\hat{R}T} + \frac{\hat{g}_{C_{(s)}}^{\,\circ}}{\hat{R}T} + \frac{\hat{g}_{H_2O}^{\,\circ}}{\hat{R}T}$$

$$\ln K_p = -(-38.881) - (-17.491) + (-1.520) + (-53.937) = 0.915$$

and

$$K_p = 2.50 \text{ at } T = 1000 \text{ K}$$

Note that solid carbon is included in the calculation of K_p, but the right hand side of Equation 3.65 does not include solid carbon because it involves only the gas phase. Therefore,

$$K_p = \frac{x_{CO}x_{H_2}}{x_{H_2O}} = 2.50 \tag{a}$$

The H to O atom mole ratio is given as 1 and can be used because they are used only in gas phase components:

$$\frac{H}{O} = \frac{2x_{H_2} + 2x_{H_2O}}{x_{H_2O} + x_{CO}} = 1 \tag{b}$$

The sum of the gaseous mole fractions is 1:

$$x_{H_2O} + x_{CO} + x_{H_2} = 1 \tag{c}$$

Solving Equations a, b, and c simultaneously gives the mole fractions in the gas phase,

$$x_{H_2O} = 0.073, x_{CO} = 0.642, x_{H_2} = 0.285$$

For the mixture of gas plus solid phases, the C/O and C/H ratios are

$$\frac{C}{O} = \frac{x'_{CO} + x'_{C_{(s)}}}{x'_{H_2O} + x'_{CO}} = 1 \tag{d}$$

$$\frac{C}{H} = \frac{x'_{CO} + x'_{C_{(s)}}}{2x'_{H_2O} + 2x'_{H_2}} = 1 \tag{e}$$

where x'_i is the molar fraction of i in the mixture, and x_i is the molar fraction of i in the gas phase. Noting the mole fraction of the gas phase is not dependent on the remaining fraction of the solid phase, we can write

$$\frac{x'_{H_2O}}{x'_{H_2}} = \frac{x_{H_2O}}{x_{H_2}} = \frac{0.073}{0.285} \tag{f}$$

The sum of the mixture fractions is 1.

$$x'_{H_2} + x'_{H_2O} + x'_{CO} + x'_{C_{(s)}} = 1 \tag{g}$$

Solving Equations d, e, f, and g simultaneously, the mole fractions for the mixture are

$$x'_{H_2} = 0.265, x'_{H_2O} = 0.068, x'_{CO} = 0.599, x'_{C_{(s)}} = 0.068$$

3.5.2 PROPERTIES OF COMBUSTION PRODUCTS

Once the mole fraction of each species has been determined, the internal energy, enthalpy, and average molecular weight of the products mixture may be obtained. For a given fuel each mole fraction of the products, x_i, is a function of T, p, and f. Because u and h are functions of x_i,

$$u = \sum_i \frac{x_i \hat{u}_i}{M_i} \tag{3.67}$$

they become functions of T, p, and f. This occurs even though each species' internal energy, u_i, is only a function of T. Thus the equilibrium composition of a reacting mixture is a function of temperature, and temperature is a function of the equilibrium composition of a reacting mixture. This coupled relationship between temperature and equilibrium composition can significantly complicate the computation of equilibrium compositions and temperatures.

Figure 3.4 shows enthalpy as a function of temperature for the products of a stoichiometric methane and air reaction. As shown, the lines of constant pressure coincide at lower temperatures where dissociation is small. At higher temperatures the changing composition due to dissociation causes the lines to separate. For a fixed temperature, dissociation is largest at low pressures and becomes quite small at very high pressures. In general, Le Chatelier's rule states that if the moles of products exceeds the moles of reactants, then an increase in pressure decreases the dissociation, as shown by Equation 3.65.

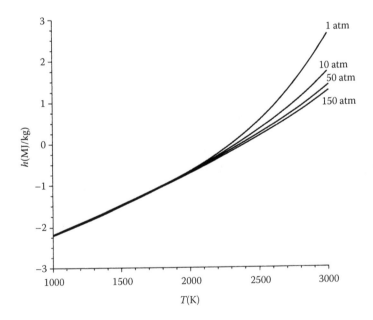

FIGURE 3.4 Absolute enthalpy of products from stoichiometric methane-air reaction.

Because the effect of dissociation is not large except at high temperatures, a starting place for computation is to determine the adiabatic flame temperature without dissociation. Then use the adiabatic flame temperature to compute the equilibrium composition of the reacting mixture with dissociation. This process can be repeated until the assumed temperature and the computed adiabatic temperature of the dissociated mixture are close.

3.6   ADIABATIC FLAME TEMPERATURE

The use of the first law of thermodynamics allows the calculation of processes that go from one equilibrium state to another. If this concept is extended to a step-by-step process in time, the assumption is that the adjustments to the equilibrium state take place very rapidly within each time step (shifting equilibrium). Several simple cases are considered in this section. First consider an adiabatic, constant pressure process during which reactants are converted to products. If the kinetic energy is small compared to the enthalpy, and there is no shaft work, the energy equation (Equation 3.12) is simply

$$h_{react} = h_{prod} \tag{3.68}$$

The problem is to determine the temperature of the products that will make this equality hold. This is the adiabatic flame temperature problem. Similarly for constant volume combustion,

$$u_{react} = u_{prod} \tag{3.69}$$

The problem is to find T and p of the products that make the equality hold. Because both p and T are now unknown, the extra equation

$$p = \frac{mRT}{V}$$

is needed. Another type of problem is the expansion of the products in a nozzle or in a piston-cylinder system. The expansion is typically not adiabatic for either of these cases. However, heat transfer can be included by means of empirical formulations.

Flames are discussed in detail in Chapter 5, but in this section we calculate the temperature rise across a flame. A flame is a rapid exothermic reaction that occurs at constant pressure, and the kinetic energy of the gases is insignificant. Hence the adiabatic flame temperature can be obtained from Equation 3.68. For stoichiometric and lean mixtures at temperatures below 1900 K, dissociation is low enough that the flame temperature may be calculated by assuming that combustion products go to completion. In this case the mole fraction of each species is known from the complete reaction. The adiabatic flame temperature is obtained by equating the enthalpy of the reactants to the enthalpy of the products. The enthalpy of the reactants is known, and by trial and error, the temperature can be determined that makes the enthalpy of the products equal to the enthalpy of the

TABLE 3.2

Adiabatic Flame Temperatures (K) of Selected Fuels with Air at 1 atm; Reactants Initially at 298 K

Fuel	Equivalence ratio, F		
	0.8	1.0	1.2
Gaseous Fuels			
Methane	2020	2250	2175
Ethane	2040	2265	2200
Propane	2045	2270	2210
Octane	2150	2355	2345
Liquid Fuels			
Octane	2050	2275	2215
Cetane	2040	2265	2195
No. 2 fuel oil	2085	2305	2260
Methanol	1755	1975	1810
Ethanol	1935	2155	2045
Solid Fuel (Dry)			
Bituminous coal	1990	2215	2120
Lignite	1960	2185	2075
Wood	1930	2145	2040
RDF[a]	1960	2175	2085
Solid Fuels (25% Moisture)			
Lignite	1760	1990	1800
Wood	1480	1700	1480
RDF[a]	1660	1885	1695

[a] Refuse-derived fuel

reactants. The procedure is shown in the following examples. For complex fuels the enthalpy of formation is first calculated from the heating value as shown in Example 3.8. Adiabatic flame temperatures for various representative fuels are shown in Table 3.2.

Example 3.13

Find the adiabatic flame temperature of carbon monoxide burning with 50% excess air at 25°C and 1 atm. Neglect dissociation. The complete reaction is

$$CO + \frac{1.5}{2}(O_2 + 3.76 \, N_2) \rightarrow CO_2 + 0.25 \, O_2 + 2.82 \, N_2$$

Solution

The energy equation (Equation 3.68) for this reaction is

$$\left[\left[\hat{h}_s + \hat{h}^\circ\right]_{CO} + 0.75\left[\hat{h}_s + \hat{h}^\circ\right]_{O_2} + 2.82\left[\hat{h}_s + \hat{h}^\circ\right]_{N_2}\right]_{react}$$

$$= \left[\left[\hat{h}_s + \hat{h}^\circ\right]_{CO_2} + 0.25\left[\hat{h}_s + \hat{h}^\circ\right]_{O_2} + 2.82\left[\hat{h}_s + \hat{h}^\circ\right]_{N_2}\right]_{prod}$$

where the subscript *s* refers to the sensible enthalpy. The reactants are at the ground state, and as a result the sensible enthalpies for the reactants is 0. From Appendix C,

$$\hat{h}^\circ_{CO} = -110.50 \text{ MJ/kgmol}$$

$$\hat{h}^\circ_{CO_2} = -393.51 \text{ MJ/kgmol}$$

$$\hat{h}^\circ_{O_2} = 0 \text{ MJ/kgmol}$$

$$\hat{h}^\circ_{N_2} = 0 \text{ MJ/kgmol}$$

Substituting and simplifying,

$$(0 - 110.53) = \left(\hat{h}_{CO_2,s} - 393.52\right) + 0.25\,\hat{h}_{O_2,s} + 0.282\,\hat{h}_{N_2,s}$$

From above, the left hand side of the equation (LHS) is equal to −110.53. The solution process is then to guess a temperature that balances the energy equation by making the right hand side of the equation (RHS) equal the LHS. Using Appendix C and guessing $T = 2000$ K,

$$RHS = (91.45 - 393.52) + 0.25(59.20) + 2.82(56.14) = -128.95$$

Guess $T = 2100$ K,

$$RHS = (97.50 - 393.52) + 0.25(62.99) + 2.82(59.75) = -111.78$$

Guess $T = 2200$ K,

$$RHS = (103.57 - 393.52) + 0.25(66.80) + 2.82(63.37) = -94.42$$

Thus, by interpolation the adiabatic flame temperature is 2107 K.
When dissociation is included the adiabatic flame temperature is 2084 K. Dissociation always reduces the flame temperature.

Example 3.14

In Example 3.5, 100 kg bituminous coal was combusted with 50% excess air at 25°C and 1 atm. The products were 5.833 kgmol CO_2, 2.778 kgmol H_2O, 3.307 kgmol O_2, and 37.303 kgmol N_2. In Example 3.8 it was found that the heat of formation of this bituminous coal is −1079 kJ/kg. Find the adiabatic flame temperature of this bituminous coal burned with 50% excess air at 25°C and 1 atm. Neglect dissociation and neglect the ash.

Solution
Following the same process as in Example 3.13, the energy equation becomes

$$\left[m\ h° \right]_{coal} = \left[N\left(\hat{h}_s + \ \hat{h}° \right) \right]_{CO_2} + \left[N\left(\hat{h}_s + \ \hat{h}° \right) \right]_{H_2O} + \left[N\left(\hat{h}_s + \ \hat{h}° \right) \right]_{O_2} + \left[N\left(\hat{h}_s + \ \hat{h}° \right) \right]_{N_2}$$

For 100 kg of coal this becomes

$$100\ kg_{coal} \left(\frac{-1079\ kJ}{kg_{coal}} \right) = 5.83\ kgmol_{CO_2} \left(\frac{\left(\hat{h}_{CO_2,s} - 395{,}520 \right) kJ}{kgmol_{CO_2}} \right)$$

$$+\ 2.78\ kgmol_{H_2O} \left(\frac{\left(\hat{h}_{H_2O,s} - 241{,}830 \right) kJ}{kgmol_{H_2O}} \right)$$

$$+\ 3.31\ kgmol_{O_2} \left(\hat{h}_{O_2,s} \frac{kJ}{kgmol_{O_2}} \right)$$

$$+\ 37.30\ kgmol_{N_2} \left(\hat{h}_{N_2,s} \frac{kJ}{kgmol_{N_2}} \right)$$

Solving iteratively using data from Appendix C,

Left Hand Side	T (K)	Right Hand Side
−107,900 kJ	1900	−137,000 kJ
−107,900 kJ	2000	58,700 kJ

Thus, by interpolation the adiabatic flame temperature is 1915 K.

As a further simplification, when only an approximate answer is desired, the enthalpy of the combustion products may be obtained from an average specific heat. Then the energy equation becomes

$$m_f \left[c_{p,f} \left(T_f - T_0 \right) + LHV \right] + m_{air} c_{p,air} \left(T_{air} - T_0 \right) = \left(m_f + m_{air} \right) c_p \left(T_{flame} - T_0 \right) \quad (3.70)$$

Solving for the adiabatic flame temperature and assuming that the air and fuel enter at the reference temperature of $25°C = 77°F$,

$$T_{flame} = T_0 + \frac{f}{1+f} \cdot \frac{LHV}{c_p}$$ (3.71)

where c_p is the average specific heat of the products. The specific heat of N_2 may be used in this approximation.

From Equation 3.71 it can be seen that adiabatic flame temperature is a function of both fuel-air ratio and lower heating value. Hence, when comparing one fuel with another, a larger heating value alone does not necessarily imply a higher stoichiometric flame temperature because the stoichiometric fuel-air ratio must also be considered. For example, carbon monoxide has one-third the heating value of natural gas, and yet the stoichiometric flame temperature is several hundred degrees higher because it has a much lower fuel-air ratio.

Example 3.15

Estimate the adiabatic flame temperature of carbon monoxide burning with 50% excess air at 25°C and 1 atm using Equation 3.71.

Solution

From Example 3.13 the complete reaction is

$$CO + \frac{1.5}{2}(O_2 + 3.76\ N_2) \rightarrow CO_2 + 0.25\ O_2 + 2.82\ N_2$$

The fuel-air ratio is

$$f_s = \frac{m_f}{m_{air(s)}} = \frac{n_f M_f}{n_{air(s)} M_{air}} = \frac{28}{0.75(4.76)(29)} = 0.2704$$

From Table 2.2,

$$LHV_{CO} = 10{,}100\ kJ/kg$$

Knowing that the flame temperature will be approximately 2200 K, we choose to use the value of c_{p,N_2} at 1000 K. From Appendix C,

$$c_{p,N_2} = \frac{32.7\ kJ}{kgmol \cdot K} \cdot \frac{kgmol}{28\ kg} = \frac{1.17\ kJ}{kg \cdot K}$$

Substituting into Equation 3.71,

$$T_{flame} = 298 + \left(\frac{0.270}{1+0.270}\right)\frac{10,100 \text{ kJ}}{\text{kg}} \cdot \frac{\text{kg} \cdot \text{K}}{1.17 \text{ kJ}} = 2133 \text{ K}$$

Thus $T_{flame} = 2133$ K. This compares favorably to $T_{flame} = 2107$ K, which was obtained in Example 3.13 using a more complete formulation. The adiabatic flame temperature for CO combustion with 50% excess air including dissociation is 2084 K. While this method is only approximate, it can provide a quick estimate of the adiabatic flame temperature.

3.7 PROBLEMS

1. Calculate (a) the stoichiometric fuel-air ratio by weight of octane (C_8H_{18}), (b) the kg of octane in one cubic meter of a stoichiometric mixture of octane and air at 300 K and 1 atm, and (c) the weight of the stoichiometric octane and air mixture at 27°C and 1 atm.

2. Wood charcoal is used as a cooking fuel in many developing countries. Generally this is locally produced, low quality wood charcoal produced at low temperature. A combustion engineer is working on designing a forced-air, charcoal cookstove with 1500 W of cooking power. To ensure that there is sufficient air to ensure complete combustion while maintaining the stove temperature high, the stove should be designed to have 10% excess air. The stove does not condense the water in the combustion products. The ultimate analysis of the locally produced charcoal is 92% C, 6% O, and 2% H (wt) and the dry, ash-free HHV is 30 MJ/kg. Determine (a) the lower heating value of the charcoal and (b) assuming 50% efficiency, the airflow rate in L/min at standard temperature and pressure.

3. A 4-cylinder, 4-stroke automobile engine is running at 4000 rpm, and each cylinder has a displacement of 0.65 L. A stoichiometric mixture of octane and air at 20°C and 1 atm is drawn into the engine. The volumetric efficiency at wide open throttle is 85%. Calculate (a) the mass of octane used per minute, (b) the volume of air drawn in per minute, and (c) the liters per hour of octane used by the engine. The volumetric efficiency is the mass of charge ingested per intake event divided by the mass of charge with volume equal to the displacement of the cylinder at density equal to that of the charge entering the port. A cylinder has 1 intake event every 2 revolutions.

4. To form a stoichiometric mixture of 27°C n-octane vapor and air, the air is brought in at 1 atmosphere and temperature T_0 and mixed with octane liquid. Assume the process is adiabatic and isobaric and calculate T_0. The octane liquid is brought in at 25°C. Use the data in Appendices A and B.

5. An engineer, Pat, calculated the stoichiometric fuel-air ratio for methane and dry air. Pat then did an experiment on a burner with room air and

methane gas. Pat measured the mass flow rate of methane and regulated the mass flow rate of air to give a stoichiometric ratio according to the calculations. Later another engineer, Fran, asked Pat if the humidity of the air had been measured. Pat said the air was 27°C and saturated, but didn't think that it mattered. Fran said, "I think the fuel-air ratio was not stoichio-metric." Pat said, "I think the error is small." Who is correct? Justify your answer by calculation.

6. Compare methane, kerosene, and bituminous coal on the basis of the mass of CO_2 produced per unit of energy released (metric tons/MJ). Use the data in the Tables 2.2, 2.7, and 2.13.

7. A burner uses methane and air. There is 3% oxygen (vol) in the dry exhaust products. Determine the excess air used for this burner.

8. Repeat Problem 7 for ethanol and air. The chemical formula for ethanol is C_2H_5OH.

9. Calculate the work required to isentropically compress 3.0 L of stoichio-metric octane-air to one tenth cubic volume starting at 27°C and 1 atm. Calculate the resulting pressure and temperature. Repeat the calculation for air only. What can you conclude about the compression work of a lean versus a stoichiometric engine? Use

$$c_{v,octane} = 0.367 + 0.00203T \, (kJ/kg \cdot K)$$

and

$$c_{v,air} = 0.634 + 0.0001287T \, (kJ/kg \cdot K)$$

where T is in K.

10. A gaseous mixture from a wood gasifier contains 15% hydrogen, 4% meth-ane, 25% carbon monoxide, 10% carbon dioxide, and 46% nitrogen by vol-ume. What is the higher heating value in $kJ/kg_{mixture}$ and $kJ/m^3_{mixture}$?

11. For the mixture of Problem 10, find the specific heat of the mixture on a mass basis and on a molar basis. The mixture is at 25°C. Does the specific heat of this mixture change with pressure?

12. Calculate the stoichiometric fuel-air mass ratio for (a) natural gas, (b) dry, ash-free bituminous coal, and (c) dry, ash-free refuse-derived fuel given the following ultimate analysis (dry, ash-free, wt%).

Element	Natural gas	Bituminous coal	Refuse-derived fuel
C	75	80	52
H	25	9	8
O	0	11	40

13. Natural gas can be simulated in the laboratory by a mixture of 83.4% meth-ane, 8.6% propane, and 8.0% nitrogen (vol). For this mixture find (a) the

stoichiometric air-fuel weight ratio and (b) the molecular weight of the combustion products at 25% excess air.

14. Using the heat of formation of water vapor, calculate the lower heating value of H_2 in units of kJ/kg and Btu/lb$_m$.

15. Using the heat of formation of methane, carbon dioxide, and water vapor at 298 K from Appendix C, calculate the lower heating value of methane.

16. A fuel with a composition of $CH_{1.8}$ has a higher heating value of 44,000 kJ/kg. Find (a) the lower heating value and (b) the heat of formation of this fuel.

17. A fuel oil is 87% carbon and 13% hydrogen (wt). Find the difference between the higher and lower heating values in kJ/kg.

18. A stoichiometric mixture of H_2 and O_2 reacts at 25 atm and 2500 K. Assume that the products are in equilibrium according to the reaction $H_2O \leftrightarrow H_2 + 1/2\ O_2$. Using the method of equilibrium constants, determine (a) the mole fractions and (b) mass fractions of the products.

19. Repeat Problem 18 using the method of minimizing Gibbs free energy.

20. Using the method of equilibrium constants, find the mole fraction and molar concentration (gmol/cm³) of atomic oxygen (O) due to dissociation of O_2 at (a) 1500 K and 1 atm, (b) 1500 K and 25 atm, (c) 2500 K and 1 atm, and (d) 2500 K and 25 atm. The reaction is $O_2 \leftrightarrow O + O$.

21. Repeat Problem 20 using the method of minimizing Gibbs free energy.

22. Repeat Problem 20 for H_2. The reaction is $H_2 \leftrightarrow H + H$.

23. Repeat Problem 20 for N_2. The reaction is $N_2 \leftrightarrow N + N$.

24. Carbon monoxide can react with water to form carbon dioxide and hydrogen if the temperature is high enough. The reaction is $CO + H_2O \leftrightarrow CO_2 + H_2$. Given a H/C ratio of 1 and an O/C ratio of 2 find the equilibrium composition at (a) 1000 K and 1 atm and (b) 2000 K, and 1 atm.

25. One mole of carbon dioxide reacts with one mole of solid carbon to form carbon monoxide. The reaction is $CO_2 + C_{(s)} \leftrightarrow 2\ CO$. Find the equilibrium composition at (a) 1000 K and 1 atm and (b) 2000 K and 1 atm.

26. Solid carbon reacts with steam to produce carbon monoxide and hydrogen. Given an H/C atom ratio of 1/1 and an O/C atom ratio of 1/1, find the equilibrium composition at 1500 K and at 1 atm. The reaction is $C_{(s)} + H_2O \leftrightarrow CO + H_2$.

27. Calculate the adiabatic flame temperature for a stoichiometric hydrogen-air flame at 1 atm. The reactants are initially at 25°C, neglect dissociation and calculate using the data in Appendix C.

28. Calculate the adiabatic flame temperature of methane and air burning with 10% excess air at 1 atm pressure. The reactants enter at 500 K. Neglect dissociation and use the data in Appendix C.

29. Repeat Example 3.13 but use 25% excess air and assume that the air enters at 440 K. The pressure is 1 atm. Find the adiabatic flame temperature neglecting dissociation. Use the data in Appendix C.

30. Bituminous coal and switchgrass mixtures are used to fuel some power plants. Determine the adiabatic flame temperature of a stoichiometric

dry, ash-free mixture of 90% coal and 10% switchgrass by energy content combusted in air. Assume the reactants enter the combustor at 25°C and are combusted at 1 atm. Use the composition and higher heating values for coal and switchgrass given in Tables 2.13 and 2.10, respectively.

31. Assume you have a computer program that calculates the equilibrium composition of a mixture and also gives the quantities R, $\partial R/\partial T$, $\partial R/\partial p$, u, $\partial u/\partial p$, and $\partial u/\partial T$. Consider the first law for a reacting mixture in a closed adiabatic system with only mechanical work (a simplified engine cylinder). The energy equation can be written as

$$m \frac{du}{dt} = -p \frac{dV}{dt}$$

and the ideal gas equation is

$$pV = mRT$$

where $u = u(p,T)$ and $R = R(p,T)$. Obtain an expression for dT/dt in terms of V, dV/dt, and the output quantities of the computer program.

Hint, differentiate the ideal gas equation with respect to time, t. Then substitute into the conservation of energy equation (the first law of thermodynamics) to eliminate dp/dt.

32. In some burners the fuel-air mixture may be stratified so that the products are reacted at different fuel-air ratios to form the final uniform equilibrium products. Consider two methane-air reactant mixtures. Both mixtures are 0.25 kg and are at 25°C and 1 atm. One mixture has an equivalence ratio (F) of 0.8 and the other has an equivalence ratio (F) of 1.2. Each methane-air mixture is burned adiabatically at constant pressure. Following this the products of the two mixtures are mixed together adiabatically at constant pressure. Determine the temperature of the mixed products, ignore dissociation, and assume complete combustion.

Hint: What equivalent homogeneous 0.50 kg mixture at 25°C and 1 atm would give the same outlet temperature?

33. To obtain a qualitative idea of the increase in adiabatic flame temperature caused by a rise in the initial reactant temperature, neglect dissociation and assume that the reactants and products each have constant (but unequal) specific heats. Show that

$$\frac{T_{af(\text{react})} - T_0}{T_{af(0)} - T_0} = C \frac{T_{react} - T_0}{T_{af(0)} - T_0} + 1$$

where $T_{af(\text{react})}$ = adiabatic flame temperature for the reactant temperature T_{react}, $T_{af(0)}$ = adiabatic flame temperature for the reactant temperature T_0,

and

$$C = \frac{c_{p,\text{react}}}{c_{p,\text{prod}}}$$

Note that dissociation causes $\partial h / \partial T$ at a given T, p, and F to increase over its non-dissociated value at the same T, p, and F. Both the increase in product species' specific heats with temperature and the increase in dissociation with temperature cause the rate of change $\partial T_{af} / \partial T_{\text{react}}$ to decrease as T_{react} is increased. This effect is partially offset by the increase in the reactants' specific heat with temperature.

REFERENCES

Bejan, A., *Advanced Engineering Thermodynamics* 3rd ed., Wiley, New York, 2006.

Cengel, Y. A. and Boles, M. A., *Thermodynamics: An Engineering Approach*, 7th ed., McGraw-Hill, New York, 2010.

Chase, M. W., ed., *NIST-JANAF Thermochemical Tables*, 4th ed., American Institute of Physics, Melville, NY, 1998.

Flagan, R. C. and Seinfeld, J. H., *Fundamentals of Air Pollution Engineering,* Prentice Hall, Englewood Cliffs, NJ, 1988.

Gordon, S. and McBride, B. J., *Computer Program for Calculation of Complex Chemical Equilibrium Compositions and Applications*, NASA RP-1311, Part I Analysis, 1994; Part II, Users Manual and Program Description, 1996.

Kee, R. J., Rupley, F. M., and Miller, J. A., *The CHEMKIN Thermodynamic Data Base*, Sandia National Laboratories Report SAND87-8215B, 1990.

Law, C. K., *Combustion Physics*, Cambridge University Press, Cambridge, UK, 2006.

Lide, D. R. and Kehiaian, H. V., *CRC Handbook of Thermophysical and Thermochemical Data,* CRC Press, Boca Raton, FL, 1994.

Moran, M. J. and Shapiro, H. N., *Fundamentals of Engineering Thermodynamics*, 6th ed., Wiley, 2007.

Powell, H. N. and Sarner, S. F., *The Use of Element Potentials in Analysis of Chemical Equilibrium*, Vol. 1, General Electric Co. report R59/FPD, 1959.

Spindt, R. S., "Air-Fuel Ratios From Exhaust Gas Analysis," SAE paper 650507, 1965.

4 Chemical Kinetics of Combustion

Thermodynamics describes the equilibrium behavior of chemical species in a mixture, but does not give the rate at which a new equilibrium is achieved when the temperature or pressure is changed. Chemical kinetics along with thermodynamics are needed to predict the reaction rates of well-mixed systems. In addition, if the system is not well mixed, mass transfer effects also come into play, as for example in non-premixed flames and solid fuel combustion.

This chapter considers basic concepts concerning elementary reactions, chain reactions, global reactions, and surface reactions.

4.1 ELEMENTARY REACTIONS

A chemical reaction may occur when two or more molecules or atoms approach one another. The type of reaction that might take place depends on the intermolecular potential forces existing during a collisional encounter, the quantum states of the molecules, and the transfer of energy. A reaction that takes place by such a collisional process is called an *elementary reaction,* to distinguish it from an overall or *global reaction,* which is the end result of many elementary reactions. For example, we might write the global reaction,

$$CO + \frac{1}{2} O_2 \rightarrow CO_2$$

which quickly summarizes the string of reactions that occur when CO and O_2 react. These elementary reactions compose the path(s) that molecules take as they move from one form (e.g., CO and O_2) to another form (e.g., CO_2). For example, in the absence of water vapor in the mixture, the elementary reactions that compose the global CO and O_2 reaction forming CO_2 are

$$1.\ CO + O_2 \rightarrow CO_2 + O$$

$$2.\ CO + O + M \rightarrow CO_2 + M$$

$$3.\ O_2 + M \rightarrow O + O + M$$

where M is a third body such as N_2 or O_2. The third body is needed to preserve the momentum and energy of the reacting molecules.

In the case of hydrocarbon fuels, hydrogen is also present. With CO and O_2, two additional elementary reactions involving CO are significant. These are

$$4.\ CO + OH \rightarrow CO_2 + H$$

$$5.\ CO + HO_2 \rightarrow CO_2 + OH$$

These additional elementary reactions greatly accelerate the oxidation of CO. Adding H also requires another 20 elementary reactions involving H, OH, HO_2, H_2O_2, O, H_2O, H_2, and O_2 molecules, as shown in Table 4.1. When hydrocarbon radicals are present, there are additional reactions that may involve CO. From this we can see that when elementary reactions are used for detailed analysis of a reaction rate, the computational modeling can become complicated very quickly.

However, elementary reactions are important because the rate of reaction for each elementary reaction stems from individual encounters between molecules and does not depend on the mixture environment. Because of this, elementary reaction rates can be determined under idealized laboratory conditions, such as low pressure and

---

TABLE 4.1

Elementary Reactions Involving Hydrogen and Oxygen

1	$H + O_2 \rightarrow O + OH$
2	$O + H_2 \rightarrow H + OH$
3	$H_2 + OH \rightarrow H_2O + H$
4	$O + H_2O \rightarrow OH + OH$
5	$H + H + M \rightarrow H_2 + M$
6	$O + O + M \rightarrow O_2 + M$
7	$O + H + M \rightarrow OH + M$
8	$H + OH + M \rightarrow H_2O + M$
9	$H + O_2 + M \rightarrow HO_2 + M$
10	$HO_2 + H \rightarrow H_2 + O_2$
11	$HO_2 + H \rightarrow OH + OH$
12	$HO_2 + H \rightarrow H_2O + O$
13	$HO_2 + OH \rightarrow H_2O + O_2$
14	$HO_2 + O \rightarrow O_2 + OH$
15	$HO_2 + HO_2 \rightarrow H_2O_2 + O_2$
16	$H_2O_2 + OH \rightarrow H_2O + HO_2$
17	$H_2O_2 + H \rightarrow H_2O + OH$
18	$H_2O_2 + H \rightarrow HO_2 + H_2$
19	$H_2O_2 + M \rightarrow OH + OH + M$
20	$O + OH + M \rightarrow HO_2 + M$

Source: Reprinted from Westbrook, C. K. and Dryer, F. L., "Chemical Kinetic Modeling of Hydrocarbon Combustion," *Prog. Energy Combust. Sci.* 10:1–57, © 1984 with permission from Elsevier Science and Technology Journals.

the presence of only reactant species, and then applied to complex situations where pressure may be high and many other species may be present. By contrast, rate data obtained for a global reaction normally cannot be applied outside the range of the experimental conditions. From the viewpoint of a physical chemist, an elementary reaction has the advantage of being amenable to evaluation by theoretical modeling. However, for engineering purposes experimental data for elementary reaction rates are almost always preferred.

There are three major types of elementary reactions:

1. Bimolecular atom exchange reactions in which two molecules (or ions) react together to form a product. For example, consider the reaction

$$AB + C \xrightarrow{k_1} BC + A$$

Molecules AB and C collide together, and in some cases they leave unchanged. In other cases there is sufficient energy and momentum that A and B are separated, and BC and A become the products.
2. Termolecular recombination reactions in which three molecules meet and form one or two new molecules. Consider the reaction

$$A + B + M \xrightarrow{k_2} AB + M$$

The third body (M) is necessary to conserve both momentum and energy during the collisional process in the reaction. The third body takes away surplus energy. Because three molecules must collide, termolecular reactions are much less likely to occur and hence the reaction rate of termolecular reactions is significantly slower than the reaction rate of bimolecular reactions.
3. Bimolecular decomposition reactions in which two molecules collide together and decompose into their constituent atoms. Consider the reaction

$$AB + M \xrightarrow{k_3} A + B + M$$

In decomposition the third body provides the energy needed to split the molecule.

To further understand the nature of these reactions, it is necessary to examine the quantum mechanical processes that give rise to them. Put simply, in each case the rate of reaction is proportional to the collision frequency. The collision frequency is in turn proportional to the product of the reactant concentrations. Only a small fraction of the molecular collisions result in a reaction. Those collisions that are energetic enough and in which the molecular orientation is favorable break a chemical bond. More information about molecular collision theory may be gained by studying the kinetic theory of gases.

For an elementary reaction, the reaction rate depends on the reaction rate constant times the concentration of each of the reactants. For example, the reaction rates of the elementary reactions 1, 2, and 3 above can be expressed as

$$-\frac{d[\text{AB}]}{dt} = -\frac{d[\text{C}]}{dt} = \frac{d[\text{BC}]}{dt} = \frac{d[\text{A}]}{dt} = k_1[\text{AB}][\text{C}] \tag{4.1}$$

$$-\frac{d[\text{A}]}{dt} = -\frac{d[\text{B}]}{dt} = \frac{d[\text{AB}]}{dt} = k_2[\text{A}][\text{B}][\text{M}] \tag{4.2}$$

$$-\frac{d[\text{AB}]}{dt} = \frac{d[\text{A}]}{dt} = \frac{d[\text{B}]}{dt} = k_3[\text{AB}][\text{M}] \tag{4.3}$$

where the factors k_1, k_2, and k_3 are the reaction rate constants for the respective reactions. Note that the reaction rate is negative for species that are being consumed in the forward reaction. In this chapter, we write the molar concentration, n, as [ ]. For example, $n_{\text{AB}} = [\text{AB}]$. Equations 4.1–4.3 give the concentration change due to chemical reactions only. The concentration can also change by system volume change and by the addition of species due to mass flow into or out of the system.

It is observed from both the kinetic theory of gases and experiments that the rate constant for an elementary reaction is an exponential function of temperature and is of the Arrhenius form

$$k = k_0 e^{-E/\hat{R}T} \tag{4.4}$$

where k_0 is the pre-exponential factor, and E is the activation energy for the reaction. The activation energy is energy required to bring the reactants to a reactive state, referred to as an *activated complex*, such that the chemical bonds can be rearranged to form products. Both E and k_0 have been determined from experiment for most elementary reactions that occur in combustion (Baulch et al. 1994).

For a reaction of the generic form,

$$a\text{A} + b\text{B} \xrightarrow{\;k_f\;} c\text{C} + d\text{D}$$

where a, b, c, and d are stoichiometric coefficients, the rate of destruction of A and B, and the rate of formation of C and D are given by

$$\frac{d[\text{A}]}{dt} = -ak_f[\text{A}]^a[\text{B}]^b \tag{4.5a}$$

$$\frac{d[\text{B}]}{dt} = -bk_f[\text{A}]^a[\text{B}]^b \tag{4.5b}$$

$$\frac{d[C]}{dt} = ck_f [A]^a [B]^b \tag{4.5c}$$

$$\frac{d[D]}{dt} = dk_f [A]^a [B]^b \tag{4.5d}$$

Elementary reactions are reversible so that the back reaction is

$$cC + dD \xrightarrow{k_b} aA + bB$$

For example, the back reaction rate of A is

$$\frac{d[A]}{dt} = ak_b [C]^c [D]^d \tag{4.6}$$

Combining the forward and backward reactions, the net reaction rate of A is

$$\frac{d[A]}{dt} = ak_b [C]^c [D]^d - ak_f [A]^a [B]^b \tag{4.7}$$

At equilibrium

$$\frac{d[A]}{dt} = 0$$

Substituting into Equation 4.7 and simplifying

$$\frac{k_f}{k_b} = \frac{[C]^c [D]^d}{[A]^a [B]^b} \tag{4.8}$$

Comparing Equation 4.8 to Equation 3.65, it follows that the ratio of the forward to backward kinetic rate constants equals the thermodynamic equilibrium constant based on concentrations K_c, that is,

$$K_c = \frac{k_f}{k_b} \tag{4.9}$$

where

$$K_c = K_p \left(\hat{R}T \right)^{a+b+c+d} \tag{4.10}$$

Thus, if only one of the rate constants is known, Equation 4.9 can be used to obtain the other rate constant.

Example 4.1

A closed chamber initially contains 1000 ppm CO, 3% O_2, and the remainder N_2 at 1500 K and 1 atm pressure. Determine the time for 90% of the CO to react, assuming only the elementary reaction

$$CO + O_2 \rightarrow CO_2 + O$$

with

$$k = 2.5 \times 10^6 e^{-24,060/T} \ m^3/gmol \cdot s$$

Solution

Initially,

$$[CO] = n_{CO} = \frac{p_{CO}}{\hat{R}T}$$

Evaluating,

$$n_{CO} = \frac{0.001(101.3) \ kPa}{1} \cdot \frac{kgmol \cdot K}{8.314 \ kPa \cdot m^3} \cdot \frac{1}{1500 \ K} = \frac{0.00816 \ gmol_{CO}}{m^3}$$

and

$$n_{O_2} = \frac{0.03}{1000 \times 10^{-6}} n_{CO} = \frac{0.245 \ gmol_{O_2}}{m^3}$$

and

$$k = 2.5 \times 10^6 e^{-24,060/1500} = 0.270 \ m^3/gmol \cdot s$$

Substituting into Equation 4.5, the rate of destruction of CO is

$$\frac{d[CO]}{dt} = -k[CO][O_2]$$

In this case because O_2 only reacts with CO and is at a much greater concentration than CO, the O_2 concentration can be assumed to be constant. Based on this, integration yields

$$\ln\left(\frac{[CO]}{[CO]_{init}}\right) = -k[O_2]t$$

Substituting and solving

$$t = -\frac{\ln(0.1)}{k[O_2]} = \frac{2.3}{(0.270 \ \text{m}^3/\text{gmol}\cdot\text{s})(0.245 \ \text{gmol}/\text{m}^3)} = 35 \ \text{s}$$

This is a very long reaction time. However, as noted above, water vapor and various radicals greatly speed up the reaction rate of CO in combustion reactions, so that the time to oxidize CO due to multiple elementary reactions is much shorter.

Most practical combustion applications involve fluid flow and heat transfer in addition to combustion. If the elementary reactions that apply to a given situation can be identified and if valid rate constants can be obtained, then the problem can be analyzed by solving the kinetic rate equations simultaneously with the equations of mass, momentum, and energy conservation. To date, the number of practical problems that have been solved following this route has been limited by both a lack of kinetic data and the numerical difficulties of solving the equations. Work is continuing on both fronts; in particular, understanding turbulent reacting flows is an ongoing area of research.

4.2 CHAIN REACTIONS

Many gas phase reactions are initiated by the formation at very low concentration of an extremely reactive species that sets off a series of reactions leading to the formation of the final products. Such a process is referred to as a chain reaction, and typically it occurs after a short induction period to allow the formation of the reactive species. The chains are the way the radicals and atoms shuffle through the set of reactions. Although difficult to apply in detail to complex systems, the concepts strictly apply to all reacting systems and are helpful in thinking about the mechanisms of reaction sets.

In *initiating reactions,* radicals are formed from stable species. Radicals are molecules with an unpaired electron such as O, OH, N, CH_3, or in general R•. For example,

$$CH_4 + O_2 \rightarrow CH_3 + HO_2$$

or

$$NO + M \rightarrow N + O + M$$

or generally

$$S \rightarrow R\bullet$$

where S is a stable species, such as a hydrocarbon fuel or nitric oxide. Radicals are often formed by the rupture of a covalent bond in a stable molecule where each fragment retains its contributing electron.

In *chain propagating* reactions, the number of radicals does not change, but different radicals are produced. For example,

$$CH_4 + OH \rightarrow CH_3 + H_2O$$

or

$$NO + O \rightarrow O_2 + N$$

or generally

$$R\cdot + S \rightarrow R\cdot + S*$$

where S* is an excited state of S or some new stable species.

In *chain branching* reactions, more radicals are produced than destroyed, for example,

$$CH_4 + O \rightarrow CH_3 + OH$$

or

$$O + H_2 \rightarrow OH + H$$

or generally

$$R\cdot + S \rightarrow \alpha R\cdot + S* \qquad \alpha > 1$$

In *terminating reactions*, radicals are destroyed by either gas phase reactions or by collisions with surfaces. For example,

$$H + OH + M \rightarrow H_2O + M$$

$$H + O_2 + M \rightarrow HO_2 \xrightarrow{\text{wall}} \frac{1}{2} H_2 + O_2$$

Reactions at the wall by which unstable species are removed are important in determining explosion limits at low pressure.

A simple chain scheme applicable to combustion can be illustrated by the following generic set:

1. Initiation		$S \rightarrow R\bullet$
2. Branching, $\alpha > 1$		$R\bullet + S \rightarrow \alpha R\bullet + S*$
3. Propagating, $\alpha = 1$		$R\bullet + S \rightarrow R\bullet + S*$
4. Terminating		$R\bullet + S \rightarrow$ (a stable product)
5. Terminating		$R\bullet \rightarrow$ destruction on the surface

Simple schemes can only be found for systems with a few species. However, the concepts of chain formalism are often useful in sorting out the reactions of larger sets of rate equations. Currently, combustion of a few fuels, such as hydrogen, methane, methanol, and ethane, can be written in terms of elementary reactions. Kinetics specialists are continuing to refine these mechanisms and to build on them to obtain schemes for more complicated hydrocarbons (Frenklach et al. 1995). As these reaction schemes are developed and evaluated, engineers can use them to better understand combustion processes. However, caution is required to ensure that all of the important elementary reactions have been included. The importance of certain reactions depends on the temperature; and thus, a reaction set that works well for flame kinetics may not properly predict ignition kinetics, which occur at much lower temperatures. Similarly, three-body collision reactions become more important at higher pressures; and thus, the kinetic mechanism can shift with pressure level.

Consider the oxidation of a hydrocarbon fuel (RH). Oxidation begins when an oxygen molecule of sufficient energy breaks a carbon–hydrogen bond to form radicals (hydrogen abstraction),

$$RH + O_2 \rightarrow R\bullet + HO_2 \tag{a}$$

For example,

$$CH_3CH_3 + O_2 \rightarrow CH_3CH_2 + HO_2$$

An alternative initiation reaction is thermally induced dissociation,

$$RH + M \rightarrow R'\bullet + R''\bullet + M \tag{b}$$

where $R'\bullet$ and $R''\bullet$ are two different hydrocarbon radicals. For example,

$$CH_3CH_2CH_3 + M \rightarrow CH_3CH_2 + CH_3 + M$$

The hydrocarbon radicals react rapidly with oxygen molecules to produce peroxy radicals,

$$R\bullet + O_2 + M \rightarrow RO_2 + M \tag{c}$$

Peroxy radicals undergo dissociation at high temperature to form aldehydes and radicals,

$$RO_2 + M \rightarrow R'CHO + R''O \tag{d}$$

The aldehydes may react with O_2,

$$RCHO + O_2 \rightarrow RCO + HO_2 \tag{e}$$

which is a branching reaction because it increases the number of free radicals. In addition to O_2 and HO_2, O and OH react with RH (the parent hydrocarbon), $R\bullet$ (hydrocarbon fragment), or RCHO (aldehydes) and with each other.

The formation of carbon monoxide initially occurs by thermal decomposition of RCO radicals,

$$RCO + M \rightarrow R\bullet + CO + M \qquad \text{(f)}$$

Oxidation of CO to CO_2 is the last step in the combustion of organic fuels. The most important CO oxidation step is the reaction with the OH radical,

$$CO + OH \rightarrow CO_2 + H \qquad \text{(g)}$$

where the OH concentration involves all of the reactions shown in Table 4.1.

The initiation Reaction (a) involves breaking a carbon–carbon bond or a carbon–hydrogen bond. The energy required for bond breakage can be estimated using the bond strengths summarized in Table 4.2. Hydrogen abstraction reactions involve breaking a carbon–hydrogen bond with a strength ranging from 364 to 465 kJ/gmol

TABLE 4.2
Typical Bond Strengths at 298 K

Diatomic Molecules	kJ/gmol
H–H	436
H–O	428
H–N	339
C–N	754
C–O	1076
N=N	945
N=O	631
O=O	498

Polyatomic Molecules	kJ/gmol
H–CH	422
$H–CH_2$	465
$H–CH_3$	438
$H–CHCH_2$	465
$H–C_2H_5$	420
H–CHO	364
$H–NH_2$	449
H–OH	498
$H–O_2$	196
$H–O_2H$	369
$HC\equiv CH$	965
$H_2C=C_2$	733
$H_3C–CH_3$	376
O=CO	532
$O–N_2$	167
O–NO	305

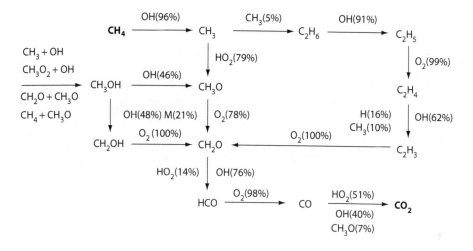

FIGURE 4.1 Reaction mechanism for methane oxidation; numbers in parentheses represent % of total species flux of the reactant through a given path. (Reprinted from Hunter, T. B., Wang, H., Litzinger, T. A., and Frenklach, M., "The Oxidation of Methane at Elevated Pressures: Experiments and Modeling," *Combust. Flame* 97(2):201–224, 1994. © 1994, with permission from Elsevier.)

and forming HO_2, leading to a net energy of reaction of 168–269 kJ/gmol. Dissociation involves breaking a carbon–carbon bond that requires 376 kJ/gmol for a single bond, 733 kJ/gmol for a double bond, and 965 kJ/gmol for a triple bond. Dissociation reactions are endothermic, and the dissociation requires a higher enthalpy of reaction than hydrogen abstraction.

Methane oxidation kinetics have been studied extensively both experimentally and theoretically because of the widespread use of natural gas. Hunter et al. (1994) obtained data from a flow reactor for lean methane-air mixtures at temperatures from 930 K–1000 K and pressures of 6–10 atm. Concentration profiles for CH_4, CO_2, and six intermediate species were obtained. A kinetic model was developed, which included 207 reactions with 40 species, of which the major trends are shown in Figure 4.1. The primary attack on CH_4 is from OH, which produces methyl radicals (CH_3) and water. CH_3 oxidation takes place primarily by reaction with hydroperoxide (HO_2) via the reaction $CH_3 + HO_2 \rightarrow CH_3O + OH$. At atmospheric pressure, other investigators have shown that methyl radical recombination and attack by oxygen atoms also play an important role. The C_2 hydrocarbon reactions appear to play a minor role for lean combustion but play an important role in fuel-rich combustion.

A flow tube reactor for obtaining chemical kinetic data is illustrated in Figure 4.2. The air is heated electrically, the fuel is mixed rapidly with the air, and the mixture flows rapidly to the lower velocity test section. A moveable sampling probe pulls and quenches a gas sample for analysis by a gas chromatograph/mass spectrometer (GC/MS). Optical access is also provided. An example of data that was obtained from another flow tube reactor is shown in Figure 4.3. Note the intermediate breakdown products of ethanol, which are formed and then consumed. The time is determined from the probe position, the flow rate, the major species, and the temperature.

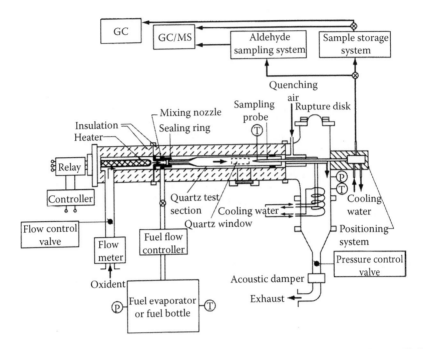

FIGURE 4.2 Flow reactor for chemical kinetic studies. (Reprinted from Hunter, T. B., Wang, H., Litzinger, T. A., and Frenklach, M., "The Oxidation of Methane at Elevated Pressures: Experiments and Modeling," *Combust. Flame* 97(2):201–224, 1994. © 1994, with permission from Elsevier.)

The kinetics of propane and higher hydrocarbons are different from that of methane because ethyl radicals (C_2H_5) are produced, which oxidize more rapidly than methyl radicals. Ethyl radicals decompose rapidly to produce C_2H_4 and H atoms. The hydrogen atoms produce chain branching from $H + O_2 \rightarrow O + OH$. Then H, O, and OH accelerate the abstraction of hydrogen from propane and the other hydrocarbons, thus producing a rapid chain reaction mechanism. A treatment of propane oxidation by Hoffman et al. (1991) includes low temperature oxidation (500–1000 K) kinetics and extends the mechanism to include 493 reactions.

Because detailed oxidation mechanisms for specific hydrocarbons can involve several hundred reactions, it is helpful to think of lean combustion reactions as proceeding along a primary path, such as

$$RH \rightarrow R\bullet \rightarrow HCHO \rightarrow HCO \rightarrow CO \rightarrow CO_2$$

For engineering purposes the oxidation of hydrocarbons is often treated by a few global reactions in order to expedite the analysis of practical combustion systems.

4.3 GLOBAL REACTIONS

For hydrocarbon fuels, the elementary reaction schemes are so complex that it is usually not feasible to consider all the chemically reacting species and their reaction

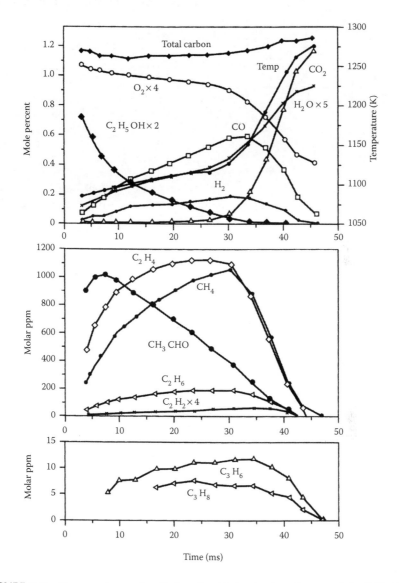

FIGURE 4.3 Flow reactor results of ethanol oxidation in air at 1 atm and $F = 0.61$. (From Norton, T. S. and Dryer, F. L., "The Flow Reactor Oxidation of C_1–C_4 Alcohols and MTBE," *Symp. (Int.) Combust.* 23:179–185, The Combustion Institute, Pittsburgh, PA, 1991, by permission of The Combustion Institute.)

rates when analyzing a practical combustion system. To simplify the chemistry, it is useful to use an overall or global reaction scheme. The basic mechanism for lean combustion is to remove hydrogen from the fuel and react it with OH and O to form water, intermediate hydrocarbons, hydrogen, and carbon monoxide. The hydrogen and carbon monoxide then oxidize to water and carbon dioxide. A global scheme for propane and other C_nH_{2n+2} fuels may be written as

$$C_nH_{2n+2} \rightarrow (n/2)\, C_2H_4 + H_2 \tag{a}$$

$$C_2H_4 + O_2 \rightarrow 2\,CO + 2\,H_2 \tag{b}$$

$$CO + \frac{1}{2}\,O_2 \rightarrow CO_2 \tag{c}$$

$$H_2 + \frac{1}{2}\,O_2 \rightarrow H_2O \tag{d}$$

Global rate equations in units of cubic centimeters, grams per mole, seconds, calories, and kelvin have been developed for each reaction species (Westbrook and Dryer 1984):

$$\frac{d[C_nH_{2n+2}]}{dt} = -10^{1/.32} \exp\left(\frac{-49{,}600}{\hat{R}T}\right) [C_nH_{2n+2}]^{0.50} [O_2]^{1.07} [C_2H_4]^{0.40} \tag{4.11}$$

$$\frac{d[C_2H_4]}{dt} = -10^{14.70} \exp\left(\frac{-50{,}000}{\hat{R}T}\right) [C_nH_{2n+2}]^{-0.37} [O_2]^{1.18} [C_2H_4]^{0.90} \tag{4.12}$$

$$\frac{d[CO]}{dt} = -10^{14.60} \exp\left(\frac{-40{,}000}{\hat{R}T}\right) [CO][O_2]^{0.25} [H_2O]^{0.50}$$
$$+ 5.0 \times 10^8 \exp\left(\frac{-40{,}000}{\hat{R}T}\right) [CO_2] \tag{4.13}$$

$$\frac{d[H_2]}{dt} = -10^{13.52} \exp\left(\frac{-41{,}000}{\hat{R}T}\right) [H_2]^{0.85} [O_2]^{1.42} [C_2H_4]^{-0.56} \tag{4.14}$$

Use of global sets of this sort should be limited to the regimes for which they have been tested, which in this case are equivalence ratios of 0.12–2, pressures of 1–9 atm, and temperatures of 960–1540 K. Note that in these units, $\hat{R} = 1.987$ cal/(gmol·K).

Even simplified schemes such as the four-reaction set given above can be too complicated for practical use in detailed computational models, e.g., numerical fluid mechanics codes. Thus, single step global rate equations are available, such as

$$\text{Fuel} + \alpha\, O_2 \rightarrow \beta\, CO_2 + \delta\, H_2O$$

where α, β, and δ are stoichiometric coefficients. The associated global rate expression is then written as

$$\hat{r}_f = \frac{d[\text{fuel}]}{dt} = -AT^n p^m \exp\left(\frac{-E}{\hat{R}T}\right) [\text{fuel}]^a [O_2]^b \tag{4.15}$$

where for many cases, one may take $n = m = 0$ for a specified range of T and p. The global reaction constants for the one-step reaction using Equation 4.15 are given in Table 4.3.

The one-step reaction overestimates the heat release rate because the products are complete, while in the actual reaction, CO oxidation continues after all the fuel has been oxidized. To improve this fault, a two-step reaction may be used.

$$\text{Fuel} + \alpha\, O_2 \rightarrow \beta\, CO + \delta\, H_2O \tag{a}$$

$$CO + \frac{1}{2} O_2 \rightarrow CO_2 \tag{b}$$

Equation 4.15 provides the rate expression for the fuel oxidation to CO (Reaction a), and the global reaction constants for the two-step reaction are given in Table 4.3. In the same way, Equation 4.13 provides the rate expression for the CO reaction to CO_2 (Reaction b). The alternative sets of data for methane and octane in Table 4.3 result from using different assumptions to fit the data.

TABLE 4.3
Global Reaction Constants for Equation 4.15[a]

Fuel	A (one-step)	A (two-step)	E (kcal/gmol)	a	b
CH_4	1.3×10^9	2.8×10^9	48.4	−0.3	1.3
CH_4	8.3×10^6	1.5×10^7	30.0	−0.3	1.3
C_2H_6	1.1×10^{12}	1.3×10^{12}	30.0	0.1	1.65
C_3H_8	8.6×10^{11}	1.0×10^{12}	30.0	0.1	1.65
C_4H_{10}	7.4×10^{11}	8.8×10^{11}	30.0	0.15	1.6
C_5H_{12}	6.4×10^{11}	7.8×10^{11}	30.0	0.25	1.5
C_6H_{14}	5.7×10^{11}	7.0×10^{11}	30.0	0.25	1.5
C_7H_{16}	5.1×10^{11}	6.3×10^{11}	30.0	0.25	1.5
C_8H_{18}	4.6×10^{11}	5.7×10^{11}	30.0	0.25	1.5
C_8H_{18}	7.2×10^{11}	9.6×10^{12}	40.0	0.25	1.5
C_9H_{20}	4.2×10^{11}	5.2×10^{11}	30.0	0.25	1.5
$C_{10}H_{22}$	3.8×10^{11}	4.7×10^{11}	30.0	0.25	1.5
CH_3OH	3.2×10^{11}	3.7×10^{12}	30.0	0.25	1.5
C_2H_5OH	1.5×10^{12}	1.8×10^{12}	30.0	0.15	1.6
C_6H_6	2.0×10^{11}	2.4×10^{11}	30.0	−0.1	1.85
C_7H_8	1.6×10^{11}	1.9×10^{11}	30.0	−0.1	1.85

Source: Reprinted from Westbrook, C. K. and Dryer, F. L., "Chemical Kinetic Modeling of Hydrocarbon Combustion," *Prog. Energy Combust. Sci.* 10:1–57, © 1984, with permission from Elsevier Science and Technology Journals.

[a] $n = 0$; $m = 0$; Units are in cubic centimeters, grams per mole, seconds, calories, and kelvin.

Example 4.2

For a stoichiometric propane-air reaction at 1 atm pressure and 1500 K, determine the fuel and CO_2 concentrations as a function of time using (a) a one-step global reaction and (b) a two-step global reaction.

Solution

The best solution approach is to develop the set of equations needed to solve for the concentrations as a function of time using one-step or two-step global reaction equations. The problem is solved using an equation solver and the data from Table 4.3. There is a core set of equations that is the same for both methods. The global reaction is

$$C_3H_8 + 5(O_2 + 3.76\,N_2) \rightarrow 3\,CO_2 + 4\,H_2O + 18.8\,N_2$$

First, we compute the number of moles of molecules initially in the stoichiometric propane-air mixture

$$n_{(init)} = \frac{p}{\hat{R}T} = \frac{gmol \cdot K}{82.05\ cm^3 \cdot atm} \cdot \frac{1\ atm}{1} \cdot \frac{1}{1500\ K} = \frac{8.12 \times 10^{-6}\ gmol}{cm^3}$$

From this we can determine the initial molar concentrations needed for the differential equations.

$$n_{f(init)} = x_{f(init)} n_{(init)} = \frac{1}{24.8} n_{(init)}$$

From the stoichiometric balance above, there is initially 1 mole of propane for every 24.8 $(1 + 5 \times 4.76)$ moles of mixture. We are now ready for the reaction rate specific equations.

Part (a) specific equation set:

For the one-step global reaction, the reaction rate equation is given by Equation 4.15, assuming that exponents $n = m = 0$

$$\hat{r}_f = -A\ \exp\left[\frac{-30,000}{1.987\ T}\right](n_f)^a (n_{O_2})^b\ gmol/cm^3 \cdot s$$

From Table 4.3

$$A = 8.6 \times 1011\ gmol/cm^3 \cdot s$$

$$a = 0.1$$

$$b = 1.65$$

The molar concentration of O_2 needs to be computed as a function of the fuel concentration for the rate equation. Because the mixture starts as a stoichiometric mixture, the ratio of fuel and O_2 remains the same throughout the reaction process. This is

$$n_{O_2} = 5n_f$$

Also we can find the molar concentration of CO_2 from a carbon balance on the one-step global reaction rate equation. That is, for every moles of fuel consumed, 3 mol of CO_2 are created.

$$n_{CO_2} = 3\left(n_{f(init)} - n_f\right)$$

We are now in a position to integrate the one-step reaction rate equation over time. The results are plotted in Figure 4.4 after normalizing n_f by the initial fuel concentration and n_{CO_2} by the final carbon dioxide concentration. Note that

$$n_{CO_2(final)} = 3n_{f(init)}$$

The conversion of the fuel and the buildup of carbon dioxide are shown, and 90% of the propane reacts to form products in 0.9 ms at 1500 K and 1 atm.

Part (b) specific equation set:
The two-step global reaction scheme is

$$C_3H_8 + 3.5\left(O_2 + 3.76\,N_2\right) \rightarrow 3\,CO + 4\,H_2O + 13.16\,N_2$$

$$CO + 0.5\left(O_2 + 3.76\,N_2\right) \rightarrow CO_2 + 1.88\,N_2$$

As with the one-step global reaction, we are using the reaction rate equation given by Equation 4.15 (assuming that $n = m = 0$).

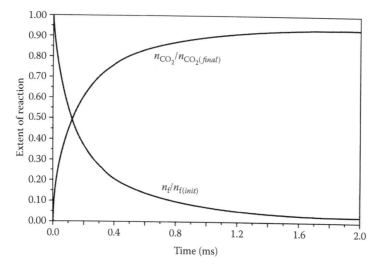

FIGURE 4.4 Calculated time history of a one-step global reaction for a stoichiometric propane-air reaction at 1 atm and 1500 K.

$$\hat{r}_f = -A \exp\left[\frac{-30{,}000}{1.987\,T}\right](n_f)^a (n_{O_2})^b \; gmol/cm^3 \cdot s$$

From Table 4.3

$$A = 1.0 \times 1012$$

$$a = 0.1$$

$$b = 1.65$$

It is simpler to integrate for CO_2 than for CO because CO_2 is only created by the second step of the two-step reaction. For the CO reaction we see that $r_{CO} = -r_{CO_2}$. From Equation 4.13

$$\hat{r}_{CO_2} = 10^{14.6} \exp\left[\frac{-40{,}000}{1.987\,T}\right](n_{CO})(n_{O_2})^{0.25}(n_{H_2O})^{0.50} - 5.0 \times 10^8 \exp\left[\frac{-40{,}000}{1.987\,T}\right]n_{CO_2}$$

The molar density of O_2 needs to be computed as a function of time. This is more complicated than in the single rate equation. The easiest way to think about this is to note that O_2 in the first rate equation is consumed at a rate of 3.5 times the fuel consumption rate. The remaining O_2 ($5.0 - 3.5 = 1.5$) can be computed as 1.5 times the initial fuel concentration and is consumed at 0.5 times the rate that CO_2 is formed. This can be written as

$$n_{O_2} = 3.5 n_f + \left(1.5 n_{f(init)} - 0.5 n_{CO_2}\right)$$

In the same way, CO is formed at three times the rate than the fuel is consumed and depleted by oxidation to CO_2 at a rate of 1 to 1. This can be written as

$$n_{CO} = 3\left(n_{f(init)} - n_f\right) - n_{CO_2}$$

H_2O is only formed in the first equation of the reaction set and can be related to the rate of consumption of the fuel. This is

$$n_{H_2O} = 4\left(n_{f(init)} - n_f\right)$$

Now we are in a position to integrate both the fuel concentration equation and the carbon dioxide concentration over time. As plotted in Figure 4.5 for this global two-step reaction, 90% of the propane reacts in 0.65 ms, and 90% of the CO_2 is formed in 0.7 ms. Carbon monoxide, which is zero initially, peaks at 0.5 ms and then decays.

4.4   NITRIC OXIDE KINETICS

Nitric oxide and nitrogen dioxide are pollutants. Combustion of hydrocarbon fuels with air produces nitric oxide (NO) and nitrogen dioxide (NO_2) in the products. Since the NO_2 to NO ratio is very small in combustion systems, the theoretical discussion

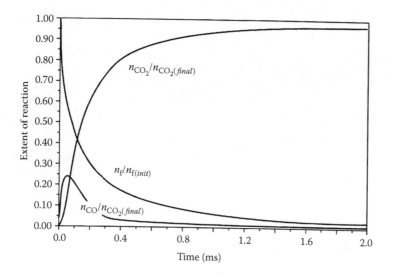

FIGURE 4.5 Calculated time history of a two-step global reaction for a stoichiometric propane-air reaction at 1 atm and 1500 K.

here will focus on nitric oxide. However, measurements of emissions typically include both NO and NO_2, which are jointly referred to as nitrogen oxides (NO_x). In ambient air, NO is converted to NO_2, which becomes the more significant pollutant.

In the atmosphere in the presence of ultraviolet sunlight, equilibrium is established between nitric oxide, nitrogen dioxide, and ozone.

$$NO_2 + O_2 \rightarrow NO + O_3$$

The addition of certain hydrocarbons unbalances the above reaction; consequently, the hydrocarbons are oxidized, and reaction products, such as nitrates, aldehydes, and PAN (peroxyacetyl nitrate), are formed. NO is converted to NO_2, and as the NO is consumed, ozone (O_3) begins to appear. Nitrogen dioxide forms a brownish haze. Nitrogen dioxide, ozone, and PAN can have adverse health effects. Nitric acid (HNO_3), which forms in the atmosphere from nitrogen dioxide, contributes to acid rain. Hence, considerable effort has been aimed at understanding and controlling the formation of nitrogen oxides from mobile and stationary combustion systems, which are major sources of nitrogen oxide emissions.

It is interesting to note that the back reaction of $NO + O_3$ produces a photon emission. Reacting a mixture containing NO with ozone and measuring the photon production is the standard method of measuring NO concentration, and this is the basis of the "chemilumenescent" analyzer.

The main sources of nitric oxide emissions in combustion are oxidation of molecular nitrogen in the post-flame zone (termed *thermal* NO), formation of NO in the flame zone (*prompt* NO), and oxidation of nitrogen-containing compounds in the fuel (*fuel-bound* NO). The relative importance of these three sources of nitrogen oxide depends on the operating conditions and the type of fuel. For adiabatic combustion with excess oxygen in the post-flame zone of a fuel containing little organic nitrogen, thermal NO formation is the main source of NO emissions.

As a starting place for understanding the basic mechanism for thermal NO production, consider the extended Zeldovich mechanism:

$$O + N_2 \rightarrow NO + N \tag{1}$$

$$N + O_2 \rightarrow NO + O \tag{2}$$

$$N + OH \rightarrow NO + H \tag{3}$$

The contribution of the third reaction pair is small for lean mixtures, but the reaction is normally included to account for rich combustion where the O_2 concentration is low. The first forward reaction controls the system, but this reaction has a high activation energy and thus is slow at low temperature. As a result, thermal NO is formed in the post-flame products. Concentrations of 1000–4000 ppm are typically observed in uncontrolled combustion systems.

From Reactions 1, 2, and 3, the rate of formation of NO is given by

$$\frac{d[NO]}{dt} = k_{+1}[O][N_2] - k_{-1}[NO][N] + k_{+2}[N][O_2]$$
$$- k_{-2}[NO][O] + k_{+3}[N][OH] - k_{-3}[NO][H] \tag{4.16}$$

To evaluate the rate of formation of NO from Equation 4.16, the O, N, OH, and H concentrations must be determined. For very high temperature applications, such as internal combustion engines, it may be safely assumed that the O, N, OH, and H concentrations remain in thermodynamic equilibrium in the post-flame zone. These values may be obtained from thermodynamic equilibrium calculations. For moderately high temperatures, such as in furnaces, N (monatomic nitrogen) does not remain in thermodynamic equilibrium. However, it can be assumed that N remains at a steady state concentration, i.e., the net rate of change of N is very small and may be set equal to zero in the following way. From Reactions 1, 2, and 3 above:

$$\frac{d[N]}{dt} = k_{+1}[O][N_2] - k_{-1}[NO][N] - k_{+2}[N][O_2]$$
$$+ k_{-2}[NO][O] - k_{+3}[N][OH] + k_{-3}[NO][H] = 0 \tag{4.17}$$

Solving for N at steady state,

$$[N] = \frac{k_{+1}[O][N_2] + k_{-2}[NO][O] + k_{-3}[NO][H]}{k_{-1}[NO] + k_{+2}[O_2] + k_{+3}[OH]} \tag{4.18}$$

Thus, the NO concentration in the post-flame combustion products may be determined as a function of time by integrating Equation 4.16. The equilibrium values of O_2, N_2, H_2O, OH, and H are obtained from a thermodynamic equilibrium program. The concentration of N is obtained either from equilibrium calculations or from Equation 4.18. The rate constants for Equation 4.16 in units of cubic centimeters per mole per second with the temperature in kelvin are as follows (Flagan and Seinfeld 1988):

$$k_{+1} = 1.8 \times 10^{14} \exp(-38{,}370/T)$$

$$k_{-1} = 3.8 \times 10^{13} \exp(-425/T)$$

$$k_{+2} = 1.8 \times 10^{10} T \exp(-4680/T)$$

$$k_{-2} = 3.8 \times 10^{9} T \exp(-20{,}820/T)$$

$$k_{+3} = 7.1 \times 10^{13} \exp(-450/T)$$

$$k_{-3} = 1.7 \times 10^{14} \exp(-24{,}560/T)$$

Calculations show that the rate of formation of NO is highly dependent on temperature, time, and stoichiometry. Let us consider two examples of NO formation that demonstrate this dependency: methane-air combustion products in a furnace and octane-air combustion products related to spark-ignition engine conditions.

Example 4.3

Methane and air initially at 298 K are burned adiabatically at 1 atm pressure and at theoretical air levels of 110%, 100%, 90%, and 80%. Calculate and plot the NO concentration as a function of time.

Solution

First determine the equilibrium concentrations, adiabatic flame temperature, and mole fractions of O_2, N_2, O, H_2O, OH, and H. The equilibrium results are

Theoretical	Mole fractions $\times 10^3$						
Air (%)	T (K)	O_2	N_2	O	H_2O	OH	H
110	2146	16.8	717	0.245	171	2.75	0.134
100	2227	4.46	708	0.215	183	2.87	0.393
90	2204	0.257	692	0.0442	189	1.30	0.679
80	2098	0.00943	667	0.00418	186	0.358	0.576

Assume these equilibrium species remain constant. Determine the N concentration from Equation 4.18 and then numerically integrate Equation 4.17. The results are plotted in Figure 4.6.

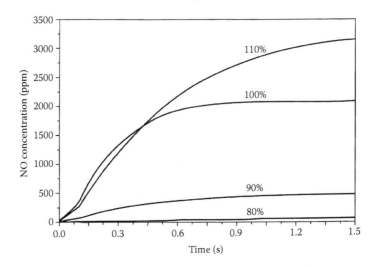

FIGURE 4.6 Calculated formation of nitric oxide in adiabatic combustion products of methane-air mixtures at 1 atm pressure and initial temperature of 300 K for 110%, 100%, 90%, and 80% theoretical air.

As shown in Figure 4.6, the lean mixture produces more NO than in the stoichiometric case though initially the rate of formation of NO is higher for the stoichiometric case. The fuel-rich cases have lower NO concentrations due to lack of oxygen and lower temperature. The NO levels reach equilibrium values in 0.5–1.5 s. At lower temperatures the approach to equilibrium is slower because the back reactions become more important. Thus, NO emissions in actual systems are highly dependent on excess air.

Example 4.4

A stoichiometric mixture of n-octane vapor and air is compressed polytropically starting at 1 atm and 298 K and then reacted adiabatically at constant pressure. For compression ratios of 6, 10, and 14, calculate and plot the NO concentration in the post-flame products as a function of time. Use a polytropic exponent of 1.3 to simulate heat loss during compression.

Solution

During compression, $pV^{1.3}$ is constant, and for an ideal gas of constant mass, pV/T is constant. Hence

$$\frac{p_2}{p_1} = \left(\frac{V_1}{V_2}\right)^{1.3}$$

and

$$\frac{T_2}{T_1} = \frac{p_2/p_1}{V_1/V_2}$$

Thus, the initial conditions for the reaction are

V_1/V_2	p_2 (atm)	T_2 (K)
6	10.3	510
10	20.0	594
14	30.9	658

Thermodynamic equilibrium calculations provide the following adiabatic flame temperature and product mole fractions ($\times 10^3$)

V_1/V_2	T_2 (K)	O_2	N_2	O	H_2O	OH	H
6	2440	5.19	726	0.239	135	3.06	0.305
10	2506	5.13	726	0.238	1124	3.19	0.293
14	2552	5.10	726	0.237	135	3.28	0.288

Use these as the input for the equation set developed in Example 4.3, solve with an equation solver, and provide the NO concentration as a function of time. The results are shown in Figure 4.7. Although the final equilibrium values of NO are only 33% higher for a compression ratio of 14 compared to 6, the initial rate of formation is much higher. For example, after 2.5 ms the NO levels are 1000, 2400, and 3600 ppm for compression ratios of 6, 10, and 14, respectively. Thus, the NO emissions from a spark engine are related to the residence time of the products in the cylinder as well as the compression ratio.

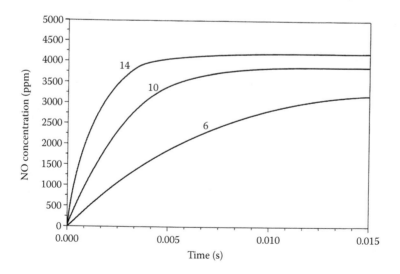

FIGURE 4.7 Calculated formation of nitric oxide in products of stoichiometric octane-air initially at 1 atm and 298 K, compressed polytropically ($n = 1.3$) at compression ratios of 6, 10 and 14, and reacted adiabatically at constant pressure.

From Examples 4.3 and 4.4 it is evident that the way to reduce NO formation is to avoid excessively high temperatures and to reduce excess oxygen in the combustion products. Another important concept to be noted is that when combustion products are cooled rapidly by expansion or heat transfer, the NO concentration tends to freeze at a certain value because the kinetic rates rapidly slow down as the temperature drops. The concept of freezing is illustrated in the following example.

Example 4.5

The combustion products of a stoichiometric n-octane-air mixture are contained in a cylinder fitted with a piston. At time $t = 0$ there is no NO in the mixture and the products are at $T_0 = 2600$ K and $p_0 = 35$ atm. Then the products expand following the polytropic law,

$$\left(\frac{p}{p_0}\right)\left(\frac{V}{V_0}\right)^{1.35} = 1.0 \quad \text{for } 0 \le t \le t_0$$

and the volume increases with time as

$$\frac{V}{V_0} = 1 + 4\left[1 + \sin\left(\frac{\pi t}{t_0} - \frac{\pi}{2}\right)\right]$$

Calculate and plot the NO concentration vs. time for $t_0 = 0.025$ s. Also plot the equilibrium NO for each temperature and show it on the same plot.

Solution
The solution procedure is similar to Example 4.4, but in this case the pressure and temperature are changing with time so that the equilibrium values of O and OH must be calculated at each time step and the new values used to solve the extended Zeldovich equations. The temperature is obtained from the ideal gas relation

$$pV = mRT$$

which implies that

$$\frac{T}{T_0} = \frac{V}{V_0} \cdot \frac{p}{p_0}$$

The details of the calculations are not shown here, but the results are plotted in Figures 4.8 and 4.9.

As shown, as the volume expands, the pressure and temperature drop. At first the NO is formed rapidly, but as the temperature drops, the reaction rates slow down, and the NO concentration freezes at a constant value of 2500 ppm. The initial theoretical equilibrium value of 4500 ppm is not attained, nor is the final equilibrium value of zero NO attained because both the forward and backward

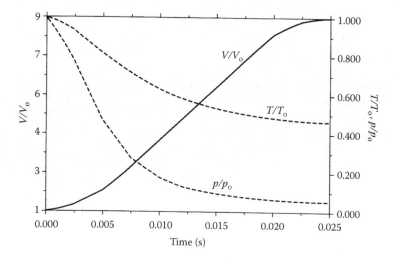

FIGURE 4.8 Decrease of pressure and temperature during polytropic expansion.

reaction rates go to very low values, and the NO concentration becomes frozen. When the expansion is faster ($t_0 = 0.01$ s), there is less time to form NO, and the concentration is frozen at a yet lower value.

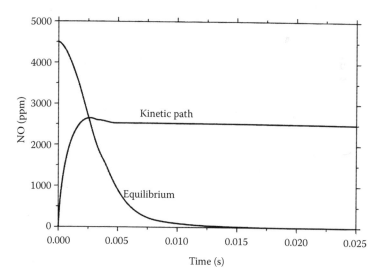

FIGURE 4.9 Nitric oxide formation for Example 4.5. The kinetic path is the correct path.

4.4.1 PROMPT NO AND FUEL-BOUND NO

Small amounts of NO (40–60 ppm) can be produced directly in a flame front. This rapidly formed or prompt NO is from reactions of hydrocarbon fragments with molecular nitrogen in the flame, such as

$$CH + N_2 \rightarrow HCN + N$$

$$CH_2 + N_2 \rightarrow HCN + NH$$

The species N, NH, and HCN then rapidly react to form NO in high temperature flame regions, where O and N can be present in excess of equilibrium concentrations. Prompt NO is usually small compared to thermally formed NO, but as thermal NO is reduced, prompt NO becomes relatively more important.

Fuel-bound NO is produced from organic nitrogen compounds in the fuel. Crude oil contains 0.1%–0.2% organic nitrogen, which is concentrated in the residual fractions after refining. Coal typically contains 1.2%–1.6% organic nitrogen. Nitrogen concentration in biomass varies depending on the source and how it was raised. Generally, wood contains 0.1%–0.2% nitrogen. Grass crops (e.g., switchgrass) can have levels of nitrogen as high as 1%. As these nitrogen-containing fuels are burned, hydrogen cyanide (HCN) is formed as an intermediate, which can further react with O, OH, and H to form NO in the post-flame zone. In combustion systems, typically 20%–50% of the fuel nitrogen is converted to NO.

4.5   REACTIONS AT A SOLID SURFACE

A homogeneous reaction is one involving a single phase, while a heterogeneous reaction involves more than one phase. The oxidation of carbon monoxide to carbon dioxide on the surface of a platinum catalyst is a heterogeneous reaction. The oxidation of char on a grate, or soot in a diesel engine, are examples of a non-catalytic heterogeneous reaction.

In each of the cases above, a gas reacts on a solid surface. The reaction rate depends on the concentration of the gas phase, the temperature, the diffusion rates, and the surface area accessible to the gas phase. For surface reaction rates to be significant, the accessible surface area per unit mass must be large, such as the solid surface of a porous platinum impregnated ceramic monolith or char or soot. Typically, the specific surface area in these examples is on the order of 100 m^2/g. As a comment to the reader, combustion of fuel sprays is not cited as an example of a surface reaction because a fuel droplet does not burn at the surface, but rather the droplet vaporizes and then burns as a gas-gas reaction.

Heterogeneous reactions at a solid surface involve a series of mass transfer and chemical reaction steps. These are

1. The reactant gas diffuses through the external gaseous boundary layer associated with the surface.
2. The reactant gas molecules diffuse into the pores of the solid while traveling to reactive surface sites.
3. Gas molecules are adsorbed on the surface, and chemical reactions between gaseous and surface molecules occur.
4. The reaction products are desorbed from the surface.

Steps 1 and 2 are external and internal mass transfer processes, whereas Steps 3 and 4 are chemical kinetic processes.

The rate of diffusion of a gaseous species such as oxygen through the external boundary layer can be obtained from computational fluid dynamics or by means of mass transfer coefficients. The concentration of a gaseous species, e.g., oxygen, due to diffusion within the pores may also be obtained from basic computations, provided a suitable description of the pore structure is available. Once the gas reaches the interior surface, it reacts with the surface if the site is active, such as oxygen reacting at a carbon edge to produce carbon monoxide. The reaction is described in terms of adsorption and desorption rate constants, which are expressed in the Arrhenius form (Equation 4.4). If the site is inactive, the reaction does not proceed at that site.

In combustion engineering the surface reaction rates of interest include soot, biomass and coal char reacting with oxygen, water vapor, carbon monoxide, carbon dioxide, and hydrogen. Recall from Chapter 2 that char is the solid material that remains after pyrolysis of a solid fuel and contains carbon, dispersed inorganic minerals, and a small amount of hydrogen. The particle surface temperature may vary due to changes in convective and radiatiative heat transfer as well as outgassing and exothermic and endothermic chemical reactions at the surface.

Char has a very porous surface, and it is useful to think of the pores as consisting of trunks and branches of trees (Figure 4.10). The porosity is established by pyrolysis (devolatilization). The diameters of the branches range from angstroms to several microns. The total surface area of each tree is several orders of magnitude greater than that of the trunk. The trunks have a distribution of sizes, and the trees overlap so that there is an interconnected porosity. Oxygen and other gases diffuse into the branches and react with the internal surfaces. During combustion, the porosity and pore surface area generally change with time and temperature. Thus,

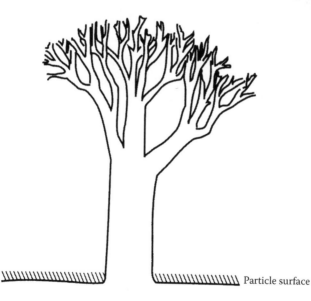

Particle surface

FIGURE 4.10 Schematic of char porosity based on the pore tree model showing continuous branching.

considering all the chemical and physical aspects of char reactivity, modeling the detailed processes occurring at the porous char surfaces from first principles is difficult.

For engineering purposes, global heterogeneous reaction rates based on a global rate constant and the external solid surface area typically are used. Global heterogeneous reaction rates are obtained from laboratory experiments of single particles of specific fuels. As with global gas phase reactions, the rate constants cannot be extrapolated beyond the range of the experiments. Use of the external surface area rather than the internal pore surface is justified at high temperatures because the bulk mass transfer through the external boundary layer is often the rate-limiting step. Global char reactions are discussed in Chapter 14.

4.6 PROBLEMS

1. Consider the stoichiometric overall reaction for octane in air

$$C_8H_{18} + 12.5 \ (O_2 + 3.76 \ N_2) \rightarrow 8 \ CO_2 + 9 \ H_2O + 47 \ N_2$$

with the global reaction rate

$$\hat{r}_f = -5.7 \times 10^{11} \exp\left(-30,000/\hat{R}T\right)\left[C_8H_{18}\right]^{0.25}\left[O_2\right]^{1.5}$$

where the units are centimeters, seconds, grams per mole, calories, and kelvin. If the reactant mixture is suddenly brought to a temperature of 2000 K and 1 atm pressure, what is the initial rate of reaction? If the temperature is held constant at 2000 K and the volume is constant, what is the rate of reaction when one-half of the original fuel has been converted to products? The reaction rate is given in units of $gmol/(cm^3 \cdot s)$.

2. Consider the following reactions from Table 4.1 for H_2-O_2 combustion. Which of the reactions are initiating, which are terminating, which are chain branching, and which are propagating?

$$H + O_2 \rightarrow O + OH \qquad\qquad (1)$$

$$H + H + M \rightarrow H_2 + M \qquad\qquad (2)$$

$$H + OH + M \rightarrow H_2O + M \qquad\qquad (3)$$

$$H_2 + OH \rightarrow H_2O + H \qquad\qquad (4)$$

$$H + O_2 + M \rightarrow HO_2 + M \qquad\qquad (5)$$

$$HO_2 + H \rightarrow H_2O + O \qquad\qquad (6)$$

$$O + H_2O \rightarrow OH + OH \qquad\qquad (7)$$

3. Reactions 1 and 5 of Problem 2 are in competition for H atoms. What is the effect of pressure on this competition?

4. The addition of a small amount of methane to an H_2–O_2 system may inhibit the overall combustion rate even though the fuel adds energy and increases the temperature. Explain this effect by comparing the reaction rates of the primary branching reaction,

$$H + O_2 \xrightarrow{\ 1\ } OH + O$$

$$k_1 = 5.13 \times 10^{16}\ T^{-0.816}\ \exp\left(-8307/T\right)$$

and the reaction,

$$H + CH_4 \xrightarrow{\ 2\ } CH_3 + H_2$$

$$k_2 = 2.24 \times 10^4\ T^3 \exp\left(-4420/T\right)$$

where T is in kelvin. (Note that in this problem the activation energy has already been divided by the universal gas constant in the rate constant expressions.)

5. Consider the prediction of NO in the products of combustion during expansion in a piston-cylinder. Find an expression for $(d[NO]/dt)$ in terms of the reaction $(d[NO]/dt)_R$ and the volume, where $V = V(t)$. Note that by definition $[NO] = n_{NO} = N_{NO}/V$.

6. A mixture of gases containing 3% O_2 and 60% N_2 by volume at room temperature is suddenly heated to 2000 K at 1 atm pressure. Assume that the O_2 and N_2 remain constant. Find the initial rate of formation of NO (ppm/s). Indicate whether the NO formation rate increases or decreases as time increases. Use the Zeldovich mechanism and the information given in Section 4.4. There is no hydrogen in the mixture.

7. Repeat Problem 6, but use 1500 K.

8. Repeat Problem 6, but use 20 atm pressure.

9. For the conditions of Problem 6, find the NO concentration (ppm) versus time (s).

10. Following the lead of Example 4.3, examine the effect of air preheated to 500 K on the NO concentration in methane-air products at 1 atm and 10% excess air.

11. Following the lead of Example 4.4, examine the effect of lean combustion on NO formation in n-octane vapor-air mixtures initially at 300 K and 1 atm, and then compressed polytropically $(n = 1.3)$ with $V_1/V_2 = 10$. Assume 180% theoretical air.

12. A lean mixture of CH_4 and O_2 react at 1300 K and 1 atm to form CO and H_2O according to the global reaction,

$$CH_4 + 3\,O_2 \rightarrow CO + 2\,H_2O + 1.5\,O_2$$

The CO reacts to form CO_2 by Equation 4.13. Using an equation solver, calculate and plot the CH_4 and CO_2 mole fractions versus time. Use the data from Table 4.3.

13. A pure carbon particle has a diameter of 0.01 μm and a density of 2000 kg/m³. How many carbon atoms does the particle contain? Assume the atoms are densely packed and use Avogadro's number.

REFERENCES

Baulch, D. L., Cobos, C. J., Cox, R. A., Frank, P., Hayman, G., Just, Th., Kerr, J. A., et al., "Summary Table of Evaluated Kinetic Data for Combustion Modeling: Supplement 1," *Combust. Flame* 98(1–2):59–79, 1994.

Benson, S. W., *The Foundations of Chemical Kinetics*, McGraw-Hill, New York, 1960.

Cowart, J. S., Keck, J. C., Heywood, J. B., Westbrook, C. K., and Pitz, W. J., "Engine Knock Predictions Using a Fully-Detailed and a Reduced Chemical Kinetic Mechanism," *Symp. (Int.) Combust.* 23:1055–1062, The Combustion Institute, Pittsburgh, PA, 1991.

Dagaut, P., Reuillon, M., and Cathonnet, M., "High Pressure Oxidation of Liquid Fuels from Low to High Temperature. 1. n-Heptane and iso-Octane," *Combust. Sci. Technol.* 95(1–6):233–260, 1994.

Flagan, R. C. and Seinfeld, J. H., "Combustion Fundamentals," Chap. 2 in *Fundamentals of Air Pollution Engineering*, Prentice Hall, Englewood Cliffs, NJ, 1988.

Frenklach, M., Clary, D. W., Gardiner, W. C., and Stein, S. E., "Detailed Kinetic Modeling of Soot Formation in Shock-Tube Pyrolysis of Acetylene," *Symp. (Int.) Combust.* 20:887–901, The Combustion Institute, Pittsburgh, PA, 1985.

Frenklach, M., Wang, H., Goldenberg, M., Bowman, C. T., Hanson, R. K., Gardiner, W. C., Lissianski, V., Smith, G. P., and Golden, D. M., *GRI-Mech -An Optimized Detailed Chemical Reaction Mechanism for Methane Combustion*, Gas Research Institute Topical Report GRI-95/0058, 1995.

Frenklach, M. and Wang, H., "Detailed Modeling of Soot Particle Nucleation and Growth," *Symp. (Int.) Combust.* 23:1559–1566, The Combustion Institute, Pittsburgh, PA, 1991.

Glassman, I. and Yetter, R., *Combustion*, 4th ed., Academic Press, Burlington, MA, 2008.

Griffiths, J. F., "Reduced Kinetic Models and Their Application to Practical Combustion Systems," *Prog. Energy Combust. Sci.* 21:25–107, 1995.

Haynes, B. S., "Soot and Hydrocarbons in Combustion," Chap. 5 in *Fossil Fuel Combustion: A Source Book,* eds. Bartok, W., and Sarofim, A. F., Wiley, New York, 1991.

Hoffman, J. S., Lee, W., Litzinger, T. A., Santavicca, D. A., and Pitz, W. J., "Oxidation of Propane at Elevated Pressures: Experiments and Modelling," *Combust. Sci. Technol.* 77(1–3):95–125, 1991.

Hunter, T. B., Wang, H., Litzinger, T. A., and Frenklach, M., "The Oxidation of Methane at Elevated Pressures: Experiments and Modeling," *Combust. Flame* 97(2):201–224, 1994.

Kee, R. J., Miller, J. A., and Jefferson, T. H., *CHEMKIN: A General-Purpose, Problem-Independent, Transportable, FORTRAN Chemical Kinetics Code Package*, Sandia National Laboratories Report, SAND 80–8003, 1980.

Kuo, K. K., "Chemical Kinetics and Reaction Mechanisms," Chap. 2 in *Principles of Combustion*, Wiley, New York, 1986.

Lide, D. R. and Haynes, W. M., eds., *CRC Handbook of Chemistry and Physics*, 90th ed., CRC Press, Boca Raton, FL, 2009.

Miller, J. A. and Fisk, G. A., "Combustion Chemistry," *Chem. Eng. News* 65(35):22–31, 34–46, 1987.

Müller, V. C., Peters, N., and Liñán, A., "Global Kinetics for n-Heptane Ignition at High Pressures," *Symp. (Int.) Combust.* 24:777–784, The Combustion Institute, Pittsburgh, PA, 1992.

Norton, T. S. and Dryer, F. L., "The Flow Reactor Oxidation of C_1–C_4 Alcohols and MTBE," *Symp. (Int.) Combust.* 23:179–185, The Combustion Institute, Pittsburgh, PA, 1991.

Siegla, D. C. and Smith, G. W., eds., *Particulate Formation During Combustion*, Plenum Press, New York, 1981.

Simons, G. A., "The Pore Tree Structure of Porous Char," *Symp. (Int.) Combust.* 19:1067–1076, The Combustion Institute, Pittsburgh, PA, 1982.

Smoot, L. D. and Smith, P. J., *Coal Combustion and Gasification,* Plenum Press, New York, 1985.

Westbrook, C. K. and Dryer, F. L., "Chemical Kinetic Modeling of Hydrocarbon Combustion," *Prog. Energy Combust. Sci.* 10:1–57, 1984.

Westbrook C. K. and Pitz, W. J., "A Comprehensive Chemical Kinetic Reaction Mechanism for Oxidation and Pyrolysis of Propane and Propene," *Combust. Sci. Technol.* 37(3–4):117–152, 1984.

Wilk, R. D., Pitz, W. J., Westbrook, C. K., and Cernansky, N. P., "Chemical Kinetic Modeling of Ethene Oxidation at Low and Intermediate Temperatures," *Sym. (Int.) Combust.* 23:203–210, The Combustion Institute, Pittsburgh, PA, 1991.

Part II

Combustion of Gaseous and Vaporized Fuels

Combustion of gaseous fuels may occur as premixed flames, diffusion flames, or radiation-dominated reactions on the surface of porous media. After a discussion of flames, the combustion processes in gas-fired furnaces and spark ignition engines are discussed. Part II ends with a discussion of gaseous detonation waves.

5 Flames

Flames are fundamental to all combustion applications. A flame is a rapid exothermic reaction between a gaseous fuel and an oxidizer that occurs over a short distance. Flames may be classified as premixed or diffusion, laminar or turbulent, and stationary or propagating.

Premixed flames arise from the combustion of gaseous reactants that are well mixed prior to combustion. *Diffusion flames* arise from the combustion of separate gaseous fuel and oxidizer streams that combust rapidly as they mix. When the gas flow is laminar, the flame is laminar and has a smooth flame front. When the gas flow is turbulent, the flame is turbulent and the flame front is wrinkled or broken up. Laminar premixed flames have a unique flame speed for a given fuel-oxidizer mixture that is less than a meter per second for hydrocarbon-air mixtures at ambient pressure and temperature. Turbulence increases the flame speed. For diffusion flames, the rate of mixing of the reactants determines the flame speed, and the reaction takes place at the interface between the fuel and the oxidizer. Stationary flames are stabilized by flow in a burner or by flow into a stagnation region. In a quiescent combustible mixture, an ignition source initiates a flame that propagates outward from the source at the laminar flame speed. Given sufficient volume, a flame will transition into a detonation, which propagates at more than 1000 m/s and creates a pressure rise (see Chapter 8).

Most applications of combustion require turbulence to produce the volumetric rates of energy release needed for compact systems. Understanding turbulent flames stems in part from concepts of laminar flames. Most smaller gas-fired systems utilize premixed turbulent flames, while larger systems use diffusion flames. Since premixed flames tend to have lower emissions, larger systems are moving to premixed burners and injectors where possible.

The laminar premixed flame is the simplest type of flame, and this chapter begins with laminar premixed flames. Then, following this discussion, turbulent premixed flames and diffusion flames are discussed.

5.1 LAMINAR PREMIXED FLAMES

A combustion reaction started by a spark or other local heat source in a quiescent fuel-air mixture will propagate into the reactants as a laminar flame. Similarly, a laminar flame can be stabilized at the tip of a burner. An exothermic chemical reaction takes place in a thin zone (the flame front) that propagates at a low velocity. For stoichiometric hydrocarbon mixtures in ambient air, the flame (that is, the chemical reaction zone) is less than 1 mm thick and moves at 20–40 cm/s. The pressure drop through the flame is very small (about 1 Pa), and the temperature in the reaction zone is high (2200–2600 K). Within the reaction zone of the flame, a multitude of active radicals are formed and diffuse upstream to attack the fuel. The fuel fragments are then converted to stable products via chemical reactions such as indicated in Chapter 4. Heat released from the reactions is conducted from the higher temperature portions to the lower temperature portions of the reaction zone to sustain the flame.

The familiar Bunsen burner, shown schematically in Figure 5.1a, provides an example of a stationary laminar premixed flame. Fuel enters under a slight positive pressure at the base of the burner and through momentum exchange entrains air. The fuel and air are mixed together in the burner tube, resulting in a homogeneous fuel-air mixture exiting the burner tube. The mixture is ignited, and a flame is stabilized at the tip of the burner as the flow expands outward. Figure 5.1b shows the streamlines relative to the flame zone, and Figure 5.1c shows isotherms and streamlines for a slot burner. A slot-burner has a rectangular cross-section that produces a tent-shaped flame similar to the cone-shaped flame of a Bunsen burner. The observed peak temperature in the flame is slightly reduced from the adiabatic temperature due to radiation losses.

The flame zone is stationary and cone shaped. The cone shape arises from the need for the local velocity of the fuel-air mixture and the flame speed to be equal. The flame speed is defined as the velocity of the flame relative to the unburned reactants. Thus, when a flame is stationary, the speed of the reacting mixture and the speed of the flame are equal. Referring to Figure 5.1b, the laminar flame speed can be measured by the normal velocity of the mixture into the flame front,

$$\underline{V}_L = \underline{V}_{tube} \sin \alpha \qquad (5.1)$$

where $\underline{V}_L$ is the laminar flame speed of the mixture, $\underline{V}_{tube}$ is the velocity of the fuel-air mixture in the tube, and α is the angle between $\underline{V}_L$ and $\underline{V}_{tube}$. A simple way to think about this problem is to realize that the flame speed represents the rate of combustion of the reactants. As the flame speed (that is, reaction rate) slows, more surface area is needed to match the reaction rate and the mass flow of the reactants. Hence, the slower the flame speed, the more pointed the cone. The flame cone is not perfectly straight, but rather is rounded at the tip and is curved at the lip due to heat transfer to the tube, which serves to stabilize the flame. Also, the velocity in the tube is not perfectly uniform due to boundary layer effects. Hence, the local velocity and angle should be used when applying Equation 5.1. However, an approximate value can be obtained simply by multiplying the mass average velocity in the tube by the ratio of the cross-sectional area of the tube to the area of the cone.

For each premixed fuel-air mixture, there is a unique laminar flame speed that depends on the fuel type, fuel-air mixture ratio, and the initial temperature and pressure of the reactants. Before considering a mathematical model of a laminar flame, some basic data on laminar burning velocities are presented.

5.1.1 Effect of Stoichiometry on Laminar Flame Speed

The effect of fuel concentration on the laminar flame speed is shown in Figure 5.2 for various fuels. The maximum flame speed occurs with slightly rich mixtures. This occurs because as discussed in Chapter 3, the highest flame temperatures occur with

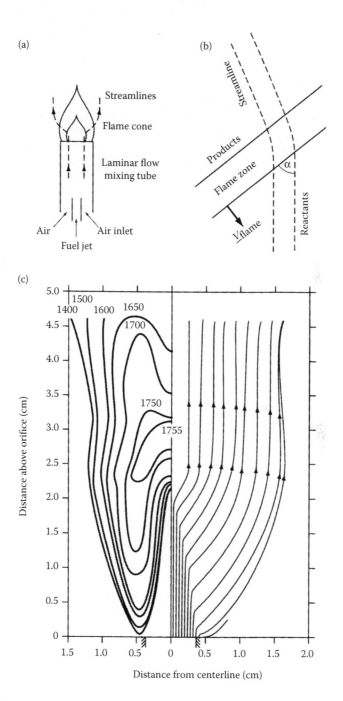

FIGURE 5.1 Bunsen burner flame: (a) schematic of burner, (b) flow diagram, (c) stream-lines and temperature (°C) for a laminar slot burner. (From Lewis, B. and Von Elbe, G., *Combustion Flames and Explosions of Gases*, 3rd ed., Academic Press, 1987. Academic Press, Ltd.)

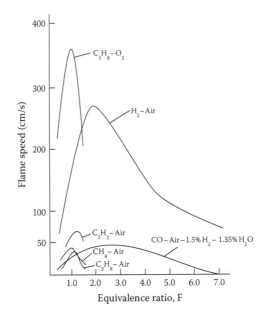

FIGURE 5.2 Laminar flame speed as a function of equivalence ratios for various fuel-air mixtures at ambient pressure and temperature. (From Strehlow, R. A., *Fundamentals of Combustion*, International Textbook Co., Scranton, PA, 1968.)

slightly rich mixtures due to dissociation. Typically, higher temperatures are associated with faster flame speeds. The rich and lean limits of flammability are indicated by the mixture curve. Laminar flames will not occur above or below these limits. Hydrogen has the highest flame speed and widest limits of flammability, while methane has the lowest flame speed and the narrowest limits of flammability. The flame temperature is highest near a stoichiometric mixture and lowest near the flammability limits (Figure 5.3).

Representative flammability limits for various fuels in air are given in Table 5.1. Most mixtures are flammable when the fuel-air volume ratio is approximately 50%–300% of the stoichiometric value though hydrogen and acetylene are exceptions. By reducing the flame temperature, non-reactive additives (for example, carbon dioxide or nitrogen) reduce the flammability limits and the laminar flame speed. For example, consider the combustion of methane (Figure 5.4). As the volume fraction of the inert component is increased, the range of flammable mixtures shrinks until the mixture is no longer flammable. As expected, as the heat capacity of diluent increases the impact on flammability limits is generally greater. For a variety of fuels, the flame temperature at the lean limit is about 1475 K–1510 K. The lean flame temperature limit may be used to estimate the effect of adding inert gases to the reactant. Inert gases with a high heat capacity have a greater effect than those with a low heat capacity for the same amount of addition.

The most common diluent addition is products of combustion. For example, in power plants, a fraction of the combustion products are sometimes recirculated with the inlet air to reduce the flame temperature and thus the amount of NO produced. Similarly, in

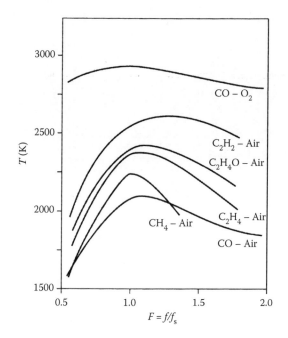

FIGURE 5.3 Flame temperature as a function of equivalence ratios for various fuel-air mixtures at ambient pressure and temperature. (From Strehlow, R. A., *Fundamentals of Combustion*, International Textbook Co., Scranton, PA, 1968.)

internal combustion engines a fraction of the residual products from the previous cycle mix with the new charge. Metghalchi and Keck (1982) showed that the addition of dry combustion products reduced the laminar flame speed by a factor of $(1-2.1 y_{prod})$, where y_{prod} is the mass fraction of products mixed with the reactants. For example, 10% exhaust gas recirculation results in a 21% reduction in the laminar flame velocity.

TABLE 5.1
Limits of Flammability in Standard Air (% by Volume)

Fuel Vapor	Stoichiometric	Lean Limit	Rich Limit
Methane	9.47	5.0	15.0
Ethane	5.64	2.9	13.0
Propane	4.02	2.0	9.5
Isooctane	1.65	0.95	6.0
Carbon monoxide	29.50	12.5	74
Acetylene	7.72	2.5	80
Hydrogen	29.50	4.0	75
Methanol	12.24	6.7	36

Source: Bartok, W. and Sarofim, A. F. eds., *Fossil Fuel Combustion: A Source Book*, © 1991, by permission of John Wiley and Sons, Inc.

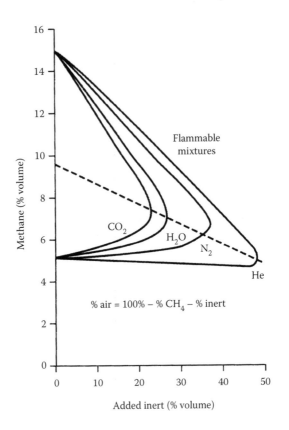

FIGURE 5.4 Limits of flammability of various mixtures of methane, air, and inert gases at 298 K and 1 atm. (From Zabetakis, M. G., "Flammability Characteristics of Combustible Gases and Vapors," Bulletin 627, Washington, U.S. Dept. of the Interior, Bureau of Mines, 1965.)

5.1.2 EFFECT OF TEMPERATURE AND PRESSURE ON LAMINAR FLAME SPEED

Laminar flame speed increases when the reactants are preheated and decreases when the reactants are pressurized. The observed temperature and pressure dependence of the reactants on laminar flame speed can be expressed as

$$\underline{V}_{L(p,T)} = \underline{V}_{L(p_0,T_0)} \left(\frac{p}{p_0} \right)^m \left(\frac{T}{T_0} \right)^n \tag{5.2}$$

where the subscript 0 refers to ambient conditions (300 K, 1 atm). The parameters for Equation 5.2 are given in Table 5.2. This correlation is plotted in Figure 5.5 for lean propane for preheat temperatures from 300 to 700 K and pressures of 1, 10, and 30 atm. Other data on the effect of reactant pressure and temperature on laminar flame speed are shown in Table 5.3. These data were obtained in a closed, constant volume chamber with a propagating spherical flame, and thus the pressure was gradually increasing. This type of data is relevant to an internal combustion engine.

TABLE 5.2
Effect of Pressure and Preheat on Laminar Flame Speed for Equation 5.2

Fuel	$\underline{V}_t$ (cm/s)	m (Pressure Exponent)	n (Temperature Exponent)
Methane ($F = 0.8$)	26	−0.504	2.105
Methane ($F = 1$)	36	−0.374	1.612
Methane ($F = 1.2$)	31	−0.438	2.000
Propane ($F = 0.8 - 1.5$)	$34 - 138(F - 1.08)^2$	$-0.16 - 0.22(F - 1)$	$2.18 - 0.8(F - 1)$
n-Heptane ($F = 1.1$)	45	–	2.39
Isooctane ($F = 1.1$)	34	–	2.19

Source: Methane data from Gu, X. J., Haq, M. Z., Lawes, M., and Woodley, R., "Laminar Burning Velocity and Markstein Lengths of Methane-Air Mixtures," *Combust. Flame* 121:41–58, 2000 (1 < p < 10 atm, 300 < T < 400 K). Other data from Metghalchi, M. and Keck, J. C., "Laminar Burning Velocity of Propane-Air Mixtures at High Temperature and Pressure," *Combust. Flame* 38:143–154, 1980 (1 < p < 50 atm, 300 < T < 700 K), by permission of Elsevier, Inc.

Clearly, the reactant preheat temperature must be less than the autoignition temperature. The *autoignition temperature* is the temperature at which a gaseous mixture self ignites without regard to the ignition delay time. Autoignition temperatures for various fuels in ambient air are shown in Table 5.4. In a flame, the ignition temperature is not precisely the autoignition temperature because of the diffusion of active species from the reaction zone to the preheat zone. Nevertheless, the autoignition temperature provides a general indication of the necessary temperature rise upstream of the flame front in order to sustain ignition and flame propagation.

Flammability limits are influenced by the temperature and pressure of the reactants. Increasing the reactant temperature increases the lean limit. The lean limit is

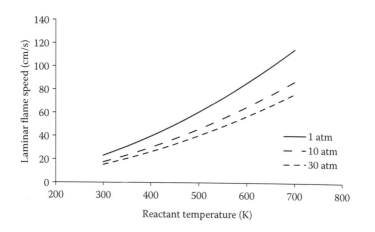

FIGURE 5.5 Laminar flame speed for a lean ($F = 0.8$) propane-air mixture at elevated temperature and pressure using data from Table 5.2.

TABLE 5.3

Empirical Fit to Laminar Burning Velocity Data in a Constant Volume Chamber for Equation 5.2

Fuel	$F = 0.8$	$F = 1.0$	$F = 1.2$
	$\underline{V}_{L(p_0,T_0)}$ (cm/s)		
Methanol	25.6	32.7	38.1
Propane	23.2	31.9	33.8
Isooctane	19.2	27.0	27.6
RMFD-303[a]	19.1	25.2	28.1
	Temperature Exponent, n		
Methanol	2.47	2.11	1.98
Propane	2.27	2.13	2.06
Isooctane	2.36	2.26	2.03
RMFD-303[a]	2.27	2.19	2.02
	Pressure Exponent, m		
Methanol	−0.21	−0.13	−0.11
Propane	−0.23	−0.17	−0.17
Isooctane	−0.22	−0.18	−0.11
RMFD-303[a]	−0.17	−0.13	−0.087

Source: Reprinted from *Combust. Flame*, Metghalchi, M. and Keck, J. C., "Burning Velocities of Mixtures of Air with Methanol, Isooctane, and Indolene at High Pressures and Temperatures," 48:191–210, © 1982 by permission of Elsevier Inc.

[a] Synthetic gasoline (45% toluene, 14% undecene, and 41% isooctane).

affected slightly by pressure, whereas the rich limit is extended markedly when the pressure is increased, as seen in Table 5.5 for natural gas.

5.1.3 STABILIZATION OF PREMIXED FLAMES

A premixed flame burning from the end of a cylindrical tube or nozzle exhibits a characteristic behavior depending on the fuel concentration and the gas velocity. Figure 5.6 shows a typical characteristic stability diagram for a premixed open burner flame. As shown, when the approach velocity to an attached open flame is decreased until the flame speed exceeds the approach velocity over some portion of the burner port, the flame flashes back into the burner (unshaded area of Figure 5.6). On the other hand, if the approach velocity is increased until it exceeds the flame velocity at every point, the flame will either be extinguished completely (that is, blowoff) or for fuel rich mixtures will be lifted until a new stable position in the gas stream above the burner is reached as a result of turbulent mixing and dilution with secondary air. The lift curve is a continuation of the blowoff curve beyond point A in Figure 5.6. The blowout curve corresponds

TABLE 5.4
Autoignition Temperature and Maximum Laminar Flame Speed at 293 K and I atm in Air

Fuel	Autoignition Temperature (K)	Maximum Laminar Flame Speed (cm/s)
Methane	810	34
Propane	743	39
n-Hexane	509	39
Isooctane	691	35
Carbon monoxide	882	39
Acetylene	578	141
Hydrogen	673	265
Methanol	658	48

Source: Bartok, W. and Sarofim, A. F. eds., *Fossil Fuel Combustion: A Source Book,* © 1991 Wiley, by permission of John Wiley and Sons, Inc.

to the velocity required to extinguish a lifted flame. Once the flame has lifted, the approach velocity must be decreased well below the lift velocity before the flame will drop back and be reseated on the burner rim. Between fuel concentrations at A and B, the blowout of a lifted flame occurs at a lower velocity than the flame blowoff from the port. Flame stabilization in burners and gas turbines will be discussed in later chapters.

5.2 LAMINAR FLAME THEORY

Consider a premixed laminar flame free from the effects of any walls. The gas flow is uniform and the flame front is planar. Let the flame be stationary so that the gas flows into the flame front as noted in Figure 5.7. The flame speed is always defined as relative to the unburned gas velocity, so in this case the flame speed is the unburned gas speed. Let subscript react refer to the mixture of reactants and subscript prod

TABLE 5.5
Limits of Flammability of Natural Gas in 293 K Air versus Pressure (% by Volume)

Pressure atm	Lean Limit	Rich Limit
1	4.50	14.2
35	4.45	44.2
70	4.00	52.9
140	3.60	59.0

Source: Jones, G. W. and Kennedy, R. E., "Inflammability of Natural Gas: Effect of High Pressure Upon the Limits," U.S. Bureau of Mines Report of Investigation 3798, 1945.

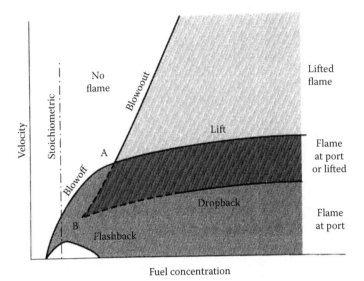

Fuel concentration

FIGURE 5.6 Characteristic stability diagram for an open burner flame. (From Wohl, K., Kapp, N. M., and Gazley, C., "The Stability of Open Flames," *Symp. (Int.) Combust.* 3:3–21, The Combustion Institute, Pittsburgh, PA, 1949, by permission of The Combustion Institute.)

refer to the mixture of products. Conservation of mass, momentum, and energy across the flame front are given by

$$\rho_{react}\, V_{react} = \rho_{prod}\, V_{prod} \tag{5.3}$$

$$p_{react} + \rho_{react}\, V^2_{react} = p_{prod} + \rho_{prod}\, V^2_{prod} \tag{5.4}$$

$$h_{react} + \frac{V^2_{react}}{2} = h_{prod} + \frac{V^2_{prod}}{2} \tag{5.5}$$

respectively, where h is the absolute enthalpy of the mixture.

In Equations 5.3 through 5.5 ρ_{react}, p_{react}, and h_{react} are known, but V_{react}, V_{prod}, ρ_{prod}, p_{prod}, and h_{prod} are unknown. The equation of state

$$p_{prod} = \rho_{prod} R_{prod} T_{prod}$$

holds and h_{prod} is a function of T. However, we are still one equation short of solving for the state of the products and the flame speed, V_{react}. If the flame speed is known from experiment or additional theory, then the state of the products can be determined from Equations 5.3 through 5.5. The burning velocities are low enough that the velocity terms in Equation 5.5 are negligible compared to the enthalpy terms. Indeed, in Chapter 3 the adiabatic flame temperature was calculated by simply assuming that the enthalpy remains constant across the flame. Similarly, the pressure change across

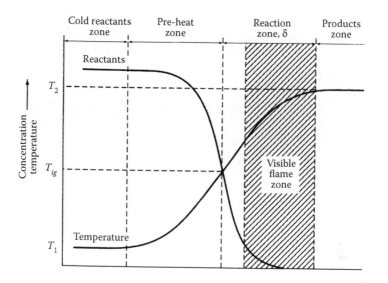

FIGURE 5.7 Laminar flame profile showing preheat and reaction zones.

the flame is very small (see Example 5.1). Hence, density decreases as the temperature increases, and by Equation 5.3 velocity increases in proportion to the temperature increase. Again, if the flame speed, V_{react} is known, then the state of the products can be directly determined, but Equations 5.3 through 5.5 alone do not yield the flame speed.

Traditionally, Equations 5.3 through 5.5 have been combined into two equations and rearranged to give the Rankine-Hugoniot equations (Williams 1985). However, no new information is obtained by doing this, and the authors prefer to work directly with the basic conservation equations. Equations 5.3 through 5.5 also apply to detonations (Chapter 8). For detonations, V_{react} is supersonic, whereas for flames, V_{react} is subsonic. This distinction is sometimes emphasized by calling the flame a deflagration.

Example 5.1

A laminar flame propagates through a propane-air mixture with an equivalence ratio of 0.9, a pressure of 5 atm, and a temperature of 300 K. The flame velocity is 22 cm/s. Find (a) the gas temperature, (b) the velocity, and (c) the pressure behind the flame.

Solution

The stoichiometric reaction is

$$C_3H_8 + 5(O_2 + 3.76\ N_2) \rightarrow 3\ CO_2 + 4\ H_2O + 18.8\ N_2$$

Accounting for the equivalence ratio of 0.9, the reacting mixture is

$$C_3H_8 + \frac{5}{0.9}(O_2 + 3.76\ N_2)$$

And so the mixture consists of 1 mol C_3H_8, 5.56 moles O_2, and 20.89 moles N_2.

Equations 5.3 through 5.5 are used to solve the problem. Since the velocity terms are much smaller than the enthalpy terms, Equation 5.5 becomes

$$h_{react} = h_{prod}$$

The easiest way to solve this problem is to use a thermodynamic equilibrium program. Several good thermodynamic programs are available online. Using a thermodynamic equilibrium program to find h_{react}, enter the moles of each of the reactants, $p_{react} = 5$ atm, and $T_{react} = 300$ K. The results are

$$h_{react} = -0.1267 \times 10^3 \text{ kJ/kg}$$

$$\rho_{react} = 5.97 \text{ kg/m}^3$$

$$M_{react} = 28.318 \text{ kg/kgmol}$$

Part (a)

In Equation 5.4 the momentum terms are much smaller than the pressure terms, and thus the pressure is constant with respect to the enthalpy. For a constant enthalpy and constant pressure process the equilibrium state can be determined from the thermodynamic equilibrium program:

$$T_{prod} = 2200 \text{ K}$$

$$\rho_{prod} = 0.784 \text{ kg/m}^3$$

$$M_{prod} = 28.23 \text{ kg/kgmol}$$

Part (b)

From Equation 5.3,

$$V_{prod} = V_{react} \frac{\rho_{react}}{\rho_{prod}} = (0.22) \frac{5.97}{0.784} = 1.67 \text{ m/s}$$

This is the velocity relative to the flame front.

Part (c)

From Equation 5.4,

$$p_{prod} - p_{react} = \rho_{prod} \left(\underline{V}_{prod}\right)^2 - \rho_{react} \left(\underline{V}_{react}\right)^2$$

$$p_{prod} - p_{react} = 5.97(0.22)^2 - 0.784(1.67)^2$$

$$p_{prod} - p_{react} = -1.90 \text{ Pa} = -0.0000187 \text{ atm}$$

and thus, $p_{prod} = 4.999981$ atm, which justifies the assumption of constant pressure across a flame.

5.2.1 LAMINAR FLAME DIFFERENTIAL EQUATIONS

Since the integral form of the conservation equations for a laminar flame (Equations 5.3 through 5.5) are insufficient to determine the flame speed, the differential form of the conservation equations through the flame zone must be used. The rate at which a laminar flame propagates through a combustible mixture is determined by the chemical reaction rate, and the heat and mass transfer rates throughout the flame zone. As a starting place this process can be modeled as a one-dimensional, steady, reacting flow problem. The governing equations consist of conservation of mass for the mixture, conservation of mass for each species, and conservation of energy for the mixture. As we saw in Example 5.1, the momentum equation is not required because the pressure is essentially constant. The full form of the laminar premixed flame equations will be presented first, and then a simplified treatment will be given to gain physical insight to the analysis of flames. A general derivation of the con-servation equations for reacting flows is beyond the scope of this text (for a detailed derivation of conservation equations of reacting flow see Chapter 5 of Law 2006).

The flow is steady-state, one-dimensional, multi-component, and laminar. Given a steady-state flame and assuming a differential element dx in the reaction zone, the net change of mass flux through the flame is zero. This can be written as

$$\frac{d(\rho \underline{V})}{dx} = 0 \qquad (5.6)$$

integrating yields

$$\rho \underline{V} = \rho_{react} \underline{V}_{react} = \text{constant} \qquad (5.7)$$

The conservation equation for each of the species, i, is complicated by two effects—mass diffusion and chemical reactions. The mass diffusion causes each species, i, to move relative to the mass average velocity of the mixture. The pri-mary cause of the diffusion velocity, $\tilde{V}_i$, is the concentration gradient dy_i/dx, which results in *ordinary diffusion*. Diffusion may also be caused by the temperature gradi-ent, resulting in *thermal diffusion*. Even in flames, where the temperature gradient is very large, the thermal diffusion velocity is only 10%–20% of the ordinary diffusion velocity and primarily affects low molecular weight species such as hydrogen. Based on this, we will neglect thermal diffusion in this discussion. The velocity of each species i is given by the overall mass average velocity plus the diffusion velocity for that species: $\underline{V}_i = \underline{V} + \tilde{V}_i$. The species continuity equation is

$$\frac{d}{dx}\left[\rho y_i \left(\underline{V} + \tilde{V}_i\right)\right] = \rho_{react} \underline{V}_{react} \frac{dy_i}{dx} + \frac{d}{dx}\left(\rho y_i \tilde{V}_i\right) = M_i \hat{r}_i \qquad (5.8)$$

The term on the right-hand side of Equation 5.8 is the net mass production rate of species i per unit volume within the differential volume due to chemical reac-tions involving species i. Recall from Chapter 4 that $\hat{r}_i$ is made up of as many terms as there are chemical reactions involving species i. The species most affecting the reaction rates are typically radicals with very small concentrations. Thus a good assumption for fuel-air flames is to approximate the diffusion in flames as a binary system consisting of the various radical species and nitrogen. For a binary system consisting of species i and j, the ordinary diffusion velocity is given by Fick's first law of diffusion

$$\tilde{V}_i = -\frac{D_{ij}}{y_i}\frac{dy_i}{dx} \qquad (5.9)$$

The binary diffusion coefficient, D_{ij}, is a property of the mixture constituents and temperature. A detailed discussion of diffusion is given in Bird, Stewart, and Lightfoot (2002).

Starting with a differential element, conservation of energy within a flame can be stated in words as

$$\begin{bmatrix} \text{the net change} \\ \text{in the convective} \\ \text{flux of enthalpy} \\ \text{of the mixture} \end{bmatrix} + \begin{bmatrix} \text{the net flux} \\ \text{of the heat} \\ \text{due to} \\ \text{conduction} \end{bmatrix} + \begin{bmatrix} \text{the net overall} \\ \text{flux of enthalpy} \\ \text{due to the} \\ \text{diffusion of mass} \end{bmatrix} = \begin{bmatrix} \text{the total heat} \\ \text{release by} \\ \text{chemical} \\ \text{reaction} \end{bmatrix}$$

The radiant energy flux is neglected here; it is small for non-luminous flames but can be significant for rich flames where carbon particles radiate, making the flame luminous. The dissipation terms in the energy equation due to work done by the viscous stresses are not included in a flame. These terms are negligible because the velocity is very small. The energy equation is

$$\rho_{\text{react}}\underline{V}_{\text{react}}c_{p(\text{react})}\frac{dT}{dx} - \frac{d}{dx}\left(\tilde{k}\frac{dT}{dx}\right) + \sum_{i=1}^{I}\rho y_i \underline{V}_i c_{p,i}\frac{dT}{dx} = \sum_{i=1}^{I}M_i\hat{r}_i h_i \qquad (5.10)$$

From Equation 5.10 we see that upstream heat conduction and diffusion of active species promote chemical reactions and sustain the flame against the convective flow of heat.

Solution of the governing equations with appropriate boundary conditions is done with numerical analysis. As an example, results from a computer solution of a stoichiometric methanol-air laminar flame are shown in Figure 5.8. Eighty-four reactions involving 26 species were used. The temperature, velocity, density, and selected species profiles within the flame zone are shown. Notice the presence of active radicals such as HO_2 and CH_2O that have diffused ahead of the flame front. Another use of detailed laminar flame calculations is to test the validity of reaction schemes at high temperatures by comparing the model results to experimentally obtained profiles of temperature, species, and flame speed.

Gottgens, Mauss, and Peters (1992) have performed laminar flame calculations for six different fuels using 82 elementary reactions. Analytical expressions for the

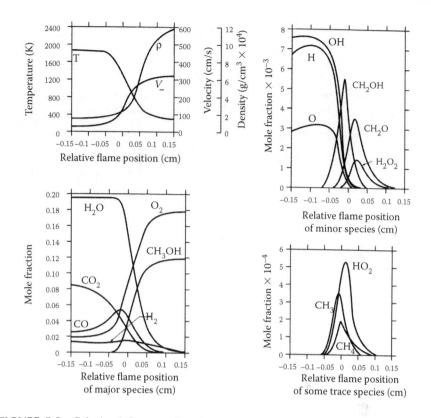

FIGURE 5.8 Calculated flame profiles for stoichiometric methanol-air, one-dimensional flame. (From Westbrook, C. K. and Dryer, F. L., "A Comprehensive Mechanism for Methanol Oxidation," *Combust. Sci. Technol.* 20:125–140, © 1979, by permission of Gordon and Breach Publishers.)

flame speed and flame thickness as a function of initial temperature and pressure, and lean equivalence ratio have been developed. Their results are shown for a laminar propane-air flame in Figure 5.9. At high pressure, such as in an internal combustion engine or gas turbine combustor, the flame speed and flame thickness are considerably reduced.

5.2.2 Simplified Laminar Flame Model

To gain an understanding of the laminar flame, let us consider a simpler, approximate flame model following the work of Mallard and Le Chatelier (1883). They reasoned that heat conduction upstream from the flame to the reactants was the rate-limiting step. They assumed that the flame consisted of a preheat zone and a reaction zone, and they neglected any diffusion of species and chemical reactions in the preheat zone. The boundary between the two zones is set by the ignition temperature, as shown in Figure 5.7. With these assumptions, the energy equation (Equation 5.10) in the preheat zone becomes

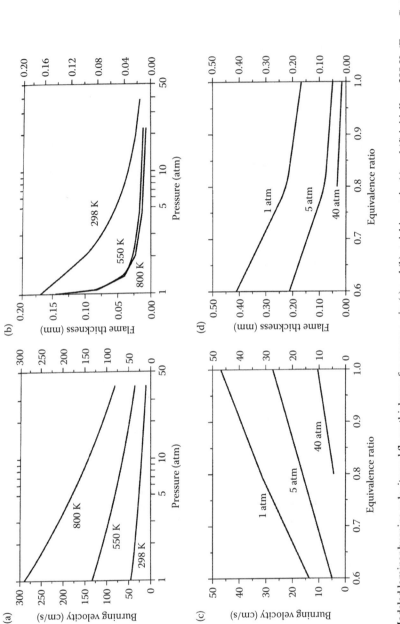

FIGURE 5.9 Modeled laminar burning velocity and flame thickness for propane-air: (a) and (b) stoichiometric, (c) and (d) initially at 298 K. (From Gottgens, J., Mauss, F., and Peters, N., "Analytical Approximations of Burning Velocities and Flame Thicknesses of Lean Hydrogen, Methane, Ethylene, Ethane, Acetylene, and Propane Flames," *Symp. (Int.) Combust.* 24:129–135, The Combustion Institute, Pittsburgh, PA, 1992, by permission of The Combustion Institute.)

$$\frac{d}{dx}\left(\tilde{k}\frac{dT}{dx}\right) = \rho_{react}\underline{V}_{react}c_p\frac{dT}{dx} \tag{5.11}$$

Solving Equation 5.11, taking $\tilde{k}$ and c_p constant

$$T = C_1 + C_2\exp\left(\frac{\underline{V}_{react}x}{\alpha}\right) \tag{5.12}$$

where α is the thermal diffusivity

$$\alpha = \frac{\tilde{k}}{\rho c_p}$$

with units of (m²/s). Assuming that a chemical reaction starts at temperature T_{ig}, where $x = 0$, the boundary conditions are

$$\text{at } x = -\infty, T = T_{react} \tag{5.13}$$

$$\text{at } x = 0, T = T_{ig} \tag{5.14}$$

and Equation 5.12 becomes

$$T = T_{react} + \left(T_{ig} - T_{react}\right)\exp\left(\frac{\underline{V}_{react}x}{\alpha}\right) \tag{5.15}$$

We desire a relation between the flame speed, $\underline{V}_{react}$, and the burned gas (products) temperature, T_{prod}. This is accomplished, following Mallard and Le Chatelier (1883) by matching the slope of the temperature–distance curve at the ignition point with the linearized temperature slope in the reaction zone. Differentiating Equation 5.15 and setting $x = 0$ yields the slope of the temperature-distance curve at the ignition point.

$$\left.\frac{dT}{dx}\right|_{x=0-} = \frac{\underline{V}_{react}}{\alpha}\left(T_{ig} - T_{react}\right) \tag{5.16}$$

Linearizing the temperature in the reaction zone gives

$$\left.\frac{dT}{dx}\right|_{x=0+} = \frac{T_{prod} - T_{ig}}{\delta} \tag{5.17}$$

Equating Equations 5.16 and 5.17 and using the laminar flame speed notation for $\underline{V}_{react}$ gives the result provided by Mallard and Le Chatelier (1883),

$$V_L = \frac{\alpha}{\delta} \left(\frac{T_{prod} - T_{ig}}{T_{ig} - T_{react}} \right) \tag{5.18}$$

This result may be extended by noting that the reaction thickness depends on the reaction rate. A characteristic reaction time, τ_{chem}, can be determined based on the average global reaction rate $\bar{r}_f$ as

$$\tau_{chem} = \frac{n_f}{\hat{r}_f} \tag{5.19}$$

The thickness of the laminar flame, δ, is roughly

$$\delta = V_L \tau_{chem} = \frac{V_L n_f}{\hat{r}_f} \tag{5.20}$$

Substituting Equation 5.20 into Equation 5.18

$$V_L = \left[\frac{\alpha \bar{r}_f}{n_f} \frac{\left(T_{prod} - T_{ig} \right)}{\left(T_{ig} - T_{react} \right)} \right]^{1/2} \tag{5.21}$$

Hence, the flame speed depends on the thermal diffusivity, the reaction rate, the initial fuel concentration, the flame temperature, the ignition temperature, and the initial temperature of the reactants. As noted previously, the actual ignition temperature in the flame is lower than the autoignition temperature because of the diffusion of active species upstream from the reaction zone. Of course, the reaction rate is a strong function of temperature. Although Equation 5.21 is incomplete because mass and heat transfer by diffusion have been neglected, it is nevertheless an interesting approximate model of the laminar flame. From Equations 5.20 and 5.21, the flame thickness is inversely proportional to the square root of the reaction rate, and when the equations are evaluated, the flame zone is indeed thin (that is, less than 1 mm; see Problem 3).

Example 5.2

Estimate the laminar flame speed of a stoichiometric propane-air mixture initially at 298 K and 1 atm pressure using the thermal flame theory with a one-step global reaction rate.

Solution
From Table 3.2,

$$T_{flame} = 2270 \text{ K}$$

From Table 5.4,

$$T_{ig} = 743 \text{ K}$$

To find the flame speed from Equation 5.21, the reaction rate is determined from Equation 4.15 with the help of Table 4.3, which yields

$$\hat{r}_f = -8.6 \times 10^{11} \exp\left(-30{,}000/1.987\,T\right)\left(n_f\right)^{0.1}\left(n_\mathbb{O}\right)^{1.65}$$

And so to determine the average reaction rate, we need to determine an appropriate average temperature, fuel concentration, and oxygen concentration. The reaction is

$$C_3H_8 + 5\left(O_2 + 3.76 \text{ N}_2\right) \rightarrow 3 \text{ CO}_2 + 4 \text{ H}_2O + 18.8 \text{ N}_2$$

Thus the initial mole fraction of fuel is

$$x_{f(init)} = \frac{1}{(5)(4.76)+1} = 0.0403$$

and the partial pressure of the fuel in the mixture is

$$P_{f(init)} = x_{f(init)}\, p = 0.0403 \text{ atm}$$

Therefore

$$n_{f(init)} = \frac{P_{f(init)}}{RT_{react}} = \frac{0.0403 \text{ atm}}{\left(82.05 \text{ cm}^3 \cdot \text{atm}/(\text{gmol} \cdot \text{K})\right)(298 \text{ K})} = \frac{1.65 \times 10^{-6} \text{ gmol}}{\text{cm}^3}$$

$$n_{O_2(init)} = 5 n_{f(init)} = 8.25 \times 10^{-6} \text{ gmol}/\text{cm}^3$$

Knowing the initial values of fuel and oxygen concentrations, the question is still: with what temperature and molar concentrations should we evaluate the average reaction rate and properties? Let us try

$$T = \frac{298 + 2270}{2} = 1284 \text{ K}$$

and assume that the fuel and oxygen mole fractions are half the initial values calculated above. These are

$$n_f = 1.91 \times 10^{-7} \text{ gmol}/\text{cm}^3$$

$$n_{O_2} = 9.55 \times 10^{-7} \text{ gmol}/\text{cm}^3$$

and the average reaction rate is

$$\bar{r}_f = -1.67 \times 10^{-4} \ \text{gmol/cm}^3 \cdot \text{s}$$

Using property values for air from Appendix B,

$$\tilde{k}_{\text{air}} = 0.0828 \ \text{W/m} \cdot \text{K}$$

$$\rho_{\text{air}} = 0.275 \ \text{kg/m}^3$$

$$c_{p,\text{air}} = 1.192 \ \text{kJ/kg} \cdot \text{K}$$

thus

$$\alpha = \frac{\tilde{k}}{\rho c_p} = 2.53 \ \text{cm}^2/\text{s}$$

From Equation 5.21,

$$\underline{V}_L = \left[\frac{\left(2.53 \ \text{cm}^2/\text{s}\right)\left(1.67 \times 10^{-4} \ \text{gmol/cm}^3 \cdot \text{s}\right)}{\left(1.65 \times 10^{-6} \ \text{gmol/cm}^3\right)} \cdot \frac{\left(2270 - 743\right) \text{K}}{\left(743 - 298\right) \text{K}} \right]^{1/2}$$

$$\underline{V}_L = 30 \ \text{cm/s}$$

The correct flame speed is 38 cm/s. Hence, the simplified thermal theory can be used to give an indication of the flame speed, but a rigorous treatment requires numerical solution of the differential Equations 5.7 through 5.10 with an appropriate full set of chemical reactions.

5.3   TURBULENT PREMIXED FLAMES

Combustion in burners and engines uses turbulence to increase the flame speed and heat release rate per unit volume. Turbulence can increase the flame speed 5–50 times the laminar flame speed. Turbulent premixed flames in burners are steady open flames, whereas flames in internal combustion engines are propagating enclosed flames. The flame front interacts with the turbulent eddies, which may have fluctuating speeds of tens of meters per second and sizes ranging from a few millimeters to a meter or more.

Turbulent flames may be categorized broadly as (a) *wrinkled flamelets*, (b) *thickened-wrinkled flames*, or (c) *thickened flames* depending on the nature of the turbulence. When the turbulence is not too intense, the turbulence distorts or wrinkles the thin laminar flame into a more brush-like appearance. Wrinkling of the flame front is conceptually an extension of the laminar flame. Moderately intense turbulence thickens the reaction zone and wrinkles the flame. Very intense, large-scale turbulence leads to a distributed reaction zone containing small lumps of reactants that are entrained and burn up as they travel through the zone.

Each of the three turbulent flame types will be discussed, but first some parameters of turbulence and turbulent reactions are discussed.

5.3.1 TURBULENCE PARAMETERS, LENGTH SCALES, AND TIME SCALES

In turbulent flow the velocity at a given point fluctuates randomly relative to the mean flow as indicated in Figure 5.10. The instantaneous velocity, $\underline{V}$, at a position is equal to the mean velocity, $\overline{\underline{V}}$, plus the fluctuating velocity, $\underline{V}'$:

$$\underline{V} = \overline{\underline{V}} + \underline{V}' \tag{5.22}$$

By definition the mean value of the fluctuating flow is zero ($\overline{\underline{V}'} = 0$), but the root mean square of the velocity perturbations is not zero and in fact is used as a measure of the intensity of the turbulence;

$$\underline{V}'_{rms} = \sqrt{\overline{\underline{V}'^2}} \tag{5.23}$$

Turbulence may be visualized as swirling eddies of various sizes. Measurements of turbulent velocity made at two nearby locations will correlate if they are inside the same eddy. The largest distance between measurements that retains a fluctuating velocity correlation is defined as the *integral scale*, L_I. Turbulent spatial scales are distributed over a range of sizes down to the smallest eddies that exist only for a very short time before they are damped out by viscous (molecular) dissipation. The smallest size eddies are characterized by the *Kolmogorov length scale*, L_K. Kolmogorov showed that

$$L_K = \left[\frac{v^3 L_I}{\left(\underline{V}'_{rms} \right)^3} \right]^{1/4} \tag{5.24}$$

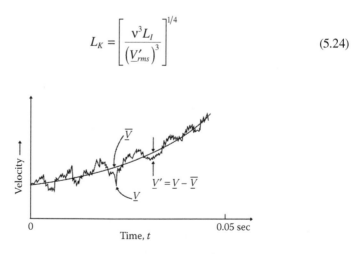

FIGURE 5.10 Velocity versus time at a point in turbulent flow.

For engineering applications concerned with the overall flow, the Reynolds number is a useful measure of turbulence. However, for flames the very local interaction of eddies with the reaction zone is of interest. A *turbulent Reynolds number* based on the turbulent intensity and the integral scale is defined by

$$\text{Re}_l = \frac{V'_{rms} L_l}{\nu} \tag{5.25}$$

Sometimes it is useful to replace L_l with L_K, giving Re_K. Recall that the Reynolds number represents the ratio of inertial to viscous forces, which can be seen by writing

$$\text{Re} = \frac{\text{inertial}}{\text{viscous}} = \frac{\rho V^2}{\mu \underline{V}/L} = \frac{VL}{\nu} \tag{5.26}$$

Thus a large Reynolds number indicates that inertia will dominate over the dissipative effects of molecular viscosity, and Re_l is a measure of how much the larger eddies are damped by viscosity. For turbulence to occur at all requires $\text{Re}_l > 1$, and in practical combustion devices Re_l is typically in the 100–2000 range.

Example 5.3

Flowing air has a temperature of 2000 K, a mean velocity of 10 m/s, and a pressure of (a) 1 atm and (b) 20 atm. The integral length scale is 1 cm, and a root mean square of the velocity fluctuations is 100 cm/s. Find the Reynolds number based on the integral scale.

Solution

Part (a)

From Appendix B

$$\rho_{air} = 0.176 \text{ kg}/\text{m}^3$$

$$\mu_{air} = 6.50 \times 10^{-5} \text{ kg}/\text{m} \cdot \text{s}$$

From this the kinematic viscosity is

$$\nu_{air} = \frac{\mu_{air}}{\rho_{air}} = \frac{6.50 \times 10^{-5} \text{ kg}}{\text{m} \cdot \text{s}} \cdot \frac{\text{m}^3}{0.176 \text{ kg}} = 3.7 \text{ cm}^2/\text{s}.$$

and from Equation 5.25

$$\text{Re}_l = \frac{100 \text{ cm}}{\text{s}} \cdot \frac{1 \text{ cm}}{1} \cdot \frac{\text{s}}{3.7 \text{ cm}^2} = 27$$

Part (b)

If the pressure is 20 atm, the density is 20 times higher and the kinematic viscosity is 20 times smaller. As a result $Re_l = 540$.

In addition to turbulent length scales, a time scale for the velocity fluctuations is also an important characteristic of turbulence. A characteristic large eddy turnover time is defined as

$$\tau_{flow} = \frac{L_I}{V'_{rms}} \tag{5.27}$$

For flames a characteristic chemical reaction time is an important time scale and can be estimated from the laminar flame thickness divided by the laminar flame speed:

$$\tau_{chem} = \frac{\delta}{V_L} \tag{5.28}$$

The ratio of the large eddy turnover time to the chemical reaction time is defined as a Damköhler number:

$$Da_I = \frac{\tau_{flow}}{\tau_{chem}} = \frac{L_I}{\delta} \cdot \frac{V_L}{V'_{rms}} \tag{5.29}$$

When $Da_I = 1$, the chemical reaction rate is comparable to the large-scale mixing rate.

In the same way, the Karlovitz number is the ratio of the smallest eddy turnover time to the chemical reaction time scale:

$$Ka = \frac{\tau_{chem}}{\tau_K} = \frac{\delta/V_L}{L_K/V'_{rms}} = \frac{\delta}{L_K} \cdot \frac{V'_{rms}}{V_L} \tag{5.30}$$

5.3.2 TURBULENT FLAME TYPES

Turbulent premixed combustion involves the interaction between the flame front and the turbulent eddies. When the laminar flame thickness is less than the Kolmogorov length ($\delta < L_K$), then the turbulence does not enter into the flame front but does wrinkle the flame front. This is referred to as a *thin flame*. Furthermore, the Karlovitz number is relevant. It has been shown that when $Ka < 1$, the turbulence does not enter the flame front. This regime is called a *wrinkled flame*. When $\delta > L_K$, the smallest scales of turbulence are able to enter the reaction zone, and more generally when $Ka > 1$, the turbulence enters and modifies the preheat zone of the flame front resulting in a *thickened wrinkled flame* regime. When $Ka > 100$, both the preheat and reaction zones are affected by the turbulence, and this is called a *thickened flame.*

To conceptually map the flame type regimes, it is helpful to use a normalized length scale (L_I/δ) versus a normalized velocity scale $\left(V'_{rms}/V_L\right)$ plot, as shown in Figure 5.11. For reference, the Da = 1 line is shown on Figure 5.11 as well. Conceptual sketches of the flame front for the three turbulent flame regimes are shown in Figure 5.12. Instantaneous images with magnification show the wrinkled flame front to consist of tiny flamelets or fluctuating tongues of flame. An idealized planar turbulent flame with a large Damköhler number (so that the reaction zone, that is, the flame sheet, is very thin relative to the turbulent eddies) is shown in Figure 5.13. Locally, the flamelet advances at the laminar flame speed, while the reactants flow into the flame front at a higher speed. Within a stream tube enclosing the flame, the wrinkling of the flame front increases the surface area of the flame front. The mass flow rate into the stream tube relative to the flame is

$$\dot{m} = \rho_{react}\underline{V}_L A_{smooth} = \rho_{react}\underline{V}_T A_{wrinkled}$$

Thus the turbulent flame speed increases in direct proportion to the increase in flame surface area.

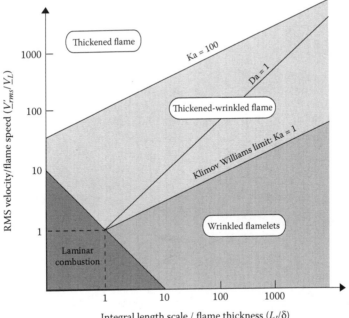

FIGURE 5.11 Turbulent premixed combustion diagram showing combustion regimes in terms of turbulent integral length/laminar flame thickness and turbulent rms velocity/laminar flame speed. (From Peters, in Poinsot, T. and Veynante, D., *Theoretical and Numerical Combustion*, 2nd ed., R.T. Edwards, Philadelphia, PA, 2005. With permission.)

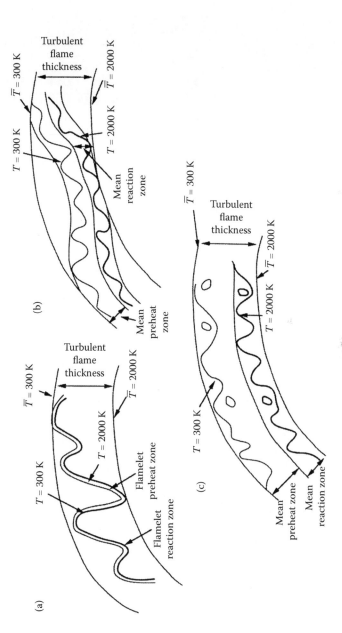

FIGURE 5.12 Diagram of a turbulent premixed flame front for (a) wrinkled flame, (b) thickened wrinkled flame, and (c) thickened flame; reactants at 300 K and products at 2000 K. (From Borghi and Destraiu, in Poinsot, T. and Veynante, D., *Theoretical and Numerical Combustion*, 2nd ed., R.T. Edwards, Philadelphia, PA, 2005. With permission.)

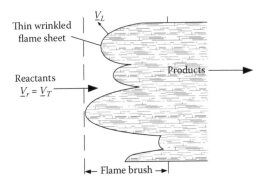

FIGURE 5.13 Idealized planar turbulent flame with a thin reaction zone compared to turbulent eddies. (It has a high Damköhler number.)

$$V_T = V_L \left(\frac{A_{\text{wrinkled}}}{A_{\text{smooth}}} \right) \tag{5.31}$$

The wrinkled area is, of course, not easy to determine. Many models of the wrinkling have been proposed, and the most basic one is a suggestion first proposed by Damköhler,

$$\frac{A_{\text{wrinkled}}}{A_{\text{smooth}}} = 1 + \frac{V'_{rms}}{V_L} \tag{5.32}$$

More intense turbulence increases the wrinkling and hence the flame speed. An example of turbulent flame speeds in a Bunsen burner is shown in Figure 5.14 as a function of turbulent intensity and pressure.

At higher turbulent intensities, Equation 5.32 does not apply as the turbulence thickens the reaction zone, and heat transfer and diffusion of the active radicals upstream further increases the flame speed. With even more intense turbulence, unreacted eddies are swept into the reaction zone, and the turbulent mixing rate rather than the chemical reaction rate determines the flame speed. With extremely intense mixing and fast chemical reactions, a flame no longer exists, and instead the combustion behaves more like a well-stirred reactor.

5.4   EXPLOSION LIMITS

When a volume of premixed gas is suddenly heated above the autoignition temperature by rapid compression, rapid mixing, or rapid heat transfer; an explosion rather than a propagating flame may occur. Under some conditions a delay (a dwell period) between the sudden heating and the rapid reaction in the form of a volumetric explosion (rather than a propagating flame) takes place. In addition, for some pressures and temperatures the mixture will not explode at all even though the mixture is flammable. Thus a plot of the limit line separating explosions from no explosions on a pressure-temperature diagram can be created for a given mixture, as shown in Figure 5.15.

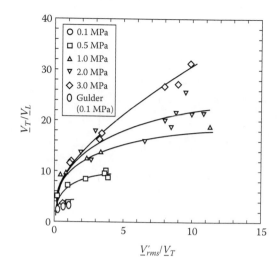

FIGURE 5.14 Turbulent premixed methane-air flame in a Bunsen burner; reactants at ambient temperature and various pressures. (From Kobayashi, H., "Experimental Study of High Pressure Turbulent Premixed Flames," *Exp. Therm. Fluid Sci.* 26(2–4):375–387, 2002, by permission of Elsevier Science & Technology Journals.)

Explosion limits are different from flame propagation limits. In flame propagation the high temperature reaction zone propagates into the reactants and is either sustained or goes out. In the explosion limits the mixture is homogeneous with no flame zone to provide a source of radicals and high temperature. Thus, for homogeneous reactions the radicals that cause the rapid reaction must be built up from reactions from within the mixture itself. If the destruction of the radicals predominates over their creation, the reactions will not trigger an explosion. If the reactions release significant energy, then a thermal self-heating mechanism causes an explosion.

Explosion limit diagrams can be complex, exhibiting various zones or limit lines. During the dwell period, most high molecular weight hydrocarbon fuels have some pressure-temperature regions where cool flames are observed. The energy release due to these reactions is very low, and thus the term cool is used. The dwell or induction period may be quite long with one or more chemiluminescent flames traveling through the mixture (Figure 5.15a). The cool flame period is followed by a thermal explosion. Two-stage ignition is not observed for methane but is observed for propane and most other higher hydrocarbons, as indicated in Figure 5.15b.

The main application of explosion limits is to internal combustion engines where the interest is in preventing knock in spark ignition engines (see Chapter 7) and promoting ignition in compression ignition engines (see Chapter 12). For a homogeneous spark ignition engine the last portion of reactants to burn (end-gas) is rapidly compressed. Our understanding of the kinetic mechanisms for complex fuels are incomplete, but some suggested synthetic two-stage processes have provided models for determining knock. Study of the end-gas temperature has indicated that for low octane fuels, considerable energy is released in the end-gas prior to knock, which indicates a two-stage reaction process.

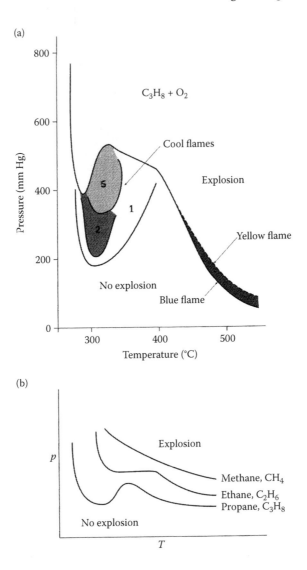

FIGURE 5.15 Explosion limits for hydrocarbons for (a) propane–oxygen, showing regions of one, two, and five cool flames and (b) methane, ethane, and propane mixtures, only showing explosion limits. (From Newitt, P. M. and Thornes, L. S., "Oxidation of Propane. Part 1, The Products of the Slow Oxidation at Atmospheric and Reduced Pressures," *J. Chem. Soc.* 1656–1665, 1937.)

5.5 DIFFUSION FLAMES

Diffusion flames take place when the source of fuel and oxidizer are physically separate. The energy release rate is limited primarily by the mixing process, and chemical kinetics plays a secondary role in the behavior of diffusion flames. Unlike premixed laminar flames, there is no fundamental flame speed. Diffusion flames commonly occur with flowing gases, vaporization of liquid fuels, and devolatilization of solid fuels.

The candle flame that was shown in Figure 1.1 is an example of a diffusion flame. The wax slowly melts, flows up the wick, and is vaporized. Air flows upward due to natural convection and diffuses inward as the vaporized fuel diffuses outward. The flame zone is the region at the intersection between the air and fuel zones. The concentric jet flow of fuel surrounded by air shown in Figure 5.16 is an example of a diffusion flame and a test setup to measure species within the flame. As the jets emerge from the tubes, the fuel diffuses outward and the air diffuses inward. On ignition, the flames stabilize near the tip of the fuel tube. If the jet flows are laminar and of similar velocities, the flame is laminar. If the jet flows have significantly different velocities, a shear layer is created, resulting in a turbulent interface between the air and the fuel that enhances diffusion. A conceptual diagram of the concentration profiles of fuel, air, and combustion products in a diffusion flame are shown in Figure 5.17.

The rest of the chapter will consider three situations: a jet of fuel flowing into quiescent air creating a flare flame; concentric, laminar flowing streams of fuel and air that form a laminar diffusion flame; and a counter flow diffusion flame. Diffusion flames from gas-fired burners, and liquid and solid fuels are considered in later chapters.

5.5.1 FREE JET FLAMES

Consider a gaseous fuel that jets upward from a nozzle of diameter d_j into stagnant air (Figure 5.18). This type of free jet is sometimes called a flare. As the velocity of the fuel jet is increased, the character of the flame changes. At a low jet velocity, the mixing rate is slow and the flame is long and smooth (laminar). The laminar flame height increases linearly with jet velocity up to a point where the flame becomes brush-like (turbulent). The flame height decreases due to more rapid turbulent

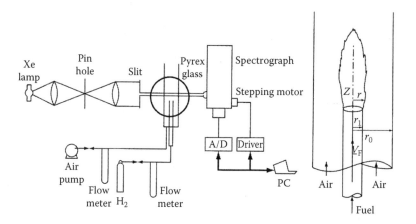

FIGURE 5.16 Diffusion flame in a concentric jet burner and test setup. (From Fukutani, S., Kunioshi, N., and Jinno, H., "Flame Structure of Axisymmetric Hydrogen-Air Diffusion Flame," *Symp. (Int.) Combust.* 23:567–573, The Combustion Institute, Pittsburgh, PA, 1991, by permission of The Combustion Institute.)

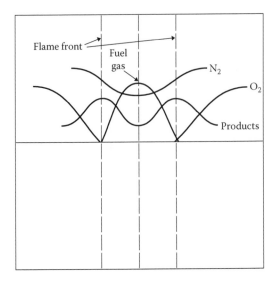

FIGURE 5.17   Concentration profiles above a concentric jet burner.

mixing. In the stable, fully developed turbulent region, the flame height is independent of the jet velocity. The turbulent flame emits more noise than a laminar flame, and the yellow luminosity due to soot formation is decreased. The flame length is proportional to $\underline{V}_j d_j{}^2$ for laminar flow and proportional to d_j for turbulent flow.

The transition to a fully developed turbulent flame is characterized by a transition Reynolds number. Interestingly, this transition Reynolds number is different for different fuels, indicating that the chemical kinetics as well as fluid mechanics plays a role in the combustion. Transition Reynolds numbers for several fuels are given in Table 5.6.

As the jet velocity (Figure 5.18) is further increased, a point is reached where the flame lifts off from the nozzle and exhibits a non-burning region at the bottom. A further increase in jet velocity causes the flame to blow off completely. A stability diagram showing the stable, lifted, and blowoff regions for an ethane-air flame is shown in Figure 5.19

TABLE 5.6
Jet Diffusion Flame Transition to Turbulent Flow

Fuel into Air	Transition Reynolds Number
Hydrogen	2,000
City gas	3,500
Carbon monoxide	4,800
Propane	9,000 – 10,000
Acetylene	9,000 – 10,000

Source: Hottel, H. C. and Hawthorn, W. R., "Diffusion in Laminar Flame Jets," *Symp. (Int.) Combust.* 3:254–266, The Combustion Institute, Pittsburgh, PA, 1949.

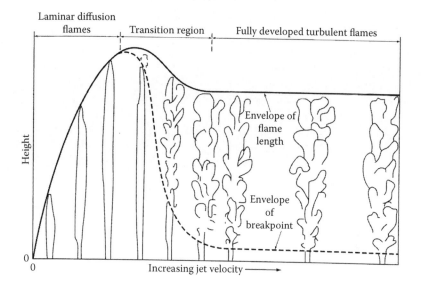

FIGURE 5.18   Free jet diffusion flame transition from laminar to turbulent flow. (From Hottel, H. C. and Hawthorn, W. R., "Diffusion in Laminar Flame Jets," *Symp. (Int.) Combust.* 3:254–266, The Combustion Institute, Pittsburgh, PA, 1949, by permission of The Combustion Institute.)

5.5.2   Concentric Jet Flames

Mixing of jets can be observed by using two concentric tubes. Fuel flows in the inner tube and exits into a larger concentric tube of flowing air (Figure 5.16). The velocity in each tube can be adjusted, and the overall fuel-air flow rate can be adjusted by selecting the tube diameters. Equal velocities produce a laminar diffusion flame, and

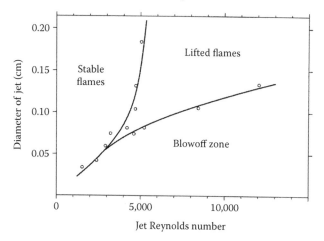

FIGURE 5.19   Stability regimes for a free jet diffusion flame of an ethylene-air mixture. (From Scholefield, D. A. and Garside, J. E., "The Structure and Stability of Diffusion Flames," *Symp. (Int.) Combust.* 3:102–110, The Combustion Institute, Pittsburgh, PA, 1949, by permission of The Combustion Institute.)

different velocities produce a shear flow that generates turbulence. The fuel diffuses outward into the air because of the concentration gradient, while the air diffuses inward. Fuel and oxygen are consumed in a flame zone when the fuel-air ratio is sufficient for combustion. The product species diffuse both inward and outward. Although this situation has been modeled theoretically beginning with Burke and Schumann in 1928, we shall consider some experimental results to gain physical insight into a diffusion flame.

Spectroscopic measurements in a methane-air laminar concentric jet diffusion flame (Figure 5.20) reveal the structure of the flame. The flame has a range of fuel to oxygen ratios within the flame zone. Just to the fuel side of the flame zone, the fuel is at a high temperature and undergoes reactions with little oxygen (pyrolysis), which produces soot particles. The flame produces CO on the fuel side, and as the CO oxidizes to CO_2, the maximum flame temperature is reached. The NO peaks at the point of maximum flame temperature. The dotted lines indicate regions of soot where the sampling probe clogged. Considerable energy is radiated by this high temperature soot, causing a reduction in peak flame temperature.

Hydrogen-air diffusion flames are free of soot and thus easier to understand. Fukutani, Kunioshi, and Jinno (1991) have used a 21-reaction scheme to compute the temperature and species concentrations in an axisymmetric hydrogen-air diffusion flame. Figure 5.21 shows some of the computed profiles and experimental data points. Hydrogen and oxygen disappear near the location of peak temperature, and the reactions are spread over a distance of several millimeters.

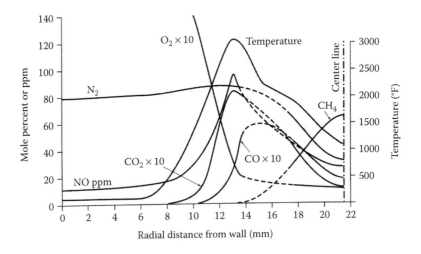

FIGURE 5.20 Methane-air diffusion flame temperature and concentration profiles, concentric jet burner (methane velocity 0.22 m/s, air velocity 0.41 m/s, overall fuel-air ratio 0.577, standard temperature and pressure initially, 17 mm above fuel inlet). (From Tuteja, A. D., "The Formation of Nitric Oxide in Diffusion Flames," Ph.D. thesis, University of Wisconsin-Madison, 1972.)

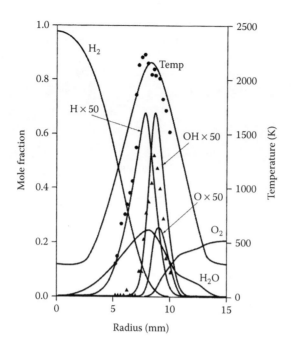

FIGURE 5.21 Concentration and temperature profiles of a hydrogen-air diffusion flame at 2 cm above the fuel inlet; the hydrogen tube had a 5 cm radius. Circles are measured temperatures and triangles are measured OH concentrations. (From Fukutani, S., Kunioshi, N. and Jinno, H., "Flame Structure of Axisymmetric Hydrogen-Air Diffusion Flame," *Symp. (Int.) Combust.* 23:567–573, The Combustion Institute, Pittsburgh, PA, 1991, by permission of The Combustion Institute.)

5.5.3 CONCENTRIC JET FLAME WITH BLUFF BODY

Most practical combustors use turbulent flow with a degree of backmixing in order to obtain high rates of combustion energy release per unit volume. A laboratory setup that achieves an intense zone of turbulent mixing of co-flowing jets of air and methane is shown in Figure 5.22. The fuel is injected at the rear of a bluff body where

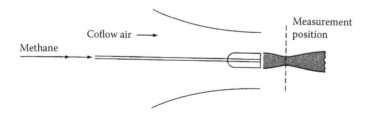

FIGURE 5.22 Schematic of a bluff-body stabilized burner setup to study diffusion flames. (From Masri, A. R., Dibble, R. W. and Barlow, R. S., "Raman-Rayleigh Measurements in Bluff Body Stabilized Flames of Hydrocarbon Fuels," *Sym. (Int.) Combust.* 24:317–324, The Combustion Institute, Pittsburgh, PA, 1992, by permission of The Combustion Institute.)

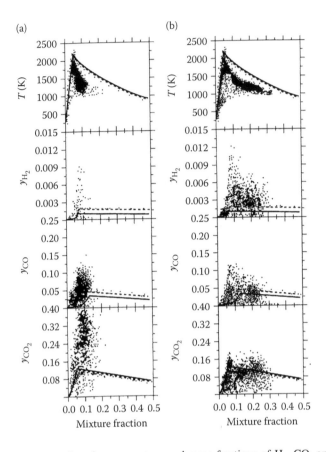

FIGURE 5.23   Scatter plots for temperature and mass fractions of H_2, CO, and CO_2 plotted versus fuel mixture fraction for the non-premixed methane-air flame of Figure 5.22. The data are collected at $x/D = 1.5$. The air velocity is 20 m/s; for (a) the fuel jet velocity is 124 m/s, and for (b) it is 155 m/s. The solid lines represent a fully burned flamelet, and the broken lines represent an intermediate flamelet. (From Masri, A. R., Dibble, R. W. and Barlow, R. S., "Raman-Rayleigh Measurements in Bluff Body Stabilized Flames of Hydrocarbon Fuels," *Sym. (Int.) Combust.* 24:317–324, The Combustion Institute, Pittsburgh, PA, 1992, by permission of The Combustion Institute.)

turbulent eddies and backmixing stabilizes the flame, as indicated by the gray area in the figure. The structure of the flame is dominated by the turbulent fluctuations. Laser Doppler velocimetry and laser light-scattering spectroscopy techniques are used to obtain instantaneous point measurements of velocity, temperature, and species concentrations (Eckbreth 1996). Scatter plots given in Figure 5.23 show how the temperature and product species are distributed as a function of fuel mixture fraction. The statistical nature of a turbulent flame is evident. Modeling turbulent flames is beyond the scope of this text.

5.6 PROBLEMS

1. A premixed stoichiometric methane-air flame has a laminar flame speed of 0.33 m/s and a temperature of 2200 K. The reactants are initially at 300 K and 1 atm. Find the velocity of the combustion products relative to the flame front and the pressure change across the flame. Assume that the reaction goes to completion and there is no dissociation.

2. Repeat Problem 1 for a premixed stoichiometric methanol-air flame. The flame speed is 0.48 m/s and the flame temperature is 2000 K.

3. The laminar flame speed of a premixed stoichiometric mixture of propane-air at 1 atm and 298 K is 0.4 m/s. Estimate the average reaction rate and the reaction zone thickness. The thermal conductivity of the mixture is $k = 0.026(T/298)^{1/2}$. Assume an appropriate average mixture temperature and ignition temperature.

4. The laminar flame speed of a premixed stoichiometric mixture of propane-air at 1 atm and 298 K is 0.4 m/s. If the nitrogen is replaced with helium, estimate the flame speed using the thermal flame theory. Assume that the temperature does not change. The thermal conductivity of the helium mixture is $\tilde{k} = 0.036(T/298)^{1/2}$. Use $c_p = 2.5R$ for helium and $c_p = 3.5R$ for nitrogen. In reality the flame temperature will increase because the specific heat of the helium is lower than that of nitrogen. Explain quantitatively how this would change the flame speed estimated above.

5. For the conditions of Problem 3, what is the approximate thickness of the preheat zone? Use Equation 5.15 and use a criterion that $T_{react} = 1.01T_1$ at the start of the preheat zone.

6. For methane, the maximum laminar flame speed in pure oxygen is 11 m/s, whereas in air the maximum flame speed is 0.45 m/s. Explain what causes this increase in the flame speed.

7. Using the data given in Table 5.3, estimate the laminar flame speed of a stoichiometric mixture of air and gasoline at 20 atm pressure and 700 K in a closed vessel. How much can turbulence increase this flame speed if $\underline{V}_T = \underline{V}_L + 2\underline{V}'_{rms}$ and the turbulent intensity is $\underline{V}'_{rms} = 1.2$ m/s?

8. When a propagating flame approaches a small hole in a plate that has reactants on both sides of the plate, there is a given hole size below which the flame will be quenched (extinguished). One approximate criterion for quenching a laminar flame is

$$d_0 < \frac{8\alpha}{\underline{V}_L}$$

where α is the thermal diffusivity of the reactants. Using this criterion, calculate the quench distance for a stoichiometric hydrogen-air flame at ambient pressure and temperature, and compare this distance to that of a propane-air flame.

REFERENCES

Bartok, W. and Sarofim, A. F., eds., *Fossil Fuel Combustion: A Source Book*, Wiley, New York, 1991.

Bird, R. B., Stewart, W. E., and Lightfoot, E. N., *Transport Phenomena*, 2nd ed., Wiley, New York, 2002.

Borghi, R. and Destriau, M., *Combustion and Flames: Chemical and Physical Principles* (transl. of *La combustion et les flammes,* 1995), TECHNIP, 1998.

Burke, S. P. and Schumann, T. E. W., "Diffusion Flames," *Ind. Eng. Chem.* 20(10):998–1004, 1928.

Cheng, R. K., Littlejohn, D., Strakey, P. A., and Sidwell, T., "Laboratory Investigations of a Low-Swirl Injectors with H_2 and CH_4 at Gas Turbine Conditions," in *Proc. Combust. Int.* 32:3001–3009, The Combustion Institute, Pittsburgh, PA, 2009.

Eckbreth, A. C., *Laser Diagnostics for Combustion, Temperature, and Species*, 2nd ed., OPA (Overseas Publishers Association), Amsterdam, 1996.

Fukutani, S., Kunioshi, N. and Jinno, H., "Flame Structure of Axisymmetric Hydrogen-Air Diffusion Flame," *Symp. (Int.) Combust.* 23:567–573, 1991.

Gottgens, J., Mauss, F. and Peters, N., "Analytical Approximations of Burning Velocities and Flame Thicknesses of Lean Hydrogen, Methane, Ethylene, Ethane, Acetylene, and Propane Flames," *Symp. (Int.) Combust.* 24:129–135, The Combustion Institute, Pittsburgh, PA, 1992.

Griffiths, J. F. and Barnard, J. A., *Flame and Combustion,* 3rd ed., Blackie Academic & Professional, London, 1995.

Gu, X. J., Haq, M. Z., Lawes, M., and Woodley, R., "Laminar Burning Velocity and Markstein Lengths of Methane-Air Mixtures," *Combust. Flame.* 121:41–58, 2000.

Gülder, Ö. L., "Laminar Burning Velocities of Methanol, Isooctane and Isooctane/Methanol Blends," *Combust. Sci. Technol.* 33:1–4, 1983.

Gülder, Ö. L., "Turbulent Premixed Flame Propagation Models for Different Combustion Regimes," *Symp. (Int.) Combust.* 23:743–750, The Combustion Institute, Pittsburgh, PA, 1990.

Hottel, H. C. and Hawthorn, W. R., "Diffusion in Laminar Flame Jets," *Symp. (Int.) Combust.* 3:254–266, The Combustion Institute, Pittsburgh, PA, 1949.

Jones, G. W. and Kennedy, R. E., "Inflammability of Natural Gas: Effect of High Pressure Upon the Limits," U.S. Bureau of Mines Report of Investigation 3798, 1945.

Kee, R. J., Grcar, J. F., Smooke, M. D., and Miller, J. A., *A Fortran Program for Modeling Steady Laminar One-dimensional Premixed Flames*, Sandia National Lab. report SAND85-8240 UC-4, 1985.

Kobayashi, H., "Experimental Study of High Pressure Turbulent Premixed Flames," *Exp. Therm. Fluid Sci.* 26(2–4):375–387, 2002.

Kolmogorov, A. N., "The Local Structure of Turbulence in Incompressible Viscous Fluid for Very Large Reynolds Numbers," *Acad. Sci. USSR* 30:301, 1941.

Law, C. K., *Combustion Physics*, Cambridge University Press, Cambridge, 2006.

Lefebvre, A. H. and Reid, R., "The Influence of Turbulence on the Structure and Propagation of Enclosed Flames," *Combust. Flame* 10:355–366, 1966.

Lewis, B. and Von Elbe, G., *Combustion Flames and Explosions of Gases*, 3rd ed., Academic Press, 1987.

Mallard, E. and Le Chatelier, H. L., "Recherches Expérimentales et Théoriques sur la Combustion des Mélanges Gazeux Explosifs," *Ann. Mines* 8:374–568, 1883.

Masri, A. R., Dibble, R. W., and Barlow, R. S., "Raman-Rayleigh Measurements in Bluff Body Stabilized Flames of Hydrocarbon Fuels," *Sym. (Int.) Combust.* 24:317–324, The Combustion Institute, Pittsburgh, PA, 1992.

Matthews, R. D., Hall, M. J., Dai, W., and Davis, G. C., "Combustion Modeling in SI Engines With a Peninsula-Fractal Combustion Model," SAE paper 960072, 1996.

Metghalchi, M. and Keck, J. C., "Laminar Burning Velocity of Propane-Air Mixtures at High Temperature and Pressure," *Combust. Flame* 38:143–154, 1980.

Metghalchi, M. and Keck, J. C., "Burning Velocities of Mixtures of Air with Methanol, Isooctane, and Indolene at High Pressures and Temperatures," *Combust. Flame.* 48:191–210, 1982.

Newitt, P. M. and Thornes, L. S., "Oxidation of Propane. Part 1, The Products of the Slow Oxidation at Atmospheric and Reduced Pressures," *J. Chem. Soc.* 1656–1665, 1937.

Poinsot, T. and Veynante, D., *Theoretical and Numerical Combustion*, 2nd ed., R.T. Edwards, Philadelphia, PA, 2005.

Scholefield, D. A. and Garside, J. E., "The Structure and Stability of Diffusion Flames," *Symp. (Int.) Combust.* 3:102–110, The Combustion Institute, Pittsburgh, PA, 1949.

Strehlow, R. A., *Fundamentals of Combustion*, International Textbook Co., Scranton, PA, 1968.

Tseng, L. K., Ismail, M. A., and Faeth, G. M., "Laminar Burning Velocities and Markstein Numbers of Hydrocarbon/Air Flames," *Combust. Flame* 95:410–426, 1993.

Tuteja, A. D., "The Formation of Nitric Oxide in Diffusion Flames," Ph.D. thesis, University of Wisconsin-Madison, 1972.

Westbrook, C. K. and Dryer, F. L., "A Comprehensive Mechanism for Methanol Oxidation," *Combust. Sci. Technol.* 20:125–140, 1979.

Williams, F. A., *Combustion Theory: The Fundamental Theory of Chemically Reacting Flow Systems*, 2nd ed., Westview Press, Boulder, CO, 1985.

Wohl, K., Kapp, N. M., and Gazley, C., "The Stability of Open Flames," *Symp. (Int.) Combust.* 3:3–21, The Combustion Institute, Pittsburgh, PA, 1949.

Zabetakis, M. G., "Flammability Characteristics of Combustible Gases and Vapors," Bulletin 627, Washington, U.S. Dept. of the Interior, Bureau of Mines, 1965.

6 Gas-Fired Furnaces and Boilers

Gaseous fuels are the easiest fuels to utilize in furnaces and boilers. No fuel preparation is necessary, the gases mix easily with air, combustion proceeds rapidly, and emissions are low compared to liquid and solid fuels. The drawbacks are that gaseous fuels can be expensive and they may not be available, particularly if a natural gas pipeline hookup is not available. Because gaseous fuels are convenient to use, they are well suited for small combustion systems such as residential, commercial, and small industrial space heating furnaces and boilers. Electric utility generating stations and district heating plants use natural gas or syngas to meet peak load demands.

Gas-fired furnaces and boilers use burners that mix the fuel and air, and stabilize the flame. The burner flames are either partially premixed or fully premixed turbulent flames. The burner design must operate safely, reliably, and efficiently, and achieve very low emissions of carbon monoxide and nitrogen oxides. In addition to using natural gas, burners are needed to combust lower heating value syngas from gasification of solid fuels. Some burners are designed for industrial processes where flame shape and temperature are the most important features. High efficiency is the most important factor for space heating furnaces and boilers, process steam plants, and steam power plants. In most cases, achieving high efficiency means operating with as little excess air as possible. Before discussing burner designs and performance, let us examine how furnace efficiency is influenced by excess air.

6.1 ENERGY BALANCE AND EFFICIENCY

The fuel and air requirements for a given heat output, volume of exhaust products, and overall efficiency of furnaces and boilers may be obtained from an energy balance on the system shown in Figure 6.1. This analysis will show that the highest efficiency is obtained by operating the burner as close to stoichiometric combustion as possible.

The fuel and air flow rates required for a given heat output are calculated from an energy and mass balance for a single control volume placed around the mixer, combustion chamber, and heat exchanger,

$$\dot{N}_{air}\hat{h}_{air} + \dot{N}_f\hat{h}_f = q + q_{loss} + \sum_{i=1}^{I} \dot{N}_i\hat{h}_i \qquad (6.1)$$

where I is the number of product species, q_{loss} is the extraneous heat loss, and q is the useful heat output. In terms of the sensible enthalpy, fuel heating value, and enthalpy of vaporization of water, the energy balance is

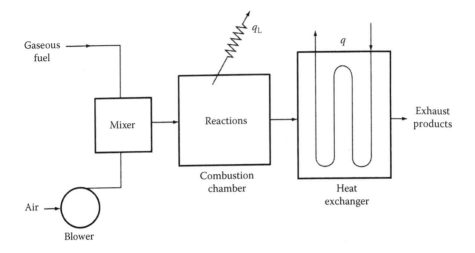

FIGURE 6.1   Schematic of a gas-fired furnace.

$$\dot{N}_{air}\hat{h}_{s,air} + \dot{N}_f\left(\hat{h}_{s,f} + \widehat{HHV}\right) = q + q_{loss} + \sum_{i=1}^{I}\dot{N}_i\hat{h}_{s,i} + \left(\dot{N}_{H_2O}\hat{h}_{fg}\right) \qquad (6.2)$$

Equation 6.2 assumes that the combustion is complete, i.e., 100% combustion efficiency, which is reasonable for a gaseous fuel. Solving for the useful heat output,

$$q = \dot{N}_f\left[\hat{h}_{s,f} + \widehat{HHV} + \left(\frac{N_{air}}{N_f}\right)\hat{h}_{s,air} - \sum_{i=1}^{I}\left(\frac{N_i}{N_f}\right)\hat{h}_{s,i} - \left(\frac{N_{H_2O}}{N_f}\right)\hat{h}_{fg}\right] - q_{loss} \qquad (6.3)$$

The moles of air and product species per mole of fuel are obtained from a chemical atom balance for the combustion process (Section 6.3). Hence, the heat output can be obtained if the fuel flow rate is specified; and conversely, if the heat output is specified, the required fuel flow rate may be calculated.

Example 6.1

Propane is burned to completion in a furnace with 5% excess air. The fuel and dry air enter at 25°C. If 10% of the heat is lost through the walls of the furnace, and the combustion products exit the furnace to the stack at 127°C, what is the useful heat output of the furnace per mass of propane?

Solution
The stoichiometric reaction is

$$C_3H_8 + 5\left(O_2 + 3.76\ N_2\right) \rightarrow 3\ CO_2 + 4\ H_2O + 18.8\ N_2$$

With 5% excess air the reaction is

$$C_3H_8 + 1.05(5)(O_2 + 3.76\ N_2) \rightarrow 3\ CO_2 + 4\ H_2O + 1.05(18.8)N_2 + 0.05(5)O_2$$

Simplifying

$$C_3H_8 + 5.25(O_2 + 3.76\ N_2) \rightarrow 3\ CO_2 + 4\ H_2O + 19.74\ N_2 + 0.25\ O_2$$

From Table 2.2,

$$HHV_{C_3H_8} = 50.4\ MJ/kg_{C_3H_8}$$

yielding

$$\widehat{HHV}_{C_3H_8} = \frac{50.4\ MJ}{kg_{C_3H_8}} \cdot \frac{44\ kg_{C_3H_8}}{kgmol_{C_3H_8}} = \frac{2218\ MJ}{kgmol_{C_3H_8}}$$

From Appendix C the enthalpy of the products is

Products (i)	N_i/N_f	$\hat{h}_{s,i}$ (MJ/kgmol$_i$)	$(N_i/N_f)\hat{h}_{s,i}$ (MJ/kgmol$_f$)
CO_2	3	4.01	12.03
H_2O	4	3.45	13.80
O_2	0.25	3.03	0.76
N_2	19.74	2.97	58.63
sum			85.22

From Equation 6.3 the useful heat output per kilogram mole of fuel is

$$q = \dot{N}_f\left[\hat{h}_{s,f} + \widehat{HHV} + \left(\frac{N_{air}}{N_f}\right)\hat{h}_{s,air} - \sum_{i=1}^{I}\left(\frac{N_i}{N_f}\right)\hat{h}_{s,i} - \left(\frac{N_{H_2O}}{N_f}\right)\hat{h}_{fg}\right] - q_{loss}$$

Noting that the fuel and the air enter at the reference state

$$\hat{h}_{s,f} = 0$$

$$\hat{h}_{s,air} = 0$$

Simplifying

$$\frac{q}{\dot{N}_f} = \widehat{HHV} - \sum_{i=1}^{I}\left(\frac{N_i}{N_f}\right)\hat{h}_{s,i} - \left(\frac{N_{H_2O}}{N_f}\right)\hat{h}_{fg} - \frac{q_{loss}}{\dot{N}_f}$$

Substituting

$$\frac{q}{\dot{N}_f} = \frac{2218 \text{ MJ}}{\text{kgmol}_f} - \frac{85.22 \text{ MJ}}{\text{kgmol}_f} - \left(\frac{4 \text{ kgmol}_{H_2O}}{\text{kgmol}_f} \cdot \frac{2.44 \text{ MJ}}{\text{kg}_{H_2O}} \cdot \frac{18 \text{ kg}_{H_2O}}{\text{kgmol}_{H_2O}}\right) - \frac{0.1q}{\dot{N}_f}$$

Solving

$$\frac{q}{\dot{N}_f} = 1779 \text{ MJ/kgmol}_f$$

and

$$\frac{q}{\dot{m}_f} = 40.4 \text{ MJ/kg}_f$$

which is 80.3% of the higher heating value.

A volumetric analysis of the products when burning natural gas is shown in Figure 6.2. The fuel for these calculations contains 83% methane and 16% ethane.

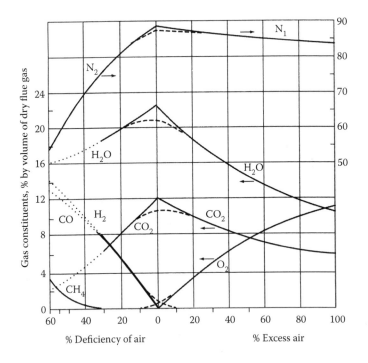

FIGURE 6.2 Volumetric analysis of natural gas combustion products as a function of fuel-air ratio. (From Reed, R. J., *North American Combustion Handbook: A Basic Reference on the Art and Science of Industrial Heating with Gaseous and Liquid Fuels,* 3rd ed., North American Manufacturing Co., Cleveland, OH, 1986, by permission of North America Mfg. Co.)

Additional external heat is required for fuel-air mixtures more than 30% deficient in air (dotted lines). Dashed lines show trends with poor mixing or wall quenching. Volumetric analysis is on a dry product basis to allow comparison with gas analyzers that require the moisture to be condensed. Note that the assumption that the combustion is complete is valid only on the lean side of stoichiometric combustion.

The mass flow rate is obtained from the molar flow rate by using the molecular weight,

$$\dot{m} = \dot{N}M \tag{6.4}$$

The volume flow rate is obtained from the molar flow rate and the equation of state,

$$p\dot{V} = \dot{N}\hat{R}T \tag{6.5}$$

The power required by the air blower is obtained from an energy balance around the blower. The result is

$$\dot{W}_{blower} = \frac{\dot{V}_{air}\Delta p_{air}}{\eta_{blower}} \tag{6.6}$$

where Δp_{air} is the pressure rise across the blower, which equals the air pressure supplied to the burner less any loss in the ducting and air preheater, and η_{blower} is the efficiency of the blower.

6.1.1 FURNACE AND BOILER EFFICIENCY

The efficiency of furnaces and boilers is defined as the ratio of the useful heat output to the energy input and by convention in the United States is always based on the higher heating value. Hence, in general for any type of fuel,

$$\eta = \frac{q}{\dot{m}_f\left(HHV\right) + \dot{W}_{blower}} \tag{6.7}$$

The furnace efficiency may be determined directly by measuring the useful heat output, q, and the fuel flow rate, or it can be determined indirectly from the products of combustion. Sometimes the fuel flow rate is not measured, so the indirect method must be used. Also, from the combustion system viewpoint, it is useful to use the indirect method of efficiency evaluation to assess where the losses arise and thus indicate how the efficiency may be improved. Substituting Equation 6.3 into Equation 6.7 and introducing the mole fraction

$$x_i = \frac{N_i}{N_{prod}}$$

where

$$N_{prod} = \sum_i N_i$$

yields

$$\eta = \frac{\hat{h}_{s,f} + \hat{h}_{s,air}\left(\dfrac{N_{air}}{N_f}\right) + \widehat{HHV} - \left(\dfrac{N_{prod}}{N_f}\right)\sum_i x_i \hat{h}_{s,i} - \left(\dfrac{N_{H_2O}}{N_f}\right)\hat{h}_{fg} - \dfrac{q_{loss}}{\dot{N}_f}}{\widehat{HHV} + \dfrac{\dot{W}_{blower}}{\dot{N}_f}} \qquad (6.8)$$

Inspection of Equation 6.8 shows that the furnace efficiency can be increased by the following measures:

1. Decreasing the temperature of the exhaust products
2. Reducing the excess air, which will reduce the moles of products per mole of fuel and reduce the blower power
3. Reducing the extraneous heat loss
4. Reducing the blower power requirements

Item (1) can be accomplished by increasing heat transfer to the load, perhaps by cleaning the heat transfer surfaces. Item (2) can be accomplished only if there is adequate mixing, and it can only be done to the point where the CO starts to increase. Item (4) indicates the motivation for aerated burners and the pulse furnace, which do not need a blower.

In order to evaluate Equation 6.8 based on the composition of the products and the ultimate analysis (i.e., elemental analysis) of the fuel, it is convenient to write.

$$\frac{N_{prod}}{N_f} = \frac{N_{prod}}{N_C}\frac{N_C}{N_f} = \frac{N_C/N_f}{x_{CO_2} + x_{CO}} \qquad (6.9)$$

where the subscripts prod, f, and C refer to products, fuel and carbon, respectively. Also note that the enthalpy of water contains the latent heat of vaporization, which is a relatively large term. Hence, if the fuel analysis is known, then the efficiency can be determined from gas sampling of the products using Equations 6.8 and 6.9. The blower power per unit fuel flow is usually small compared to the heating value and hence, the efficiency can be determined without knowing the fuel and air flow rates. Of course, if the fuel and air flow rates are known, the indirect method for efficiency determination may be compared with the direct method.

Example 6.2

Stack gas analysis of a natural gas-fired furnace gave the following volumetric analysis: 4% O_2, 10% CO_2, 17% H_2O and 86% N_2, all on a dry basis (the water vapor in the gas sample was condensed before entering the measuring instruments). The fuel was 84% CH_4 and 16% C_2H_6 by volume, and the higher heating value was 54.15 MJ/kg. The fuel and air entered the furnace at 25°C, and the stack gas temperature was 127°C. No blower was used and heat losses were negligible. Find (a) the percent excess air and (b) the operating efficiency of this furnace based on the higher heating value.

Solution

Part (a)

From Equation 3.45 the percent excess air based on the combustion products is

$$\% \text{ excess air} = 100 \left(\frac{x_{O_2}\big|_{prod}}{\dfrac{x_{N_2}\big|_{prod}}{3.76} - x_{O_2}\big|_{prod}} \right) = 100 \left(\frac{4}{\dfrac{86}{3.76} - 4} \right) = 21.2\%$$

Part (b)

Use Equation 6.8 to determine the furnace efficiency. The wet product analysis is needed for this equation and by volume is

$$\frac{0.04}{1.17} = 3.42\% \ O_2$$

$$\frac{0.10}{1.17} = 8.55\% \ CO_2$$

$$\frac{0.17}{1.17} = 14.53\% \ H_2O$$

$$\frac{0.86}{1.17} = 73.50\% \ N_2$$

Using Appendix C at 400 K the enthalpy of the products is

$$\hat{h}_{s,prod} = \sum_i x_i \hat{h}_{s,i} = 0.0342(3.03) + 0.0855(4.01) + 0.1453(3.45) + 0.7350(2.97)$$

$$\hat{h}_{s,prod} = 3.13 \ \text{MJ/kgmol}_{prod}$$

Next use Equation 6.9 to find the moles of products formed per mole of fuel,

$$\frac{N_{prod}}{N_f} = \frac{N_C/N_f}{x_{CO_2} + x_{CO}} = \frac{1(0.84) + 2(0.16)}{0.0855 + 0} = \frac{13.6 \ \text{kgmol}_{prod}}{\text{kgmol}_f}$$

The enthalpy of the water vapor condensation term of Equation 6.8 is

$$\left(\frac{N_{H_2O}}{N_f}\right)\hat{h}_{fg} = \frac{0.1453\ \text{kgmol}_{H_2O}}{\text{kgmol}_{prod}} \cdot \frac{13.6\ \text{kgmol}_{prod}}{\text{kgmol}_f} \cdot \frac{2.44\ \text{MJ}}{\text{kg}_{H_2O}} \cdot \frac{18\ \text{kg}_{H_2O}}{\text{kgmol}_{H_2O}}$$

$$\left(\frac{N_{H_2O}}{N_f}\right)\hat{h}_{fg} = \frac{86.8\ \text{MJ}}{\text{kgmol}_f}$$

The molecular weight of the fuel is

$$M_f = \frac{0.84\ \text{kgmol}_{CH_4}}{\text{kgmol}_f} \cdot \frac{16\ \text{kg}_{CH_4}}{\text{kgmol}_{CH_4}} + \frac{0.16\ \text{kgmol}_{C_2H_6}}{\text{kg}_{C_2H_6}} \cdot \frac{30\ \text{kg}_{C_2H_6}}{\text{kgmol}_{C_2H_6}} = \frac{18.24\ \text{kg}_f}{\text{kgmol}_f}$$

And the molar higher heating value is

$$\widehat{HHV} = \frac{54.15\ \text{MJ}}{\text{kg}_f} \cdot \frac{18.24\ \text{kg}_f}{\text{kgmol}_f} = \frac{987.7\ \text{MJ}}{\text{kgmol}_f}$$

Substituting into Equation 6.8, noting that the enthalpy of the fuel and air is zero at 25°C, and remembering that there is no blower and that the heat losses are negligible, the furnace efficiency based on the HHV is

$$\eta = \frac{0 + 0 + 987.7 - 13.6(3.13) - 86.6}{987.7 + 0} = 0.869$$

The furnace efficiency is 86.9%.

6.2 FUEL SUBSTITUTION

Substitution of one gaseous fuel for another more readily available or sustainable fuel is sometimes desirable for a particular burner. Perhaps a manufactured gas from biomass gasification is to be substituted for natural gas. When a new gaseous fuel is being used, the burner should be adjusted to preserve the heat rate and flame stability and shape where possible. If the substitute fuel is similar, as in the case of substituting propane for methane, it may be sufficient simply to adjust the airflow to achieve the proper equivalence ratio.

If the substitute fuel is dissimilar, the fuel flow rates may have to be adjusted by changing the fuel pressure or orifice size to maintain the heat rate. The heat rate, q, equals the volumetric flow rate of the fuel times the heating value of the fuel per unit volume.

$$q = \dot{V}_f\left(HHV\right)$$

The volumetric flow rate of the fuel is given by

$$\dot{V}_f = \sqrt{\frac{2\Delta p}{\rho_f}} A_f \qquad (6.10)$$

where

Δp is the pressure drop across the fuel orifice,
A_f is the effective area of the fuel orifice, and
ρ_f is the density of the fuel.

Hence, for a fixed orifice size, fuel pressure and temperature, the heat rate is given by

$$q = K\frac{HHV}{\sqrt{sg_f}} = K(WI) \qquad (6.11)$$

where K is the system constant and WI is the Wobbe Index, which is a measure of the interchangeability of fuels, defined by

$$WI = \frac{HHV}{\sqrt{sg_f}} \qquad (6.12)$$

If the Wobbe Index of the substitute fuel is significantly different from the design fuel, the burner should be modified. In addition, the flame length, flashback, and blow-off characteristics should be considered. The density of a fuel cannot be changed appreciably by increasing the pressure because this will increase the flow rate, moving the burner out of the stable design region. For low heating value fuels such as from biomass gasification, the fuel flow rate must be considerably greater than for natural gas for the same heat output. The volume of products will also be greater, which may require larger ductwork.

Example 6.3

Compare the Wobbe Index for wood producer gas to the index for natural gas and comment on the interchangeability of these fuels.

Solution

From Equation 6.12 the ratio of the Wobbe Index for producer wood gas to natural gas is

$$\frac{WI\big|_{producer}}{WI\big|_{natural}} = \left(\frac{HHV}{\sqrt{\rho}}\right)_{producer}\left(\frac{\sqrt{\rho}}{HHV}\right)_{natural} = \frac{HHV_{producer}}{HHV_{natural}}\cdot\sqrt{\frac{\rho_{natural}}{\rho_{producer}}}$$

From Table 2.2 the volumetric higher heating values of wood producer gas and natural gas are 4.8 MJ/m³ and 38.3 MJ/m³, respectively. The ratio of the specific gravities are proportional to the ratios of the density of the gases and hence ratio is proportional to the molecular weight as follows:

$$\frac{\sqrt{sg_{\text{natural}}}}{\sqrt{sg_{\text{producer}}}} = \sqrt{\frac{\rho_{\text{natural}}}{\rho_{\text{producer}}}} = \sqrt{\frac{\left[pM/\hat{R}T\right]_{\text{natural}}}{\left[pM/\hat{R}T\right]_{\text{producer}}}} = \sqrt{\frac{M_{\text{natural}}}{M_{\text{producer}}}}$$

From Table 2.1 the molecular weight of wood producer gas is approximately 23 and the molecular weight for natural gas is about 17. Thus,

$$\frac{WI\big|_{\text{producer}}}{WI\big|_{\text{natural}}} = \frac{HHV_{\text{producer}}}{HHV_{\text{natural}}}\sqrt{\frac{M_{\text{natural}}}{M_{\text{producer}}}} = \frac{4.8}{38.3}\sqrt{\frac{17}{23}} = 0.11$$

Thus, not only will the flow rate of fuel need to be increased to maintain the same heat output, but the burner will need to be modified to assure stability because the Wobbe Index of the producer gas is approximately 1/10th that of natural gas.

6.3 RESIDENTIAL GAS BURNERS

As noted above, the burner should operate as close to stoichiometric combustion as possible to realize high efficiency while maintaining emissions as low as possible. The shape and distribution of the flame should match the furnace combustion chamber. In general, for smaller systems it is desirable that the flame should fill the combustion chamber, but not impinge on the furnace or boiler walls. Domestic or residential burners are either partially premixed or fully premixed.

Older residential burners are partially premixed and do not use an air blower. Rather, combustion air from the ambient surroundings is entrained by the fuel flow in a manner similar to a Bunsen burner. The fuel enters at low pressure (0.5–15 kPa) and at high velocity through an orifice, as shown in Figure 6.3. Primary air is drawn in through shutter openings due to the momentum of the fuel jet and is controlled by adjusting inlet shutters. A venturi throat improves the entrainment. Mixing occurs in the extended tube leg and the throat. The mixture flows through the burner head and burns as an attached flame at the burner ports. The momentum of the flame entrains

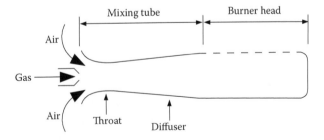

FIGURE 6.3 Multiple port, partially premixed atmospheric gas burner.

secondary air that completes the combustion. Entrained air burners of this type typi-cally operate at 40% excess air overall.

Ports in the burner head are typically 1–2 mm in diameter and spaced 4–8 mm apart edge to edge with a port loading of 10 W/mm² and 50% primary aeration (% theoretical air). Port load is defined as the fuel flow rate times the lower heating value divided by the port area.

$$q''_{port} = \frac{\dot{m}_f \, LHV_f}{A_{port}} \tag{6.13}$$

Burner stability depends on the primary aeration, burner port load, and port spacing. If the flow rate of the reactants is too high compared to the flame speed, the flame will become unstable and lift off (Figure 6.4). At low primary aeration the array of ports burn with a single cone, but as the air is increased each port burns as an individual cone. The exact size and location of the stable area is dependent on the particular burner configuration. The port diameter must be small enough to prevent flashback of the flame (i.e., 1–2 mm). Gases that have significant hydrogen content have a different stability diagram because of the high flame speed. Because of this, flames containing higher levels of hydrogen have a greater tendency to flash back and a lesser tendency to lift off and so the minimum quench diameter for these types of fuels is 0.8 mm.

Partially premixed burners with entrained secondary air do not meet today's stan-dards for efficiency and emissions of carbon monoxide and nitrogen oxides. New residential burners use a fan to supply forced air and fully premix the fuel and air without the need for secondary air (Figure 6.5). In this way the burner can be oper-ated near stoichiometric conditions with low excess air. However, the nitrogen oxide emissions tend to be well above 100 ppm. By recirculating some of the flue gas back

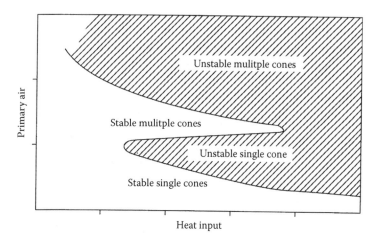

FIGURE 6.4 Typical stability diagram for a multiple port, partially premixed atmospheric gas burner showing the unstable liftoff limit. (From Jones, H. R. N., *The Application of Combustion Principles to Domestic Gas Burner Design*, Taylor Francis, Boca Raton, FL, 1990. British Gas.)

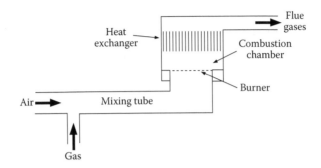

FIGURE 6.5   Premixed gas burner using fan supplied air (residential size).

in with the inlet air, the flame temperature is lowered and the nitrogen oxide emissions reduced to less than 20 ppm while maintaining a good heat input per unit area.

6.4   INDUSTRIAL GAS BURNERS

Industrial burners serve a wide range of process heating applications beyond space heating where gas temperature, heat transfer and flame shape are important. For example, the premixed tunnel burner (Figure 6.6) meets some needs, but as the heat input and range of operating conditions increases, the limitations of flashback in the premix manifold and liftoff from the nozzle (the grey area in Figure 6.6) can necessitate a non-premixed burner.

Safety and flexibility considerations led to the development of nozzle mix burners where the fuel and air are introduced separately and mixed in a refractory nozzle. By re-radiating heat from the flame, the refractory nozzle stabilizes the flame. Figure 6.7 shows a two-stage nozzle mix burner. Single stage nozzle mix burners use only the inner refractory nozzle, but the nitrogen oxide emissions tend to be high. By using two stages, the first stage can burn rich and the second stage can burn lean, thereby avoiding peak temperatures in the flame. By changing the amount and ratio of primary and secondary air and by using different shaped refractory nozzles, a wide range of flame shapes can be achieved for different applications and different fuels. Two stage nozzle mix burners have a wide turndown range.

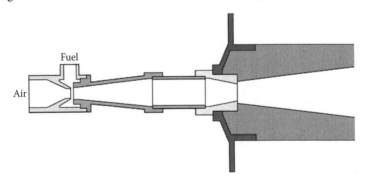

FIGURE 6.6   Premixed tunnel burner (industrial size).

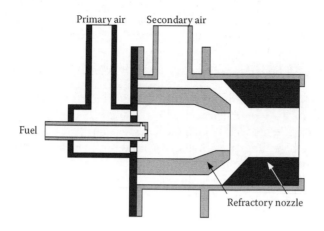

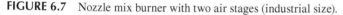

FIGURE 6.7 Nozzle mix burner with two air stages (industrial size).

For small and medium size industrial burners, an interesting concept is the ceramic fiber burner, which uses a porous matrix of ceramic and metallic fibers formed into a cylindrical tube (Figure 6.8.) Premixed fuel and air flow uniformly through the porous material and, aided by catalytic reaction, burn within the fiber matrix. The low conductivity of the fibers and convective cooling by the out-flowing reactants allow the burner to operate safely with no flashback tendencies at surface velocities below the mixture flame speed. The surface temperature is maintained below 1370 K for low NO_X formation and above 1250 K to completely react the CO. The surface temperature depends in part on the velocity of the reactants through the surface. Heat release rates up to 800 kW/m^2 are possible with a turndown ratio of 6 to 1. Typically 70% of the heat release is transferred by radiation. The fiber matrix burner reduces nitrogen oxide emissions to a level as low as 15 ppm. Because the heat release per unit burner area is reduced, the fiber matrix burner area must have a

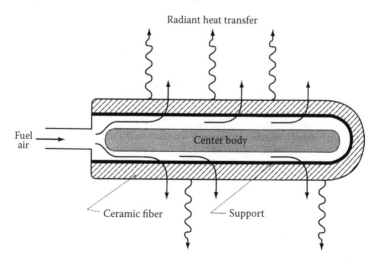

FIGURE 6.8 Cross section of a fiber matrix burner.

larger surface area than a flame, which is a limitation for some applications. Excess air may be as low as 10% without increasing CO or hydrocarbon emissions.

6.5   UTILITY GAS BURNERS

Gas burners for electrical utility and district heating boilers operate with low excess air to achieve high system efficiency while maintaining very low carbon monoxide and nitrogen oxide emissions. A high flame temperature near stoichiometric operation forms excessive nitric oxide, so the flue gas is mixed with the inlet air to reduce the flame temperature. However, without additional measures, flue gas recirculation (FGR) causes the flame to be less stable. By introducing *swirl* to the flow, the flame tends to be stabilized. Swirl is a rotating flow around the central axis of the nozzle. Swirl is imparted by vanes in the airflow passage that provide angular momentum to the axial flow. The swirling flow spreads the flame, enhances mixing, and stabilizes the flame. Figure 6.9 shows how the flame shape changes as the amount of swirl is

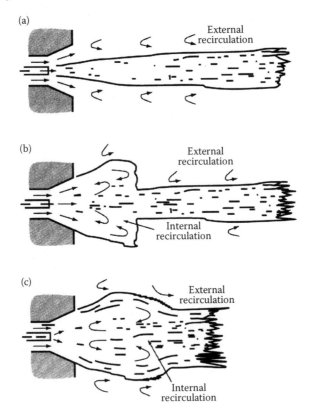

FIGURE 6.9   Effect of burner swirl on flame type: (a) long jet flame, no swirl; (b) combination jet flame and partial internal recirculation zone, intermediate swirl; (c) flame with closed internal recirculation zone, high swirl. (From Weber R. and Dugue, J., "Combustion Accelerated Swirling Flows in High Confinements," *Prog. Energy Combust. Sci.* 18(4):349–367, 1992, by permission from Elsevier Science Ltd.)

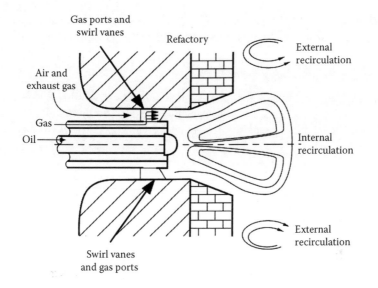

FIGURE 6.10 Utility size high swirl burner for gas or oil. (From Baukal, C. E. and Schwartz, R. E., eds., *The John Zink Combustion Handbook*, CRC Press, Boca Raton, FL, 2001, p. 582, by permission.)

increased. With high swirl a closed internal recirculation zone is formed by a positive radial pressure gradient that is caused by the swirl, and this creates backmixing. In this way the flame is stable even with large amounts of flue gas recirculation.

A high swirl burner where the fuel is injected from the tip of the swirl vanes is shown in Figure 6.10. This design achieves very low nitrogen oxide emissions (below 10 ppm at 3% oxygen in the products) even with 260°C preheat (Figure 6.11). This burner is stable with up to 50% FGR at heat input levels up to 12 MW (thermal). For

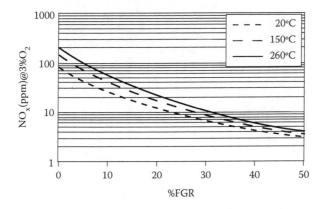

FIGURE 6.11 Emissions from a high swirl burner with natural gas-firing as a function of flue gas recirculation (FGR) for three air preheat levels. (From Baukal, C. E. and Schwartz, R. E., eds., *The John Zink Combustion Handbook*, CRC Press, Boca Raton, FL, 2001, p. 584, by permission.)

heat inputs above 12 MW, an outer ring of vanes is placed where additional fuel is injected. The vanes in the outer ring are straight and outer swirl is not used.

6.6 LOW SWIRL GAS BURNERS

Recently, a low swirl burner that uses swirl in a different way has been developed and shown to operate stably and with low emissions. In this burner an annular vane swirler surrounds a cylindrical center channel (Figure 6.12). A perforated plate is

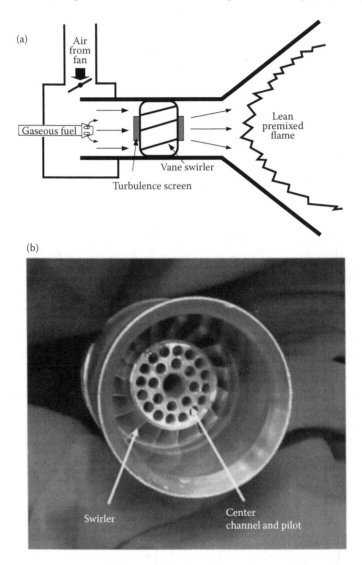

FIGURE 6.12 Low swirl premixed burner (a) side view, (b) end view. The center hole provides a provisional pilot/central burner. (Courtesy of R. K. Cheng, U.S. DOE Lawrence Berkeley National Laboratory, Berkeley, CA.)

placed in the flow path of the center channel to establish a controlled level of turbulence. Flow through the center channel has no swirl. Gaseous fuel is injected from a spray nozzle in the mixing zone. The centrifugal forces of the swirling flow act on the non-swirling center core to create flow divergence downstream of the burner exit. Because of the flow divergence, the velocity beyond the exit plane decreases, and the axial velocity near the centerline decreases to zero. Because the amount of swirl is low and the center core of the burner is open, there is no recirculation zone. At the exit plane of the burner, the velocity of the reactants exceeds the turbulent flame speed, but there is a standoff distance where the flow velocity is equal and opposite to the turbulent flame speed. Because of the divergent flow pattern, the turbulent flame brush of the reaction zone has a ring-like shape, and the streamlines flow through the ring without recirculation.

A high swirl burner typically has a solid center body, and the flame is stabilized at the burner face (the refractory nozzle in Figure 6.10). However, with the low swirl burner the center is open and the burner operates with a stable lifted flame. As the load (i.e., heat input rate) is increased, the turbulence intensity of the reactants increases, but the flame position is relatively unchanged. Low swirl burners have been implemented with diameters of 2.5–71 cm with heat inputs of 5 kW to 44 MW, and the turndown range is better than 10 to 1.

Scaling rules for low swirl burner design have been developed. The ratio of the center channel radius to the outer swirler radius is 0.4–0.55. The mass flow through the center channel is approximately equal to the mass flow through the swirl annulus. The mass flow ratio is adjusted by changing the number of holes in the turbulence screen, which alters screen blockage from 60% to 80%. The angle of the swirl vanes is 37°–45°. For methane, the average velocity in the burner is 5–25 m/s. For methane, flashback in these burners occurs at 1.7 m/s, and blow-off occurs well above 25 m/s.

When operating at 100% capacity on natural gas, the emissions are about 5 ppm NO_X and 3 ppm CO at 3% O_2 in the flue gas. At 20% capacity NO_X is 6 ppm and CO is essentially zero. In general for a range of commercial sizes, the NO_X and CO are 4–7 ppm in the flue gas at 3% O_2. The swirl burner has low NO_X without FGR because the residence time in the hottest part of the flame is short compared to high swirl burners with an internal recirculation zone.

Tests in the low swirl burner have shown that the turbulent flame speed is up to 30 times faster than the laminar flame speed for methane and up to 50 times faster than the laminar flame speed for hydrogen. Turbulent flame speed increases linearly with the intensity of the turbulence in these burners (Cheng et al. 2000). For methane-air the flame speed correlation is

$$\frac{V_T}{V_L} = 1 + 1.73\frac{V'_{rms}}{V_L} \tag{6.14}$$

and for a hydrogen-air flame in a low swirl burner,

$$\frac{V_T}{V_L} = 1 + 3.15\frac{V'_{rms}}{V_L} \tag{6.15}$$

6.7 PROBLEMS

1. The useful heat output of a residential furnace is 30 kW. The fuel is methane and heat losses are 10% of the useful output. The stack temperature is 400 K and 40% excess air is used. Both air and fuel enter the furnace at 298 K and 1 atm. Combustion is assumed to be complete. Calculate:

   a. Flow rate of the methane (std m³/min)
   b. Flow rate of the inlet air (std m³/min)
   c. Flow rate of the exhaust products (actual m³/min)
   d. Furnace efficiency

2. Repeat Problem 1, but assume that combustion products leave at 310 K and the moisture is condensed in the heat exchanger.

3. Repeat Problem 1, but assume that combustion products leave at 310 K, heat losses are negligible, and excess air is 5%.

4. An entrained air burner uses methane gas. The air and fuel enter at 298 K. The stack temperature is 400 K, and the CO_2 concentration is 9% by volume. The CO concentration is 10 ppm by volume. Assuming the extraneous heat losses are negligible, what is the efficiency of the furnace? The higher heating value of methane is 55.5 MJ/kg.

5. A residential furnace is rated at 30 kW heat input using propane. Assume that the burner head consists of 1 mm diameter holes, offset with 4 mm between hole centers on an equilateral triangular lattice. How many holes are required and what is the area of the burner head in square centimeters? Assume a burner loading of 30 W/mm². What is the heat release rate per centimeter square of burner head? The temperature of the burner head is 323°C. Assuming 80% primary aeration, what is the velocity in the burner ports? How does this compare to the laminar flame speed? From Table 2.2 the lower heating value of propane is 83.6 MJ/m³. Use the data in Table A.3.

6. A furnace is rated at 30 kW heat input using methane. If a low swirl burner is used in the furnace, what is an appropriate burner diameter? Assume the channels occupy 50% of the center area, a stoichiometric mixture of fuel and air, and an average burner velocity of 17 m/s.

7. A utility generates 500 MW of electricity while operating on natural gas. The overall thermal efficiency of the plant is 33%. Air is supplied to the burners at 2.0 kPa and 300 K. Find the cost of natural gas per hour assuming the cost of natural gas is $20 per MWh. Also, find the power required to deliver the burner air across a 2 kPa pressure rise at 300 K. Assume that 10% excess air is required for clean combustion, and the fan efficiency is 90%. Assume that the stoichiometric air-fuel mass ratio is 17.2.

8. Producer gas from a wood gasifier consists of 20% CO, 13% H_2, 3% CH_4, 10% CO_2, 10% H_2O and 44% N_2 by volume. The wood gasifier is used to retrofit an industrial natural gas-fired furnace rated at 150 kW on an input basis. Calculate the standard volume flow rate of air and fuel (scmh) for both methods of firing assuming 10% excess air in both cases. How would

the burner need to be modified for the retrofit? How does the standard volume flow rate of combustion products compare to natural gas firing for this case? The natural gas is 84% CH_4 and 16% C_2H_6 (vol). Use Table 2.2 for the heating value data. Standard cubic meters per hour (scmh) is determined at 1 atm and 15°C.

9. Outside air is brought directly into a gas furnace through a well-insulated duct. The furnace burner was originally set to give a stoichiometric mixture for 294 K air. On a 260 K winter day, what will be the equivalence ratio of the burner?

10. Determine the adiabatic flame temperature and equilibrium NO concentration for a methane-air flame that has 15% excess air by volume and 20% of the combustion products recirculated back into the inlet of the burner. The inlet air and recirculated products are at 250°C and 1 atm.

REFERENCES

Bartok, W. and Sarofim, A. F., eds., *Fossil Fuel Combustion: A Source Book*, Wiley, New York, 1991.

Baukal, C. E., ed., *Industrial Burners Handbook*, CRC Press, Boca Raton, FL, 2004.

Baukal, C. E. and Schwartz, R. E., eds., *The John Zink Combustion Handbook*, CRC Press, Boca Raton, FL, 2001.

Breen, B. P., Bell, A. W., Bayard de Volo, N., Bagwell, F. A., and Rosenthal, K., "Combustion Control for Elimination of Nitric Oxide Emissions for Fossil-Fuel Power Plants," *Symp. (Int.) Combust.* 13:391–401, The Combustion Institute, Pittsburgh, PA, 1970.

Cheng, R. K., Yegian, D. T., Miyasato, M. M., Samuelson, G. S., Benson, C. E., Pellizzari, B., and Loftus, P., "Scaling and Development of Low-Swirl Burners for Low Emission Furnaces and Boilers," *Symp. (Int.) Combust.* 28:1305–1313, The Combustion Institute, Pittsburgh, PA, 2000.

Faulkner, E. A., *Guide to Efficient Burner Operation: Gas, Oil and Dual Fuel*, 2nd ed., Fairmont Press, Atlanta, GA, 1987.

Griffiths, J. F. and Barnard, J. A., *Flame Combust.*, 3rd ed., Blackie Academic & Professional, London, 1995.

Jones, H. R. N., *The Application of Combustion Principles to Domestic Gas Burner Design*, Taylor Francis, Boca Raton, FL, 1990.

Kamal, M. M. and Mohamed, A. A., "Combustion in Porous Media," *Proc. Inst. Mech. Eng., Part A* 220(5):387–508, 2006.

Reed, R. J., *North American Combustion Handbook: A Basic Reference on the Art and Science of Industrial Heating with Gaseous and Liquid Fuels*, 3rd ed., North American Manufacturing Co., Cleveland, OH, 1986.

Rhine, J. M. and Tucker, R. J., *Modeling Gas-Fired Furnaces and Boilers: And Other Industrial Heating Processes,* McGraw-Hill, New York, 1990.

Sayre, A., Lallemant, N., Dugue, J., and Weber, R., "Effect of Radiation on Nitrogen Oxide Emissions From Nonsooty Swirling Flames of Natural Gas," *Symp. (Int.) Combust.* 25:235–242, The Combustion Institute, Pittsburgh, PA, 1994.

Toqan, M. A., Beér, J. M., Jansohn, P., Sun, N., Testa, A., Shihadeh, A., and Teare, J. D., "Low NO_X Emission from Radially Stratified Natural Gas-Air Turbulent Diffusion Flames," *Symp. (Int.) Combust.* 24:1391–1397, The Combustion Institute, Pittsburgh, PA, 1992.

United States Environmental Protection Agency, *Compilation of Air Pollutant Emission Factors, Vol. 1: Stationary Point and Area Sources*, 5th ed., EPA-AP-42, 5th ed., 1995.

Weber, E. J. and Vandaveer, F. E., "Gas Burner Design," Chap. 12 in *Gas Engineers Handbook*, ed. C. G. Segeler, The Industrial Press, New York, 1965.

Weber, R. and Dugue, J., "Combustion Accelerated Swirling Flows in High Confinements," *Prog. Energy Combust. Sci.* 18(4):349–367, 1992.

Weber, R. and Visser, B. M., "Assessment of Turbulent Modeling for Engineering Prediction of Swirling Vortices in the Near Burner Zone," *Int. J. Heat Fluid Flow* 11(3):225–235, 1990.

7 Premixed-Charge Engine Combustion

Automobiles, light duty trucks, off-road vehicles, and small utility engines have internal combustion engines that use volatile liquid fuels, especially gasoline. The fuel rapidly vaporizes and mixes with air prior to combustion in the engine, hence the term "premixed-charge engine." The discussion here is limited to automotive spark ignition engines, and the emphasis is on the combustion aspects of the engine rather than overall performance or mechanical design of the vehicle. A compilation of combustion engine terminology is given at the end of this chapter to aid the reader.

7.1 INTRODUCTION TO THE SPARK IGNITION ENGINE

From a combustion perspective an internal combustion engine consists of multiple piston-cylinders, each with separate intake and exhaust valves and a spark plug. After the spark plug fires, a turbulent flame front propagates outward from the spark plug. This flame front then creates elevated pressure against the piston and connecting rod, thus creating torque on the drive shaft. The piston travels in a repetitive cycle between being fully inserted at top dead center (TDC) and being fully withdrawn at bottom dead center (BDC). The four-stroke engine sequence for a cylinder (Figure 7.1) is as follows: (1) with the intake valve(s) open and exhaust valve(s) closed, a mixture of air and fuel is drawn into the cylinder as the piston moves downward toward BDC; (2) with all valves closed, the fuel-air charge is compressed as the piston moves upward toward TDC; (3) the spark discharges when the fuel-air mixture is nearly fully compressed, and the power stroke occurs as the charge combusts rapidly forcing the piston downward toward BDC due to the high pressure in the cylinder; and (4) as the piston nears BDC the exhaust valve opens and the piston then moves upward toward TDC while expelling the combustion products. The cycle is repeated continuously with the power stroke occurring during every other revolution of the drive shaft.

In newer engines, the spark plug is centrally located to minimize flame travel distance, and there are usually two intake valves and two exhaust valves to ensure good circulation and control of the intake process. The piston displaces a certain volume, and at TDC there is a clearance volume. For example, in a 2.4 L engine with four cylinders with a bore and stroke of 91.4 mm, each cylinder displaces a volume of 0.6 L. The clearance volume is 6.7 cm^3. If the compression ratio is 10, the volume at TDC is 0.0667 L (66.7 cm^3).

The torque produced by the engine is a function of the pressure force exerted on the pistons. The pistons then exert force on the drive shaft. The power produced by the engine is the torque times the shaft revolutions per unit time. Engine torque is

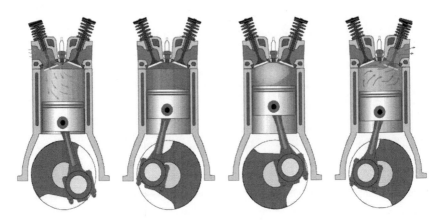

FIGURE 7.1 Four-stroke engine cycle showing crankshaft, piston, and valve positions during intake, compression, power, and exhaust strokes. The crankshaft rotates clockwise 720 CA° for one complete cycle. (Courtesy of K. Hoag, University of Wisconsin-Madison.)

controlled by means of a throttle plate in the main inlet air manifold. With a wide open throttle the volumetric efficiency is nearly 100% (typically 90–96%), but with decreasing throttle setting, the pressure drop across the throttle increases and the mass of air (and corresponding amount of fuel) drawn into the cylinder is reduced. At low throttle positions the volumetric efficiency is as low as 30%.

The proper amount of fuel relative to the amount of air is injected into the inlet manifold of each cylinder by means of fuel injectors. In older vehicles a carburetor was used to feed the fuel. Because of the catalytic exhaust system used to control pollutant emissions, it is necessary to maintain a near stoichiometric fuel-air mixture for all throttle settings. An oxygen sensor in the exhaust manifold provides feedback control to the fuel injectors to maintain the near stoichiometric fuel-air ratio, which can be done more accurately with fuel injectors than a carburetor.

Automotive vehicles typically operate from 500 revolutions per minute (rpm) at idle to a maximum of 5000 rpm. The time available to complete combustion varies greatly depending on the rpm. Some residual combustion products remain in the cylinder after blowdown, and this is referred to as inherent exhaust gas recirculation (EGR). Depending on the rpm and valve timing, inherent EGR varies from 5% to 20% by mass.

Combustion is not instantaneous, but occurs over time as a flame front propagates outward from the spark plug and sweeps over the unburned mixture. This typically occurs over 30–60 crank angle degrees (CA°) out of the 360 CA° crank angle motion (about 8 ms at 1400 rpm engine speed). At the time of spark ignition, the average gas temperature and pressure are about 700 K and 700 kPa. During combustion the pressure can reach 2 MPa and the temperature can reach 2400 K. Timing (crank angle setting) of the spark is set to ignite before top dead center (BTDC) of the piston cycle and is determined by the burning rate of the reactants and the engine speed. Slow flame speed causes late burnout, which means that more of the combustion energy is lost in the exhaust. Additionally, the emissions are increased unless the spark is advanced (set to ignite sooner); however, this may reduce peak pressure and power. Spark timing is adjusted so that peak pressure takes place 5–20 CA° after top dead center (ATDC) based on best efficiency, lowest emissions, and maximum measured torque. Fortunately, as the engine rpm increases, the flame speed tends to increase because of increased turbulence in the cylinder.

Before proceeding further, the reader may find that it is helpful to study the terminology in Example 7.1 and the terminology summary at the end of this chapter.

Example 7.1

The cylinder pressure-volume diagram is shown for a spark ignition engine. As discussed earlier, a single combustion cycle consists of two strokes of the piston (720 CA°).

Part (a)

Match the cycle events below to the numbers shown on the figure. Note that during the intake stroke from point 6′ to point 1 the pressure is below ambient because of the throttle.

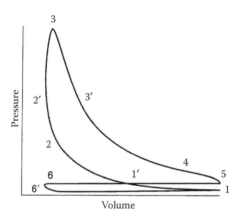

Cycle Events

(a) Maximum cylinder volume
(b) Intake valve closes
(c) Spark discharge starts
(d) Clearance volume
(e) Peak pressure of combustion
(f) End of combustion heat release
(g) Exhaust valve opens
(h) BDC of expansion stroke
(i) Intake valve opens
(j) Exhaust valve closes

Part (b)

Describe each of the process parameters below in terms of the cycle events above and/or the numbers shown on the figure.

Process parameters

(k) Displacement volume
(l) Compression ratio
(m) Closed cycle portion of the engine cycle
(n) Gas exchange portion of the engine cycle
(o) Combustion duration
(p) Trapped mass in the cylinder
(q) Theoretical mass in the cylinder
(r) Volumetric efficiency (%)
(s) Blowdown portion of the exhaust
(t) Valve overlap period
(u) Pumping work
(v) Indicated work
(w) Indicated mean effective pressure
(x) Net indicated work

Solution

Part (a)

(a) 1; (b) 1'; (c) 2; (d) 2'; (e) 3; (f) 3'; (g) 4; (h) 5; (i) 6; (j) 6';

Part (b)

(k) from 1 to 2'; (l) $d/(d + k)$; (m) from 1' to 4; (n) from 4 to 1'; (o) from 2 to 3'; (p) the mass in the cylinder at 1'; (q) $\rho_0(d + k)$; (r) 100 (p/q); (s) from 4 to 5; (t) from 6 to 6';

(u) $\int_5^1 pdV$; (v) $\int_1^5 pdV$; (w) v/k; (x) $u + v$.

Notes: (d) this is the minimum cylinder volume; (t) some of the products (residuals) in the cylinder blow back into the intake port; (u) defined this way even though the valve events are not at TDC and BDC; the pumping work is typically large at part throttle (about 40% of indicated work at highway load conditions), but small (<10%) at wide open throttle; (x) larger than the engine shaft work (brake work) because of friction and work used to drive accessories.

7.2   ENGINE EFFICIENCY

The ideal cycle (the Otto cycle), given in thermodynamic textbooks for this type of engine, consists of an isentropic compression, an instantaneous constant volume combustion, an isentropic expansion, and an instantaneous constant volume exhaust. It can be shown for this cycle that the ideal thermal efficiency depends only on the compression ratio, CR, and the ratio of specific heats, γ,

$$\eta_{t,ideal} = 1 - (CR)^{1-\gamma} \tag{7.1}$$

where

$$\eta_t = \frac{\text{Work out}}{\text{Fuel energy in}}$$

For example, for a compression ratio of 8 and specific heat ratio of 1.3, the ideal engine thermal efficiency is 0.464. For a compression ratio of 10 and specific heat ratio of 1.3, the ideal thermal efficiency is 0.536. The ideal thermal efficiency is less than 1.0 because of heat expelled in the exhaust and the piston work needed to compress the charge.

For a given compression ratio the actual engine thermal efficiency is less than the ideal thermal efficiency because of (1) finite combustion time before and after TDC; (2) heat loss to the piston head and cylinder walls; (3) friction between the piston rings and cylinder; (4) crevice volume loss and blowby between the piston and cylinder; (5) incomplete combustion; (6) back pressure during the exhaust stroke and negative pressure during the intake stroke, especially when the throttle is actuated, and (7) mechanical losses in the system. Engine thermal efficiency, η_t, is measured in the laboratory by using a dynamometer to measure the output torque (τ), rotation rate (ω) of the engine output shaft, and fuel flow rate as follows:

$$\eta_t = \frac{\dot{W}_b}{\eta_{comb}\dot{m}_f\text{LHV}} \tag{7.2}$$

where the brake power, $\dot{W}_b$, is

$$\dot{W}_b = 2\pi\tau\omega$$

The combustion efficiency, η_{comb}, accounts for incomplete combustion and is typically 0.95–0.98. To determine the combustion efficiency, the engine is treated as a steady flow device, the mass flow of fuel and air are measured, and the enthalpy of the gases entering and leaving the engine are evaluated at ambient temperature. This is

$$\eta_{comb} = \frac{\dot{m}_f h_f + \dot{m}_{air}h_{air} - \dot{m}_{out}h_{out}}{\dot{m}_f\text{LHV}} \tag{7.3}$$

The overall engine efficiency is the product of the thermal efficiency and the combustion efficiency. As a rule of thumb, roughly one-third of the energy in the fuel provides the net power to the piston, one-third of the energy is lost in heat transfer to the piston and cylinder walls, and one-third of the energy is expelled as heat in the exhaust. Of the energy transferred to the piston, part goes to power the wheels, part goes to power auxiliaries such as the water pump, fuel pump, and air conditioning, and part is lost to friction in the driveline.

Knocking combustion is the primary limit to increasing the compression ratio. As the flame propagates outward from the spark plug, the last portion of the gas to burn (the end-gas), which experiences the most preheat, may undergo rapid reaction, thus causing autoignition of the entire remaining unburned gas volume. Rapid autoignition causes pressure waves to develop that produce noise and cause the engine to vibrate, and heat loss to increase. Adjustments to the engine can be made to avoid knock, such as retarding the spark (setting the spark to ignite the fuel-air mixture later, e.g., closer to TDC), but this tends to increase emissions. A more fundamental approach to reducing knock depends on understanding fuel chemistry and flame propagation in the engine. During the last 15 years, understanding this phenomenon has made it possible to increase the compression ratio from 8 to 10 while still keeping the fuel-air mixture needed for the three-way catalyst in the exhaust to maintain low emissions of hydrocarbons, carbon monoxide, and nitrogen oxides. Before the advent of tight emission controls, spark engines achieved higher compression ratios without knocking by running lean. However, the three-way catalytic converter needed to meet today's emission controls requires essentially no free oxygen in the exhaust gas. As a result, stoichiometric combustion is needed in today's spark ignition engines.

Knock causes reduced efficiency, increased heat transfer, and if severe, engine damage. Engines with higher compression ratios typically require a higher octane gasoline to avoid knocking. The tendency to knock is accentuated by increasing the compression ratio, advancing the ignition timing (i.e., moving the spark timing to allow a longer combustion time before the piston reaches top dead center), opening the throttle to wide open, slightly rich mixtures, increasing the inlet air temperature, increasing the coolant temperature, and a buildup of deposits on the cylinder walls. Knock is discouraged by part throttle, lean or over-rich mixtures, increased engine speed, and decreased inlet pressure. Reducing the time for the end-gas reaction by the chamber shape design and spark plug location, and by locating the exhaust valve away from the end-gas, also discourages knock.

7.3 ONE-ZONE MODEL OF COMBUSTION IN A PISTON-CYLINDER

The one-zone model of combustion in a cylinder is a simplified model in which the mass within the closed cylinder is considered to be a single homogeneous mass that is pressurized, combusted, and heated uniformly. While the assumption of homogenous combustion is not strictly true, the one-zone model is useful for calculating the heat release rate and the average temperature in the cylinder versus crank angle.

Consider a piston-cylinder with the valves closed and containing a known mass, m, of homogeneous fuel and air mixture that is trapped when the intake valve is closed (Figure 7.2). The volume and pressure in the cylinder vary with time and are determined from measurements. The energy equation for the trapped mass is

$$m\frac{du}{dt}=q_{chem}-q_{loss}-p\frac{dV}{dt} \tag{7.4}$$

where q_{chem} is the rate of heat release from combustion, q_{loss} is the rate of heat loss to the piston head, cylinder walls, and valves, and the $p(dV/dt)$ is that rate of work transferred to the piston. Heat loss to the walls can be estimated from correlations based on measurements. Assuming constant molecular weight and an ideal gas mixture

$$pV = mRT \tag{7.5}$$

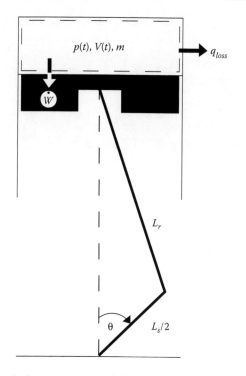

$p(t), V(t), m$

q_{loss}

$\dot{W}$

L_r

θ

$L_s/2$

FIGURE 7.2 Schematic for a one-zone model of combustion in a cylinder showing the connecting rod, the stroke, and the crank angle.

differentiating and rearranging

$$mR\frac{dT}{dt} = \frac{d(pV)}{dt}$$

substituting

$$du = c_v dT \qquad (7.6)$$

and using

$$R = c_p - c_v = c_v(\gamma - 1) \qquad (7.7)$$

it follows that

$$m\frac{du}{dt} = \frac{1}{\gamma - 1}\frac{d(pV)}{dt} \qquad (7.8)$$

Expanding the right hand side in Equation 7.8 and substituting it into Equation 7.4 yields

$$q_{chem} = \frac{\gamma}{\gamma-1} p \frac{dV}{dt} + \frac{1}{\gamma-1} V \frac{dp}{dt} + q_{loss} \tag{7.9}$$

The ratio of specific heats is estimated for the gaseous mixture and typically may be taken as 1.26 during combustion and 1.40 during intake.

The heat transfer is modeled using a convective heat transfer coefficient and an effective area,

$$q_{loss} = \tilde{h} A (T - T_w) \tag{7.10}$$

where a correlation such as that by Woschni (1967) is used.

$$\tilde{h} = 3.26 L_b^{-0.2} p^{0.8} T^{-0.55} \left(2.28 \overline{V}_p + 0.00324 \frac{V_d T_1}{p_1 V_1} (p - p_{motored}) \right)^{0.8} \tag{7.11}$$

where L_b is the cylinder bore and $\overline{V}_p$ is the mean piston speed. T_1, V_1 and p_1 are reference values at intake valve closing, and V_d is the displacement volume. The units of $\tilde{h}$ are $kW/m^2 \cdot K$. The exposed surface area at any particular crank angle is

$$A_{wall} = A_{piston\ head} + A_{cyl\ head} + \frac{\pi L_b L_s}{2} \left(r_c + 1 - \cos\theta - (r_c^2 - \sin^2\theta)^{1/2} \right) \tag{7.12}$$

where r_c is the ratio of the connecting rod length to the crank radius. An example of engine wall heat loss measurements is shown in Figure 7.3.

The volume in Equation 7.9 is determined from knowing the cylinder bore, rod and crank dimensions and crank angle, which is related to time from the measured revolutions per minute (see the summary of the engine equations at the end of this chapter). The pressure, which is essentially uniform throughout the cylinder volume, is measured as a function of crank angle and is filtered for noise and averaged over 100 or more cycles to determine an accurate time derivative. Equation 7.9 is solved by finite difference using the relation between crank angle and time,

$$\Delta\theta = \frac{360° }{rev} \cdot \frac{(rpm)\ rev}{min} \cdot \frac{min}{60\ s} \cdot \frac{\Delta t\ s}{1} = 6(rpm)\Delta t \tag{7.13}$$

The fuel burning rate is determined from

$$\frac{dm_f}{dt} = \frac{q_{chem}}{\eta_{comb} LHV} \tag{7.14}$$

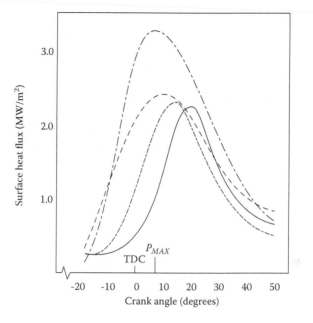

FIGURE 7.3 Heat flux measurements at four locations on a cylinder head; the engine speed is 2000 rpm and the air-fuel ratio is 18. (From Alkidas, A. C. and Myers, J. P., "Transient Heat-Flux Measurements in the Combustion Chamber of a Spark-Ignition Engine," *J. Heat Transfer* 104(1):62–67, 1982, with permission of ASME.)

The mass fraction of fuel burned is calculated from the sum of the heat release at that crank angle divided by the total heat release between intake valve closing (IVC) and exhaust valve opening (EVO).

Application of the one-zone model to a set of measurements using a single cylinder engine is shown in Figure 7.4. The pressure is measured as a function of crank angle, and the volume is calculated from the crank angle. The temperature (from Equation 7.5) and burning rate (from Equations 7.9 and 7.14) were calculated and plotted versus crank angle for the specific set of engine conditions given in Table 7.1. The engine performance parameters are given in Table 7.2. For the higher compression ratio case, the burn rate is faster, peak pressure is higher, and the thermal efficiency is slightly higher. The peak temperature is similar for the three cases. The net engine efficiency for a compression ratio of 10 is 27%. The terminology "indicated" is used in Table 7.2 because measurements of the cylinder pressure and crank angle (volume) were used to determine the power rather than dynamometer measurements.

Determining the temperature and burning rate from the one-zone model is useful for evaluating the effects of poor mixing, non-optimal spark and valve timing, and the behavior of different fuels. In practice the trapped mass may contain 5–10% residual combustion products from the previous cycle and correction for this can be made. An additional small correction can also be made for heat and mass loss due to the crevice volume between the piston and cylinder, which is normally 1–2% of the

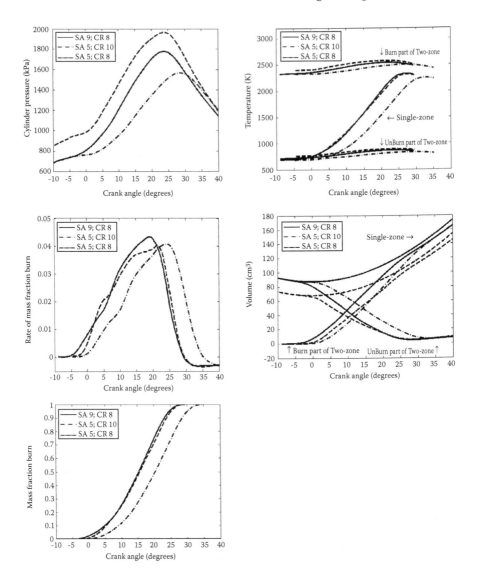

FIGURE 7.4 One-zone model calculations of temperature, fuel mass fraction burned, and fuel burning rate from cylinder pressure and volume measurements for three spark advance and compression ratio cases. Two-zone model calculations of unburned and burned volumes, and unburned and burned temperatures are included. The engine parameters are given in Tables 7.1 and 7.2. (Courtesy of Naber and Yeliana, Michigan Technological University, Houghton, MI.)

clearance volume. In some engines exhaust gas recirculation (EGR) is deliberately added to the inlet manifold to lower the temperature and nitric oxide formation, and this can also be taken into account.

While the one-zone model is useful for engineering purposes, it does have limitations, especially because the temperature at any given time is a cylinder average

TABLE 7.1

Engine Test Parameters for Table 7.2 and Figure 7.4

Bore (mm)	82.55
Stroke (mm)	114.3
Connecting rod length (mm)	254
Number of cylinders	1
Displacement volume (cm^3)	611.7
Compression ratio	8, 8, 10
Spark advance (CA°)	5, 9, 5
Engine speed (rpm)	900
Intake pressure (kPa)	50
Intake valve open (IVO) timing (CA°)	−350
Intake valve closed (IVC) timing (CA°)	−146
Exhaust valve open (EVO) timing (CA°)	140
exhaust valve closed (EVC) timing (CA°)	375
Air-fuel ratio	14.54
Internal EGR (%)	12.9, 13.0, 10.8
External EGR (%)	0
Ethanol content (%)	0
Gasoline LHV (MJ/kg)	43.46
Fuel flow rate (mg/cycle)	300

Source: Courtesy of Naber and Yeliana, Michigan Technological University, Houghton, MI.

Note: TDC is at 0 CA°

temperature. Because the flame propagates outward from the spark plug as the combustion progresses, the trapped mass in the cylinder consists of a burned zone and an unburned zone. Although the pressures are the same in both zones, the temperatures are quite different from each other in the burned and unburned zones. This is important in modeling nitric oxide formation in the cylinder and thus provides strong motivation for us to examine the two-zone model that will be described in the next section. As the reader may anticipate, given today's computational resources, a many zone, reacting, computational fluid dynamic (CFD) model of the combustion is even more useful for detailed design, optimization, and research. However, a discussion of CFD for the reacting flows within a cylinder is beyond the scope of this book. Readers interested in pursuing this topic further may wish to consult the book by Poinset and Veynante (2005).

7.4 TWO-ZONE MODEL OF COMBUSTION IN A PISTON-CYLINDER

In the two-zone model of engine combustion, the premixed-charge in the cylinder is treated as being composed of two zones: an unburned zone (u) and a burned zone (b). The unburned zone and the burned zone are separated by a very thin turbulent

TABLE 7.2

Engine Test Data from Pressure Versus Crank Angle Measurements

Performance Parameter	CR = 8 SA = 5	CR = 8 SA = 9	CR = 10 SA = 5
Peak pressure (kPa)	1568	1777	1966
Peak pressure location (CA°)	28.5	23.6	23.6
Gross IMEP (kPa)	371	374	378
Gross indicated power (W)	1702	1715	1735
Gross indicated torque (J)	18.1	18.2	18.4
Gross indicated work (J)	227	229	231
Gross indicated efficiency (%)	28	29	30
Net IMEP (kPa)	333	335	336
Net indicated power (W)	1530	1538	1539
Net indicated torque (J)	16.2	16.3	16.3
Net indicated work (J)	204	205	205
Net indicated efficiency (%)	25	26	27
10% of burn time (CA°)	13.1	13.0	9.6
50% of burn time (CA°)	24.6	23.8	20.1
90% of burn time (CA°)	33.2	31.9	28.6

Source: Courtesy of Naber and Yeliana, Michigan Technological University, Houghton, MI.

Note: CR (compression ratio), SA (spark advance in the negative direction from TDC). Gross values exclude the exhaust-intake stroke (i.e., the pumping loop). Net values include the exhaust-intake stroke. Duration values are determined from a two-zone model.

flame (Figure 7.5). The equation set for the two-zone model consists of conservation of mass and energy for the unburned and burned zones, mass and volume constraints, and the ideal gas equations of state for the unburned and burned zones. Because the flame is thin, the mass and enthalpy transport from the unburned zone to the burned zone is assumed to be equal to the enthalpy transport into the burned zone. The cylinder pressure, $p(t)$, is uniform throughout and the mass is constant. The heat transfer rate out of each zone to the piston-cylinder walls is denoted by q_u and q_b and is modeled from measurements. The governing equations for the two-zone model are

Conservation of mass

$$\frac{dm_u}{dt} = -\dot{m}_u \tag{7.15}$$

$$\frac{dm_b}{dt} = \dot{m}_u \tag{7.16}$$

Conservation of energy

$$\frac{d(m_u u_u)}{dt} = -h_u \dot{m}_u - p\frac{dV_u}{dt} - q_u \tag{7.17}$$

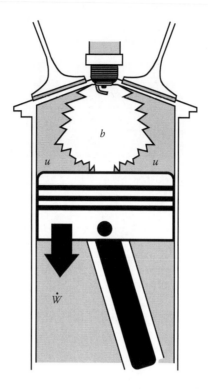

FIGURE 7.5 Schematic for two-zone model of combustion in cylinder showing burned (b) and unburned (u) zones

$$\frac{d\left(m_b u_b\right)}{dt} = h_b \dot{m}_b - p \frac{dV_b}{dt} - q_b \tag{7.18}$$

Mass constraint

$$m_u + m_b = m \tag{7.19}$$

Volume constraint

$$V_u + V_b = V \tag{7.20}$$

Equations of state

$$p = \frac{m_u R_u T_u}{V_u} \tag{7.21}$$

$$p = \frac{m_b R_b T_b}{V_b} \tag{7.22}$$

First substitute Equations 7.15 and 7.16 into Equations 7.17 and 7.18, expand the left hand sides, and use the specific heat. The energy equations become

$$\left(h_u - u_u\right)\frac{dm_u}{dt} = m_u c_{v,u}\frac{dT_u}{dt} + p\left(\frac{dV_u}{dt}\right) + q_u \tag{7.23}$$

$$\left(h_b - u_b\right)\frac{dm_b}{dt} = m_b c_{v,b}\frac{dT_b}{dt} + p\left(\frac{dV_b}{dt}\right) + q_b \tag{7.24}$$

In the six equations to be solved (Equations 7.19–7.24), the unknowns are m_u, m_b, T_u, T_b, V_u, and V_b; $V(t)$ and $p(t)$ are known from measurement; R_u, R_b, and m are known from the fuel specification and the fuel-air ratio; q_u and q_b are modeled from heat transfer correlations; and u, h, and c_v are known and are functions of T_u and T_b. The heat transfer is the sum of the heat transfer from each zone:

$$q = q_u + q_b \tag{7.25}$$

The heat transfer from each of the zones is modeled using a convective film coefficient

$$q_u = \tilde{h}A_u\left(T_u - T_{wall}\right) \tag{7.26}$$

$$q_b = \tilde{h}A_b\left(T_b - T_{wall}\right) \tag{7.27}$$

To determine the cylinder areas A_u and A_b in applying Equations 7.26 and 7.27, it is necessary to model the shape of V_u and V_b, for example by assuming a spherical expansion outward from the spark.

Initial conditions are set at the time of the spark. A small but finite volume of products at the adiabatic flame temperature is taken as the initial burned zone. The six coupled equations are solved using finite differences.

In Figure 7.4, the temperature and volume plots show the results of a two-zone model solution for the same engine data cited above. The combustion products temperature, T_b, peaks at about 2500 K, while the unburned mixture temperature is less than 1000 K. (The temperature computed by the one-zone model is the mass average of the burned and unburned temperatures.) Knowing the burned temperature and volume is useful in understanding nitric oxide (NO) formation. Figure 7.6 shows a calculation using the extended Zeldovich NO kinetic mechanism discussed in Section 4.4. Mass that combusts during early combustion forms more NO because the temperature is higher for a longer time than for mass that burns near the end of combustion. In both cases the rate of reaction is too slow to follow shifting equilibrium.

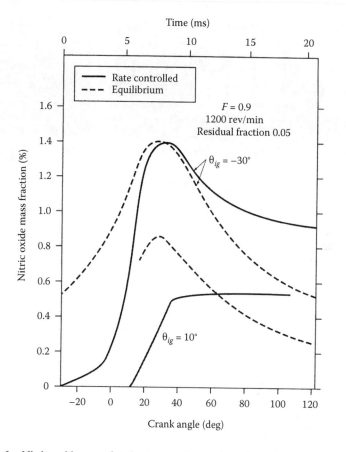

FIGURE 7.6 Nitric oxide mass fraction versus time and crank angle in burned gas for the first part of charge to burn (θ_{ig} = −30 CA°) and when one-third of the charge is burned (θ_{ig} = 10 CA°). (From Lavoie, G. A., Heywood, J. B. and Keck, J. C, "Experimental and Theoretical Study of Nitric Oxide Formation in Internal Combustion Engines," *Combust. Sci. Technol.* 1(4):313–326, 1970. © 1970, with permission of Taylor & Francis, Inc.)

7.5   IN-CYLINDER FLAME STRUCTURE

The flame front in the cylinder is not smooth but rather is a wrinkled flame. The brush-like nature of the flame surface is due to persistent turbulence formed by flow through the inlet manifold and valves. Under magnification the chemical reaction zone (flame front) appears instantaneously as a thin laminar flame. Macroscopically, the flame propagates outward from the spark plug as a brush-like, wrinkled flame sheet. For the first 10% of the burn near the spark, the flame speed is independent of the engine rpm and is much slower. Then the flame propagates at a near con-stant speed of up to 10 m/s depending on the rpm before slowing down near the walls. Laser photographs of in-cylinder flame propagation through quartz windows (Figure 7.7) show the highly wrinkled nature of the reaction sheet. The turbulent

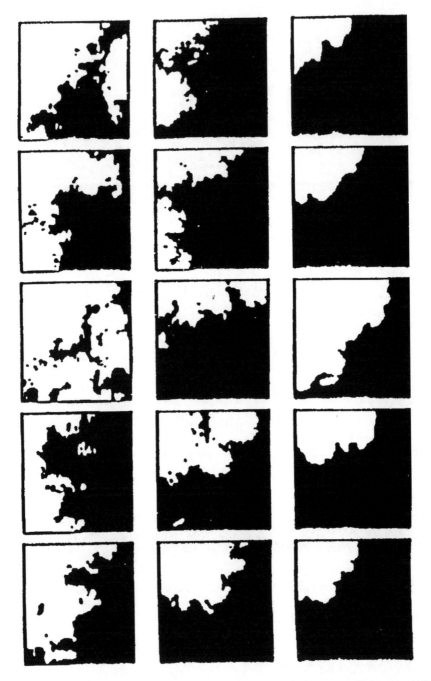

FIGURE 7.7 In-cylinder flame images at three engine speeds: top row at 300 rpm, middle row at 1200 rpm, bottom row at 2400 rpm. White is burned zone and black is unburned zone. (From Bracco, F. V., "Structure of Flames in Premixed-Charge IC Engines," *Combust. Sci. Technol.* 58:209–230, 1988. © 1988, with permission of Taylor & Francis, Inc.)

Reynolds number in the cylinder varies from 10 to 6000 over the full range of engine speeds (500–5000 rpm) from idle to full load. The Damköhler number, which is the ratio of the time to traverse a turbulent eddy to the chemical reaction time, varies from 1 to 1500. The flame is highly wrinkled, and for some conditions parcels of unburned gas occur within the reaction sheet.

The reaction sheet area is much larger than the flame envelope area (e.g., the ratio of the reaction sheet to the flame envelope area is typically 10 at 2000 rpm), and the propagation rate is much faster than laminar flame speed. Interestingly, the flame propagation speed is proportional to the engine rpm, which is proportional to the turbulent intensity in the cylinder. It is convenient to define a flame speed ratio (FSR) as the actual flame speed divided by the laminar flame speed relative to the unburned gas,

$$\text{FSR} = \frac{V_{flame}}{V_L} \tag{7.28}$$

The FSR varies from 3 at a low rpm to 35 or more at a high engine rpm. The FSR is related to the turbulent intensity by

$$\text{FSR} = \sqrt{1 + \left(\frac{V'}{V_L}\right)^2} \tag{7.29}$$

Of course, if the actual flame speed did not increase with rpm, it would be impossible to have low emissions at a high engine speed. As the rpm increases and the flow rate through the manifold and valves increases, turbulence in the cylinder becomes more intense. Turbulence is due to large-scale swirl and tumble induced in the curved inlet manifold and the smaller scale turbulence generated as the charge flows through the valves.

7.6 COMBUSTION CHAMBER DESIGN

Modern combustion chambers have tended toward a more compact design, which provides for a larger flame area and a faster burn rate. A centrally located spark plug (Figure 7.1) provides the largest flame area and keeps the hot products away from the walls for as long as possible so that heat transfer is reduced. This is important because a 10% reduction in heat transfer gives about a 3% increase in mean effective pressure and thus power. A strong tumble flow is also often used not only to produce more turbulence and higher flame speed, but also to reduce cycle-to-cycle variations. The reduction of cyclic variability allows the spark timing to be optimum over a greater fraction of the cycles. Fast burn designs are also more tolerant to higher values of exhaust gas recirculation (EGR), which facilitates a higher compression ratio without knock. Obtaining swirl without reduction in volumetric efficiency at wide-open throttle is important for maximum power. Most designs use four valves per cylinder to allow better breathing.

Avoiding engine knock is an important consideration in chamber design. The end-gas (last part of the gas to burn) should be in a cool part of the chamber (away from the exhaust valve) and should undergo as little time as possible in the compressed condition. Fast burn geometry with a central spark location combined with swirl and tumble allows for higher compression ratios before knock occurs.

Today's engines have been greatly improved by port fuel injection, electronic controls, high energy ignition systems, fast burn chamber designs, and the use of three-way catalysts with closed loop control to maintain a stoichiometric mixture. Further engine improvements may be possible by reducing throttling losses, decreasing heat transfer losses, and improving breathing to produce desirable turbulence in the cylinder with less cycle-to-cycle variations or increases in heat transfer and to provide for variable inlet and exhaust valve timing. Other engine concepts are being investigated, such as having a stratified rather than a homogeneous charge in the cylinder, and compression ignition rather than a spark plug, and the use of alternative fuels.

7.7   EMISSION CONTROLS

In the early 1960s in Los Angeles, automobile emissions were identified as the main contributor to photochemical smog. Photochemical smog is the brown haze present in major cities around the world when the sun is shining. In the presence of sunlight, hydrocarbons, nitric oxide and nitrogen dioxide in the urban atmosphere react photochemically to produce ozone, more nitrogen dioxide, aldehydes, and aerosol particulates. Smog is an eye irritant and respiratory hazard and is harmful to some plants.

Starting in 1968, automobiles driven in the United States were required to meet emissions standards set by Congress and administered by the Environmental Protection Agency. The first step was the installation of a positive crankcase ventilation valve to prevent venting of hydrocarbon fumes from lubricating oil to the atmosphere. Over the last 40 years the emission standards for hydrocarbons (HC), carbon monoxide (CO), and oxides of nitrogen (NO_x) have been progressively tightened to protect human health and the environment as knowledge of the effects of these emissions has become better understood (Table 7.3).

Emission standards for light duty vehicles (passenger cars, vans, sport utility vehicles and light trucks) are expressed in terms of grams of a specified pollutant emitted per unit distance driven (g/mi or g/km) during specified urban and highway driving cycles. The U.S. urban driving test cycle lasts 31 minutes at vehicle speeds up to 56 mph and includes 23 stops; the highway test cycle lasts 12.5 minutes and includes higher speeds and no stops. The entire vehicle is tested on a dynamometer in the laboratory, and emissions are collected in a bag. The emission standards apply equally to small and large vehicles rather than using a standard of grams of pollutant per gram of fuel.

Before the advent of catalytic converters in the exhaust manifold, engines were run slightly lean to reduce HC and CO emissions. In addition, vehicle manufacturers

TABLE 7.3

History of U.S. Emission Standards for Passenger Vehicles and Light Duty Trucks (g/mi)

Model Year	CO	NO$_X$	NMHC[a]	Comment
1960	(84)	(4.1)	(10.6)	Typical before standards
1970[b]	34	–	4.1	U.S. Clean Air Act begins
1972[b]	28.0	3.1	3.0	
1975	15.0	3.1	1.5	Oxidizing catalyst used
1978	15.0	2.0	1.5	
1980	7.0	2.0	0.41	
1981	3.4	1.0	0.41	Three-way catalyst used
1997	3.4	0.4	0.25	
2008[c]	2.1	0.05	0.075	Vehicle fleet average from 7 categories called bins

[a] Non-methane hydrocarbons.
[b] Adjusted to account for changes in federal engine test procedures that began in 1975.
[c] Applies only to regions of the United States where federal ambient air quality standards are exceeded; also includes particulate emission standard of 0.01 g/mi and formaldehyde emission standard of 0.015 g/mi in these regions; requires sulfur in gasoline <30 ppm.

focused their efforts on better fuel-air mixture preparation, better mixture control during starting and idling to reduce HC emissions, and on reducing crevices in the cylinders. Fuel trapped in crevices around the pistons, spark plugs, valves, and quench layers near cylinder walls is incompletely burned and contributes to HC and CO emissions. However, in spite of these efforts, it has not been possible to reduce CO, HC, and NO$_X$ sufficiently without aftertreatment of the exhaust. Lean combustion was insufficient to meet the more stringent emission standards, and furthermore, lean combustion tends to increase NO$_X$ emissions. Very lean burn reduces NO$_X$, but the flame speed becomes too slow. Additionally, advancing the spark timing, which results in a longer burn time, excessively decreases engine efficiency.

In 1975, a two-way catalytic converter was used on some spark engine vehicles to reduce CO and HC. To meet the emission controls, a three-way catalyst that oxidizes CO and HC while simultaneously reducing NO$_X$ was introduced in 1981. The three-way catalyst requires that the fuel-air ratio be maintained close to stoichiometric over all engine speeds and load conditions. A stoichiometric mixture is maintained by replacing the carburetor with a fuel injector at the air intake manifold and an oxygen sensor in the exhaust manifold with feedback control to the fuel injectors.

With the advent of catalytic converters, it became necessary to use unleaded gasoline because the catalyst was rendered ineffective by lead in the gasoline. Because no equally cheap, effective, and environmentally acceptable substitute was found for lead, the octane number of regular unleaded gasoline decreased. With the removal of the lead additive starting in 1975 compression ratios dropped from 9–9.5 to 8–8.5 to avoid knock. The reduction in the compression ratio plus timing adjustments caused an estimated 15% decrease in engine efficiency. Today, with better engine design and tuning, compression ratios are approaching 10.

Figure 7.8 provides an example of how gasoline is used in a typical warmed-up spark ignition automobile engine. The right side path shows the mechanisms by which 9% of the fuel escapes normal combustion. Fuel can follow this path either directly as fuel or as fuel-air mixtures. Approximately one-third of this 9% escapes directly as fuel and two-thirds escapes as fuel-air mixtures. Approximately 1.8%

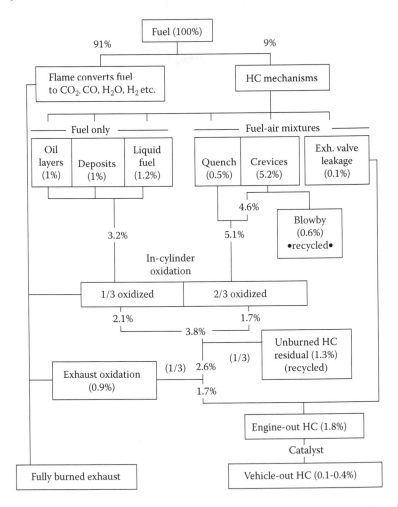

FIGURE 7.8 Flowchart for gasoline fuel that enters each cylinder: the normal combustion process is on the left side, and unburned or partially burned HC mechanisms are on the right side. Numbers in parenthesis denote % of gasoline entering the cylinder for each step in the total process. (Cheng, W. K., Hamrin, D., Heywood, J. B., Hochgreb, S., Min, K. and Norris, M., "An Overview of Hydrocarbon Emissions Mechanisms in Spark-Ignition Engines," SAE paper 932708, 1993, reprinted with permission from SAE paper 932708. © 1993, SAE International.)

of the fuel escapes from the engine and enters the catalytic converter. The catalyst reduces the vehicle-out HC emissions to 0.1–0.4% of the fuel used by the engine. It is estimated that the net fuel economy loss is about 6% due to the 9% of fuel that does not burn in the normal combustion process. For cold starting and the 15 minutes or so of driving before warmed-up conditions are reached, the amount of unburned fuel is higher, contributing to reduced fuel economy and higher HC emissions until the catalyst reaches its warmed up state. As noted in Table 7.3, the latest emission standards require a further tightening of HC, CO, NO_x in regions of the country where the ambient air ozone standards are exceeded in an effort to reduce photochemical smog. In the U.S., vehicles operating in these non-attainment areas are required to use reformulated gasoline.

7.8 ETHANOL CONSIDERATIONS

The Energy Independence and Security Act of 2007 set the goal of producing 36 billion gallons (136 billion liters) of ethanol by 2022. In 2007 in the U.S. 6.5 billion gallons of ethanol were produced from 3.0 billion bushels of corn. In the future, cellulose rather than corn sugars most likely will be used to produce ethanol. The significant differences in fuel properties between ethanol and gasoline are:

- On a volumetric basis the lower heating value of ethanol is 66% that of gasoline. As a result the miles per gallon is lower for ethanol.
- The combustion efficiency is slightly higher for ethanol blends because of lower combustion variability and slightly higher initial burning rate.
- The octane number of ethanol is 100 (average of research and motor octane numbers) compared to 86–93 for gasoline, and thus the compression ratio can be slightly increased when using ethanol.
- Ethanol has a stoichiometric air-fuel ratio of 9.0 versus 14.7 for gasoline.
- The oxygen content of ethanol is 34.7% by weight compared to about 2.7% for gasoline.
- Ethanol has better dilution tolerance compared to gasoline and thus more EGR can be used.
- The boiling point of ethanol is 78°C compared to 30°C–225°C for gasoline. As a consequence, gasoline vaporizes more readily at the lower temperatures experienced during cold start. By contrast, pure ethanol does not perform well during a cold start. However, E85 does perform well in a cold start.

Engine tests comparing gasoline and E85 run at a stoichiometric air-fuel ratio and constant indicated mean effective pressure (IMEP), valve and spark timings show little difference between the two fuels. The peak pressure occurs at the same crank angle, and the 90% burn crank angle is about 7% faster for ethanol. Ethanol has low HC emissions, although the HC emissions do contain acetaldehyde, which is photochemically active in the atmosphere. Emissions of nitrogen oxides from ethanol are not significantly different from gasoline.

7.9 REVIEW OF TERMINOLOGY FOR PREMIXED GAS, FOUR-STROKE ENGINES

A. Geometry

CR	Compression ratio	Maximum cylinder volume/minimum cylinder volume.

$$\mathrm{CR} = \left(V_d + V_c\right)/V_c$$

L_b	Cylinder bore	Inside diameter of a cylinder liner.
L_r	Connecting rod length	Connecting rod connects the crank of length $L_s/2$ to the piston.
L_s	Stroke length	$0 \le L_{s,\theta} \le L_s$

$$L_{s,\theta} = \frac{L_s}{2}\cos\theta + \left[L_r^2 - \left(\frac{L_s}{2}\right)^2 \sin^2\theta \right]^{1/2}$$

r_c	Ratio of the connecting rod length to the crank radius	$r_c = 2L_r/L_s$ Typically r_c is 3–4
rpm	Engine speed	Revolutions per minute
V_θ	Cylinder volume at θ	$V_\theta = V_{d,\theta} + V_c = x_c V_d + V_c$ $V_\theta = \left[x_\theta\left(CR - 1\right) + 1 \right]V_c$
V_c	Clearance volume	Minimum cylinder volume, which occurs at TDC.
V_d	Displacement volume	Swept volume as a piston moves over its stroke, L_s from TDC to BDC.

$$V_d = \frac{\pi L_b^2 L_s}{4}$$
$$V_{d,\theta} = x_c V_d$$
$$V_{d,\theta} = \left(\frac{L_s}{2} + L_r - L_{s,\theta}\right)\left(\frac{\pi}{4}\right)L_b^2 + V_c$$

V_r	Crevice volume	Volumes that fill with mixture but are inaccessible to normal combustion, includes the space between the piston head and top piston ring.
$\underline{V}_P$	Piston speed	$\underline{V}_p = \dfrac{dL_{s,\theta}}{d\theta}\left(\dfrac{d\theta}{dt}\right)$

$$= \bar{\underline{V}}_p \frac{\pi\sin\theta}{2}\left[1 + \frac{\cos\theta}{\left(4\left(L_r/L_s\right)^2 - \sin^2\theta\right)^{1/2}} \right]$$

$\bar{\underline{V}}_p$	Mean piston speed	$\bar{\underline{V}}_p = 2L_s\omega_e$
ω_e	Engine rotational speed	

| x_θ | Fraction of the stroke | $x_\theta = \dfrac{1}{2}\left[1 - \cos\theta + r_c - r_c\sqrt{1 - \left(\dfrac{\sin\theta}{r_c}\right)^2}\,\right]$ |

B. Timing

θ	Crank angle	θ is expressed in CA° before top dead center (BTDC) and after top dead center (ATDC) with $\theta = 0$ CA° at TDC of the compression stroke. $$\Delta\theta = \frac{360(\text{rpm})\Delta t}{60}$$
θ_s	Spark timing	Crank angle when spark discharge begins. MBT timing is that which gives maximum brake torque for a given operating condition. Timing is "retarded" from MBT if it is moved closer to TDC and "advanced" if it is moved further back in the compression stroke away from TDC.
	Valve timing	Intake and exhaust valve openings and closing crank angles are abbreviated as IVO, IVC, EVO, EVC. The interval during which both intake and exhaust are open is called the "valve overlap period."

C. Gas Exchange

m_t	Trapped mass	The mass of gases in the cylinder at IVC.
η_v	Volumetric efficiency	Ratio expressed as a percentage of trapped mass to mass of intake air at inlet density, ρ_0, that would fill the cylinder at BDC. $$\eta_v = \frac{100 m_t}{\rho_0(V_c + V_d)}$$
X_r	Residual fraction	The mass fraction of products from the previous cycle retained in the trapped mass.
EGR	Exhaust gas recirculation	Some exhaust products are recirculated with the fresh charge to reduce emissions.
	Blowback	Products that flow from the cylinder to the intake port near the start or end of the intake valve open period.
	Blowby	Gas that escapes from the cylinder, mainly due to leakage past the piston rings.
	Blowdown	The period of rapid exhaust between EVO and BDC of the expansion stroke.

D. Performance Parameters

Note: Engine performance is expressed by indicated values that are based on the cylinder gas as a system, and brake values, which are based on engine dynamometer measurements.

| W_i | Indicated work | By convention, the net work computed from the *pdV* integral from BDC of the compression stroke to BDC of the expansion stroke. |

W_p	Pumping work	By convention, the net work computed from the pdV integral from BDC of exhaust stroke to BDC of intake stroke.
W_{in}	Net indicated work	$W_{in} = W_i + W_p$ Net work done by the cylinder gas against the moving piston during two revolutions of the cycle.
IMEP	Indicated mean effective pressure	$\text{IMEP} = W_i/V_d$ for each cylinder; IMEP normalizes the work output by the engine size.
PMEP	Pumping mean effective pressure	$\text{PMEP} = W_P/V_d$
$\dot{W}_i$	Indicated power	$\dot{W}_i = n(\text{IMEP})\omega/2,$ where $\omega/2$ is the number of power strokes per unit time. For an n-cylinder engine the IMEP must be the average value.
ISFC	Indicated specific fuel consumption	$\text{ISFC} = \dot{m}_f/\dot{W}_i$
W_b	Brake work	Based on the measured engine shaft output and is less than the indicated work because of pumping, friction and accessory power.
BMEP	Brake mean effective pressure	$\text{BMEP} = \dfrac{W_b}{nV_d}$ where n = the number of cylinders that contribute to W_b.
$\dot{W}_b$	Brake power	$\dot{W}_b = 2\pi\omega\tau,$ where τ is the measured shaft torque exerted by the engine. Brake power is the rate at which the engine does work against a load.
BSFC	Brake specific fuel consumption	$\text{BSFC} = \dot{m}_f/\dot{W}_b$
η_c	Combustion efficiency	$\eta_c = \dfrac{\dot{m}_f h_f + \dot{m}_{air} h_{air} - \dot{m}_p h_p}{\dot{m}_f \text{LHV}}$ The engine is treated as a steady flow device with fuel and air entering at ambient T_0 and exhaust gases leaving engine at T_0. This definition accounts for incomplete conversion of products and for lean to stoichiometric mixtures being $0.95 - 0.98$.
η_t	Thermal efficiency	$\eta_t = \dfrac{\dot{W}_b}{\eta_c \dot{m}_f \text{LHV}}$

7.10 PROBLEMS

1. Calculate the dew point temperature for a stoichiometric mixture of octane and air at atmospheric pressure. Use the data in Appendix A.
2. Consider the impact of compression of the air-fuel mixture on turbulence. Assume that an idealized turbulent eddy is represented by a gas disc 3

mm in diameter, d, and 0.5 mm thick, l, at atmospheric pressure and temperature. The disk is spinning around its axis at 1000 rpm. What is the rotational speed of the disk after it is adiabatically compressed to 25 atm pressure? The gas disc is compressed equally in both the radial and axial direction; that is $d_1/d_2 = l_1/l_2$. Discuss your result in terms of turbulence in engines.

3. Assume that the temperature of a combustion chamber can be represented with a two-zone model. Zone 1 is the hot core of the gas and Zone 2 is the cool boundary layer next to the cylinder wall. For the following data, estimate the temperature of the core gas in a pancake-shaped combustion chamber with a 100 mm bore and 10 mm height, mass average gas temperature = 2500 K, pressure = 1800 kPa, average surface temperature = 500 K, average boundary layer thickness = 0.5 mm. Assume

$$T_2 = \left(T_{wall} + T_1\right)/2.$$

4. A fairly general plot of mass burned fraction (m_b/m) as a function of the volume (V_b/V) burned fraction is shown below. Plot the flame radius normalized by the cylinder radius (r_f/r_c) as a function of the mass fraction burned for a pancake-shaped chamber. Approximate the flame shape as a cylinder with the same axis and depth as the combustion chamber (this creates a simple 2D geometry) and assume that the flame starts in the center of the cylinder.

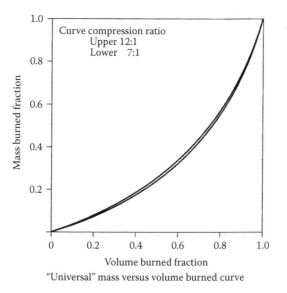

"Universal" mass versus volume burned curve

5. During combustion in spark ignition engines the ratio of the density of the cooler unburned fuel-air mixture to the density of the hot burned products (ρ_u/ρ_b) is about 4. Using this, show that

$$\frac{V_b}{V} = \frac{4m_b/m}{1+3m_b/m}$$

Compare mass burned fraction as function (m_b/m) of volume burned fraction (V_b/V) from this formula to the data shown in the figure given in Problem 4.

6. It has been suggested that the turbulence level in the end-gas could be increased by a piston with a raised rim at the radius corresponding to 80% volume burned as indicated in the sketch below. Discuss what effects this might have on flame propagation.

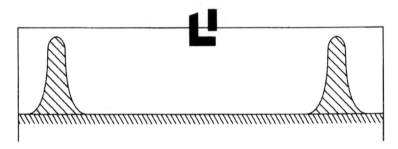

7. Consider a central spark location in a "double-hemi" chamber (see sketch). Would a long reach spark plug location (B) decrease the heat transfer significantly over a conventional location (A)?

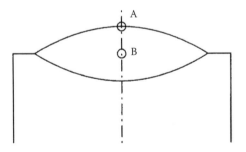

8. Return to Problem 4 and sketch to scale the flame and cylinder wall at the 80% mass burned point. Use a view similar to that of Problem 6 but without the raised rims. Assume the engine has a 125 mm bore and 125 mm stroke with an 8:1 compression ratio. The crank angle at 80% mass burned is 20 CA° after TDC. At this crank angle the volume is at 1.25 of its TDC value. Remember the wrinkled flame covers a thickness of about 5–6 mm.

9. Sketch to scale a wrinkled flame as a 0.1 mm thick sheet that is in the form of isosceles triangles. The wrinkled flame thickness is 6 mm and its flame speed ratio (FSR) is 20.

10. Assume you are given the detailed output from a two-zone model of combustion in a cylinder of a spark ignition engine. Explain in words how you would go about calculating the nitrogen oxide concentration versus crank

angle similar to that in Figure 7.6. Explain this in as much detail as you can (words only, no equations).

11. Discuss the reasons for the trends of knock promotion given in the text, giving physical arguments.

12. Show the effect of engine operating variables on ignition and flame propagation by filling in the blanks in the table using the symbols provided and discuss your answers in terms of the governing phenomena.

	Effect	
	Ignition	Flame Propagation
Ignition System Variables		
Increased spark energy		
Hotter spark		
Engine Variables		
Increased compression		
Increased mixture swirl		
Increased residual gas		
Increased flame propagation distance		

Symbols: helps ↑, hinders ↓, no effect –.

13. The following factors influence cycle-to-cycle variability and the best condition for minimizing the variability. In each case discuss briefly the mechanism you think is operative and if you think the factor is likely to be important.

Factor	Best Conditions for Minimizing Cycle-to-Cycle Variability
1. Fuel	Fuel that provides the best burning velocity
2. Fuel-air ratio	Slightly rich of stoichiometric
3. Spark plug	High breakdown energy; long duration, wide spark gap, thin or sharp pointed electrode, gap discharge direction perpendicular to the mean flow, which is 4 m/s.
4. Turbulence	Small scale at time of ignition, with high turbulent intensity over the whole cylinder volume.

14. Typically when the peak pressure of cycle $n + 1$ is plotted as a function of the peak pressure of cycle, n, there is no correlation; that is, there is no effect of past cycle pressures on future cycle pressure and the variations are random. However, for very lean fuel-air ratios at part load or idle conditions, it is observed that a very low peak pressure on cycle n is followed by a high peak pressure on cycle $n + 1$. This is called "prior-cycle effect." Discuss why the typical case should be expected to be random and what might cause the prior-cycle effect for lean fuel-air ratios.

15. The selection of the distillation curve for automotive gasoline is a compromise between various operating requirements. For the distillation curve below, match the following operating characteristics to the position on the curve: (a) poor hot starting, vapor lock, and high evaporative losses; (b) combustion deposits and oil dilution; (c) poor long-trip economy; (d) poor warm up, rough acceleration, and poor short trip economy; (e) increased icing; (f) poor cold starting.

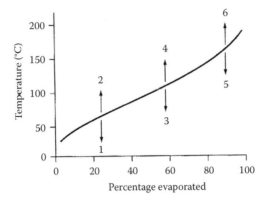

16. In view of the issues noted in Problem 15, discuss the problems that arise from a single component fuel such as ethanol when used as an alternative to gasoline.

REFERENCES

Abraham, J., Williams, F. A., and Bracco, F. V., "A Discussion of Turbulent Flame Structure in Premixed Charges," *SAE paper 850345*, 1985.

Alkidas, A. C. and Myers, J. P., "Transient Heat-Flux Measurements in the Combustion Chamber of a Spark-Ignition Engine," *J. Heat Transfer* 104(1):62–67, 1982.

Annand, W. J. D., "Heat Transfer in the Cylinders of Reciprocating Internal Combustion Engines," *Proc. Inst. Mech. Eng.* 177(36):973–996, 1963.

Arcoumanis, C., Hu, Z., Vafidis, C., and Whitelaw, J. H., "Tumbling Motions—A Mechanism of Turbulence Enhancement in Spark-Ignition Engines," *SAE paper 900060*, 1990.

Benson, R. S. and Whitehouse, N. D., *Internal Combustion Engines: A Detailed Introduction to the Thermodynamics of Spark and Compression Ignition Engines, Their Design and Development*, Pergamon Press, Oxford, 1979.

Bianco, Y., Cheng, W. K., and Heywood, J. B., "The Effects of Initial Flame Kernel Conditions on Flame Development in SI Engine," *SAE paper 912402*, 1991.

Borman, G. and Nishiwaki, K., "Internal-Combustion Engine Heat Transfer," *Prog. Energy Combust. Sci.* 13(1):1–46, 1987.

Bracco, F. V., "Structure of Flames in Premixed-Charge IC Engines," *Combust. Sci. Technol.* 58:209–230, 1988.

Catania, A. E., Misul, D., Mittica, A., and Spessa, E., "A Refined Two-Zone Heat Release Model for Combustion Analysis in SI Engines," *JSME Int. J. Ser. B* 46(1):75–85, 2003.

Cheng, W. K., Hamrin, D., Heywood, J. B., Hochgreb, S., Min, K. and Norris, M., "An Overview of Hydrocarbon Emissions Mechanisms in Spark-Ignition Engines," *SAE paper 932708*, 1993.

Chun, K. M. and Heywood, J. B., "Estimating Heat Release and Mass-of-Mixture Burned from Spark-Ignition Engine Pressure Data," *Combust. Sci. Technol.* 54:133–143, 1987.

Eriksson, L. E., "Requirements for and a Systematic Method for Identifying Heat-Release Model Parameters," SAE paper 980626, 1998.

Ferguson, C. R. and Kirkpatrick, A. T., *Internal Combustion Engines: Applied Thermosciences*, 2nd ed., Wiley, New York, 2001.

Foster, D. E., "An Overview of Zero-Dimensional Thermodynamic Models for IC Engine Data Analysis," SAE paper 852070, 1985.

Fox, J. W., Cheng, W. K., and Heywood, J. B., "A Model for Predicting Residual Gas Fraction in Spark-Ignition Engines," SAE paper 931025, 1993.

Groff, E. G. and Matekunas, F. A., "The Nature of Turbulent Flame Propagation in a Homogeneous Spark-Ignited Engine," SAE paper 800133, 1980.

Heywood, J. B., *Internal Combustion Engine Fundamentals*, McGraw-Hill, New York, 1988.

Heywood, J. B., Higgins, J. M., Watts, P. A., and Tabaczynski, R. J., "Development and Use of a Cycle Simulation to Predict SI Engine Efficiency and Emissions," SAE paper 790291, 1979.

Krieger, R. B. and Borman, G. L., "The Computation of Apparent Heat Release for Internal Combustion Engines," ASME 66-WA/DGP-4, 1966.

Kummer, J. T., "Catalysts for Automobile Emission Control," *Prog. Energy Combust. Sci.* 6(2):177–199, 1980.

Lavoie, G. A., Heywood, J. B., and Keck, J. C, "Experimental and Theoretical Study of Nitric Oxide Formation in Internal Combustion Engines," *Combust. Sci. Technol.* 1(4):313–326, 1970.

Mattavi, J. N., Groff, E. G., Lienesch, J. H., Matekunas, F. A., and Noyes, R. N., "Engine Improvements Through Combustion Modeling," in *Combustion Modeling in Reciprocating Engines,* J. N. Mattavi and C. A. Amann, eds., 537–587, Plenum Press, New York, 1980.

Muranaka, S., Takagi, Y., and Ishida, T., "Factors Limiting the Improvement in Thermal Efficiency of S.I. Engine at Higher Compression Ratio," SAE paper 870548, 1987.

Obert, E. F., *Internal Combustion Engines and Air Pollution*, 3rd ed., Intext Education Publishers, New York, 1973.

Poulos, S. G. and Heywood, J. B., "The Effect of Chamber Geometry on Spark-Ignition Engine Combustion," SAE paper 830334, 1983.

Poinsot, T. and Veynante, D., *Theoretical and Numerical Combustion*, 2nd ed., Edwards, Philadelphia, PA, 2005.

Reitz, R. D., "Assessment of Wall Heat Transfer Models for Premixed-Charge Engines Combustion Computations," SAE paper 910267, 1991.

Woschni, G., "Universally Applicable Equation for the Instantaneous Heat Transfer Coefficient in the Internal Combustion Engine," SAE paper 670931, 1967.

Yeliana, Y., Cooney, C., Worm, J., and Naber, J. D., "The Calculation of Mass Fraction Burn of Ethanol-Gasoline Blended Fuels Using Single and Two-Zone Models," SAE paper 2008-01-0320, 2008.

Zur Loye, A. O. and Bracco, F. V., "Two-Dimensional Visualization of Premixed-Charge Flame Structure in an IC Engine," SAE Paper 870454, 1987.

8 Detonation of Gaseous Mixtures

A detonation is a combustion wave that travels at supersonic speeds producing high temperatures and pressures for short periods of time. Detonation is an important combustion topic because given sufficient volume and time, a propagating flame (deflagration) can change into a detonation. Detonations often create dangerous situations, such as an explosion due to leakage of natural gas in a building, a coal mine explosion due to the presence of methane, or a hydrogen-oxygen explosion in an overheated nuclear reactor. The study of detonations can lead to ways to prevent such accidents.

Just as a laminar premixed flame has a unique flame speed, each fuel and air mixture has a unique detonation speed. However, the detonation speed is faster than the speed of sound of the quiescent mixture; consequently, the reaction front includes shock waves. Typically detonations propagate about 1000 times faster than a laminar flame in ambient air. Whereas the pressure across a flame is essentially constant, the pressure rises by 10–30 times across a gaseous detonation.

As will be discussed later in this chapter, attempts have been made to utilize the intense combustion of a detonation for energy applications, but practical energy uses of detonations have not yet been developed. Let us first consider the transition from a flame to a detonation and then consider the features of steady state detonations.

8.1 TRANSITION TO DETONATION

Consider a combustible mixture of gaseous fuel and air in a long tube. The mixture is ignited at the closed end of the tube. A flame forms and begins to propagate along the tube at laminar flame speed. The propagating flame gradually loses its smooth shape and becomes wrinkled. As a result of the increase in effective flame surface, the flame accelerates with respect to the unburned gas. The wrinkled, fluctuating flame front generates turbulence and weak pressure pulses that run ahead of the flame front and gradually preheat the gas ahead of the flame, causing the flame to speed up. High-speed schlieren photography (a light scattering technique sensitive to density gradients) of the transition from a flame to a detonation shows that as the flame accelerates, the pressure pulses become stronger, coalesce, and further preheat the gas ahead of the flame. Eventually, a pocket of gas ahead of the flame reaches its autoignition temperature and produces a local explosion. The rapidly expanding gases produce a shock wave that interacts with the walls, sending a forward propagating shock that rapidly ignites the fuel ahead and a backward moving shock that dies out. The forward moving shock-combustion complex is a detonation, and the rearward moving shock is called a retonation (Figure 8.1). Note that the velocity can be obtained from the slope of the various lines indicated in Figure 8.1.

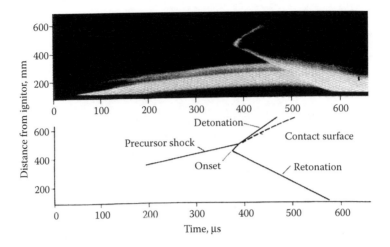

FIGURE 8.1 Streak self-light photograph and interpretation of the onset of detonation. The onset of detonation occurs at 375 microseconds after ignition.

A transition to detonation also occurs in large diameter tubes and in spherical geometries, such as focused laser ignition at the center of a spherical chamber or a hot jet of gas. The transition distance from ignition to detonation depends on the strength of the ignition source and the fuel mixture. Typically, a distance of 1–4 m and times of 2–200 ms are required for spark initiation of a detonation of gaseous fuels initially at standard temperature and pressure. A transition from a flame to a detonation can be accelerated by induced turbulence ahead of the flame, such as from rough or irregular walls or jets.

8.2 STEADY-STATE DETONATIONS

A detonation may be viewed macroscopically as a shock wave followed by combustion. The heat release is coupled to the shock and drives the shock. The propagation velocity depends on the heat release due to combustion. Representative detonation velocities, which are given in Table 8.1, are about 1000 times greater than the corresponding laminar flame speed. The detonation velocity is higher for oxygen-fuel mixtures than for air-fuel mixtures, as is the case with flames. Detonation causes a pressure jump of 10–30 times the initial pressure, which can be very destructive, although it is transient. The detonation velocity also depends on the fuel-air ratio and exhibits rich and lean mixture limits that are somewhat narrower than the flammability limits.

Detailed measurements of the structure of gaseous detonation waves show that the picture is more complicated than a normal shock wave followed by rapid combustion. Indeed, it was shown theoretically (after detailed experimental observations in the 1960s) that combustion behind a normal shock wave is unstable and warps the shock in a cellular manner. The shock front is composed of a three-dimensional grid of shock intersections that move transverse to the wave front as the front moves forward. This view helps to explain the rapid reactions that occur in a detonation because the local temperatures and pressures are higher than what the one-dimensional model would suggest. Experimentally, the three-dimensional, cellular

TABLE 8.1

Detonation Velocities for Various Premixed Gases Initially at 1 atm and 25°C

Mixture	Detonation Velocity (m/s)
Stoichiometric H_2–O_2	2840
Stoichiometric H_2–air	1970
1.12 Stoichiometric H_2–O_2	3390
0.37 Stoichiometric H_2–O_2	1760
Stoichiometric CH_4–O_2	2320
Stoichiometric CH_4–air	1800
1.5 Stoichiometric CH_4–O_2	2530
1.2 Stoichiometric CH_4–O_2	2470
Stoichiometric C_2H_2–air	1870
Stoichiometric C_2H_2–O_2	2430
Stoichiometric CO–O_2	1800
Stoichiometric C_3H_8–O_2	2350
Stoichiometric C_3H_8–air	1800

Source: Soloukhin, R. I., *Shock Waves and Detonations in Gases*, Mono Book, Baltimore, MD, 1966.

nature of the detonation front may be observed best at pressures below atmospheric pressure where the cells are larger. Mixtures near the limits of detonability have larger cells than stoichiometric mixtures, and at the limit take on a spinning motion normal to the direction of propagation.

High-speed, self-luminous photographs of the structure of an acetylene-oxygen detonation are shown in Figure 8.2. The luminous zones are caused by triple-shock intersections that sweep across the detonation front consuming the fuel. Imprints of a soot-blackened wall from a reflected detonation, as in Figure 8.3, can be used to investigate the cellular nature of the detonation wave. Specially designed pressure transducers have been used to measure the pressure spikes in a detonation wavefront, and pressures more than twice that predicted by normal shock theory have been observed.

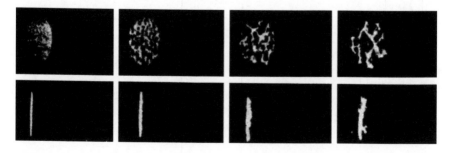

FIGURE 8.2 Self-luminous photographs of an acetylene-oxygen detonation wave. The top row is taken at right angles to the front, and the bottom row is at an oblique angle. (From Soloukhin, R. I., *Shock Waves and Detonations in Gases*, Mono Book, Baltimore, MD, 1966.)

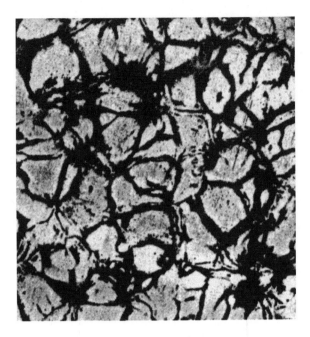

FIGURE 8.3 Imprints from a reflection of a detonation off a soot covered end wall. (From Shchelkin, K. I. and Troshin, Ya, K., *Gas Dynamics of Combustion* translated from "Gazodinamika Goreniya," Moscow Izd. Akad. Nauk SSSR, 1963, NASA Technical Translation NASA-TT-F-231, Washington, DC, 1964.)

The cellular nature of the detonation front is caused by the rapid heat release that warps the front, thus causing curved shocks, which interact by means of triple-shock interactions, as indicated in Figure 8.4. (It can be shown that double shock interactions cannot exist.) The triple-shock structure involves incident, transmitted, and reflected shock waves. The incident shock is oblique and directs the flow towards the point of intersection. The transmitted shock, called a Mach-stem shock, with which the incident shock intersects is nearly normal to the flow and hence is stronger. The

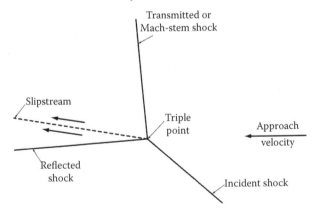

FIGURE 8.4 Schematic diagram of a triple shock configuration in a detonation front.

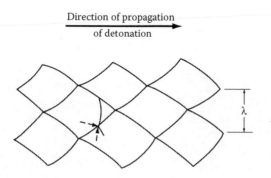

FIGURE 8.5   Soot track formed by the triple points of a detonation sweeping along a side wall. The characteristic cell size is λ. A triple shock intersection (arrows) etches the soot track on the foil.

reflected shock balances the pressure, and the slipstream includes regions of different temperatures but constant pressure. The triple points propagate along curved paths and periodically collide, leaving a pattern on smoked foil as indicated ideally in Figure 8.5. The incident and reflected shock waves extend above the trajectory of the triple point, and the transmitted shock and slipstream extend below. The incident shock is curved and extends to another triple point. The transmitted and reflected shocks are also curved and are intensified where the triple points collide and decay away from the intersections. Rayleigh light scattering photographs of the detonation structure are compared to schlieren photographs and numerical analysis in Figure 8.6. Rayleigh scattering uses a pulsed laser and measures the gas density, whereas schlieren images measure density gradients and are sensitive to shock waves and turbulence. The Rayleigh scattering images show the high density triple point regions behind the leading shock front.

Most of the heat release occurs near the triple points. The distance between triple points defines a detonation cell size (λ). The detonation cell size is a fundamental characteristic length of a detonation wave that can be used to correlate mixture limits as well as initiation energy and quenching behavior. Cell size data for various fuels and equivalence ratios at atmospheric pressure are shown in Figure 8.7. The solid lines are based on the theoretical correlation that the cell size (λ) equals a constant multiplied by the distance for the fuel to react according to a one-dimensional model calculation. The constant (A) is fitted at stoichiometric conditions and is different for each fuel (e.g., $A = 10.1$ for C_2H_2, $A = 52.2$ for H_2). Cell sizes for detonations in pure oxygen are smaller than in air, and diluents such as argon or carbon dioxide alter the cell size.

The fuel-air mixture limits to sustain a detonation can be estimated from Figure 8.7. Near the rich and lean limits the detonation cell size becomes so large that the shock waves decay, and eventually the detonation reverts to a flame. For hydrogen-air mixtures the rich and lean limits are approximately $F = 0.4$–3.5; for acetylene-air they are $F = 0.4$–3.0; and for ethane-air, propane-air, and butane-air they are $F = 0.7$–2.0. Detonation limits in small diameter tubes are narrower because of interaction with the walls. Experimental observations have shown that the tube diameter must be approximately 13 times greater than the characteristic cell size to sustain detonation.

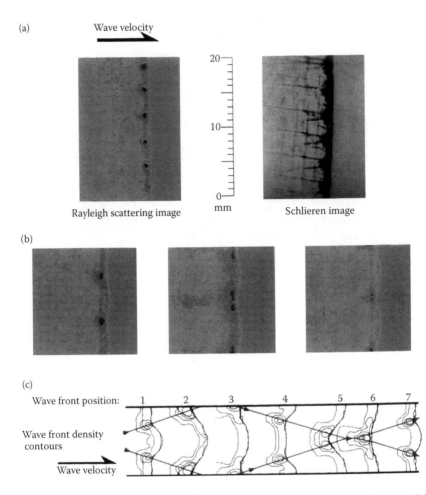

FIGURE 8.6 Structure of a detonation wave in a hydrogen-oxygen-argon mixture at 0.374 atm: (a) Rayleigh and schlieren images acquired simultaneously from the same wave front; (b) comparison of cell structures from Rayleigh images with (c) density contours from the numerical analysis of Kailasanath et al. The straight lines represent triple point trajectories. (From Anderson, T. J. and Dabora, E. K., "Measurements of Normal Detonation Wave Structure Using Rayleigh Imaging," *Symp. (Int.) Combust.* 24:1853–1860, The Combustion Institute, Pittsburgh, PA, 1992, with permission of The Combustion Institute.)

Direct initiation of detonation means that a strong blast wave such as from a solid explosive or a focused laser is used to start the process and that the energy decays asymptotically to a steady detonation wave of constant cell size. If the ignition energy is less than a certain critical value, the reaction zone progressively decouples from the blast wave as it decays, and a deflagration (flame) results. Experiments have shown that the ignition energy required for direct initiation of detonation is proportional to the cube of the cell size. Hence, cell size is the key indicator of detonability.

If the ignition energy is less than the critical value, it does not mean that detonation cannot occur. Rather, it means that indirect instead of direct initiation occurs.

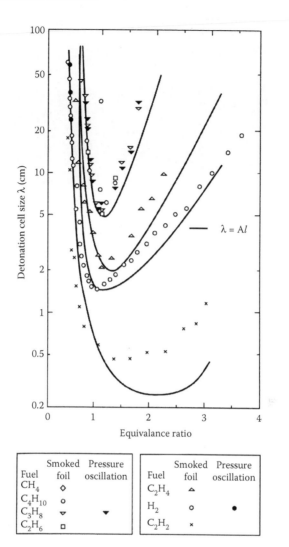

FIGURE 8.7 Detonation cell sizes of fuel-air mixtures at 1 atm. (From Lee, J. H. S., "Dynamic Parameters of Gaseous Detonations," *Annu. Rev. of Fluid Mech.* 16:311–336, 1984, with permission of Annual Reviews, Inc.)

Indirect initiation is the mechanism by which a flame accelerates to a detonation, as discussed in Section 8.1. This requires a much longer transition distance than direct initiation.

When a detonation wave in a tube of diameter d suddenly emerges from the tube into an unconfined volume containing the same mixture, quenching of the detonation will occur when

$$d < 13\lambda \tag{8.1}$$

where λ is the cell size. Quenching here means reversion to a deflagration (a flame). This correlation holds for a wide range of fuels and mixture ratios including flow through an orifice. For noncircular orifices,

$$d_{eff} < 13\lambda \tag{8.2}$$

where d_{eff} is the average of the smallest and largest openings. For example, for a square orifice with sides of length L, the effective diameter is

$$d_{eff} = 0.5\left(L + \sqrt{2}L\right) = 1.2L \tag{8.3}$$

Since quenching here means quenching of the detonation but not the flame, it is always possible that given sufficient distance the flame can again accelerate into a detonation.

8.3 ONE-DIMENSIONAL MODEL FOR PROPAGATION VELOCITY, PRESSURE, AND TEMPERATURE RISE ACROSS A DETONATION

Because detonations have a three-dimensional microstructure, a one-dimensional model of the structure of a detonation is not realistic for analysis of the chemical kinetics. Nevertheless, a one-dimensional gas dynamic model is reasonably accurate for predicting detonation velocity and post reaction pressure and temperature, and thus is useful for engineering purposes. By writing one-dimensional equations for conservation of mass, momentum, and energy across the detonation wave, expressions for the detonation velocity, average pressure rise, and average temperature rise across the detonation can be obtained.

Consider a plane detonation wave that travels into a premixed combustible gas that is at rest. The gas can be in a very large volume, or it can be in a tube. In either case, the reaction zone is thin; consequently, the area of the reaction zone exposed to the wall is negligible and losses to the walls may be neglected. The detonation wave sweeps over the reactants, converts them to products at elevated pressure and temperature, and sets the products into motion. At a fixed position in space (or in a tube), we have an unsteady problem. Hence, it is convenient to transform the velocities to a wave-fixed coordinate system by superimposing a velocity equal in magnitude and in the opposite direction as the detonation velocity, V_D, onto the gas in the room (or tube) as indicated in Figure 8.8. The transformation between the wave-fixed (standing wave) coordinate system and room-fixed (traveling wave) coordinate system is

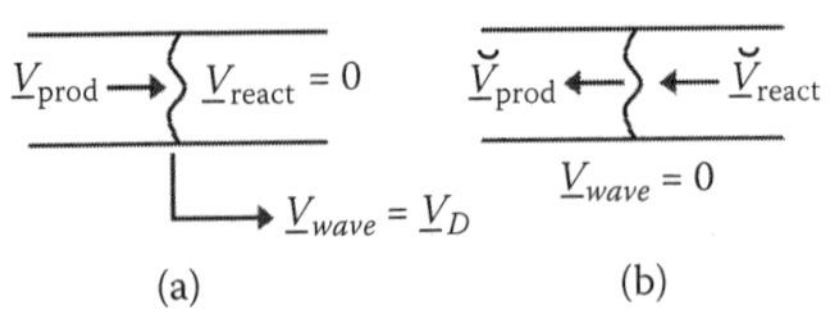

FIGURE 8.8 Coordinate transformation between a traveling wave (a) and a standing wave (b).

$$\breve{V}_{react} = \underline{V}_D \tag{8.4}$$

$$\breve{V}_{prod} = \underline{V}_D - \underline{V}_{prod} \tag{8.5}$$

where $\breve{V}$ denotes velocities in the standing wave coordinate system, and $\underline{V}$ indicates velocities in the traveling wave coordinate system. The conservation equations for mass, momentum, and energy across a detonation in the standing wave coordinate system are the same as Equations 5.3 through 5.5 for a laminar flame. As a side note, these conservation equations also describe the conditions across a normal shock wave. A flame is a rapid change in subsonic flow due to heat addition. A shock wave (in wave fixed coordinates) is an abrupt change from supersonic to subsonic flow without heat addition, and a detonation is an abrupt change from supersonic to subsonic flow with heat addition. The conservation equations across a detonation, shock, or flame formulated in wave fixed coordinates are

$$\rho_{react}\, \breve{V}_{react} = \rho_{prod}\, \breve{V}_{prod} \tag{8.6}$$

$$p_{react} + \rho_{react}\, \breve{V}^2_{react} = p_{prod} + \rho_{prod}\, \breve{V}^2_{prod} \tag{8.7}$$

$$h_{react} + \frac{\breve{V}^2_{react}}{2} = h_{prod} + \frac{\breve{V}^2_{prod}}{2} \tag{8.8}$$

The problem for a detonation is generally stated as "given p_{react}, T_{react}, ρ_{react}, and h_{react} (the absolute enthalpy per unit mass of reactants), find $\breve{V}_{react}$, $\breve{V}_{prod}$, p_{prod}, T_{prod}, and ρ_{prod}." Using the equation of state $p = \rho RT$ and the enthalpy tables, it is apparent that, as with the laminar flame, we are again short one equation. For a detonation this difficulty is overcome by asserting that the velocity of the products, $\breve{V}_{prod}$, is at sonic velocity (relative to the wave front and with respect to the hot gas). This is called the Chapman-Jouguet condition. It is a justifiable assumption in that it accurately predicts the observed detonation velocity, $\breve{V}_{react}$. Physically it is reasonable to view the heat addition as choking the flow; that is, driving the flow of the products to $\mathrm{Ma}_{prod} = 1$. With this assumption, the solution of the problem can now be completed.

Across a detonation the species composition changes, and it is assumed that the products are in chemical equilibrium. The speed of sound of the products, a_{prod}, is

$$a_{prod} = \sqrt{\left(\frac{\partial p_{prod}}{\partial \rho_{prod}} \right)_s} \tag{8.9}$$

where *s* refers to isentropic. In general the speed of sound is not equal to $\sqrt{\gamma_{prod}R_{prod}T_{prod}}$ for a reacting mixture. Similarly, in general the isentropic exponent for a reacting mixture is not equal to the ratio of specific heats. To include variable molecular weight and specific heats, Equations 8.6 through 8.8 can be rewritten in the Rankine-Hugoniot form for a reacting mixture (presented in Kuo 2005). Because these effects are relatively minor, it is instructive to proceed with the solution of a Chapman-Jouguet detonation by assuming constant molecular weights and constant ratio of specific heats.

As a starting place it is convenient to rewrite Equations 8.6 through 8.8 in terms of pressure, temperature, and Mach number. Then the Mach number is given by

$$\breve{Ma}^2 = \frac{\breve{V}^2}{\gamma RT} = \frac{\rho \breve{V}^2}{\gamma p} \tag{8.10}$$

Rearranging Equation 8.10 yields

$$\rho = \frac{\gamma p \left(\breve{Ma}^2 \right)}{\breve{V}^2}$$

and

$$\breve{V}^2 = \gamma RT \left(\breve{Ma}^2 \right)$$

Substituting into Equations 8.6 and 8.7 and simplifying, the conservation of mass and momentum equations become

$$\frac{p_{react}^2 \left(\breve{Ma}_{react}^2 \right)}{T_{react}} = \frac{p_{prod}^2 \left(\breve{Ma}_{prod}^2 \right)}{T_{prod}} \tag{8.11}$$

$$p_{react} \left(1 + \gamma \breve{Ma}_{react}^2 \right) = p_{prod} \left(1 + \gamma \breve{Ma}_{prod}^2 \right) \tag{8.12}$$

Combining energy and continuity (Equations 8.6 and 8.8),

$$\left[c_{p,prod} T_{prod} + \frac{\breve{V}_{prod}^2}{2} \right] - \left[c_{p,react} T_{react} + \frac{\breve{V}_{react}^2}{2} \right] = \frac{q}{\dot{m}} \tag{8.13}$$

where $q/\dot{m}$ is the heat of reaction per unit mass of reactants. Dividing by $c_{p,react}$ and noting that

$$c_p = \gamma R / (\gamma - 1)$$

Equation 8.13 becomes

$$T_{react}\left[1+\frac{\gamma-1}{2}\breve{Ma}_{react}^2\right]+\frac{q}{\dot{m}c_{p,react}}=T_{prod}\left[1+\frac{\gamma-1}{2}\breve{Ma}_{prod}^2\right] \qquad (8.14)$$

Now obtain the pressure and temperature jump in terms of the detonation Mach number, $\breve{Ma}_{react}=Ma_D$, and using the Chapman-Jouguet condition that $\breve{Ma}_{prod}=1$. From Equation 8.12,

$$\frac{p_{prod}}{p_{react}}=\frac{1+\gamma\breve{Ma}_{react}^2}{1+\gamma} \qquad (8.15)$$

Combining Equations 8.11 and 8.15,

$$\frac{T_{prod}}{T_{react}}=\frac{\left(1+\gamma\breve{Ma}_{react}^2\right)^2}{\breve{Ma}_{react}^2\left(1+\gamma\right)^2} \qquad (8.16)$$

Dividing Equation 8.14 by T_{react}, substituting with Equation 8.16, and rearranging yields

$$\frac{\left(1+\gamma\breve{Ma}_{react}^2\right)^2}{2\,\breve{Ma}_{react}^2\left(1+\gamma\right)}-\left[1+\frac{\left(\gamma-1\right)}{2}\breve{Ma}_{react}^2\right]=\frac{q}{\dot{m}c_{p,react}T_{react}} \qquad (8.17)$$

For $\breve{Ma}_{react}^2\gg1$ Equation 8.17 may be simplified to

$$\breve{Ma}_{react}=Ma_D=\left[\frac{2\left(\gamma+1\right)}{c_{p,react}T_{react}}\left(\frac{q}{\dot{m}}\right)\right]^{1/2} \qquad (8.18)$$

From Equation 8.18, it is apparent that the detonation propagation rate depends primarily on the heat release per unit mass of reactants, which in turn depends on the heating value of the fuel and the air-fuel ratio. The effect of the initial temperature and initial pressure is to influence the heat release per unit mass of reactants. Preheating the initial gas mixture, which causes more final dissociation, decreases the detonation velocity, while increasing the initial pressure decreases dissociation and increases the detonation velocity. Thermodynamic equilibrium programs are used to calculate Chapman-Jouguet detonation properties, and some representative results are shown in Table 8.2. The effect of the fuel-air ratio and initial pressure on a methane-air detonation is shown in Figure 8.9.

TABLE 8.2

Calculated Detonation Properties for Several Gaseous Mixtures Initially at 298 K and 1 atm

Reactants

Fuel (1 mol)	C_2H_2	C_2H_2	CO	H_2	H_2	CH_4	C_3H_8
O_2 (mol)	2.5	2.5	0.5	0.5	0.5	2	5
N_2 (mol)	0	9.32	0	0	1.88	7.52	18.8

Detonation Products (Mole Fraction)

CO_2	0.0930	0.0880	0.4033	0	0	0.0696	0.0836
H_2O	0.0872	0.0615	0	0.5304	0.2943	0.1721	0.1384
O_2	0.1167	0.0221	0.1659	0.0486	0.0078	0.0098	0.0116
CO	0.3463	0.0660	0.3813	0	0	0.0235	0.0300
OH	0.1157	0.0146	0	0.1370	0.0183	0.0097	0.0099
O	0.1288	0.0056	0.0495	0.0386	0.0021	0.0012	0.0015
H	0.0746	0.0039	0	0.0811	0.0060	0.0017	0.0017
NO	0	0.0169	0	0	0.0078	0.0072	0.0085
H_2	0.0370	0.0062	0	0.1641	0.0317	0.0085	0.0072
N_2	0	0.7152	0	0	0.6319	0.6967	0.7076

Detonation Parameters

V_D (m/s)	2425	1867	1799	2841	1971	1804	1801
Ma_D	7.36	5.41	5.24	5.28	4.84	5.11	5.31
T_{prod} (K)	4214	3113	3525	3682	2949	2780	2823
p_{prod}/p_{react}	33.87	19.13	13.98	18.85	15.62	17.20	18.27
γ_{prod}	1.152	1.157	1.125	1.129	1.163	1.169	1.166
a_{prod} (m/s)	1317	1027	977	1545	1092	999	994
M_{prod}	23.3	28.4	34.5	14.5	23.9	27.0	27.7

Source: Soloukhin, R. I., *Shock Waves and Detonations in Gases*, Mono Book, Baltimore, MD, 1966.

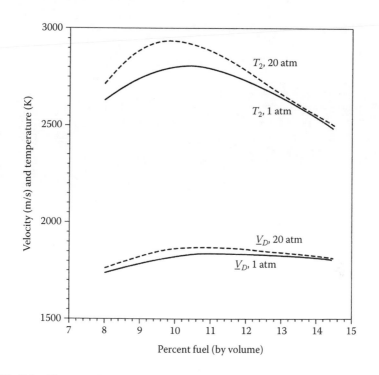

FIGURE 8.9 Chapman-Jouquet detonation velocities and products' temperatures for methane-air mixtures initially at 298 K and 1 and 20 atm between lean and rich limits. A stoichiometric methane-air mixture is 9.51% methane.

Example 8.1

A large volume contains a stoichiometric mixture of methane and air at 1 atm and 25°C. Using the detonation velocity given in Table 8.1, calculate an approximate value for the pressure, temperature, and gas velocity immediately behind the detonation wave.

Solution

The stoichiometric balance equation for a methane-air reaction is

$$CH_4 + 2(O_2 + 3.76\,N_2) \rightarrow CO_2 + 2\,H_2O + 7.52\,N_2$$

From this the molecular weight of a stoichiometric methane-air mixture is

$$M_{react} = \frac{\left[\dfrac{1\,\text{kgmol}_{CH_4}}{1} \cdot \dfrac{16\,\text{kg}_{CH_4}}{\text{kgmol}_{CH_4}} + \dfrac{9.52\,\text{kgmol}_{air}}{1} \cdot \dfrac{29\,\text{kg}_{air}}{\text{kgmol}_{air}} \right]}{10.52\,\text{kgmol}_{react}}$$

$$M_{react} = 27.8\ \text{kg/kgmol}$$

From Table 8.1,

$$V_D = \breve{V}_{react} = 1800 \text{ m/s}$$

The speed of sound of the reactants is

$$a_{react} = \left(\gamma R_{react} T_{react}\right)^{1/2} = \left[1.4\left(\frac{8314 \text{ kg} \cdot \text{m}^2/\text{s}^2}{\text{kgmol} \cdot \text{K}} \cdot \frac{\text{kgmol}}{27.8 \text{ kg}}\right)(298 \text{ K})\right]^{1/2} = 353 \text{ m/s}$$

and the Mach number is

$$\breve{\text{Ma}}_{react} = \frac{1800}{353} = 5.10$$

Using Equations 8.15 and 8.16,

$$p_{prod} = (1 \text{ atm})\left[\frac{1+(1.4)(5.10)^2}{1+1.4}\right] = 15.6 \text{ atm}$$

$$T_{prod} = (298 \text{ K})\frac{\left[1+(1.4)(5.10)^2\right]^2}{5.10^2(1+1.4)^2} = 2784 \text{ K}$$

The pressure p_{prod} is lower than the more exact calculation given in Table 8.2, and the temperature T_{prod} is slightly higher. Note that if $\gamma_{prod} = 1.169$ from Table 8.2 is used in the denominator of Equation 8.14, then the correct value of p_{prod} is predicted. The speed of sound of the products is

$$a_{react} = \left[1.4\left(\frac{8314 \text{ kg} \cdot \text{m}^2/\text{s}^2}{\text{kgmol} \cdot \text{K}} \cdot \frac{\text{kgmol}}{27.8 \text{ kg}}\right)(2784 \text{ K})\right]^{1/2} = 1080 \text{ m/s}$$

Now use the Chapman-Jouguet condition and Equation 8.5 to get the velocity behind the detonation in room-fixed coordinates

$$V_{prod} = V_D - a_{prod} = 1800 - 1080 = 720 \text{ m/s}$$

Thus, the Mach number behind the wave in room-fixed coordinates is

$$\text{Ma}_{prod} = \frac{720 \text{ m/s}}{1080 \text{ m/s}} = 0.667$$

Example 8.2

A large volume contains a stoichiometric mixture of methane and air at 1 atm and 25°C. Using the lower heating value of methane, calculate the detonation Mach number of this mixture and compare the results to Table 8.2.

Solution

From Table 2.2 the lower heating value of methane is 50 MJ/kg. The stoichiometric balance equation for a methane-air reaction is

$$CH_4 + 2(O_2 + 3.76 N_2) \rightarrow CO_2 + 2 H_2O + 7.52 N_2$$

Neglecting dissociation (although it is surely significant), the heat of reaction is estimated to be

$$\frac{q}{\dot{m}} = \frac{50 \text{ MJ}}{kg_{CH_4}} \cdot \frac{16 \text{ kg}_{CH_4}}{(16 + 9.52(29.0)) \text{ kg}_{react}} = 2739 \text{ kJ/kg}_{react}$$

Using Equation 8.18 with $\gamma = 1.4$ and $c_{p,react} = 1.0$ kJ/kg · K, we get

$$\breve{Ma}_{react} = \left[\frac{2(1.4+1)}{(1.0 \text{ kJ/kg} \cdot \text{K}) 298 \text{ K}} (2739 \text{ kJ/kg}) \right]^{1/2} = 6.6$$

Using Equation 8.17, $\breve{Ma}_{react} = 6.8$. However, Table 8.2, which is based on thermochemical calculations of reacting mixtures, gives $\breve{Ma}_{react} = Ma_D = 5.1$. The error occurs because we have chosen a heat release that is too large due to dissociation of the products. In addition Equations 8.17 and 8.18 assume constant molecular weights and a constant specific heats ratio, γ. The constant specific heat detonation model with heat release to completion overestimates the detonation Mach number but gives the correct trends.

8.4 MAINTAINED AND PULSE DETONATIONS

Flames can be fixed in position by flame holders or by special flow patterns such as swirl stabilization. Similarly, devices have been built to stabilize detonations behind standing normal shock waves. Three such designs are shown in Figure 8.10. A question arises as to whether these are true detonations because three-dimensional transverse waves are not observed. However, the onset of detonation does force a readjustment of the normal shock waves to a new upstream position, and hence, they are considered detonations. Attempts have been made to develop chemical lasers using standing detonation waves.

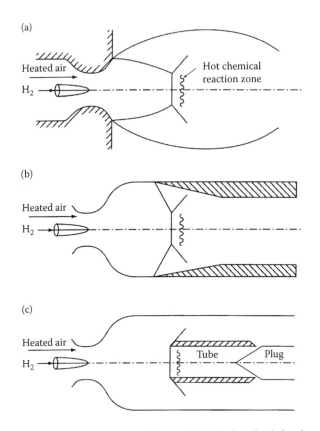

FIGURE 8.10 Standing detonation experiments (a) behind a shock bottle of an overexpanded jet, (b) on a Mach stem of two oblique shocks, and (c) behind a normal shock wave held on the lip of a tube by a choking plug. (From Strehlow, R. A., Liaugminas, R., Watson, R. H. and Eyman, J. R., "Transverse Wave Structure in Detonations," *Symp. (Int.) Combust.* 11:683–692, The Combustion Institute, Pittsburgh, PA, 1967.)

The concept of a maintained detonation wave moving around an annulus that is continuously replenished with fresh combustibles has been investigated in England, the former Soviet Union, and the United States. Consider an annular chamber as shown in Figure 8.11. A gaseous fuel such as methane or hydrogen, and either air or oxygen are introduced into manifolds which feed two rings of nozzles. The nozzles impinge, mix, and supply a detonable mixture to the combustion chamber. A detonation wave is initiated by a pulse of hot, high pressure gas which is introduced tangentially into the annulus. The detonation wave propagates around the annulus. The high pressure behind the wave front gradually decays, allowing fresh reactants to flow into the combustion chamber and exhaust products to flow out of the chamber.

Because detonations travel several thousand times faster than flames, the increase in heat release per unit volume would be enormous if a rotating detonation wave could be sustained. Suggestions have been made to use a nozzle at the end

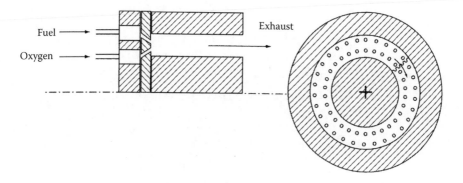

FIGURE 8.11 Concept drawing for a rotating detonation combustor. (From Nicholls, J. A. and Dabora, E. K., "Recent Results on Standing Detonation Waves," *Symp. (Int.) Combust.* 8:644–655, The Combustion Institute, Pittsburgh, PA, 1961, with permission of The Combustion Institute.)

of the annular combustion chamber and make a rotating detonation wave rocket engine. Others have suggested a rotating detonation wave gas turbine combustor. The high frequency, high velocity swirling flow of combustion products emitted from the annular chamber could perhaps be effectively mixed with dilution air in a short distance. Such a design could possibly reduce circumferential variability in temperature at the entry to the turbine while also reducing the size of the combustor.

The rotating detonation wave combustor is not without problems, however. There is a tendency to form multiple waves, which makes it impossible to supply enough fresh fuel; consequently, the combustion decays to a deflagration. The only proven way to prevent the formation of multiple waves, which prevent sustained operation, is to operate the annulus at sub-atmospheric pressure, which greatly reduces the power.

Pulsed detonations in a tube are under investigation as a possible thrust-producing device that could operate at high speed without the need of a compressor. For example, consider a 50 mm diameter tube 1 m long with inlet valves for gaseous fuel and air, a spark plug at the closed end, and the other end open. The valves are opened briefly, the tube is filled with a mixture of hydrogen-air or propane-air, and the spark plug fired. After a short induction distance a detonation wave is formed and propagates down the tube. If the inlet valves are cycled at 10 to 50 Hz, a semi-steady thrust is produced. Experiments have been done in such a tube where the fill flow rate was varied (Schauer, Stutrud, and Bradley 2001). For a 50% purge ratio at 16 Hz with a stoichiometric hydrogen-air mixture, the detonation exit speed was 2000 m/s, and a thrust of 4 lb was produced. The specific impulse (thrust/fuel flow rate) was an encouraging 5000 s.

8.5 PROBLEMS

1. Use the information given in Table 8.2 to verify that Equation 8.8 holds for a stoichiometric hydrogen-air detonation. The sensible enthalpy of each of the products at 2949 K is

Product	h_s (MJ/kgmol)
H_2O	123.52
O_2	96.06
OH	87.71
O	30.84
H	30.63
NO	93.06
H_2	86.85
N_2	90.85

2. Use the detonation Mach number in Table 8.2 for a stoichiometric hydrogen-air mixture to compare Equations 8.15 and 8.16 with p_{prod}/p_{react} and T_{prod} in Table 8.2.

3. Use the detonation Mach number in Table 8.2 for a stoichiometric propane-air mixture to compare Equations 8.15 and 8.16 with p_{prod}/p_{react} and T_{prod} in Table 8.2.

4. For a methane and air mixture at standard initial conditions, calculate the detonation velocity at stoichiometric and at the rich and lean limits using Equation 8.18. Comment on the validity of Equation 8.18. The rich and lean limits for a methane-air detonation are shown in Figure 8.9.

5. Calculate the detonation velocity for a stoichiometric propane-air mixture at standard initial conditions using Equation 8.18. Comment on the validity of Equation 8.18.

6. Use the information for the propane-air mixture given in Table 8.2 to calculate the velocity of the gas behind the detonation with respect to laboratory coordinates. Is it subsonic or supersonic?

7. A stoichiometric gasoline vapor-air mixture, which is initially at 300 K and 1 atm, is compressed isentropically to 1/8th of its initial volume. Estimate the peak pressure and temperature that could be generated if a detonation occurred after compression. Use the data for gasoline given in Tables 2.4 and 3.1. Use the specific heat of N_2 given in Appendix C. Assume that $\gamma = 1.4$.

8. For methane-air detonation at the lean detonation limit, estimate the peak pressure that could be generated. Assume that the reactants are at 20°C and 1 atm, and use the one-dimensional model. Comment on the validity of the one-dimensional model. The lean limit for a methane-air detonation is shown in Figure 8.9. Use the data for methane given in Table 2.2. Use the specific heat of N_2 given in Appendix C. Assume that $\gamma = 1.4$.

9. What pressure is required to burst a 0.5 m diameter by 3 mm wall steel pipe? Assume a yield stress for steel of 250 MPa. If the initial pressure in the tube is 1 atm, what Mach number will generate this pressure based on the pressure at the end of the reaction zone of a detonation. For a thin-wall tube

$$\sigma = \frac{pr_0}{t}$$

where σ is the stress, p is the pressure inside the tube, r_0 is the tube radius and t is the thickness of the tube. Assume that $\gamma = 1.4$.

REFERENCES

Anderson, T. J. and Dabora, E. K., "Measurements of Normal Detonation Wave Structure Using Rayleigh Imaging," *Symp. (Int.) Combust.* 24:1853–1860, The Combustion Institute, Pittsburgh, PA, 1992.

Bowen, J. R., Ragland, K. W., Steffes, F., and Loflin, T., "Heterogeneous Detonation Supported by Fuel Fogs or Films," *Symp. (Int.) Combust.* 13:1131–1139, The Combustion Institute, Pittsburgh, PA, 1971.

Cullen, R. E., Nicholls, J. A., and Ragland, K. W., "Feasibility Studies of a Rotating Detonation Wave Rocket Motor," *J. Spacecr. Rockets* 3(6):893–898, 1966.

Edwards, B. D., "Maintained Detonation Waves in an Annular Channel: A Hypothesis Which Provides the Link Between Classical Acoustic Combustion Instability and Detonation Waves," *Symp. (Int.) Combust.* 16:1611–1618, The Combustion Institute, Pittsburgh, PA, 1977.

Gordon, S. and McBride, B. J., *Computer Program for Calculation of Complex Chemical Equilibrium Compositions, Rocket Performance, Incident and Reflected Shocks and Chapman-Jouguet Detonations,* NASA-SP-273, 1976.

Kailasanath, K., Oran, E. S., Boris, J. P., and Young, T. R., "Determination of Detonation Cell Size and the Role of Transverse Waves in Two-Dimensional Detonations," *Combust. Flame* 61(3):199–209, 1985.

Knystautas, R., Guirao, C., Lee, J. H., and Sulmistras, A., "Measurement of Cell Size in Hydrocarbon-Air Mixtures and Predictions of Critical Tube Diameter, Critical Initiation Energy, and Detonability Limits," *Prog. Astronaut. Aeronaut.* 94:23–37, 1984.

Kuo, K. K., "Detonation and Deflagration Waves of Premixed Gases" in *Principles of Combustion,* 354–435, 2nd ed., John Wiley & Sons, Hoboken, NJ, 2005.

Law, C. K., "Combustion in Supersonic Flows" in *Combustion Physics,* 634–686, Cambridge University Press, Cambridge, UK, 2006.

Lee, J. H. S., "Dynamic Parameters of Gaseous Detonations," *Annu. Rev. of Fluid Mech.* 16:311–336, 1984.

Nicholls, J. A. and Dabora, E. K., "Recent Results on Standing Detonation Waves," *Symp. (Int.) Combust.* 8:644–655, The Combustion Institute, Pittsburgh, PA, 1961.

Oppenheim, A. K., Urtiew, P. A., and Weinberg, F. J., "The Use of Laser-Light Sources in Schlieren-Interferometer Systems," *Proc. Royal Soc. A* 291(1425):279–290, 1966.

Schauer, F., Stutrud, J., and Bradley, R., "Detonation Initiation Studies and Performance Results for Pulsed Detonation Engine Applications," 39th AIAA Aerospace Sciences Meeting, Reno, NV, 2001.

Shchelkin, K. I. and Troshin, Ya., K., *Gas Dynamics of Combustion* translated from "Gazodinamika Goreniya," Moscow Izd. Akad. Nauk SSSR, 1963, NASA Technical Translation NASA-TT-F-231, Washington, DC, 1964.

Soloukhin, R. I., *Shock Waves and Detonations in Gases,* Mono Book, Baltimore, MD, 1966.

Strehlow, R. A., Liaugminas, R., Watson, R. H., and Eyman, J. R., "Transverse Wave Structure in Detonations," *Symp. (Int.) Combust.* 11:683–692, The Combustion Institute, Pittsburgh, PA, 1967.

Strehlow, R. A., "Detonations," Chap. 9 in *Combustion Fundamentals,* McGraw-Hill, New York, 1984.

Taki, S. and Fujiwara, T., "Numerical Simulation of Triple Shock Behavior of Gaseous Detonation," *Symp. (Int.) Combust.* 18:1671–1681, The Combustion Institute, Pittsburgh, PA, 1981.

Williams, F. A., "Detonation Phenomena," Chap. 6 in *Combustion Theory*, Westview Press, Boulder, CO, 2nd ed., 1994.

Part III

Combustion of Liquid Fuels

Oil-fired furnaces, gas turbine combustors, and diesel engines involve spray combustion. Analogous to gaseous detonation, liquid fuels can also combust in a detonation mode. First, sprays are considered, and then spray combustion in furnaces, gas turbines, and diesel engines are discussed. The reader may wish to review the material on liquid fuels in Chapter 2 before proceeding with Part III.

9 Spray Formation and Droplet Behavior

Combustion of liquid fuels differs from the combustion of gaseous fuels in that a liquid fuel must be vaporized and then combusted. This additional step adds a significant complication to the combustion process. In the analysis of gaseous fuel combustion systems, we were concerned about the energy density of the fuel, the reaction rate, the heat release rate, the flame temperature, and the flame speed—all of which are coupled together. In the analysis of liquid fuel combustion systems, we are again concerned about the energy density of the fuel, the reaction rate, the heat release rate, the flame temperature, and the flame speed; but the rate controlling phenomenon is the evaporation of the fuel.

Students who have followed the text closely may at this point wonder about the discussion of spark ignition engines in Part II, "Combustion of Gaseous and Vaporized Fuels." Gasoline is a liquid that is vaporized and then combusted. How is the vaporization and combustion of gasoline in a spark ignition engine different from the case we are discussing here? The difference is that in a spark ignition engine the vaporization of the gasoline and combustion of the gasoline are separate processes that are uncoupled. That is, the gasoline is first vaporized and mixed with air, then it is introduced to the combustion environment (the cylinder). In the liquid fuel combustion systems we will examine in Part III of this text, the liquid fuel is admitted to the combustion chamber and is then evaporated as a part of the combustion process.

In almost all practical cases of liquid fuel combustion, the liquid fuel is broken up into a fine spray of liquid droplets as it is introduced to the combustion chamber. Oil-fired furnaces and boilers, gas turbines, and diesel engines utilize sprays of liquid fuel droplets to increase the fuel surface area and thus increase the vaporization and combustion rate. For example, breaking up one 3 mm drop into 30 μm droplets results in one million droplets and increases the surface area of the fuel by 100 times. Because the droplet burning rate is approximately inversely proportional to the diameter squared, the burning rate increases ten thousand times in this example, assuming the droplets burn under the same ambient conditions. Thus the need to use sprays is quite clear. Good atomization (creation of fine and uniform sprays) is the key to complete combustion and low emissions.

On injection of the liquid fuel into a combustion chamber, the liquid is broken up into a large number of droplets of various sizes and velocities. Depending on the density of the spray and the ambient conditions, some of the droplets may continue to shatter, and some may recombine in droplet collisions. As soon as the fuel is broken up into droplets, it begins to vaporize. The fuel vapor mixes with the surrounding gas, and either because of the high temperature of the ambient air or because of an existing flame front, combustion of the vapor-air mixture occurs. The hot products of combustion mix with the uncombusted fuel-air vapor and droplets, and given sufficient time or combustor length, the fuel is completely converted to combustion products.

This chapter will consider spray formation mechanisms, droplet size distributions, types of spray nozzles, and droplet vaporization, to provide a basis for examining liquid fuel combustion in furnaces, boilers, gas turbines, and diesel engines in the following chapters.

9.1 SPRAY FORMATION

Sprays can be formed in a number of ways. Most commonly, liquid fuel sprays are formed by pressurized jet atomization or air blast atomization. In pressurized jet atomization a spray is formed by pressurizing a liquid and forcing it through an orifice at a high velocity relative to the surrounding air or gas. Alternatively, air blast atomization produces a spray by impinging a high velocity air flow on a relatively slow-moving liquid jet.

As a liquid jet emerges from an orifice into a gas, the breakup mechanism may be visualized sequentially beginning with stretching or narrowing of the liquid followed by the appearance of ripples, protuberances, and ligaments in the liquid, which leads to the rapid collapse of the liquid into droplets. Further breakup then occurs due to vibration and shear of the droplets, and finally some agglomeration of the droplets occurs due to collisions if the spray is not dilute. For visual examples of spray formation and droplet breakup see the photographs in Saminy et al. (2004) and Van Dyke (1982).

The spray formation process is characterized by three dimensionless groups. These are

- Jet Reynolds number (the ratio of inertia force to viscous force)

$$\mathrm{Re}_{jet} = \frac{\rho_\ell V_{jet} d_{jet}}{\mu_\ell} \tag{9.1}$$

- Jet Weber number (the ratio of inertia force to surface tension force)

$$\mathrm{We}_{jet} = \frac{\rho_g V_{jet}^2 d_{jet}}{\sigma} \tag{9.2}$$

- Ohnesorge number (the ratio of viscous force to surface tension force)

$$\mathrm{Oh} = \frac{\mu_\ell}{\sqrt{\rho_\ell \sigma d_{jet}}} = \frac{\left[\left(\rho_\ell/\rho_g\right)\mathrm{We}_{jet}\right]^{1/2}}{\mathrm{Re}_{jet}} \tag{9.3}$$

The subscripts *jet*, ℓ, and *g* refer to the incoming jet, the properties of the liquid being atomized, and the properties of the gas into which the liquid jet is being atomized, respectively. For example, the parameter d_{jet} is the diameter of the undistributed jet, and σ is the surface tension of the liquid.

When swirl is induced in the liquid as it flows into an orifice, the jet forms a wider conical sheet and breaks up in a similar wave-like manner as in a plain jet (Figure 9.1). The spray from a plain or swirl type orifice penetrates a certain distance before coming to rest in quiescent air. The three dimensionless numbers above are useful in formulating empirical relationships for droplet size, spray angle, and penetration. For the case of liquid jet breakup due to aerodynamic shattering (air blast atomization), the Reynolds number and Weber number are defined from the droplet rather than the jet point of view, and air density, droplet velocity relative to the gas, and droplet diameter are used. That is, the droplet Reynolds number and the droplet Weber number are

$$\mathrm{Re}_{drop} = \frac{\rho_g V_{drop} d_{drop}}{\mu_g} \tag{9.4}$$

$$\mathrm{We}_{drop} = \frac{\rho_g V_{drop}^2 d_{drop}}{\sigma} \tag{9.5}$$

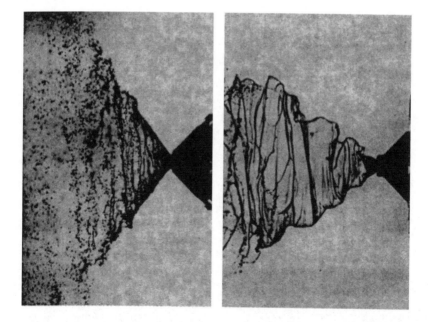

FIGURE 9.1 Orifice with swirl forms a thin conical jet in quiescent air; the disturbances grow until the sheet disintegrates into droplets. (From Van Dyke, M., *An Album of Fluid Motion*, Parabolic Press, 10th ed., Stanford, CA, 1982. Courtesy of H. E. Fiedler, Technical University of Berlin.)

respectively. For $We_{drop} > 12$, droplet breakup occurs. For large values of We_{drop}, small droplets are stripped from the parent droplet as the parent droplet deforms into an elliptical shape with its major axis perpendicular to the airflow. This micro-droplet shedding regime occurs when

$$\frac{We_{drop}}{\sqrt{Re_{drop}}} > 0.7 \qquad (9.6)$$

Observations from experiments with falling droplets of liquid CO_2 in helium at high pressures and temperatures (well above the liquid critical point) show that as the ambient pressure and temperature are increased, droplets first break up, and then for even higher pressures and temperatures they totally disintegrate. This phenomenon indicates that the droplet surface mixture has reached its thermodynamic critical point and that the surface tension has become negligible.

Example 9.1

Consider a 15 μm fuel droplet traveling at 200 m/s relative to the surrounding air. The fuel density is 850 kg/m³ and the surface tension, σ, is 0.031 N/m. The air is compressed to 6.2 MPa and 864 K. The density of the air is 25 kg/m³. Calculate the droplet Weber number and the droplet Reynolds number. Will the fuel droplet break up?

Solution

$$We_{drop} = \frac{\rho_g V_{drop}^2 d_{drop}}{\sigma} = \frac{25 \text{ kg}}{\text{m}^3} \cdot \left(\frac{200 \text{ m}}{\text{s}}\right)^2 \cdot \frac{15 \times 10^{-6} \text{ m}}{1} \cdot \frac{\text{m}}{3.1 \times 10^{-2} \text{ N}} = 500$$

From linear interpolation of Appendix B

$$\mu_g = 3.819 \times 10^{-5} \text{ kg/ms}$$

$$Re_{drop} = \frac{\rho_g V_{drop} d_{drop}}{\mu_g} = \frac{25 \text{ kg}}{\text{m}^3} \cdot \frac{\text{m} \cdot \text{s}}{3.819 \times 10^{-5} \text{ kg}} \cdot \frac{200 \text{ m}}{\text{s}} \cdot \frac{15 \times 10^{-6} \text{ m}}{1} = 1960$$

$$\frac{We_{drop}}{\sqrt{Re_{drop}}} = \frac{500}{\sqrt{1960}} = 11$$

The droplet will break up because

$$We_{drop} > 12$$

and

$$\frac{We_{drop}}{\sqrt{Re_{drop}}} > 0.7$$

Remembering that the Weber number is the ratio of the inertia force to the surface tension force, we see that the surface tension force is very weak in this case. Based on this, we expect a catastrophic type of breakup.

9.2 DROPLET SIZE DISTRIBUTIONS

Droplet size measurements in sprays are made using various optical techniques. For example, a short pulse laser can be used to penetrate the spray and illuminate a high-resolution digital camera screen. Digital images from the camera are then transferred to a computer, and particle sizing software is used to analyze the images obtained in order to build up a distribution of diameters. By double pulsing the laser, both particle size and velocity distributions may be obtained. A spatial distribution is obtained by counting the droplets in a given volume at a given instant. A temporal distribution is obtained by counting all droplets passing through a given surface.

Droplet size distribution measurements are typically plotted as a histogram, such as represented in Figure 9.2, where ΔN_i is the fraction of droplets counted in size interval Δd_i. As the size intervals, Δd_i, become smaller and smaller, the histogram takes the form of a differential number distribution, $dN_i/d(d)$, as a function of diameter. Another way to examine droplet size distributions is to plot the

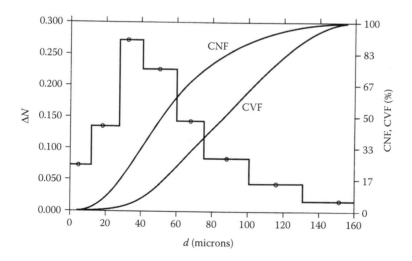

FIGURE 9.2 Droplet size distributions from Example 9.2.

cumulative number fraction, CNF_k, the fraction of droplets smaller than a given droplet diameter, d_k.

$$CNF_k = \frac{\sum_{i=1}^{k}(d_i \Delta N_i)}{\sum_{i=1}^{\infty}(d_i \Delta N_i)} \tag{9.7}$$

The cumulative volume fraction, CVF, is the volume fraction of all droplets less than size d_k.

$$CVF_k = \frac{\sum_{i=1}^{k}(d_i^3 \Delta N_i)}{\sum_{i=1}^{\infty}(d_i^3 \Delta N_i)} \tag{9.8}$$

The histogram can be replaced by a continuous smooth curve, and the summation replaced by integration to give

$$CVF = \int_0^d \left[d^3 \frac{dN}{d(d)} \right] d(d) \tag{9.9}$$

The average droplet volume is given by

$$\bar{V}_{drop} = \frac{\pi}{6} \sum_i (d_i^3 \Delta N_i) \tag{9.10}$$

There are five different measurements of diameter that are commonly used to describe the average size of a distribution of droplets in a simple way. These are the most probable diameter, the mean diameter, area mean diameter, the volume mean diameter, and the Sauter mean diameter.

- The most probable droplet diameter is the droplet diameter with the largest fraction of droplets.
- The mean diameter $(\bar{d}_1)$ is the average diameter of the group of droplets based on the fraction of droplets at each diameter.

$$\bar{d}_1 = \sum_{i=1}^{\infty}(d_i \Delta N_i) \tag{9.11}$$

- The area mean diameter (AMD or $\bar{d}_2$) is the average diameter based on the fraction of droplets with a given surface area.

$$\text{AMD} = \bar{d}_2 = \left[\sum_{i=1}^{\infty} \left(d_i^2 \Delta N_i \right) \right]^{1/2} \qquad (9.12)$$

- The volume mean diameter (VMD or $\bar{d}_3$) is the average diameter based on the fraction of droplets with a given volume.

$$\text{VMD} = \bar{d}_3 = \left[\sum_{i=1}^{\infty} \left(d_i^3 \Delta N_i \right) \right]^{1/3} \qquad (9.13)$$

- The Sauter mean diameter (SMD or $\bar{d}_{32}$) is used in a number of spray models. The SMD is the VMD divided by the AMD.

$$\text{SMD} = \bar{d}_{32} = \frac{\displaystyle\sum_{i=1}^{\infty} \left(d_i^3 \Delta N_i \right)}{\displaystyle\sum_{i=1}^{\infty} \left(d_i^2 \Delta N_i \right)} \qquad (9.14)$$

Example 9.2

The following droplet size distribution data has been obtained experimentally:

Size Interval (μm)	Number of Droplets
0–10	80
11–25	150
26–40	300
41–60	250
61–75	160
76–100	95
100–130	50
131–170	20
171–220	0
Total (0–220)	1105

Determine the size distribution, the cumulative number fraction distribution, and the cumulative volume fraction distribution. Determine the most probable diameter, the mean diameter, the area mean diameter, the volume median diameter, and the Sauter mean diameter.

Solution

Column	1 $d_i(\mu m)$	2 ΔN_i	3 $d_i \Delta N_i$	4 CNF (%)	5 $d_i^2 \Delta N_i$	6 $d_i^3 \Delta N_i$	7 CVF (%)
	5	0.072	0.36	0.7	1.8	9	0.003
	18	0.136	2.44	5.8	44.0	791	0.3
	33	0.272	8.96	24.3	295.7	9,757	3.8
	50.5	0.226	11.42	47.8	576.9	29,132	14.4
	68	0.145	9.85	68.0	669.6	45,530	31.0
	88	0.086	7.56	83.6	665.7	58,586	52.2
	115.5	0.045	5.23	94.4	603.6	69,721	77.6
	150.5	0.018	2.74	100	410.0	61,700	100
	195.5	0.000	0	100	0	0	100
Sum		1.000	48.5	–	3267	275,226	–

The size distribution is given in Column 2, which is the number of droplets in a specific size interval divided by the total number of droplets (1105). The CNF and CVF are given in Columns 4 and 7, respectively. The size distribution, the CNF, and the CVF are shown in Figure 9.2.

- From Column 2, the *most probable diameter* = 33 μm
- From the sum of Column 3, *the mean diameter* = 48 μm
- The *area mean diameter* is the square root of the sum of Column 5,

$$d_2 = \sqrt{3267} = 57 \text{ μm}$$

- The *volume mean diameter* is the cube root of the sum of Column 6,

$$d_3 = \sqrt[3]{275,226} = 65 \text{ μm}$$

- The *Sauter mean diameter* is the sum of Column 6 divided by the sum of Column 5,

$$\overline{d}_{32} = 275,226/3267 = 84 \text{ μm}$$

Various equations have been used to describe the droplet-size distribution functions. A general function often used dates to the early work of Nukiyama and Tanasawa (1939); a, b, c, and q are empirically derived parameters.

$$\frac{d(\text{VF})}{d(d)} = ad^b \exp(-cd^q) \tag{9.15}$$

where VF is the volume fraction. If the liquid density is constant, VF also represents the mass fraction.

The cumulative volume distribution is often represented by the Rosin-Rammler distribution (Rosin and Rammler 1933), which was originally developed for pulverized powders:

$$\text{CVF}_i = 1 - \exp\left[-(d/d_0)^q\right] \tag{9.16}$$

The reference diameter, d_0, is chosen such that

$$\text{CVF}_i = 1 - \exp(-1) = 63.2\%$$

The exponent q can be found by nonlinear curve fit of the data to

$$\left(d_i/d_0\right)^q = \ln\left[\left(1-\text{CVF}_i\right)^{-1}\right] \tag{9.17}$$

In Example 9.2 it can be deduced from linear interpolation of Column 7 that $d_0 = 100\ \mu\text{m}$. A nonlinear curve fit of the data to Equation 9.17 provides $q = 2.7$. A chart for graphically estimating the cumulative volume fraction is shown in Figure 9.3.

9.3 FUEL INJECTORS

A fuel injector should provide good penetration and dispersion of small droplets and allow for good mixing with the combustion air. The degree of atomization required depends primarily on the time available for vaporization, mixing, and combustion. First we will consider steady flow injectors and then intermittent injectors.

9.3.1 STEADY FLOW INJECTORS

The simplest injector is the plain orifice with a hole of length L and diameter d_{orf} injecting into still air. The spray cone angle lies between 5° and 15° depending on

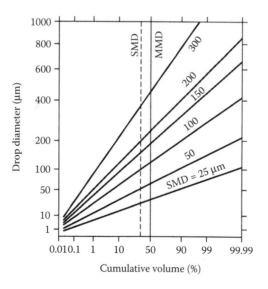

FIGURE 9.3   Plot for estimating the cumulative volume fraction below a given droplet size. (From Simmons, H. C., "The Correlation of Drop-Size Distributions in Fuel Nozzle Sprays. Parts I and II.," *J. Eng. Power* 99(3):309–319, 1977, by permission of ASME.)

fluid viscosity and surface tension. The correlation for the Sauter mean diameter (SMD) for a plain orifice, suggested by Tanasawa and Toyoda (1955), is

$$\text{SMD} = 47 \frac{d_{orf}}{V_{jet}} \left(\frac{\sigma}{\rho_g}\right)^{0.25} \left(1 + 331 \frac{\mu_\ell}{(\sigma \rho_\ell d_{orf})^{0.5}}\right) \tag{9.18}$$

All variables in Equation 9.18 are given in standard SI units, e.g., meters, seconds, meters per second, kilograms per cubic meter, and newtons per meter. It should be noted that Equation 9.18 (as well as many of the equations that follow) is not based on dimensionless numbers but represents a curve fit to data using a particular set of units. The user needs to take care in each of these cases to ensure that the correct set of units is used.

The "simplex" or swirl atomizer shown in Figure 9.4 achieves a considerable improvement in droplet dispersion over the plain orifice. Tangential slots in the atomizer impart angular momentum to the fluid causing the fluid to form a hollow cone as it emerges from the orifice, thus forming an air core vortex inside the cone. The spray angle can be quite large—up to 90°. The ratio of d_s/d_{orf} should be about 3.3 to obtain the highest discharge coefficient. The L/d_{orf} ratio of the orifice should be as small as possible, but practical considerations limit it to the 0.2–0.5 range.

The Sauter mean diameter for a pressure-swirl atomizer depends on liquid surface tension and viscosity, the mass flow rate of the liquid, and the pressure drop across the atomizer. The correlation by Radcliffe (see Lefebvre 1989) is

$$\text{SMD} = 7.3 \sigma^{0.6} v_\ell^{0.2} \dot{m}_\ell^{0.25} \Delta p^{-0.4} \tag{9.19}$$

where the SMD is in microns and all other parameters are in standard SI units.

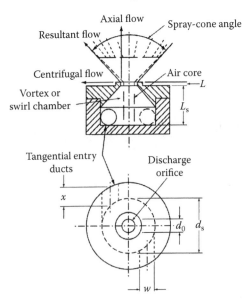

FIGURE 9.4 Simplex pressure-swirl atomizer.

Figure 9.5 shows a measured droplet size distribution for mass and number density obtained from a typical pressure-swirl type nozzle. The average SMD for this atomizer is about 45 μm. At 50 mm downstream from the exit, the spray is 10 cm in diameter, and the SMD varies from 80 μm at the edge to 10 μm at the axis. The volume fraction peaks at about 4 cm from the axis.

The flow rate of the swirl atomizer is proportional to the square root of the injection pressure differential. For a 20:1 range of flow rates, as might be required in an aircraft turbine, the higher flow rate would require a pressure of about 400 atm to be sure that satisfactory operation would take place at the lowest flow rate. Such pressures represent the lower end of diesel injection pressures, but are too high for the large steady flows of a gas turbine or a furnace. To solve the problem of fuel turndown in gas turbines and oil-fired furnaces, various forms of air-blast atomizers are used.

An air blast atomizer, such as shown in Figure 9.6, requires relatively low fuel pressure and produces fine droplets. The additional air creates good mixing and reduces soot formation. An example of a formula for determining the SMD of prefilming air-blast atomizers is given by Lefebvre (1989),

$$
SMD = 3.33 \times 10^{-3} \left(\frac{\sigma \rho_\ell d_p}{\rho_{air}^2 V_{air}^2} \right)^{0.5} \left(1 + \frac{\dot{m}_\ell}{\dot{m}_{air}} \right)
$$

$$
+ 13 \times 10^{-3} \left(\frac{\mu_\ell^2}{\sigma \rho_\ell} \right)^{0.425} \left(1 + \frac{\dot{m}_\ell}{\dot{m}_{air}} \right)^2 d_p^{0.575}
$$

(9.20)

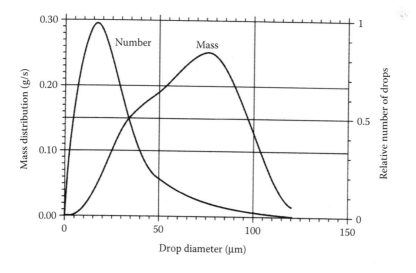

FIGURE 9.5 Measured drop size distribution for a pressure-swirl atomizer with an 80° cone angle; the pressure difference across the atomizer was 689 kPa; the fuel was aircraft gas turbine test fuel type II. (Derived from data of Dodge, L. G. and Schwalb, J. A., "Fuel Spray Evolution: Comparison of Experiment and CFD Simulation of Nonevaporating Spray," *J. Eng. Gas Turbines Power* 111(1):15–23, 1989, by permission of the ASME.)

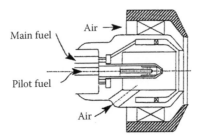

FIGURE 9.6   Airblast atomizer with pilot fuel for low turndown. The crosses denote swirl vanes.

where the SMD is in meters, and all quantities are in SI units, and where d_p is the diameter of the pintle. The first term of Equation 9.20 accounts for the momentum of the air and the surface tension of the droplet. Higher velocities and higher air densities act to create smaller droplets. Higher surface tensions, more dense liquids, and larger droplets are more effective at resisting breakup. The second term in Equation 9.20 balances the viscosity of the liquid and the surface tension. More viscous fluids resist deformation and result in ligament breakup further downstream and larger droplets. Increasing the air-to-liquid ratio above five has little effect.

Example 9.3

Compute and plot the SMD for a kerosene spray from a prefilming air blast nozzle over a range of air flow to liquid fuel flow ratios from 1 to 10. The air is at standard temperature and pressure. Let the air velocity be a constant value of 50, 75, 100, and 125 m/s. The pintle diameter is 36 mm.

Solution
The kerosene properties are

$$\sigma = 0.0275 \ \text{N/m} \qquad\qquad \text{Appendix A.6}$$

$$\rho_\ell = 825 \ \text{kg/m}^3 \qquad\qquad \text{Table 2.7}$$

$$\nu_\ell = 1.6\times10^{-6} \ \text{m}^2/\text{s} \qquad\qquad \text{Table 2.7}$$

$$\mu_\ell = \rho_\ell \nu_\ell = \frac{1.6\times10^{-6} \ \text{m}^2}{\text{s}} \cdot \frac{825 \ \text{kg}}{\text{m}^3} = 0.0132 \ \text{kg/m}\cdot\text{s}$$

From Appendix B

$$\rho_{\text{air}} = 1.177 \ \text{kg/m}^3$$

Equation 9.20 can be simplified to

$$\text{SMD} = A\left(1+\frac{\dot{m}_\ell}{\dot{m}_{\text{air}}}\right) + B\left(1+\frac{\dot{m}_\ell}{\dot{m}_{\text{air}}}\right)^2$$

where

$$A = 3.33 \times 10^{-3} \left(\frac{\sigma \rho_\ell d_p}{\rho_{air}^2 V_{air}^2} \right)^{0.5}$$

$$A = 3.33 \times 10^{-3} \left[\frac{0.0275 \text{ N}}{\text{m}} \cdot \frac{825 \text{ kg}}{\text{m}^3} \cdot \frac{0.036 \text{ m}}{1} \cdot \frac{\text{kg} \cdot \text{s}^2}{\text{N} \cdot \text{m}} \right]^{0.5} \frac{\text{m}^3}{1.177 \text{ kg}} \cdot \frac{\text{s}}{V_{air} \text{ m}} \cdot \frac{10^6 \text{ μm}}{\text{m}}$$

$$A = \frac{3009}{V_{air}} \text{ μm}$$

and

$$B = 13 \times 10^{-3} \left(\frac{\mu_\ell^2}{\sigma \rho_\ell} \right)^{0.425} d_p^{0.575}$$

$$B = 13 \times 10^{-3} \left[\left(\frac{0.00132 \text{ kg}}{\text{m} \cdot \text{s}} \right)^2 \frac{\text{m}}{0.0275 \text{ N}} \cdot \frac{\text{m}^3}{825 \text{ kg}} \cdot \frac{\text{N} \cdot \text{s}^2}{\text{kg} \cdot \text{m}} \right]^{0.425} \frac{(0.036 \text{ m})^{0.575}}{1} \cdot \frac{10^6 \text{ μm}}{\text{m}}$$

$$B = 1.820 \text{ μm}$$

The SMD as a function of ratio of the fuel flow to airflow and air velocity is

$$\text{SMD} = \frac{3009}{V_{air}} \left(1 + \frac{\dot{m}_\ell}{\dot{m}_{air}} \right) + 1.820 \left(1 + \frac{\dot{m}_\ell}{\dot{m}_{air}} \right)^2 \text{ μm}$$

This is plotted below. As shown, the atomization quality starts to decline when the air-to-liquid ratio falls below about 2. When the air-to-liquid ratio exceeds about 3, only slight improvement in atomization quality is gained by the addition of more air.

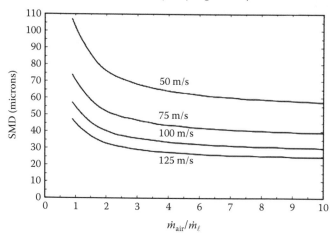

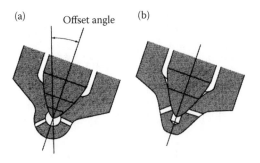

FIGURE 9.7　Multi-hole nozzle for diesel: (a) nozzle with sac volume, (b) sacless nozzle.

9.3.2 Intermittent Injectors

Injectors for direct injection into the intake port or cylinder of an internal combustion engine operate with a short pulse once per combustion cycle. Until recently, direct injection has been primarily used in diesel engines; however, it has recently been used in a number of spark ignition engines. The diesel injection system uses high pressures (300–1400 atm) to achieve good atomization and electronic control of the injection.

Diesel fuel injectors are typically of the plain orifice type in which the needle is inwardly opening. Figure 9.7 shows a multiple hole diesel nozzle, (a) showing a sac volume below the needle, which causes dribble at the end of injection, and (b) with the sac volume eliminated, which reduces exhaust smoke. Figure 9.8 shows standard and throttling pintle nozzles, which are typical of single hole nozzles. The throttling pintle nozzle tends to prevent weak injection at the start of injection and shuts faster to prevent dribble at the end of injection.

Table 9.1 gives three older correlations for the Sauter mean diameter, which are useful for the lower injection pressures typically used prior to the 1990s in heavy-duty diesels. Note that the positive exponent on air density of the last two formulas may be the result of agglomeration effects. These formulas give trends but should not be applied rigorously to spray modeling.

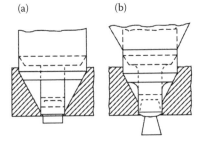

FIGURE 9.8　Pintle nozzle: (a) standard and (b) throttling.

TABLE 9.1
Equations for the Sauter Mean Diameter of Diesel Spray

Source	SMD (μm)
Elkotb (1982)	$\left(3.08 \times 10^6\right) v_f^{0.385} \sigma^{0.737} \rho_f^{0.737} \rho_{air}^{0.06} \Delta p^{-0.54}$
Knight (1955)	$\left(1.605 \times 10^6\right) \Delta p^{-0.458} \dot{m}_f^{0.209} v_f^{0.215} \left(A_{orf}/A(t)_{eff}\right)^{0.916}$
Hiroyasu and Kadota (1974)	$\left(2.33 \times 10^6\right) \Delta p^{-0.135} \rho_{air}^{0.121} V_f^{0.131}$

Where

Δp	Pressure difference across injector	(Pa)
ρ_{air}	Density of air	(kg/m³)
ρ_f	Density of fuel	(kg/m³)
v_f	Kinematic viscosity of fuel	(m²/s)
σ	Surface tension	(N/m)
V_f	Injection fuel volume	(m³/stroke)
$\dot{m}_f$	Injection rate	(kg/s)
A_{orf}	Area of nozzle holes	(m²)
A_{eff}	Effective area of nozzle holes at Δp	(m²)

Source: Elkotb, M. M., "Fuel Atomization for Spray Modeling," *Prog. Energy Combust. Sci.* 8(1):61–91, 1982; Knight, B. E., Communication on A. Radcliffe, "The Performance of a Type of Swirl Atomizer," *Proc. Inst. Mech. Eng.* 169:104–105, 1955; Hiroyasu H. and Kadota, T., "Fuel Droplet Size Distribution in Diesel Combustion Chamber," SAE paper 740715, 1974.

Example 9.4

Using the following data for a single hole of a multi–hole diesel injector, calculate the SMD from the formulas of Table 9.1.

d_{jet} = 0.0003 m (0.3 mm)
Δp = 35.5 MPa (based on a peak injection pressure and an air pressure of 6.2 MPa)
ρ_{air} = 25 kg/m³
ρ_f = 850 kg/m³
v_f = 2.82 × 10⁻⁶ m²/s
σ = 0.03 N/m
V_f = 1 × 10⁻⁸ m³ (based on a 2 ms duration)
$\dot{m}_f$ = 8.68 × 10⁻³ kg/s
A_{orf} = 7.1 × 10⁻⁸ m²
A_{eff} = 3.5 × 10⁻⁸ m² (includes the effect of needle motion)

Solution

For Knight's formula

$$SMD = 1.605 \times 10^6 \left(35.5 \times 10^6\right)^{-0.458}$$

$$\times \left(8.68 \times 10^{-3}\right)^{0.209} \left(2.82 \times 10^{-6}\right)^{0.215} \left(\frac{7.1 \times 10^{-8}}{3.5 \times 10^{-8}}\right)^{0.916}$$

$$\text{SMD} = 25.4 \ \mu\text{m}$$

For Hiroyasu and Kadota's formula

$$\text{SMD} = 2.33 \times 10^3 \left(35.5 \times 10^6\right)^{-0.135} \left(25\right)^{0.121} \left(1 \times 10^{-8}\right)^{0.131}$$

$$\text{SMD} = 29.5 \ \mu\text{m}$$

For Elkotb's formula

$$\text{SMD} = 3.08 \times 10^6 \left(2.82 \times 10^{-6}\right)^{0.385} \left(0.03\right)^{0.737} \left(850\right)^{0.737} \left(25\right)^{0.06} \left(35.5 \times 10^6\right)^{-0.54}$$

$$\text{SMD} = 24.8 \ \mu\text{m}$$

Although the three formulas give very similar results for these data, the trends are not the same for all variables. For example, Hiroyasu and Kadota's formula does not include the effect of viscosity, while Knight's formula does not include the effect of air density. Typically it has been found that the increase in the SMD with air density is due to increased droplet agglomeration with increased air density. Of course, agglomeration effects increase as the location of the SMD measurement is moved downstream away from the injector orifice.

Hiroyasu, Arai, and Tabata (1989) have given two formulas for the SMD based on data obtained from Fraunhofer diffraction measurements applied to an injector with pressures of 3.5–90 MPa. Data for the lower injection velocities were fit by Equation 9.21a and data for the higher injection velocities were fit by Equation 9.21b. The SMD is the larger result of the two equations.

$$\text{SMD} = 4.12 \, \text{Re}_{jet}^{0.12} \, \text{We}_{jet,\ell}^{-0.75} \left(\frac{\mu_\ell}{\mu_{air}}\right)^{0.54} \left(\frac{\rho_\ell}{\rho_{air}}\right)^{0.18} d_{jet} \tag{9.21a}$$

$$\text{SMD} = 0.38 \, \text{Re}_{jet}^{0.25} \, \text{We}_{jet,\ell}^{-0.32} \left(\frac{\mu_\ell}{\mu_{air}}\right)^{0.37} \left(\frac{\rho_\ell}{\rho_{air}}\right)^{-0.47} d_{jet} \tag{9.21b}$$

When using Equations 9.21a and 9.21b, it is important to note that in these equations the jet Weber number is defined differently from Equation 9.2. In this case, ρ_ℓ is being used to define the Weber number instead of ρ_g. We identify this Weber number as $\text{We}_{jet,\ell}$.

$$\text{We}_{jet,\ell} = \frac{\rho_\ell V_{jet}^2 d_{jet}}{\sigma} \tag{9.22}$$

The Ohnesorge number can then be written in terms of $\mathrm{We}_{jet,\ell}$.

$$\mathrm{Oh} = \frac{\left[\mathrm{We}_{jet,\ell}\right]^{1/2}}{\mathrm{Re}_{jet}} \tag{9.23}$$

Example 9.5

Consider a 0.2 mm diameter nozzle hole with a flow rate of diesel fuel of 0.00531 kg/s. The fuel density is 850 kg/m^3. The diesel fuel sprays into compressed air at 25 kg/m^3 density and 6.2 MPa pressure. Prior to compression the air was at 1 atm and 25°C. Use Equations 9.21a and 9.21b to calculate the SMD and select the larger value.

Solution

First calculate Re_{jet}, $\mathrm{We}_{jet,\ell}$, and Oh.

$$V_{jet} = \frac{\dot{m}_\ell}{\rho_\ell A_{jet}} = \frac{0.00531 \ \mathrm{kg}}{\mathrm{s}} \cdot \frac{\mathrm{m}^3}{850 \ \mathrm{kg}} \cdot \frac{4}{\pi(0.0002)^2 \ \mathrm{m}^2} = 199 \ \mathrm{m/s}$$

The viscosity of diesel fuel is

$$\mu_\ell = 2.4 \times 10^{-3} \ \mathrm{kg/m \cdot s} \qquad \text{Appendix A.5}$$

The jet Reynold's number is then

$$\mathrm{Re}_{jet} = \frac{\rho_\ell V_{jet} d_{jet}}{\mu_\ell} = \frac{850 \ \mathrm{kg}}{\mathrm{m}^3} \cdot \frac{0.2 \times 10^{-3} \ \mathrm{m}}{1} \cdot \frac{199 \ \mathrm{m}}{\mathrm{s}} \cdot \frac{\mathrm{m \cdot s}}{2.4 \times 10^{-3} \ \mathrm{kg}} = 14{,}100$$

From Appendix A.6

$$\sigma = 3 \times 10^{-2} \ \mathrm{N/m}$$

$$\mathrm{We}_{jet,\ell} = \frac{\rho_\ell V_{jet}^2 d_{jet}}{\sigma}$$

$$\mathrm{We}_{jet,\ell} = \frac{850 \ \mathrm{kg}}{\mathrm{m}^3} \cdot \frac{0.2 \times 10^{-3} \ \mathrm{m}}{1} \cdot \left(\frac{199 \ \mathrm{m}}{\mathrm{s}}\right)^2 \cdot \frac{\mathrm{m}}{3 \times 10^{-2} \ \mathrm{N}} \cdot \frac{\mathrm{N \cdot s}^2}{\mathrm{kg \cdot m}}$$

$$\text{We}_{jet,\ell} = 224 \times 10^3$$

and

$$\text{Oh} = \frac{\left[\text{We}_{jet}\right]^{1/2}}{\text{Re}_{jet}} = \frac{\left[224 \times 10^3\right]^{1/2}}{14,100} = 0.0336$$

From the ideal gas law

$$T_{air} = \frac{p_{air}}{p_{air,0}} \cdot \frac{\rho_{air,0}}{\rho_{air}} T_{air,0} = \frac{6.2}{0.1} \cdot \frac{1.177}{25} 298 = 870 \text{ K}$$

From Appendix B

$$\mu_{air} = 3.836 \times 10^{-5} \text{ kg/m} \cdot \text{s}$$

Thus

$$\frac{\text{SMD}}{d_{jet}} = 4.12(14,100)^{0.12} \left(224 \times 10^3\right)^{-0.75} \left(\frac{2.4 \times 10^{-3}}{3.836 \times 10^{-5}}\right)^{0.54} \left(\frac{850}{25}\right)^{0.18} = 0.0222$$

$$\frac{\text{SMD}}{d_{jet}} = 0.38(14,100)^{0.25} \left(224 \times 10^3\right)^{-0.32} \left(\frac{2.4 \times 10^{-3}}{3.836 \times 10^{-5}}\right)^{0.37} \left(\frac{850}{25}\right)^{-0.47} = 0.0708$$

Selecting the larger value,

$$\text{SMD} = (0.0708)(200) = 14 \text{ μm}$$

For an impulsive liquid spray the entrainment of air varies with time and position along the spray. Figures 9.9a and 9.9b show the progression of the spray and the entrainment pattern as a function of time. From Figure 9.9a note that air is at first entrained into the spray, and then the flow goes outward after a length l_c as designated on the spray outline shown for $t = 1.05$ ms. Other work has shown that a recirculation pattern exists within the spray at the tip region. The mass of entrained air per mass of fuel injected up to a given time, m_{air}/m_f, can be calculated from

$$1 + \frac{m_{air}}{m_f} = \alpha_\varepsilon \left(\frac{x}{d_{orf}}\right) \left(\frac{\rho_{air}}{\rho_f}\right)^{1/2} \tag{9.24}$$

where x is the axial distance from the nozzle tip and α_ε is the entrainment coefficient. After about 1 ms, α_ε is approximately constant and has a value of 0.25 ± 0.05. The formula is based on data taken for the injection of fuel into a pressurized bomb

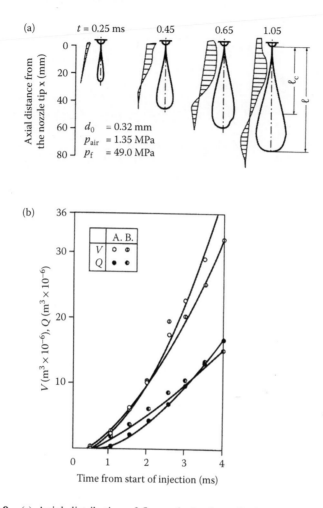

FIGURE 9.9 (a) Axial distribution of flow velocity for a liquid jet. (b) Time history of entrainment gas quantity Q and overall volume V. Based on a single hole (0.32 mm) nozzle with fuel at 22.6 MPa and ambient air at 1.35 MPa. (Ha, J.-Y., Iida, N., Sato, G. T., Hayashi, A., and Tanabe, H., "Experimental Investigation of the Entrainment into a Diesel Spray," SAE paper 841078, 1984, reprinted with permission from SAE paper 841078. © 1984, SAE International.)

at ambient temperature and thus does not include the effects of vaporization or air motion other than that induced by the spray.

Empirical observations of thick sprays formed by pressure atomization have been carried out for more than 60 years. The basic measurements are spray cone angle and tip penetration distance. There are several dozen formulas for the penetration of diesel sprays into stagnant air. For a short time, t_b, during the early development of the spray, the tip moves linearly with time; after that the spray length is proportional to the square root of time. For the initial linear portion ($t \le t_b$) the penetration distance L is given by Arai et al. (1984),

$$\frac{L}{L_b} = 0.0349 \left(\frac{\rho_{air}}{\rho_\ell} \right)^{1/2} \left(\frac{\Delta p}{\rho_\ell} \right)^{1/2} \left(\frac{t}{d_{orf}} \right) \qquad (9.25a)$$

where

$$t_b = 28.65 \left(\frac{\rho_\ell}{\rho_{air}} \right)^{1/2} \left(\frac{\rho_\ell}{\Delta p} \right)^{1/2} d_{orf}$$

and

$$L_b = 15.8 d_{orf} \left(\frac{\rho_\ell}{\rho_{air}} \right)^{1/2}$$

For $t \geq t_b$ the penetration distance is proportional to the 1/4 power of the pressure difference and to the square root of the hole diameter,

$$L = 2.95 \left[d_{orf} t \left(\frac{\Delta p}{\rho_{air}} \right)^{1/2} \right]^{1/2} \qquad (9.25b)$$

It should be noted that these formulas give a discontinuity in slope at time t_b. Specifically, the slope of the initial linear portion up to time t_b is twice that of the slope given by Equation 9.25b for the time after t_b.

Example 9.6

Calculate the penetration distance and time for a diesel fuel injector with the following data: six-hole nozzle with each hole having a 0.2 mm diameter, and 75 mm³ of fuel is injected over a duration of 0.002 s. The air density inside the cylinder is 25 kg/m³, and the air pressure is 6.2 MPa. The nominal fuel density is 850 kg/m³ and the fuel-air density ratio is 34. From experiment the flow coefficient of the nozzle holes is 0.7. That is,

$$A_{eff} = 0.7 A_{actual}$$

Solution

$$L_b = 15.8 (0.2) \left(\frac{850}{25} \right)^{1/2} = 18.4 \text{ mm}$$

To calculate t_b first find Δp.

$$\dot{m}_{f,total} = \frac{75 \text{ mm}^3}{1} \cdot \frac{850 \text{ kg}}{\text{m}^3} \cdot \frac{\text{m}^3}{10^9 \text{ mm}^3} \cdot \frac{1}{0.002 \text{ s}} = 0.0319 \text{ kg/s}$$

or

$$\dot{m}_{f,hole} = 0.00531 \text{ kg/s}$$

The area of each injector hole is

$$A = \frac{\pi d^2}{4} = \frac{\pi (0.0002)^2 \text{ m}^2}{4} = 0.0314 \times 10^{-6} \text{ m}^2$$

Using the flow coefficient of 0.7, the injection velocity is

$$V_f = \frac{\dot{m}_f}{\rho_f(0.7A)} = \frac{0.00531 \text{ kg}}{\text{s}} \cdot \frac{\text{m}^3}{850 \text{ kg}} \cdot \frac{1}{(0.7)(0.0314 \times 10^{-6} \text{ m}^2)} = 284 \text{ m/s}$$

Remembering from fluid mechanics that the pressure drop across an orifice with a constant Δp is

$$p = \frac{\rho_f V_f^2}{2} = \frac{1}{2} \cdot \frac{850 \text{ kg}}{\text{m}^3} \cdot \left(\frac{284 \text{ m}}{\text{s}}\right)^2 \cdot \frac{\text{Pa} \cdot \text{m} \cdot \text{s}^2}{\text{kg}} \cdot \frac{\text{MPa}}{10^6 \text{ Pa}} = 34.3 \text{ MPa}$$

This is the average Δp. Thus the average injection pressure is 40.53 MPa and

$$t_b = 28.65 \left(\frac{\rho_\ell}{\rho_{air}}\right)^{1/2} \left(\frac{\rho_\ell}{p}\right)^{1/2} d_{orf}$$

$$t_b = 28.65 \left(\frac{850}{25}\right)^{1/2} \left(\frac{850 \text{ kg}}{\text{m}^3} \cdot \frac{1}{34.3 \times 10^6 \text{ Pa}} \cdot \frac{\text{Pa} \cdot \text{m} \cdot \text{s}^2}{\text{kg}}\right)^{1/2} 0.0002 \text{ m}$$

$$t_b = 0.166 \text{ ms}$$

which is 8.3% of the spray duration

$$\frac{0.166 \text{ ms}}{2 \text{ ms}} = 8.3\%$$

From Equation 9.25b, the spray penetration is

$$L = 2.95 \left[d_{orf} t \left(\frac{p}{\rho_{air}} \right)^{1/2} \right]^{1/2}$$

$$L = 2.95 \left[(0.0002 \text{ m})(0.002 \text{ s}) \left(\frac{34.3 \times 10^6 \text{ Pa}}{1} \cdot \frac{\text{m}^3}{25 \text{ kg}} \cdot \frac{\text{kg}}{\text{Pa} \cdot \text{m} \cdot \text{s}^2} \right)^{1/2} \right]^{1/2}$$

$$L = 0.0639 \text{ m}$$

Thus the spray tip will be about 64 mm from the nozzle at the end of the injection period.

A correction factor for the effect of a cross flow velocity is given by Arai et al. (1984). The cross flow is taken as a solid body rotation with the spray traveling radially outward from the center of the rotation. The penetration with cross flow, L_{cf}, is then given by

$$\frac{L_{cf}}{L} = \frac{1}{1 + 2\pi\omega L / \underline{V}_\ell} \tag{9.26}$$

where ω is the rotational speed of air and $\underline{V}_\ell$ is the velocity of the liquid at the orifice.

Example 9.7

Consider the previous example above with $L = 64$ mm and $\underline{V}_\ell = 200$ m/s. For an engine with a speed of 1800 rpm and a swirl rate of 3, find the penetration of the spray. It should be noted that cavitation can change the density of the discharge fluid and thus the velocity is uncertain; however, this effect is neglected here.

Solution

$$\omega = \frac{3(1800) \text{ rev/min}}{60 \text{ min/s}} = 90 \text{ s}^{-1}$$

Using Equation 9.26,

$$\frac{L_{cf}}{L} = \frac{1}{1 + 2\pi\omega L / \underline{V}_\ell} = \frac{1}{1 + 2\pi(90)64 / (200,000)} = 0.847$$

and

$$L_{cf} = 54 \text{ mm}$$

Thus the swirl causes the fuel to be distributed in a volume of about 2/3 that of the non-swirl case but improves fuel-air mixing within that smaller volume.

The total cone angle, θ (expressed in degrees), of the spray is a function of hole geometry, but for simple orifices can be approximated by

$$\theta = 0.05 \left(\frac{\Delta p d_{orf}^2}{\rho_{air} v_{air}^2} \right)^{1/4} \tag{9.27}$$

Cone angle taken from measurements is sometimes used to approximately calculate the amount of entrained air.

Example 9.8

Consider a 0.2 mm diameter nozzle hole with Δp of 34.3 MPa. The diesel fuel sprays into compressed air at 25 kg/m³ density and 6.2 MPa pressure. Prior to compression the air was at 1 atm and 25°C. Calculate the total cone angle θ of the spray.

Solution

From the ideal gas law

$$T_a = \frac{p_{air}}{p_{air,0}} \frac{\rho_{air,0}}{\rho_{air}} T_{air,0} = \frac{6.2}{0.1} \cdot \frac{1.177}{25} (298 \text{ K}) = 870 \text{ K}$$

From Appendix B

$$\mu_{air} = 3.836 \times 10^{-5} \text{ kg/m} \cdot \text{s}$$

so that

$$v_{air} = \frac{3.836 \times 10^{-5} \text{ kg}}{\text{m} \cdot \text{s}} \cdot \frac{\text{m}^3}{25 \text{ kg}} = 1.534 \times 10^{-6} \text{ m}^2/\text{s}$$

Substituting into Equation 9.27,

$$\theta = 0.05 \left[\frac{34.3 \times 10^6 \text{ Pa}}{1} \cdot \frac{\text{m}^3}{25 \text{ kg}} \left(\frac{0.0002 \text{ m}}{1} \right)^2 \left(\frac{\text{s}}{1.534 \times 10^{-6} \text{ m}^2} \right)^2 \cdot \frac{\text{kg}}{\text{Pa} \cdot \text{m} \cdot \text{s}^2} \right]^{1/4}$$

$$\theta = 19.5°$$

9.4   VAPORIZATION OF SINGLE DROPLETS

Droplet vaporization rate and the time to completely evaporate a single droplet are of significant interest in the design of combustion systems. In many combustion systems evaporation is the rate-limiting step. The vaporization rate depends on the latent heat of vaporization of the droplet, the boiling point of the droplet, the size of the droplet, the gas temperature and pressure, and the relative velocity between the droplet and the gas. In most instances the relative velocity between the droplet and the surrounding gas is small and not a factor. Pure fuels have a well-defined boiling point, but mixtures such as diesel and jet fuel have a range of boiling points. High pressure causes a decrease in the latent heat, and at the approach to the critical point, the latent heat goes to zero.

Droplet vaporization rates have been measured extensively by suspending a single droplet on a fine thermocouple wire and photographing the change of droplet diameter with time. After an initial transient period, it is observed that the droplet diameter squared decreases linearly with time:

$$d^2 = d_0^2 - \beta t \tag{9.28}$$

where β is the vaporization constant. Substituting $d = 0$ and rearranging Equation 9.28, the vaporization time for a droplet of size d_0 is

$$t_v = \frac{d_0^2}{\beta} \tag{9.29}$$

As shown in Figure 9.10, droplet evaporation is a two-step process. Initially, heat transfer from the environment heats the droplet and there is little vaporization. As the droplet heats up, the vapor pressure of the liquid builds up, and the evaporation rate starts to increase. Evaporation removes heat from the droplet, and an energy balance is quickly reached where the energy lost from the droplet due to evaporation is equal to the heat transfer to the droplet. At that point the evaporation rate is limited by the rate of heat transfer to the droplet. For a liquid sphere (droplet) the mass is

$$m_\ell = \frac{\rho_\ell \pi d^3}{6} \tag{9.30}$$

Differentiating the mass loss from the droplet gives

$$\dot{m}_\ell = \frac{dm_\ell}{dt} = \frac{d}{dt}\left(\frac{\rho_\ell \pi d^3}{6}\right) \tag{9.31}$$

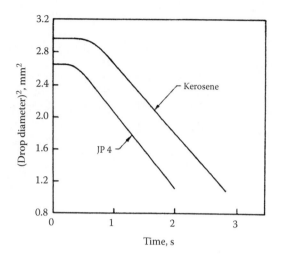

FIGURE 9.10 Vaporization rate curves for kerosene and jet fuel. (From Wood, B. J., Wise, H., and Inami, S. H., "Heterogeneous Combustion of Multicomponent Fuels," *Combust. Flame* 4:235–242, 1960.)

Assuming the density of the liquid remains constant, it follows that the steady state vaporization rate is given by

$$\dot{m}_\ell = \frac{\rho_\ell \pi}{6} \frac{d(d^3)}{dt} = \frac{\rho_\ell \pi}{2} d^2 \frac{d(d)}{dt} \tag{9.32}$$

Rearranging,

$$\dot{m}_\ell = \frac{\rho_\ell \pi}{4} d \frac{d(d^2)}{dt} \tag{9.33}$$

Substituting Equation 9.28 into Equation 9.33 yields

$$\dot{m}_\ell = \frac{\rho_\ell \pi}{4} d \frac{d(d_0^2 - \beta t)}{dt} = -\frac{\rho_\ell \pi \beta d}{4} \tag{9.34}$$

Some values of the vaporization constant, β, are given in Table 9.2. The vaporization constant increases with ambient pressure and temperature. As an example, measurements of the vaporization constant for n-heptane are shown in Figure 9.11 as a function of the surrounding temperature and pressure.

For spherical droplets of a pure liquid fuel and assuming radiation heat transfer is negligible, heat and mass transfer theory can be used to show that the evaporation constant, β, is given by

TABLE 9.2

Vaporization Constant for Selected Fuels in Air at 1 atm Pressure

Fuel	Air Temperature (K)	β (mm²/s)
Gasoline	971	1.06
Gasoline	1068	1.49
Kerosene	923	1.03
Kerosene	971	1.12
Kerosene	1014	1.28
Kerosene	1064	1.47
Diesel fuel	437	0.79
Diesel fuel	971	1.09

Source: Bartok, W. and Sarofim, A. F., *Fossil Fuel Combustion: A Source Book*, © 1991,
by permission of John Wiley & Sons, Inc.

$$\beta = \left(\frac{8\tilde{k}_g}{\rho_\ell c_{p,g}}\right) \ln(1+B) \tag{9.35}$$

where B is the mass transfer driving force. At the high temperatures usually encountered in combustion (around 1400–2000 K)

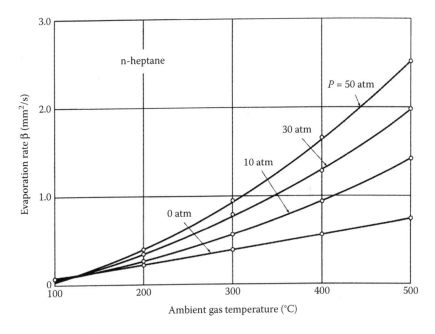

FIGURE 9.11   Vaporization rate, β, for large n-heptane droplets in stagnant nitrogen vs. ambient pressure and temperature. (From Hiroyasu H. and Kadota, T., "Fuel Droplet Size Distribution in Diesel Combustion Chamber," SAE paper 740715, 1974, reprinted with permission from SAE paper 740715. © 1974, SAE International.)

$$B = \frac{c_{p,g}\left(T_\infty - T_{boil}\right)}{h_{fg}}$$

(9.36)

where T_{boil} is the boiling point temperature of the liquid.

9.5 PROBLEMS

1. Consider the following very simple distribution of droplet sizes:

   1/4 of the droplets have a diameter of 10 μm

   1/2 of the droplets have a diameter of 20 μm

   1/4 of the droplets have a diameter of 30 μm.

   Find the mean diameter $(\bar{d}_1)$, the area mean diameter $(\bar{d}_2)$, the volume mean diameter $(\bar{d}_3)$, and the Sauter mean diameter $(\bar{d}_{32})$ for this droplet distribution.

2. Using the plot of Figure 9.3, make a table of cumulative volume fractions versus droplet diameters for SMDs of 50 and 25 μm. What is the mean diameter $(\bar{d}_1)$, the area mean diameter $(\bar{d}_2)$, and the volume mean diameter $(\bar{d}_3)$ of each of the size distributions?

3. For the prefilming airblast atomizer shown in Figure 9.6, use the following data and Equation 9.20. Compute the SMD for fuels A and B. The diameter of the pintle is 1 cm.

Property	Fuel A	Fuel B
σ (kg/s^2)	27×10^{-3}	74×10^{-3}
ρ_ℓ (kg/m^3)	784	1000
μ_ℓ (kg/m·s)	1.0×10^{-3}	53×10^{-3}
p_{air} (kPa)	100	100
T_{air} (K)	295	295
V_{air} (m/s)	100	100
$\dot{m}_{air}/\dot{m}_\ell$	3	3

   If the air velocity is doubled, but the mass flow rate of air $(\dot{m}_{air})$ is kept constant, what is the effect on the SMD for each of the fuels. Discuss how the mechanisms of atomization differ for these two fuels.

4. Assume that the shape of a liquid fuel spray can be modeled as a cone of angle θ with a hemisphere at the tip; it will look like a single scoop ice cream cone with half of the ice cream inside the cone. Using the angle and penetration formulas, find an expression for the spray volume versus time. The spray is in air at 13.5 atm pressure and 20°C temperature. The injector orifice diameter is 0.32 mm, the liquid mass flow rate is constant at 0.014 kg/s, and the injection pressure is 22.6 MPa. The spray is injected into a large volume of air and after 4 ms the injection ends. The fuel is dodecane. Plot the average fuel-air ratio in the spray volume for $t \leq 4$ ms. Assume that all of the fuel remains in the spray volume and neglect vaporization.

What effect would droplet vaporization have on the validity of the method? Compare with Figure 9.9.

5. Consider the following for a droplet of dodecane which has been partially vaporized and transported to a region where it is moving at the local bulk velocity and is vaporizing at a steady state temperature of 450 K. (a) The temperature of the air is 1500 K and the pressure is 1 atm. Compute the vaporization constant and vaporization time of a 30 μm diameter droplet. See Appendix A.9 for fuel properties. (b) The turbulent eddies in the region where the droplet is located have an average size of 3000 μm and are turning at 300 rev/s. If the 30 μm droplet travels around the outer edge of an eddy with zero relative velocity, how many times must it go around the circular 3000 μm path before it is vaporized? (c) Given the same conditions as in Part (a) what size droplet will travel just one eddy revolution in its lifetime?

6. For a particular simplex nozzle the Sauter mean diameter of the spray is given by

$$SMD = 2.25\sigma^{0.25}\mu_\ell^{0.25}\dot{m}_\ell^{0.25}\Delta p^{-0.5}\rho_{air}^{-0.25}$$

a. Rewrite this equation in terms of σ, μ_ℓ, d_{orf} (the orifice diameter), Δp, and ρ_{air}. That is, derive a relation between $\dot{m}_\ell$ and d_0, and substitute. Here $\dot{m}_\ell$ is the mass flow rate of liquid. The flow coefficient for the orifice is K. Hint: review the Bernoulli equation.

b. Using the above expression for SMD, if the SMD is decreased by a factor of 2 while everything is held constant except the pressure drop, how much does the power of the fuel pump increase?

7. For a kerosene drop with an initial diameter of 1 mm in air at 700°C and 1 atm absolute pressure, (a) find the time to vaporize 50%, 90% and 100% of initial mass of the droplet, and (b) what is the mass flow rate from the droplet when the droplet is 50% and 90% vaporized?

8. For diesel fuel in air at 971 K and 1 atm, find the time required to completely vaporize 100, 50, 25, and 10 μm droplets. Also find the initial mass flow rate of the vapor from each of these drop sizes and the mass flow rate when 50% of the drop is vaporized. Use the data in Table 9.2 and 2.4.

REFERENCES

Arai, M., Tabata, M., Hiroyasu, H., and Shimizu, M., "Disintegrating Process and Spray Characterization of Fuel Jet Injected by a Diesel Nozzle," SAE paper 840275, 1984.

Bartok, W. and Sarofim, A. F., *Fossil Fuel Combustion: A Source Book*, Wiley, New York, 1991.

Bosch, *Diesel Fuel Injection*, SAE International, 1997.

Curtis, E. W. and Farrell, P. V., "A Numerical Study of High-Pressure Droplet Vaporization," *Combust. Flame* 90(2):85–102, 1992.

Dodge, L. G. and Schwalb, J. A., "Fuel Spray Evolution: Comparison of Experiment and CFD Simulation of Nonevaporating Spray," *J. Eng. Gas Turbines Power* 111(1):15–23, 1989.

Elkotb, M. M., "Fuel Atomization for Spray Modeling," *Prog. Energy Combust. Sci.* 8(1):61–91, 1982.

Ha, J.-Y., Iida, N., Sato, G. T., Hayashi, A., and Tanabe, H., "Experimental Investigation of the Entrainment into a Diesel Spray," SAE paper 841078, 1984.

Hiroyasu H. and Kadota, T., "Fuel Droplet Size Distribution in Diesel Combustion Chamber," SAE paper 740715, 1974.

Hiroyasu H., Arai, M., and Tabata, M., "Empirical Equations for the Sauter Mean Diameter of a Diesel Spray," SAE paper 890464, 1989.

Knight, B. E., Communication on A. Radcliffe, "The Performance of a Type of Swirl Atomizer," *Proc. Inst. Mech. Eng.* 169:104–105, 1955.

Kadota, T. and Hiroyasu, H., "Evaporation of a Single Droplet at Elevated Pressures and Temperatures," *Bull. JSME* 19(138):1515–1521, 1976.

Kong, S.-C. and Reitz, R. D., "Spray Combustion Processes in Internal Combustion Engines," in *Recent Advances in Spray Combustion: Spray Combustion Measurements and Model Simulation* Vol. 2, ed. K. K. Kuo, 171:395–424, ed. P. Zarchan, American Institute of Aeronautics and Astronautics, 1996.

Lefebvre, A. H., *Atomization and Sprays*, Taylor & Francis, Boca Raton, FL, 1989.

Nukiyama S. and Tanasawa Y., "An Experiment on the Atomization of Liquid. III. Distribution of the Size of Drops," *Trans. Jpn. Soc. Mech. Eng.* 5(18):63–67, 1939.

Reitz, R. D. and Bracco, F. V., "Mechanisms of Breakup of Round Liquid Jets," in *Encyclopedia of Fluid Mechanics,* 233–249, ed. N. P. Cheremisnoff, Gulf Publishing, Houston, TX, 1986.

Rosin, P. and Rammler, E., "The Laws Governing the Fineness of Powdered Coal," *J. Inst. Fuel* 7:29–36, 1933.

Saminy, M., Breuer, K. S., Leal, L. G., and Steen, P. H., *A Gallery of Fluid Motion*, Cambridge University Press, Cambridge, UK, 2004.

Sirignano, W. A., *Fluid Dynamics and Transport of Droplets and Sprays*, 2nd ed., Cambridge University Press, Cambridge, UK, 2010.

Simmons, H. C., "The Correlation of Drop-Size Distributions in Fuel Nozzle Sprays. Parts I and II.," *J. Eng. Power* 99(3):309–319, 1977.

Stiesch, G., *Modeling Engine Spray and Combustion Processes*, Springer Verlag, Berlin, 2003.

Tanasawa, Y. and Toyoda, S., "On the Atomization of a Liquid Jet Issuing from a Cylindrical Nozzle," *Technol. Rep. Tohoku Univ.*, Japan 19:135, 1955.

———, "On the Atomization Characteristics of Injection for Diesel Engines," *Technol. Rep. Tohoku Univ.*, Japan 21:117, 1956.

Van Dyke, M., *An Album of Fluid Motion*, Parabolic Press, 10th ed., Stanford, CA, 1982.

Wood, B. J., Wise, H., and Inami, S. H., "Heterogeneous Combustion of Multicomponent Fuels," *Combust. Flame* 4:235–242, 1960.

10 Oil-Fired Furnace Combustion

In this chapter, we first consider the system aspects of oil-fired furnaces and boilers. Following this, combustion of sprays is discussed and a simple plug flow model of droplet combustion is developed. Finally, emissions from oil-fired furnaces and boilers are examined.

In industrial applications, fuel oil furnaces and boilers are used for process heat and steam as well as a number of other applications. Distillate fuel (No. 2) oil is used in residential, commercial, and industrial furnaces and boilers, while residual fuel (No. 6) oil is used in power plants and large industrial boilers. Because of the need to optimize crude oil in refineries for gasoline, Nos. 3, 4, and 5 fuel oil grades are rarely available. Due to their high cost, distillate fuel oil is rarely used in large burners.

Fuel oils are atomized by means of a spray nozzle in order to achieve rapid combustion. The exception is kerosene, which vaporizes rapidly as a warmed liquid. Because of this, small kerosene space heaters are made with a wick or a combustion pot where heat from the kerosene flame heats and vaporizes the fuel. No. 2 fuel oil is sprayed at ambient temperature (even in cold climates), but No. 6 fuel oil must be heated to 100°C to ensure proper pumping and atomization.

10.1 OIL-FIRED SYSTEMS

A schematic diagram of an oil-fired furnace is shown in Figure 10.1. Fuel oil is pumped through a spray nozzle, the fuel droplets mix in a burner with air from a blower, and the two-phase mixture burns out in the combustion chamber. The fuel and air flow requirements for a given heat output are calculated from an energy balance around the combustion chamber and heat exchanger. The resulting equation set is the same as for the energy balance around a gas-fired furnace and is given in Equations 6.1 through 6.6. The only significant difference is the work associated with the oil fuel pump. The power required by the fuel pump is obtained from an energy balance around the pump. The result is

$$\dot{W}_{pump} = \frac{\dot{V}_f \Delta p_f}{\eta_{pump}} \tag{10.1}$$

where Δp_f is the pressure rise across the fuel pump, which essentially equals the fuel pressure supplied to the fuel nozzle, and η_{pump} is the efficiency of the pump. The efficiency of the furnace or boilers is then defined as the ratio of the useful heat output

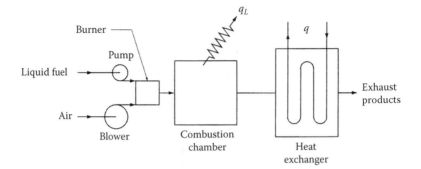

FIGURE 10.1 Schematic drawing of an oil-fired furnace, where q is the useful heat output.

to the energy input, and by convention in the United States efficiency is based on the higher heating value.

$$\eta = \frac{q}{\dot{m}_f\left(\text{HHV}\right) + \dot{W}_{blower} + \dot{W}_{pump}} \qquad (10.2)$$

where the work of the blower is given by Equation 6.6.

In small oil-fired systems, the combustion chamber is refractory lined to promote combustion. Refractory is high temperature (generally ceramic) material that provides structural strength and insulation in furnace applications. In large oil-fired systems, the burners fire directly into a large volume chamber that contains the heat exchanger surfaces. *Fire tube* boilers, where combustion products pass through tubes immersed in a tank of water, are typically used for small applications, and *water tube* boilers, where water flows in tubes exposed to combustion products, are used for large systems.

As with gas-fired furnaces and boilers, the highest efficiency is obtained by completely burning the fuel with as little excess air as possible. This requires fine atomization of the fuel and rapid penetration of the air into the spray. Older domestic distillate oil-fired furnaces and boilers require 40%–60% excess air and have efficiencies of 75%–80%. Older commercial oil burners require 30% excess air, while industrial oil burners may require 15% excess air. Utility boilers using residual fuel oil require as little as 3% excess air. Attempts to reduce the excess air with inadequate mixing result in increased particulate (smoke) emissions. Newer systems have improved mixing, lower excess air, and reduced particulate emissions.

Atomization of fuel oil is typically accomplished by one of three methods: single fluid atomizers, twin fluid atomizers, or rotary cup atomizers. The burner design should include a fuel filter and fuel pump, fuel atomizer, and inlet air blower. The burner should provide a stable flame with low excess air, produce low emissions, and be durable and reliable.

The single fluid pressure jet atomizer is used widely for residential, commercial, and small industrial furnaces burning No. 2 fuel oil. In this type of application, the burner is referred to as a high-pressure gun-type burner (Figure 10.2). Fuel oil is pressurized to 7 atm and forced through a nozzle that atomizes the liquid into small droplets, which typically have a mass mean diameter of 40 μm, and 10% of the

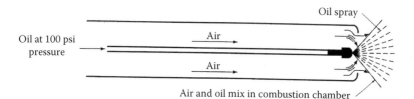

FIGURE 10.2   High-pressure gun-type oil burner.

droplets are larger than 100 μm. Air from a low-pressure (about 25 cm H_2O gauge pressure) blower flows around the fuel nozzle and mixes with the spray. To obtain good mixing of the fuel droplets and air, guide vanes are mounted near the spray nozzle.

The liquid jet from the fuel nozzle breaks up into a spray due to the high axial velocity of the liquid relative to the surrounding air and due to the tangential velocity that is imparted to the fuel jet by means of swirl. As shown in Figure 10.3, the fuel oil flows tangentially into a small swirl chamber at the tip of the nozzle before exiting the nozzle. The angle of the external spray cone can be varied from 30° to 90° by varying the angle of the interior cone, which changes the ratio of the axial velocity to rotational velocity of the jet. An 80° hollow-cone nozzle is typical of domestic use. Electrodes located near the upstream edge of the spray provide a continuous source of ignition.

Residential oil burners typically use 2–10 L/h. Commercial and small industrial burners use nozzles that provide up to 130 L/h per nozzle. Fuel droplet size is distributed between 50 and 300 μm. Turndown cannot be achieved by reducing the oil pressure because this increases the droplet size, which causes incomplete burnout in small furnaces. Control of the combustion rate is achieved by shutting off the oil pressure to the nozzle. Residential units have only one nozzle and thus on-off control is used.

Small oil burner designs have both improved the mixing of fuel and air and reduced the excess air by adding swirl to the air and by adding a flame retention device (Figure 10.4). Air swirl is achieved with guide vanes mounted to the inside of the burner can. The flame retention device is a metal cone that is mounted a short distance from the spray nozzle. The gap between the nozzle and the cone, and slots and holes in the cone allow for air penetration into the spray. The cone acts as an air

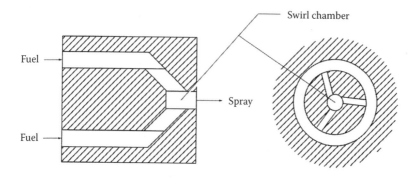

FIGURE 10.3   Pressure atomization nozzle for distillate fuel oil.

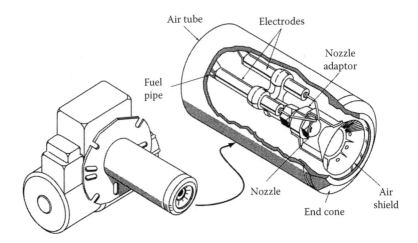

FIGURE 10.4 Residential oil burner with flame retention air shield. (From Offen, G. R., Kesselring, J. P., Lee, K., Poe, G. and Wolfe, K. J., *Control of Particulate Matter from Oil Burners and Boilers*, Environmental Protection Agency EPA-450/3-76-005, 1976.)

shield that stabilizes the flame, creating a more compact and intense flame. Swirl creates more mixing beyond the flame retention cone. In this way, excess air can be reduced and efficiency increased.

The single hole spray burner cannot be fired below 2 L/min, which represents about 75 MJ/h, because it is not practical to reduce the hole size. There are numerous applications where a smaller firing rate is needed. One option is pre-vaporizing the fuel in a pot burner; however, this tends to yield a sooty flame. The Babbington burner, which operates on a different principle, can burn cleanly at much lower flow rates. This burner uses an "inside-out" nozzle shown in Figure 10.5. The liquid fuel flows over the outside of

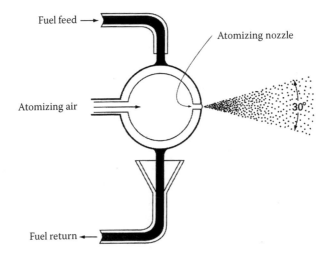

FIGURE 10.5 The Babbington atomization principle (burner). (From Babbington, R. S., McLean, Virginia, personal communication, 1993.)

a hollow sphere and drains off into a fuel return tube. Low pressure air at 70 kPa flows into the sphere and through a small hole that atomizes a thin film of fuel. The atomization air is about 1% of the combustion air, and the mass mean droplet size of No. 2 fuel oil is about 15 μm with less than 1% of the droplets greater than 70 μm. This small droplet size facilitates good mixing and low emissions. The spray is formed by rupturing the fuel film, and the droplet size depends on surface tension but not viscosity. Turndown is achieved by reducing the fuel flow, which reduces the thickness of the fuel film on the sphere. Fuel flow rates from 4 L/h to as low as 2 L/week have been demonstrated.

Utility and large industrial burners operating with heavy fuel oil use single fluid or twin fluid atomizers. In a single fluid pressure jet atomizer for heavy fuel oil, shown in Figure 10.6a, the fuel oil pressure is typically about 30 atm with fuel rates up to 4700 L/h per nozzle. Liquid enters a swirl chamber with a tangential component of velocity and passes through a discharge orifice as an annular film that then disintegrates into a spray with additional assistance from a central core air jet. The volumetric size distribution for heavy fuel oil is shown in Figure 10.6b. The mass mean diameter is 155 μm. The spill flow pressure jet atomizer shown in Figure 10.7 is a modification of the atomizer shown in Figure 10.6a to allow for turndown and shut-off. A return flow path for the oil is provided inside the nozzle so that turndown of up to 3 to 1 is achieved by applying a backpressure to the return flow. However, the central air core is sacrificed.

Twin fluid atomizers use a jet of steam or air to impinge on the fuel jet. A steam atomizing nozzle of the Y-jet type is shown in Figure 10.8. Typically, the steam pressure is only 1–3 atm above the oil pressure. A two-phase mixture is formed in the exit port and expands out of the port to form a spray. The flow rate of steam should not exceed 10% of the flow rate of oil. Of course, this requires makeup boiler water and reduces the boiler efficiency slightly. If steam is not available, compressed air can be used. The droplet size distribution is similar to single fluid atomizers. The main advantage of twin fluid atomizers is that the fuel pressure is reduced.

The rotary cup atomizer was introduced as a means of further reducing the required oil pressure. As shown in Figure 10.9, oil is fed to a conical cup through a tube inside a hollow rotating shaft. The shaft and cup rotate at 3500 rpm. Oil flows along the inside of the cup in a sheet and is thrown by centrifugal force into the surrounding air stream. The droplet sizes formed are considerably larger than for the other two types of atomizers mentioned, which tends to cause smoking. Thus, in recent years rotary cup atomizers have been used less frequently than other types of atomizers.

10.2 SPRAY COMBUSTION IN FURNACES AND BOILERS

The process by which a liquid fuel spray burns depends on the number density of the spray, the degree of turbulent mixing, and the fuel volatility. If the number density is low, the mixing is high, and the fuel volatility is relatively low, then the spray burns as individual droplets surrounded by individual flames. If the number density of the spray is high (i.e., a dense spray), the mixing is relatively low, and the fuel volatility is high, then the droplets vaporize, but the flame occurs at the outer edge of the spray. This is termed *external combustion* because the local fuel-air ratio is too high for combustion except at the outer edge of the spray. An intermediate case, which is termed *group combustion*, occurs when a group of droplets within the spray burns in

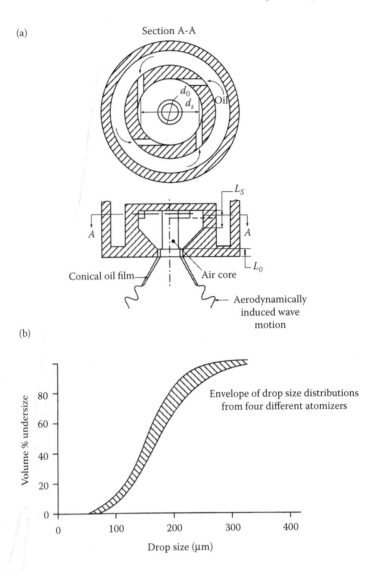

FIGURE 10.6 Pressure jet atomizer with swirl for heavy fuel oil: (a) atomizer and (b) drop-let size distribution. (From Lawn, C. J., ed., *Principles of Combustion Engineering for Boilers*, Academic Press, London, 1987, by permission of author.)

the manner of a large single drop. The external combustion flame tends to exhibit a blue flame, whereas the droplet combustion flame tends to be yellow, indicating rich combustion with the consequent production of soot.

The spray processes for multicomponent fuel oil combustion can be summarized as follows:

1. Heating of the droplet and vaporization of the low boiling point components
2. Ignition of the volatiles surrounding the droplet

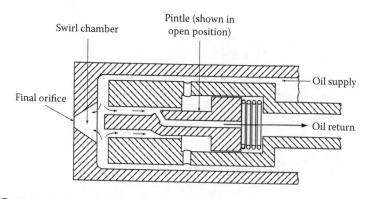

FIGURE 10.7 Spill flow pressure jet atomizer with shut-off. (From Lawn, C. J., ed., *Principles of Combustion Engineering for Boilers*, Academic Press, London, 1987, by permission of author.)

3. Thermal decomposition, disruptive boiling, and swelling of the droplet
4. Continued thermal decomposition of the droplet as the volatile flame continues
5. Burning of the carbonaceous residue on the remaining droplet surface at about one-tenth the initial burning rate of the droplet

Fuel oils that contain water exhibit enhanced boiling and swelling prior to ignition. Extensive disruptive boiling of the droplets effectively results in finer atomization by ejection of smaller satellite droplets from the parent droplet. The satellite droplets may ignite before the parent droplet.

The fluid dynamic flow of the air dominates the shape of the combustion zone. As noted earlier, most small oil-fired burners use a flame retention cone that stabilizes the initial flame zone, provides mixing of air into the droplets, and imparts a small amount of swirl. The droplets move on a set trajectory in this cone region and then flow away from the cone in a more nearly uniform flow, all the while burning primarily as a collection of individual droplets. For light fuel oil the droplet vaporization rate is proportional to the droplet diameter (Equation 9.34). Because vaporization is the rate-limiting

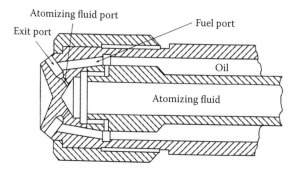

FIGURE 10.8 Y-jet atomizer. (From Lawn, C. J., ed., *Principles of Combustion Engineering for Boilers*, Academic Press, London, 1987, by permission of author.)

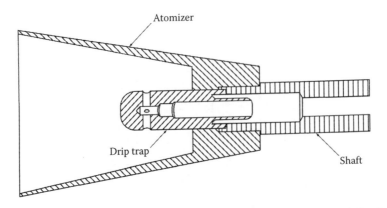

FIGURE 10.9 Rotary cup atomizer. (From Lawn, C. J., ed., *Principles of Combustion Engineering for Boilers*, Academic Press, London, 1987, by permission of author.)

step in droplet combustion, the total droplet burning time is proportional to the diameter squared (Equation 9.29). At a furnace temperature of 1000 K, the vaporization constant, β, is about 1.25 mm^2/s. For heavy fuel oil, the "diameter-squared-law" does not hold because the droplet swells, and a carbonaceous residue forms on the surface.

In larger nozzle type burners the spray is injected axially with high momentum, and the air is blown in with varying amounts of swirl. Combustion air comes from a windbox through movable registers that facilitate adjusting the flame length and shape. A vaned stabilizer (slotted turbulator) attached to the fuel nozzle causes mixing of the air and spray, and provides flame stabilization, as shown in Figure 10.10a. With low swirl the flame is long and narrow, as shown in Figure 6.9a. Gases within

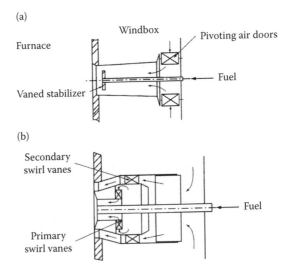

FIGURE 10.10 Industrial burners showing an airflow path: (a) swirl air with a vaned stabilizer (slotted turbulator), (b) dual swirl burner. (From Lawn, C. J., ed., *Principles of Combustion Engineering for Boilers*, Academic Press, London, 1987, by permission of author.)

the combustion chamber near the flame are entrained by the momentum of the jet, thus causing recirculation and mixing at the flame boundary. The droplets travel in a straight line and burn out as a combination of individual droplet flames and an external flame. If a divergent fuel spray is used and the air is blown in with swirl (Figure 10.10b), then a short brush-like flame, which is stabilized on the surface of the refractory nozzle of the burner (called the quarl), is produced. With high swirl there is a closed internal recirculation zone that promotes a short, high intensity flame. The different types of flames are used for different applications. Long jet flames are used in corner-fired boilers and with industrial processes where radiant heat transfer is used. Swirl flames are used in wall-fired boilers and industrial heaters. Temperature contours near the burner of Figure 10.10b are shown in Figure 10.11. There was no evidence of a closed recirculation zone in this case.

Large industrial and utility burners typically burn residual fuel oil, and the droplets burn quite differently and more slowly than distillate droplets. When a residual oil droplet is subjected to a high temperature gas, say 1300°C as in Figure 10.12, it heats up and starts to evaporate at about 200°C with slight swelling until the onset of boiling, which is accompanied by distortion and swelling. Droplet expansions alternate with contractions in rapid succession, the overall diameter increasing and reaching about twice the initial diameter. Puffs of vapor are ejected at this stage with considerable force and over distances up to 10 diameters from the parent droplet. Sometimes satellites are expelled, and in exceptional cases the entire droplet disintegrates. Generally, volatilization proceeds by distillation (boiling-off of increasingly

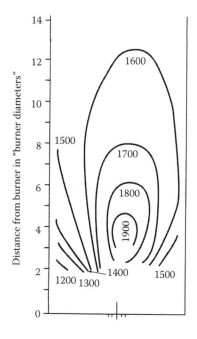

FIGURE 10.11 Temperature (K) contours for a swirl pressure jet nozzle using heavy fuel oil. (From Lawn, C. J., ed., *Principles of Combustion Engineering for Boilers*, Academic Press, London, 1987, by permission of author.)

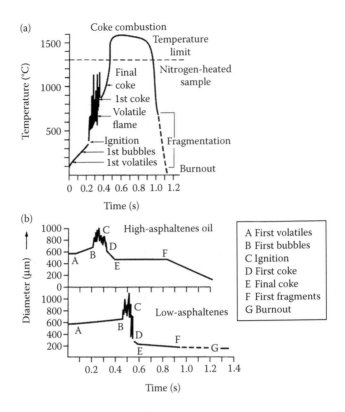

FIGURE 10.12 Temperature and size history of a single residual fuel oil droplet in 1300°C air. (From Lightman, P. and Street, P. J., "Single Drop Behaviour of Heavy Fuel Oils and Fuel Oil Fractions," *J. Inst. Energy* 56(426):3–11, 1983. © 1983, with permission of Maney Publishing.)

higher boiling point compounds) and by pyrolysis (breaking of molecular bonds to form lower molecular weight compounds). As the volatiles are released, 0.02–0.2 μm soot particles are formed, which can agglomerate into filaments up to several thousand microns long before burning out in leaner regions of the flame. Towards the end of the volatile loss phase, the droplets become very viscous and solidify into porous coke particles. Fuels that are high in asphaltenes tend to form coke. Coke from a residual oil spray is a very porous, carbonaceous particle that burns out with decreasing density until fragmentation occurs towards the end of burnout, as shown in Figure 10.12b. Solid particle burnout will be considered in detail in Chapter 14.

10.3 PLUG FLOW MODEL OF A UNIFORM FIELD OF DROPLETS

The combustion of an oil burner spray typically involves a complex three-dimensional flow of droplets and air with turbulent mixing, droplet vaporization, and combustion. However, it is instructive to simplify the problem by assuming a one-dimensional, constant pressure, constant area stream tube containing a uniform flow of *monodispersed* droplets in air. Monodispersed means uniformly sized droplets of uniform spacing. The droplets and gas have the same velocity. This case is referred to as plug flow, and it applies to a low swirl flow.

Vaporization and ignition start at $x = 0$, as indicated in Figure 10.13. The object is to find the temperature as a function of distance, the length of the reaction zone, and the overall combustion intensity (the heat release per unit volume). It is reasonable to assume that the mixing and chemical reaction time is short compared to the droplet vaporization time. The spray is dilute so that the spray does not occupy significant volume, but it does add mass. At $x = 0$, the number of droplets per unit volume, n'_0, at a specified fuel-air ratio can be determined as follows. Given the differential volume, $A dx$, initially the mass of the fuel is

$$m_{f,0} = n'_0 \rho_\ell \frac{\pi d_0^3}{6} A dx \tag{10.3}$$

where the subscript 0 indicates the position $x = 0$. Similarly, the mass of the air is

$$m_{air,0} = m_{total,0} - m_{f,0} = \left(\rho_0 - n'_0 \rho_\ell \frac{\pi d_0^3}{6} \right) A dx \tag{10.4}$$

where ρ_0 is the initial density of the air-droplet mixture. The initial fuel-air ratio is found by dividing Equation 10.3 by Equation 10.4.

$$f = \frac{\left(n'_0 \rho_\ell \pi d_0^3 / 6 \right) A dx}{\left[\rho_0 - \left(n'_0 \rho_\ell \pi d_0^3 / 6 \right) \right] A dx} \tag{10.5}$$

Solving for n'_0,

$$n'_0 = \frac{f}{1+f} \cdot \frac{\rho_0}{\rho_\ell} \cdot \frac{6}{\pi d_0^3} \tag{10.6}$$

Moving downstream the temperature increases, the density decreases, and the velocity increases. Recalling that the area is constant, conservation of mass is

$$\rho_0 \underline{V}_0 = \rho \underline{V} \tag{10.7}$$

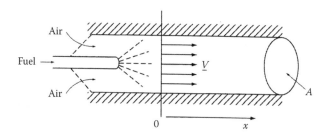

FIGURE 10.13 Plug flow model of spray combustion.

Combustion Engineering, Second Edition

where ρ is the density of the air-droplet mixture. The number of droplets remains the same, but due to the expansion of the gas, the number of droplets per unit volume decreases,

$$n_0' \underline{V}_0 = n' \underline{V} \tag{10.8}$$

Combining Equations 10.7 and 10.8 and simplifying,

$$n' = n_0' \rho / \rho_0 \tag{10.9}$$

From Chapter 9, a droplet at steady state conditions vaporizes according to

$$\dot{m}_{drop} = -\frac{\rho_\ell \pi \beta d}{4} \tag{9.34}$$

and droplet diameter squared decreases linearly with time:

$$d^2 = d_0^2 - \beta t \tag{9.28}$$

where β is the vaporization rate constant.

The energy transport by conduction and diffusion is small in comparison to the energy transport by convection for the extended reaction zone (Figure 10.13). Based on this conservation of energy for the differential volume, $A dx$, is

$$\rho \underline{V} A c_p \frac{dT}{dx} dx = q''' A dx \tag{10.10}$$

where q''' is the heat release rate per unit volume due to combustion. Canceling the volume element $A dx$ and noting that $\underline{V} = dx/dt$, the energy equation becomes

$$\rho c_p \frac{dT}{dt} = q''' \tag{10.11}$$

The heat release rate per unit volume in the differential element is

$$q''' = -n' \dot{m}_{drop} (\text{LHV}) \tag{10.12}$$

Substituting Equations 10.6, 10.9, 9.34, and 10.12 into Equation 10.11 and simplifying,

$$\frac{dT}{dt} = \frac{3}{2} \cdot \frac{f}{1+f} \cdot \frac{\beta(\text{LHV})}{c_p d_0^3} \cdot d \tag{10.13}$$

Differentiating Equation 9.28 yields

$$dt = -\frac{2d}{\beta} d(d) \tag{10.14}$$

Substituting Equation 10.14 into Equation 10.13,

$$dT = -3\left(\frac{f}{1+f}\right)\frac{(LHV)d^2}{c_p d_0^3} \cdot d(d)$$

(10.15)

Integrating,

$$T = T_0 + \frac{f}{1+f} \cdot \frac{LHV}{c_p}\left[1 - \left(\frac{d}{d_0}\right)^3\right]$$

(10.16)

From Chapter 3 the adiabatic flame temperature is

$$T_{flame} = T_0 + \frac{f}{1+f} \cdot \frac{LHV}{c_p}$$

(3.71)

Substituting Equation 3.71 into Equation 10.16,

$$T = T_0 + \left(T_{flame} - T_0\right)\left[1 - \left(\frac{d}{d_0}\right)^3\right]$$

(10.17)

The reaction zone length is given by

$$L_v = \int_0^{t_v} V dt$$

(10.18)

Substituting Equation 10.7 and the ideal gas law (remember that the pressure is constant) into Equation 10.18,

$$L_v = V_0 \int_0^{t_v} \frac{T}{T_0} dt$$

(10.19)

Substituting Equation 10.17 into Equation 10.19 yields

$$L_v = V_0 \int_0^{t_v}\left[1 + \left(\frac{T_{flame}}{T_0} - 1\right)\left(1 - \frac{d^3}{d_0^3}\right)\right] dt$$

(10.20)

Substituting Equation 10.14 into Equation 10.20 and integrating yields

$$L_v = \frac{V_0 d_0^2}{\beta}\left(\frac{2}{5} + \frac{3}{5}\frac{T_{flame}}{T_0}\right)$$

(10.21)

Finally, the combustion intensity, which is the heat release per unit volume, can be estimated by assuming that the mixing and the chemical reactions occur much faster than the vaporization. This yields

$$I = \frac{\dot{m}_f\left(\text{LHV}\right)}{AL_v} \tag{10.22}$$

where $\dot{m}_f$ is the fuel feed rate to the burner. It can be easily shown that

$$\dot{m}_f = \dot{m}_{total}\frac{f}{\left(1+f\right)} = \rho_0\underline{V}_0 A \frac{f}{\left(1+f\right)} \tag{10.23}$$

Then

$$I = \frac{\rho_0\underline{V}_0}{L_v}\cdot\frac{f}{1+f}\text{LHV} \tag{10.24}$$

or substituting Equation 3.71 into Equation 10.24

$$I = \frac{\rho_0\underline{V}_0}{L_v}c_p\left(T_{flame}-T_0\right) \tag{10.25}$$

These relationships are explored in the example below.

Example 10.1

A monodispersed spray moves in a plug flow of air with a fuel-air ratio of 0.077, initial droplet diameter of 100 µm, flame temperature of 2100 K, burning rate constant of 0.25 mm²/s, initial velocity of 1 m/s, and an initial air temperature of 500 K. The droplet density is 900 kg/m³ and the pressure is 1 atm. Find the initial droplet number density, the vaporization time, the length of the reaction zone, and the combustion intensity.

Solution

$$\rho_{air,0} = 0.706 \text{ kg/m}^3 \qquad\qquad\qquad\qquad\qquad\text{(Appendix B)}$$

$$\rho_0 = \rho_{air,0} + \rho_{\ell,0} = \rho_{air,0}\left(1+f\right) = \left(0.706 \text{ kg/m}^3\right)\left(1.077\right) = 0.760 \text{ kg/m}^3$$

From Equation 10.6 the initial droplet number density is

$$n_0' = \left(\frac{0.077}{1.077}\right)\left(\frac{0.760 \text{ kg/m}^3}{900 \text{ kg/m}^3}\right)\left(\frac{6}{\pi\left(10^{-2} \text{ cm}\right)^3}\right) = 115/\text{c m}^3$$

From Equation 9.29 the vaporization time is

$$t_v = \frac{\left(100 \times 10^{-3} \text{ mm}\right)^2}{0.25 \text{ mm}^2/\text{s}} = 0.04 \text{ s} = 40 \text{ ms}$$

Evaluating Equation 10.21,

$$L_v = \frac{\left(1 \text{ m/s}\right)\left(0.10^2 \text{ mm}^2\right)}{\left(0.25 \text{ mm}^2/\text{s}\right)} \left[\frac{2}{5} + \frac{3}{5} \cdot \frac{2100 \text{ K}}{500 \text{ K}}\right] = 0.11 \text{ m} = 11 \text{ cm}$$

The specific heat is an average value appropriate for the temperature range. Based on this, we will choose to use the average temperature (1300 K). Noting that air is the primary component of the mixture, from Appendix B

$$c_p = 1.195 \text{ kJ/kg} \cdot \text{K}$$

Substituting into Equation 10.25

$$I = \frac{\left(0.760 \text{ kg/m}^3\right)\left(1 \text{ m/s}\right)}{0.11 \text{ m}} \left(1.195 \text{ kJ/kg} \cdot \text{K}\right)\left(2100 - 500\right) \text{ K} = 13.2 \text{ MW/m}^3$$

This calculation assumes that the mixing and chemical reaction rates are very rapid compared to the vaporization rate such that the true reaction zone will be somewhat longer than 11 cm. Using these relationships, the velocity and temperature profiles versus distance along the reaction zone may be determined (Problem 5). Of course, when swirl is used, the flow field becomes more complex.

10.4   EMISSIONS FROM OIL-FIRED FURNACES AND BOILERS

Particulates and nitrogen oxides are the emissions of primary concern from distillate fuel oil furnaces, while sulfur oxides must also be considered when burning residual fuel oil. Fuel oil contains a small but significant percentage of noncombustible ash. Number 2 fuel oil contains approximately 0.1% ash, while No. 6 fuel oil can contain up to 1.5% ash. A representative composition of ash from No. 6 fuel oil is given in Table 10.1. Particulate emissions from residual fuel oil burners are due to three sources: ash, soot, and sulfates. Approximately 5% of the sulfur in the oil is converted to sulfur trioxide, while the remainder goes to sulfur dioxide. The SO_3 readily goes to sulfuric acid in the presence of water or combines with calcium, magnesium, and sodium to form sulfates. Vanadium pentoxide, which is present in some oils to a significant extent, acts as a catalyst to convert SO_2 to SO_3, and hence increased vanadium means increased particulate emissions. Sulfate compounds increase opacity (light scattering) and particulate emissions. Also, in the upper part of a stack, the flue gas temperature can drop below the dew point of sulfuric acid. If this happens, the acid condenses on the fly ash, making it sticky and prone to deposition on the walls of the stack.

TABLE 10.1

**Representative Ash
from Residual Fuel Oil**

Compound	Weight %
SiO_2	1.7
Al_2O_3	0.3
Fe_2O_3	3.8
CaO	1.7
MgO	1.1
NiO	1.9
V_2O_5	7.9
Na_2O	31.8
SO_3	42.3

Both sodium and vanadium in fuel oil may form sticky ash compounds inside the boiler that can lead to fouling of heat exchanger surfaces. Frequent soot blowing is needed to clean these surfaces. Soot blowing uses forced air, water, or steam to clean heat exchanger surfaces. Fuel oil additives such as alumina, dolomite, and magnesia have been found effective in reducing fouling and corrosion. Additives either produce higher melting point ash deposits that do not fuse together, or form refractory sulfates which are more easily removed in soot blowing. Other fuel oil additives can reduce smoke, such as organometallic compounds of manganese, iron, and cobalt that have a catalytic influence on soot oxidation. Additives often increase the emissions of particulates; however, they turn sulfuric acid mist into a dry powder. Whereas soot can be controlled with good burner design, acid mist is controlled with additives.

Hydrocarbon fuels, such as fuel oil, have a tendency to produce soot in fuel rich areas of the combustion zone. During volatilization at high temperature, the fuel oil tends to crack or split into compounds of lower molecular weight and then in fuel rich regions to polymerize or build up again with the elimination of hydrogen. Fresh soot has the empirical formula C_8H. During the polymerization process, fuel ions nucleate to an atomic mass of about 10,000 and then form crystallites and finally soot spheres. The final stage of soot particle formation is the aggregation of the spheres into filaments. Soot particles in the furnace can be identified by a bright white-yellow radiation (continuum radiation). Hence, soot effectively transfers heat to the boiler walls. However, soot is more difficult to burn than gaseous volatile species. Thus, to avoid soot emissions, especially in smaller furnaces and boilers, it is important to bring air into direct contact with the volatilizing spray. For example, distillate oil burners exhibit a decrease in particulate emissions with increasing excess air due to a reduction in soot.

The Bacharach Smoke number is used to characterize particulate emissions. In this method a small volume of flue gas is manually pulled through a filter, and the darkness of the spot on the filter is matched by eye to a scale of 0 to 10. In practice, the excess air is decreased (to increase the efficiency) until the smoke number exceeds 2. Design improvements, such as the flame retention head discussed earlier, enable reduced excess air. However, the flame retention head has higher NO_X emissions than the standard burner configuration due to the hotter flame zone.

Nitric oxide is formed from nitrogen bound to the organic fuel molecules as well as nitrogen in the air. No. 6 fuel oil contains 0.1%–0.5% fuel bound nitrogen, while No. 2 fuel oil contains approximately 0.01% fuel bound nitrogen. When fuel is burned, 10%–60% of the fuel nitrogen is oxidized to NO. This fraction depends on the amount of oxygen available after the fuel molecules decompose. If the combustion zone is fuel rich, the fuel molecules crack, and much of the nitrogen forms N_2. Reduced excess air helps to lower thermal NO and SO_3 formation. In utility boilers using residual fuel oil burners with reduced airflow, over-fire air ports above the burner reduce NO emissions without excessive soot formation. Low NO_X oil burners, such as the one shown in Figure 10.14, for utility boilers have been developed that internally stage the air. This burner is similar to some gas burners except that the fuel nozzle is surrounded by a flame stabilizer disk that improves the flame stability and turndown. Dual air zones with multistage swirl vanes regulate the combustion air. The flow rate and degree of swirl influence the mixing of the air into the fuel-rich core of the flame. The extended flame zone reduces peak temperatures.

Emission factors for generic residential, commercial, industrial, and utility burners have been developed by the United States Environmental Protection Agency by testing many furnaces and boilers (Table 10.2). These values can vary widely between models, and they depend on how well the units are maintained and adjusted. The general equation for estimating emissions is

$$E = AR \cdot EF \left(1 - ER/100\right) \tag{10.26}$$

where E is the emissions rate, AR is the activity rate, EF is the emissions factor, and ER is the percent emissions reduction required to meet the emission standards. The federal emission standards for new large sources (greater than 260 GJ/h) are shown in Table 10.3. Frequently, these standards can be met without external controls for distillate fuels but require external equipment such as electrostatic precipitators and scrubbers for residual fuels.

Example 10.2

A large industrial burner proposes to use residual fuel oil containing 3% sulfur and 0.3% nitrogen. The higher heating value is 42,000 kJ/L. What emissions reduction will be required to meet United States Environmental Protection Agency emission standards?

Solution

Using the factors from Table 10.2 and Equation 10.26, the uncontrolled particulate emissions ($ER = 0$) are

$$\frac{E}{AR} = \frac{\left(1.09(3) + 0.38\right) kg}{10^3 \, L} \times \frac{1000 \, g}{kg} \times \frac{10^3 \, L}{42 \times 10^6 \, kJ} = \frac{86.9 \, g}{10^6 \, kJ}$$

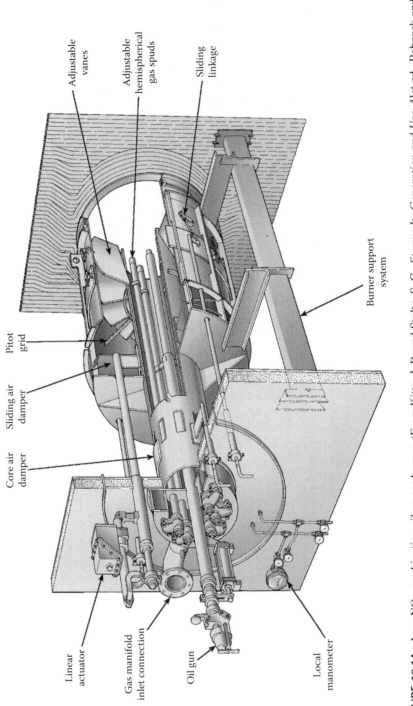

FIGURE 10.14 Low NO$_X$ combination oil or gas burner. (From Kitto, J. B. and Stultz, S. C., *Steam: Its Generation and Use*, 41st ed., Babcock and Wilcox, Barberton, OH, 2005. With permission.)

TABLE 10.2
Uncontrolled Emission Factors for Fuel Oil Combustion in Units of kg/10³ L of Oil

Type	Particulates	SO$_2$	NO$_X$ as NO$_2$
Residential (D)	0.3	17 S	2.2
Commercial and industrial (D)	0.24	18 S	2.4
Industrial and utility (R)	1.09 S + 0.38	19 S	2.5 + 12.5 N

Source: US Environmental Protection Agency, *Compilation of Air Pollution Emission Factors; Vol. I – Stationary Point Sources and Area Sources*, 5th ed., EPA-AP-42, 1995.

D means distillate fuel oil; R means residual fuel oil #6; S means multiply by percent sulfur in oil (3% means S = 3); N means multiply by percent nitrogen in oil (0.2% means N = 0.2).

For sulfur dioxide the uncontrolled emissions are

$$\frac{E}{AR} = \frac{19(3)\,kg}{10^3\,L} \cdot \frac{1000\,g}{kg} \cdot \frac{10^3\,L}{42 \times 10^6\,kJ} = \frac{1357\,g}{10^6\,kJ}$$

For nitrogen dioxide the uncontrolled emissions are

$$\frac{E}{AR} = \frac{\left(2.5 + 12.5(0.3)\right)kg}{10^3\,L} \cdot \frac{1000\,g}{kg} \cdot \frac{10^3\,L}{42 \times 10^6\,kJ} = \frac{149\,g}{10^6\,kJ}$$

Using Table 10.3 and rearranging Equation 10.26, the required emission reduction for particulates is

TABLE 10.3
Federal Emission Standards for Large New Residual Fuel Oil Combustion Sources in Units of g/10⁶ kJ

Emissions	Utility	Industrial
Particulates[a]	13	43
SO$_2$[b]	344	344
NO$_X$ as NO$_2$	130	172

Source: US Environmental Protection Agency, *Compilation of Air Pollution Emission Factors; Vol. I – Stationary Point Sources and Area Sources*, 5th ed., EPA-AP-42, 1995.

[a] Also requires a limit of 20% opacity.

[b] Also requires a 90% reduction but not lower than 86 g/10⁶ kJ.

$$ER = \left[1 - \frac{E}{AR \cdot EF}\right]100 = \left[1 - \frac{13}{86.9}\right]100 = 85\%$$

The required emission reduction for sulfur dioxide is $344 \text{ g}/10^6$ kJ or 90% but not lower than 86 g/10^6 kJ (see footnote b in Table 10.3). A 90% reduction of sulfur dioxide emissions is 136 g/10^6 kJ. The required reduction of sulfur dioxide is 90%. The required emission reduction for nitrogen dioxide controls is

$$ER = \left[1 - \frac{344}{1357}\right]100 = 75\%$$

In this case, particulate control would require a baghouse or electrostatic precipitator, sulfur dioxide control would require a scrubber, and nitrogen dioxide control would require combustion modification.

10.5 PROBLEMS

1. Find the auxiliary power needed for a burner rated at 180,000 MJ/h input (based on the HHV) using No. 6 fuel oil pressurized to 7 MPa and sprayed through an atomizing nozzle. The air pressure drop across the burner is 125 mm H$_2$O, and the inlet air is 400 K. Assume 10% excess air is used. The efficiency of the fuel pump is 90%, and the efficiency of the air blower is 70%. Calculate the power required by the fuel pump and by the air blower. Use the data provided in Tables 2.7 and 3.1 for No. 6 fuel oil. Use Appendix B for air properties.

2. An industrial boiler produces 68,000 kg/h of saturated steam at 4.5 MPa using residual fuel oil (No. 6 fuel oil). The ambient air temperature is 255 K, which is also the temperature of the oil storage tank. There is no air preheater but the oil is preheated to 373 K. The oil is pressurized to 5 MPa. There is a 50 mm water pressure drop across the air ducts and windbox. An economizer heats the feed water to 400 K. The stack temperature is 470 K. The heat loss due to incomplete combustion is 0.5% of the total heat input, and the radiation heat transfer losses are 5% of the total heat input. The excess air is measured to be 5%. The efficiency of the pump is 90%, and the efficiency of the blower is 70%. Calculate the boiler efficiency. Assume that properties of the products are the same as the properties of air (Appendix B). Use the data provided in Table 2.7, Table 3.1, and Appendix A for the properties of No. 6 fuel oil. The enthalpy of water at 400 K is 535 kJ/kg, and the enthalpy of saturated steam at 4.5 MPa is 2794 kJ/kg. What recommendations can you make to attempt to improve the boiler efficiency?

3. A commercial oil-fired furnace is rated at 300,000 kJ/h input and uses No. 2 fuel oil. An atomizing nozzle is used with a pressure of 700 kPa and a flow coefficient of 0.8 ($A_{eff} = 0.8A$). Find the fuel flow rate and the diameter of the nozzle. Use the data given in Table 2.7 for the properties of No. 2 fuel oil.

4. Approximately how long does it take 50 μm and 300 μm distillate fuel oil (No. 2 fuel oil) droplets to burn out in a furnace? Assume $\beta = 1.75$ mm²/s. Does this result depend on the flame temperature, excess air, or turbulence? Explain.

5. For Example 10.1 calculate and plot the velocity and the temperature versus distance along the reaction zone. This can be done either analytically or using a numerical equation solver.

6. What is the maximum percent sulfur in residual fuel oil (No. 2 fuel oil) that can be burned without exceeding the federal emission standards?

REFERENCES

Babbington, R. S., McLean, Virginia, personal communication, 1993.

Burkhardt, C. H., *Domestic and Commercial Oil Burners: Installation and Servicing,* 3rd ed. Glencoe/McGraw-Hill, New York, 1969.

Kitto, J. B. and Stultz, S. C., *Steam: Its Generation and Use,* 41st ed., Babcock and Wilcox, Barberton, OH, 2005.

Lawn, C. J., ed., *Principles of Combustion Engineering for Boilers,* Academic Press, London, 1987.

Lightman, P. and Street, P. J., "Single Drop Behaviour of Heavy Fuel Oils and Fuel Oil Fractions," *J. Inst. Energy* 56(426):3–11, 1983.

Offen, G. R., Kesselring, J. P., Lee, K., Poe, G., and Wolfe, K. J., *Control of Particulate Matter from Oil Burners and Boilers,* Environmental Protection Agency EPA-450/3-76-005, 1976.

Sayre, A. N., Dugue, J., Weber, R., Domnick, J., and Lindenthal, A., "Characterization of Semi-Industrial-Scale Fuel-Oil Sprays Issued from a Y-Jet Atomiser," *J. Inst. Energy* 67(471):70–77, 1994.

US Environmental Protection Agency, *Compilation of Air Pollution Emission Factors; Vol. I - Stationary Point Sources and Area Sources,* 5th ed., EPA-AP-42, 1995.

Williams, A., *Combustion of Liquid Fuel Sprays,* Butterworth-Heinemann, Burlington, MA, 1990.

Williams, A., "Fundamentals of Oil Combustion," *Prog. Energy Combust. Sci.* 2:167–79, 1976.

11 Gas Turbine Spray Combustion

Gas turbine engines are used to produce thrust for aircraft and electrical power for stationary applications. As shown schematically in Figure 11.1, the basic components of a gas turbine engine consist of a rotary compressor, a combustor, a turbine that drives the compressor, and a load such as an electric generator. In aircraft applications the combustion products are expanded through the turbine and exhausted at high velocity to produce thrust. Gas turbines have a high energy output per unit volume compared to piston engines and hence are well suited for aircraft applications. In industrial and utility applications the turbine generates shaft power, and the exhaust can be used in a heat recovery steam generator for process heat or to drive a steam turbine in a combined cycle. In a combined cycle, electric power is produced from both the gas turbine and the steam turbine. Aircraft gas turbines use liquid distillate fuels, and stationary gas turbines use both gaseous fuels and liquid fuels.

In this chapter we first consider the operating parameters for gas turbine combustors. Following this combustor design, combustion processes, heat transfer considerations, and emissions are examined in more detail.

11.1 GAS TURBINE OPERATING PARAMETERS

Gas turbine combustors operate at pressures from 3 atm for small simple turbines to more than 40 atm for new aircraft engine turbines. New industrial turbines have pressure ratios ranging from 17:1 to 35:1. The combustor inlet temperature depends on the compressor pressure ratio and whether or not regenerative heating is used. Combustor inlet air temperatures range from 200°C–450°C without regenerative heating, and 400°C–600°C with regeneration. Combustor outlet temperatures are set by the metallurgical requirements of the turbine blades and range from 1000°C to 1500°C for industrial turbines and 1300°C–1700°C for aircraft turbines. As with all heat engines, the highest gas turbine cycle efficiency is achieved with the highest feasible turbine inlet temperature. Because the stoichiometric flame temperature of distillate fuels is over 2000°C, significant excess air must be used to prevent overheating of the first several stages of the turbine blades. An example of design parameters for a large industrial gas turbine is given in Table 11.1.

To further consider the relationship between combustion pressure, temperature, and cycle efficiency, consider the ideal air cycle gas turbine or Brayton cycle, shown in the pressure-volume and temperature-entropy diagrams of Figure 11.2. Ideal flow through the compressor (1–2) and expansion through the turbine (3–4) is isentropic. Pressure drop through the combustor is usually about 3% but will be neglected here. The ideal net power output of the simple cycle is

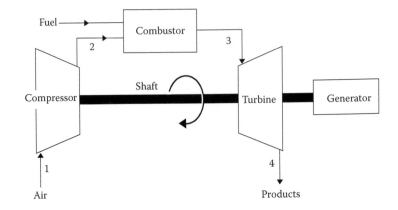

FIGURE 11.1 Schematic diagram of an open cycle gas turbine system.

$$\dot{W}_{net} = \dot{m}\left[\left(h_3 - h_4\right) - \left(h_2 - h_1\right)\right]$$ (11.1)

For an ideal gas with constant specific heat, it follows that

$$\frac{p_2}{p_1} = \frac{p_3}{p_4}$$

TABLE 11.1
Heavy-Duty Industrial Gas Turbine Design Parameters

Simple cycle gas turbine:	
Power output	136 MW
Heat rate	10,390 Btu/kWh
Combined cycle with gas turbine and steam turbine:	
Power output	200 MW
Heat rate	6828 Btu/kWh
Compressor:	
Number of stages	18
Overall pressure ratio	13.5:1
Airflow rate	287 kg/s
Turbine:	
Number of stages	3
Inlet temperature	1260°C
Outlet temperature	593°C
Combustors:	
Number of chambers	14
Number of fuel nozzles per chamber	6

Source: Brandt, D. E., "Heavy Duty Turbopower: The MS7001F,"
           Mech. Eng. 109(7):28–37, 1987.

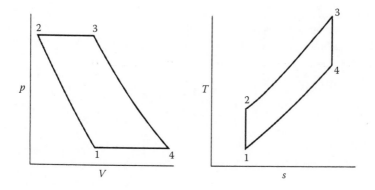

FIGURE 11.2   Ideal Brayton cycle pressure-volume and temperature-entropy diagrams.

and

$$\frac{T_2}{T_1} = \frac{T_3}{T_4} = T_{ratio}$$

Substituting into Equation 11.1

$$\dot{W}_{net} = \dot{m}c_p\left[T_3\left(1 - \frac{1}{T_{ratio}}\right) - T_1\left(T_{ratio} - 1\right)\right] \tag{11.2}$$

By inspection of Equation 11.2, there is an optimum temperature ratio, T_{ratio}, for maximum power per unit mass at a fixed inlet temperature. Differentiating Equation 11.2 with respect to T_{ratio} and setting it equal to zero yields

$$T_3 = \frac{T_2^2}{T_1} \tag{11.3}$$

or

$$\left(\frac{T_2}{T_1}\right)_{opt} = \left(\frac{T_3}{T_1}\right)^{1/2} \tag{11.4}$$

For isentropic compression the pressure ratio for maximum power is

$$\left(\frac{p_2}{p_1}\right)_{opt} = \left(\frac{T_2}{T_1}\right)_{opt}^{\gamma/(\gamma-1)} = \left(\frac{T_3}{T_1}\right)^{\gamma/2(\gamma-1)} \tag{11.5}$$

For example, if the inlet air temperature is 300 K and the combustor outlet temperature is 1200 K, the optimum pressure ratio across the compressor is 11. If the combustor outlet temperature is 1800 K, then the optimum pressure ratio across the compressor is 23. The turbine efficiency, on the other hand, continues to increase with an increasing compression ratio.

The gas turbine cycle efficiency is the net power output divided by the heat input to the combustor,

$$\eta = \frac{\left(h_3 - h_4\right) - \left(h_2 - h_1\right)}{h_3 - h_2} \tag{11.6}$$

Assuming constant specific heats,

$$\eta = 1 - \frac{T_1}{T_2} = 1 - \frac{T_4}{T_3} \tag{11.7}$$

Thus for an ideal gas with no flow losses

$$\eta = 1 - \frac{1}{\left(p_2/p_1\right)^{(\gamma-1)/\gamma}} \tag{11.8}$$

The heat rate (*HR*) is the heat input from the fuel divided by the net power output, which is the inverse of the cycle efficiency. In power engineering in the United States it is customary to use British thermal unit per hour input and kilowatt output, thus

$$HR = \frac{q_{in}}{\dot{W}_{net}} \frac{\text{Btu}}{\text{kWh}}$$

Remembering that efficiency is dimensionless

$$\eta = \frac{\dot{W}_{net}}{q_{in}}$$

heat rate and efficiency can be related as follows,

$$HR = \frac{3413}{\eta} \frac{\text{Btu}}{\text{kWh}} \tag{11.9}$$

Example 11.1

For compressor inlet air at 27°C and 1 atm, and for a combustor outlet temperature of 1260°C, find the combustor pressure for maximum work. At this optimum pressure, find the ideal compressor outlet temperature, the ideal cycle efficiency, and the ideal heat rate. Assume a simple, ideal air cycle gas turbine and refer to Figure 11.2.

Solution

Given $T_1 = 300$ K, $T_3 = 1260°C$, and $p_1 = 1$ atm; the specific heat ratio of an ideal gas is 1.4; and rearranging Equation 11.5 we find the optimum pressure;

$$p_{2,opt} = p_1 \left(\frac{T_3}{T_1} \right)^{\gamma/2(\gamma-1)}$$

$$p_{2,opt} = 1 \text{ atm} \left(\frac{(1260+273)}{300} \right)^{1.4/2(1.4-1)}$$

$$p_{2,opt} = 17.4 \text{ atm}$$

Remembering that for isentropic compression,

$$\frac{p_2}{p_1} = \left(\frac{T_2}{T_1} \right)^{\gamma/(\gamma-1)}$$

We find that the ideal compressor outlet temperature, T_2, is

$$T_2 = T_1 \left(\frac{p_2}{p_1} \right)^{(\gamma-1)/\gamma} = 300 \text{ K} \left(\frac{17.4}{1} \right)^{(1.4-1)/1.4} = 679 \text{ K}$$

From Equation 11.8

$$\eta = 1 - \frac{1}{\left(p_2/p_1 \right)^{(\gamma-1)/\gamma}} = 1 - \frac{1}{(17.4/1)^{(1.4-1)/1.4}} = 0.558$$

And from Equation 11.9 the heat rate for our ideal turbine is

$$HR = \frac{3413}{\eta} = \frac{3413}{0.558} = 6114 \text{ Btu/kWh}$$

11.2 COMBUSTOR DESIGN

Gas turbine combustors should have low pressure loss, high combustion efficiency, wide stability limits (i.e., be able to operate stably over a wide range of pressures and airflow rates), a uniform outlet temperature, and low emissions. In addition, gas turbines used in aircraft must be compact and lightweight. To meet these requirements, the velocity in the combustor cannot be too high or the pressure loss will be excessive. Moreover, mixing of fuel and air must be excellent, the fuel-air ratio must be maintained within certain limits, the combustor residence time must be sufficient to complete combustion, the temperatures must be compatible with metal durability, and pollutant emissions must be low.

A conventional gas turbine combustor (Figure 11.3) consists of an inlet diffuser section to slow the velocity, a fuel injector, a primary combustion zone with an air swirler to maintain back-mixing, an intermediate combustion zone to complete combustion of the dissociated products, and a dilution zone to meet turbine inlet temperature requirements. Secondary air flows through holes and slots in the combustion chamber liner. The flow pattern through the combustor is shown in Figure 11.4. The diffuser section reduces the flow velocity and splits the flow between the primary combustion zone (15%–20%) and the liner air (80%–85%). The reduced velocity reduces the pressure loss due to heat addition and also reduces the chance for blow-off of the flame. The liner mixes air into the intermediate combustion zone and the dilution zone.

Liquid fuel injectors used in gas turbine combustors are typically of the pressure-swirl or airblast types (Chapter 9). Typical Sauter mean diameters are 30–60 µm, and fuel nozzle pressure drops are in the range of 5–20 atm. Aircraft gas turbines use liquid fuels such as JP-4, while industrial gas turbines use kerosene, fuel oil, or natural gas. Natural gas injectors have plain orifices with venturi nozzles.

There are three basic configurations of gas turbine combustors: annular, cannular, and silo. Figure 11.5 shows schematic drawings of these configurations. In the *annular* configuration, there is one annular combustion chamber liner mounted concentrically inside an annular casing. Older aircraft gas turbines and heavy-duty industrial gas turbines use the *cannular* design in which up to 18 tubular combustion "cans" are mounted in an annular plenum. Some industrial gas turbines use the single *silo* combustor design.

The aircraft gas turbine propulsion system shown in Figure 11.6 is a turbofan engine that generates 274 kN thrust. The high pressure turbine drives the fan and compressor; the low pressure turbine and fan provide the thrust. The annular combustor, which is very compact and lightweight, delivers nearly uniform hot gas to power the turbine. An aircraft derivative industrial engine of cannular design is similar to the one shown in Figure 11.6 except that there is no fan, low pressure compressor, or

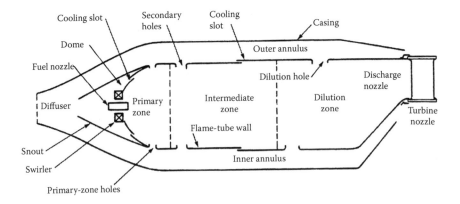

FIGURE 11.3 Main components of a gas turbine combustor. (From Lefebvre, A. H., *Gas Turbine Combustion*, 2nd ed., Taylor & Francis, Philadelphia, PA, 1999. © 1999, with permission of Taylor & Francis.)

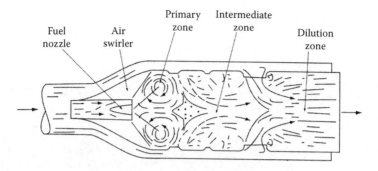

FIGURE 11.4 Flow pattern in a combustor created by swirl vanes and radial jets (20% of air added to the primary zone, 30% to the intermediate zone, and 50% to the dilution zone).

exhaust nozzle. An electrical generator is connected to the turbine shaft, the engine runs at constant speed, and the load varies. A heavy duty gas turbine engine with a silo type combustor, which can use harder-to-burn fuels, is shown in Figure 11.7.

The combustion process is similar for the turbines shown in Figures 11.6 and 11.7. Fuel is sprayed through an atomizing nozzle into the primary combustion zone. A spark plug is used for ignition. Flame stabilization is provided by swirl imparted to the primary inlet air. Swirl vanes curving 45° to the flow and located around the fuel nozzle give a tangential velocity component to the primary air. The flow expands into the combustion chamber and mixes with the fuel spray. As shown in Figure 11.4, a recirculation pattern is set up by means of the vortex motion created by the inlet swirl because the pressure is lower in the center of the vortex and because of the increasing axial pressure gradient in the primary combustion zone due to the addition of secondary air through the walls of the flame tube. This provides aerodynamic stabilization of the flame by back-mixing the hot combustion products. The primary combustion zone is maintained at near-stoichiometric conditions.

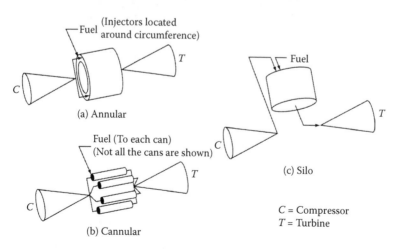

FIGURE 11.5 Aircraft gas turbine combustor types: (a) annular, (b) cannular, and (c) silo. "C" indicates the combustor side and "T" indicates the turbine side.

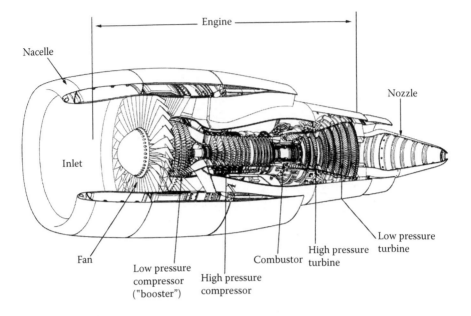

FIGURE 11.6 Aircraft gas turbine engine with an annular combustion chamber. This engine has a mass flow of 812 kg/s, a pressure ratio of 31.9, an overall length of 4 m, a fan diameter of 2.4 m, a weight of 4420 kg, and a takeoff thrust of 274 kN. (Courtesy of General Electric Co., Schenectady, NY.)

Secondary and dilution air flows between the casing and the liner and is admitted gradually into the combustion chamber through holes and slots. In this way, cooling of the liner is achieved, combustion is completed, and combustion products are cooled to the required turbine inlet temperature. Approximately 30% of the air flows through the liner into the intermediate combustion zone, and the remaining 50% is added in the dilution zone. The size and number of holes in the liner is a compromise between a large number of small holes to give fine scale mixing and a smaller number of large holes to give better penetration and to cool the center core before entry into the turbine. By overlapping sections of the liner, thin slots are formed to provide film cooling of the liner wall. Liners for aircraft combustors are typically only 1 mm thick, and hence the design of the slots is critical. The flow pattern shown in Figure 11.4 is also applicable to annular combustors, which use a ring of fuel nozzles and air swirlers that discharge into the annular space.

The size of the combustor for a given power output depends on the fuel type, droplet size distribution, and other factors. The cross-sectional area may be roughly sized by noting that aircraft turbines have an average flow velocity of 25–40 m/s, whereas industrial combustors typically have velocities of 15–25 m/s. The length of the combustor also varies, but tubular combustors typically have a length to diameter ratio of 3 to 6 based on the liner. More space and better liner cooling is needed for No. 2 fuel oil because the fuel oil droplet vaporization rate is lower and radiation heat transfer is higher. Inorganic compounds in the fuel must be carefully limited to avoid erosion and corrosion of the turbine blades.

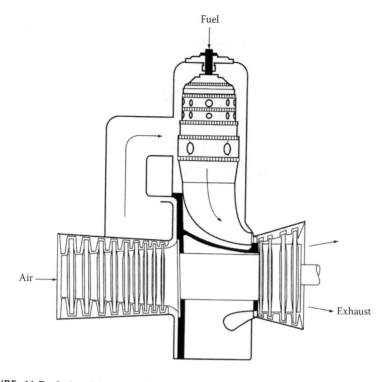

Fuel

Air

Exhaust

FIGURE 11.7 Industrial gas turbine combustor. (From Lefebvre, A. H., *Gas Turbine Combustion*, 2nd ed., Taylor & Francis, Philadelphia, PA, 1999. © 1999, with permission of Taylor & Francis.)

11.2.1 IGNITION

Ignition and relight are of obvious importance in aircraft gas turbines. Typically, ignition is caused by a spark plug using short-duration pulses at a continuous rate of 60–250 pulses per minute. Surface discharge igniters that utilize a thin film of semiconductor material to separate the electrodes at the firing end are also used. The electrical resistance of the semiconductor falls with increasing temperature, giving a rapid discharge. The surface igniter gives optimal performance when a thin layer of fuel is on the plug face.

The ignition process itself begins with the formation of a flame kernel. The flame must now spread through the chamber, and for tubular chambers it must also spread to the other chambers Generally, good flame spread conditions are similar to the conditions needed for good stability: reduced primary zone velocity, increased pressure and temperature, and near stoichiometric fuel-air ratios. Short, large flow area interconnectors with low heat loss facilitate flame spread between chambers.

For mists of uniform-sized fuel droplets in quiescent atmospheres, the flame quench distance is proportional to the droplet diameter and to the square root of fuel density. The minimum ignition energy is proportional to the cube of quench distance, and thus changes in the spray droplet size have an important effect on ignition. The low vaporization rate of large size droplets does not provide the fuel-air mixture needed for ignition. Figure 11.8 shows the effect of droplet size on minimum ignition energy for several fuels.

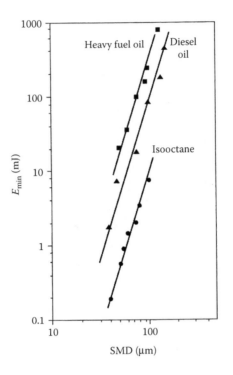

FIGURE 11.8 Effect of mean droplet size on the minimum ignition energy for conditions of 1 atm, 15 m/s, and $f = 0.65$. (From Ballal, D. R. and Lefebvre, A. H., "Ignition and Flame Quenching of Flowing Heterogeneous Fuel-Air Mixtures," *Combust. Flame* 35:155–168, 1979. © 1979, with permission of Elsevier.)

11.2.2 FLAME STABILIZATION

For the flame to be stable, the flame velocity must be equal and opposite to the reactant velocity at each point along the flame. Flame holding is achieved by creating high inlet swirl, which causes back-mixing of hot products. Experiments to determine flame stability for a given design are performed at constant inlet air temperature, pressure, and velocity. The fuel flow is varied until the lean and rich extinction limits are determined. The resulting data obtained by repeating the experiment at different airflow rates is plotted to give a stability loop diagram, similar to that shown in Figure 11.9. Interestingly, the shape of the curve is similar to the curve for blowout data in a well-stirred reactor (to be discussed in the next section). In general the blowout velocity and stability limits can be extended by the following:

- Reducing the primary stream velocity and turbulence intensity
- Increasing the inlet temperature, pressure, and swirl
- Keeping the primary combustion zone near stoichiometric

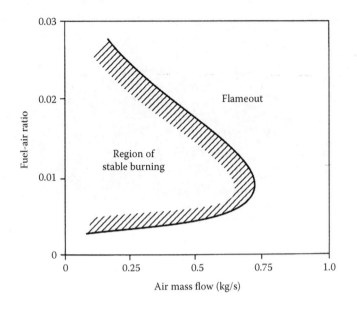

FIGURE 11.9 Combustion chamber stability plot for constant inlet pressure. (From Lefebvre, A. H., *Gas Turbine Combustion*, 2nd ed., Taylor & Francis, Philadelphia, PA, 1999. © 1999, with permission of Taylor & Francis.)

- Using a higher volatility fuel
- Decreasing the droplet size

Maximum stability is achieved by injecting primary air through a small number of large holes. The basic idea is to produce large-scale circulation patterns. Fuel effects are not important at high power conditions, but are important at idle conditions where the flame stabilization becomes vaporization limited. For air blast atomizers the mixture is nearly homogeneous, and stirred reactor correlations can be applied to predict the lean limit.

11.2.3 A Specific Combustor Design

Figure 11.10 shows one of the 14 combustion chambers (combustors) for the heavy-duty cannular industrial gas turbine design parameters indicated in Table 11.1. Figure 11.11 shows the cap and liner, and transition piece for the combustor. The 13° tilt of the combustors allows for a more compact combustor and gas turbine design. The combustion system consists of a fuel nozzle assembly, a flow sleeve, a cap and liner assembly, and a transition piece. The combustion system incorporates advanced liner cooling technology to permit firing temperatures of 1560 K. Ignition is achieved with a spark plug in 2 of the 14 combustion chambers with cross-firing connections to ignite the balance of the chambers. Successful ignition is sensed by four ultraviolet flame detectors located opposite the spark plugs.

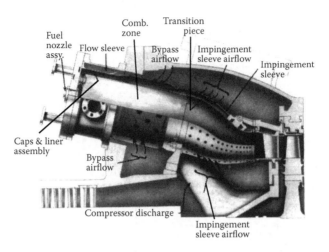

FIGURE 11.10 Gas turbine combustor showing components. (Claeys, J. P., Elward, K. M., Mick, W. J. and Symonds, R. A., "Combustion System Performance and Field Test Results of the MS7001F Gas Turbine," *J. Eng. Gas Turbines Power* 115(3):537–546, 1993. © 1993, with permission of ASME.)

The airflow paths are as follows. After leaving the diffuser area of the compressor, air flows past the transition pieces and moves forward toward the head end of the liner. The air splits between two parallel paths before entering the liner. Some of the air flows through the holes provided in the impingement sleeve of the transition piece for cooling; the balance of the air passes through an array of multiple holes in the flow sleeve. Air outside the liner is used for dilution, mixing, and cooling as it enters holes in the liner sleeve where the air and fuel react and flow downstream through the transition piece before entering the turbine section.

The fuel nozzle assembly is designed to use either natural gas or distillate fuel, and to transfer from one fuel to the other during operation. Six individual, dual-fuel nozzles are attached to an internally manifolded cover. Each fuel nozzle consists of a swirl tip, an atomizing air tip, and a distillate spray nozzle. Gaseous fuel is injected through metering holes in the swirler. Continuous atomizing air is used with the distillate fuel spray. The multi-nozzle design offers two significant advantages over single-nozzle designs. First, it allows more thorough mixing and control of the fuel and air in the reaction zone, resulting in a shorter flame length and improved combustor performance. Second, it produces significantly lower noise levels than single-nozzle combustors and thereby improving combustor component wear.

The cap and liner assembly, shown in Figure 11.11a, consists of a slot film cooled, 0.35 m diameter, 0.76 m long liner sleeve and a multi-nozzle cap assembly. The relatively short liner length is another benefit of the multi-nozzle design in that it produces a shorter reaction zone than single nozzle combustors. The shorter liner has less surface area to cool. The cap is film cooled in the center and impingement cooled in the region surrounding the fuel nozzles. The surfaces of the cap and liner exposed to hot gas are coated with a ceramic thermal barrier to further limit metal temperatures and reduce the effects of thermal gradients.

The transition piece, shown in Figure 11.11b, directs the hot gases from the exit of the liner sleeve to the inlet of the turbine. The transition piece is surrounded by a sleeve that has an array of holes. Air flows through the holes and impinges on the backside of the transition piece, thereby providing an effective impingement-type cooling of the surface. The use of impingement cooling allows for more precise control of cooling the transition piece while having only a small adverse effect on overall

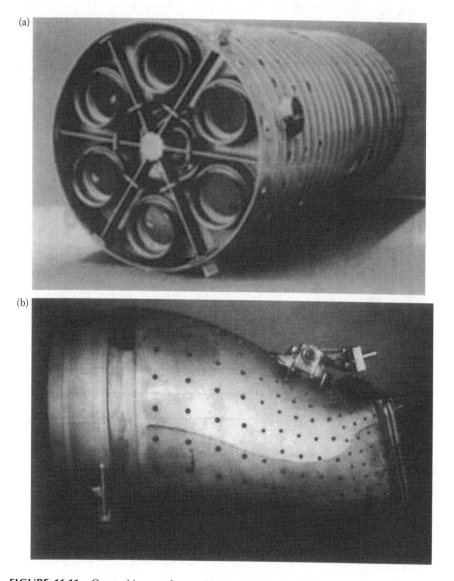

(a)

(b)

FIGURE 11.11   Gas turbine combustor (a) cap and liner, and (b) transition piece. (From Claeys, J. P., Elward, K. M., Mick, W. J. and Symonds, R. A., "Combustion System Performance and Field Test Results of the MS7001F Gas Turbine," *J. Eng. Gas Turbines Power* 115(3):537–546, 1993. © 1993, with permission of ASME.)

pressure drop. The internal surfaces of the transition piece have a ceramic thermal barrier coating that limits the maximum metal temperature.

11.3 COMBUSTION RATE

Aircraft gas turbines require that the combustion chamber be as small as possible and have the highest feasible combustion rate. Gas turbines used for stationary power need not meet such tight size constraints; nonetheless, gas turbine engines derived from aircraft engines are sometimes used for stationary power generation. The combustor designer needs to know how the operating conditions, such as inlet pressure and temperature, control the combustion rate and how to scale the size of the combustor for different fuels. Because of the difficulty of observing processes inside the combustor, for many years gas turbine combustor technology advanced by using intuitive design based on years of experience and extensive testing. More recently, computational modeling has become an important tool in the design and engineering of gas turbines.

Because of the complexity of combustion systems and the time required to build and analyze complex multi-dimensional models for reacting systems, computational modeling is typically used to explore one aspect or area of combustion system but not to design or engineer the entire system (Liu and Niksa 2004). For example, detailed models of a complete gas turbine, including the compressor, combustion, heat transfer, and the turbine, are not available. One area of modeling that can be applied to practical design is the use of simplified models to explore and design combustion systems. Referred to as reduced-order models or surrogate models, these models use physics-based methods or non-physics-based methods. The non-physics-based methods develop a mathematical relationship between the inputs and the outputs. These methods are similar to (but generally far more complex than) linear regression and interpolation. These methods include orthogonal decomposition, neural networks, and other similar techniques.

The physics-based approaches start with simplified models of the underlying physics and then assemble these models together to build systems of models. These models are commonly used in the design and engineering of process systems and are widely used in analyzing liquid and solid fuel systems. In this approach, the goal is to identify zones within an energy system that share common characteristics and can be modeled with simpler models. In many cases these simplified models are then tuned using CFD or experimental results. If needed, after tuning and validation these simplified models can be knit together into a multizonal system of models that describe the combustion/energy system or process. Although this process of model development and integration to create a system of models is not a part of this book, we have already encountered in this book a number of simplified models as a means to describe and understand the processes occurring in combustion systems. For example, in Chapter 10 we developed a plug flow model of spray flow combustion to describe the initial combustion region of an oil burner. While this model did not include complex three-dimensional flows, turbulent mixing, and details of droplet interaction, it did provide a starting place for understanding the phenomena occurring in an oil burner. These simplified models are useful for both the qualitative and quantitative understandings needed in engineering design, but the student should remember that tuning and validation are a critical part of the process when they are applied to a specific application.

For example, in a gas turbine combustor we can identify several distinct zones that can be modeled separately. The air coming from the compressor is split into two streams. The smaller of the two streams enters the primary combustion zone with the fuel, while the second stream is added downstream in the intermediate and dilution zones to complete the combustion process before products enter the turbine. Due to the high temperature and turbulent convective and radiant heat transfer, in the primary combustion zone the droplet vaporization rate is very rapid. Mixing between the fuel vapor and the gas in the primary combustion chamber occurs rapidly due to the swirling inlet flow, jets from the liner ports, and local turbulence on the scale of the droplet spacing. Because the vaporization and mixing rates are fast, and because the high swirl causes back-mixing, it is instructive as a first step to model the primary zone of the combustor (Figure 11.3) as a *well-stirred reactor.* A well-stirred reactor, shown schematically in Figure 11.12, incorporates a way to rapidly achieve a well-mixed state. Fuel and air flow through opposing jets into the center of a spherical, insulated volume, causing combustion products to flow outward through many holes in the sphere. Because the mixing is intense, the temperature and species composition are assumed to be uniform throughout the reactor.

Conservation of mass, conservation of mass of each species, and conservation of energy are applied to the reactor. The inlet mass flow rate of air and fuel, $\dot{m}$, is constant and equal to the outflow of products. The species conservation equation for each species in the well-stirred reactor is

$$\dot{m}\left(y_i - y_{i,in}\right) = \hat{r}_i M_i V \tag{11.10}$$

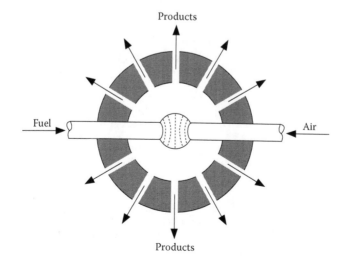

FIGURE 11.12 Physical model of a well-stirred reactor.

where y_i is the mass fraction of species i, V is the reactor volume, $\hat{r}_i$ is the molar reaction rate of species i per unit volume, and M_i is the molecular weight of species i. Conservation of energy for the well-stirred reactor is

$$\dot{m}\sum_{i=1}^{I}\left(y_i h_i - y_{i,in}h_{i,in}\right)+q=0 \tag{11.11}$$

where q is the rate of heat loss through the walls and h_i is the absolute enthalpy of species i. The nominal residence time can be obtained by

$$\tau=\frac{\rho V}{\dot{m}} \tag{11.12}$$

where

$$\rho=\frac{p\bar{M}}{RT} \tag{11.13}$$

Equations 11.10 and 11.11 can involve many species and reaction steps. To simplify the analysis, let us consider a one-step global chemical reaction where the only species are fuel and air reacting to produce products. From Equation 4.15 the molar reaction rate of the fuel is

$$\hat{r}_f=-AT^n p^m \exp\left(-\frac{E}{RT}\right)\left(n_f\right)^a \left(n_{O_2}\right)^b \tag{11.14}$$

Equations 11.10 through 11.14 provide a sufficient set of equations to determine the maximum heat release in a given volume as well as the combustion efficiency. Let us explore this in Example 11.2.

Example 11.2

Consider the primary combustion zone of a gas turbine as a well-stirred reactor with a volume of 900 cm³. Kerosene and air at 298 K with an equivalence ratio of 0.80 flow into the well-stirred reactor, which operates at 10 atm and 2000 K. Assume that mixing and droplet vaporization are rapid relative to combustion. To keep it simple, neglect dissociation and heat loss, and assume the products go to completion. The lower heating value of the kerosene is 42.5 MJ/kg. Assume a one-step global kinetic reaction with $A = 5 \times 10^{11}$, $E = 30,000$ cal/gmol, $a = 0.25$, $b = 1.5$, and $m = n = 0$ in Equation 11.14. Assume that kerosene is $C_{12}H_{24}$. Find the mass fraction of the kerosene burned before it flows out of the reactor, the fuel flow rate, and the heat output of the reactor.

Solution

From Equation 3.42

$$\text{excess air} = \frac{1-F}{F} = \frac{1-0.80}{0.80} = 0.25$$

The reaction is

$$C_{12}H_{24} + (1.25)(18)(O_2 + 3.76\ N_2) \rightarrow$$

$$12\eta\ CO_2 + 12\eta\ H_2O + (1-\eta)C_{12}\ H_{24} + \left[(1.25)(18) - 12\eta - 6\eta\right]O_2$$

$$+ (1.25)(18)(3.76)\ N_2$$

where η is the fraction of fuel that is consumed in the reactor (the progress variable):

$$\eta = 1 - \frac{y_f}{y_{f,in}}$$

Simplifying,

$$C_{12}H_{24} + 22.5\ O_2 + 84.6\ N_2 \rightarrow$$

$$12\eta\ CO_2 + 12\eta\ H_2O + (1-\eta)C_{12}\ H_{24} + (22.5 - 18\eta)\ O_2 + 84.6\ N_2$$

Using a basis of 1 kgmol of fuel, we determine the mass fraction of species flowing into the reactor as follows

Species	N_i (kgmol)	M_i (kg/kgmol)	m_i (kg)	$y_{i,\ in}$
Fuel	1	168	168	0.0514
O_2	22.5	32	720	0.2202
N_2	84.6	28.16	2382	0.7284
Total	108.1		3270	1.000

The mass of fuel plus air flowing into the reactor equals the mass of products flowing out,

$$\dot{m}_{air} + \dot{m}_f = \dot{m}_{prod}$$

Because the water in the products is not condensed, the lower heating value of the fuel consumed in the reactor is equal to the difference between the sensible enthalpies of the products flowing out of the reactor and the sensible energies of the fuel and air flowing into the reactor. This is

$$\dot{m}_f (\eta) LHV = \dot{m}_{prod} h_{prod} - \left(\dot{m}_{air} h_{air} + \dot{m}_f h_f\right)$$

where all enthapies are sensible enthalpies. Because the fuel and air enter at the reference state in this example, $h_{air} = h_f = 0$. Rearranging, using conservation of total mass, and simplifying,

$$\eta = \frac{\dot{m}_{prod}}{\dot{m}_f} \cdot \frac{h_{prod}}{LHV} = \frac{h_{prod}}{y_{f,in}(LHV)}$$

For simplicity, h_{prod} is taken from the nitrogen tables. The enthalpy of the products becomes

$$h_{prod} = \frac{56.14\ MJ}{kgmol} \cdot \frac{kgmol}{28\ kg} = 2.00\ MJ/kg$$

Substituting values,

$$\eta = \frac{2.00}{0.0514(42.5)} = 0.916$$

Therefore, in this case we see that 8.4% of the fuel flows out of the reactor without reacting.

Now using a 1 kgmol of fuel basis, determine the moles and mole fraction of the species in the well-stirred reactor per mole of fuel for the above reaction:

Species	N_i (kgmol)	x_i	M_i (kg/kgmol)	m_i (kg)
CO_2	10.86	0.096	44	478
H_2O	10.86	0.096	18	195
Fuel	0.095	0.00084	168	16
O_2	6.21	0.055	32	199
N_2	84.6	0.751	28.16	2382
Total	112.6	1.00		3270

To determine the fuel reaction rate from Equation 11.14, the moles of fuel and oxygen in the reactor are needed. The total moles in the reactor are

$$n = \frac{p}{\hat{R}T} = \frac{10\ atm}{1} \cdot \frac{gmol \cdot K}{82.05\ cm^3 \cdot atm} \cdot \frac{1}{2000\ K} = 6.09 \times 10^{-5}\ gmol/cm^3$$

and the moles of fuel and oxygen in the reactor are

$$n_f = x_f n = (0.00084)(6.09 \times 10^{-5}) = 5.12 \times 10^{-8}\ gmol/cm^3$$

$$n_{O_2} = x_{O_2} n = (0.055)(6.09 \times 10^{-5}) = 3.35 \times 10^{-6}\ gmol/cm^3$$

The molar reaction rate of the fuel can now be evaluated as

$$\hat{r}_f = -5 \times 10^{11} \exp\left[\frac{-30{,}000}{(1.987)(2000)}\right]\left(5.12 \times 10^{-8}\right)^{0.25}\left(3.35 \times 10^{-6}\right)^{1.5}$$

$$\hat{r}_f = -0.0243 \ \text{gmol}/\left(\text{cm}^3 \cdot \text{s}\right)$$

From fuel continuity (Equation 11.10) the mass flow rate of fuel into the reactor is obtained,

$$\dot{m}_{f,\,in} - \dot{m}_{f,\,out} = -\hat{r}_f M_f V$$

$$\dot{m}_{f,in}\left(\frac{\dot{m}_{f,in} - \dot{m}_{f,out}}{\dot{m}_{f,in}}\right) = \dot{m}_{f,in}\left(\eta\right) = -\hat{r}_f M_f V$$

Thus

$$\dot{m}_{f,in} = -\frac{\hat{r}_f M_f V}{\eta}$$

$$\dot{m}_{f,in} = \frac{0.0243 \ \text{gmol}}{\text{cm}^3 \cdot \text{s}} \cdot \frac{168 \ \text{g}}{\text{gmol}} \cdot \frac{\text{kg}}{1000 \ \text{g}} \cdot \frac{900 \ \text{cm}^3}{0.914}$$

$$\dot{m}_{f,in} = 4.02 \ \text{kg/s}$$

The heat output of the reactor is

$$\eta\left(\dot{m}_{f,in} \text{LHV}\right) = (0.914)\frac{4.02 \ \text{kg}}{\text{s}} \cdot \frac{42.5 \ \text{MJ}}{\text{kg}} = 156 \ \text{MW}$$

The heat output per unit volume is 174 kW/cm³. While this heat output is large, it should be noted that it is of similar magnitude to that of a laminar flame, which is approximately 1 mm thick.

Example 11.3

Consider the primary combustion zone of a gas turbine as a well-stirred reactor with a volume of 900 cm³ and the conditions in Example 11.2. In this case, assume that combustion and mixing are rapid relative to droplet vaporization. Assume a mono-sized distribution of droplets with a diameter of 10 μm. Use the droplet vaporization relationships given in Chapter 9. Find the vaporization time, the fuel flow rate, and the heat output of the reactor. The heat of vaporization of kerosene is 250 kJ/kg and the boiling point temperature is 500 K.

Solution

In a well-stirred reactor, temperature and pressure are constant. From Equation 9.29 the vaporization time is

$$t_v = \frac{d_0^2}{\beta}$$

β can be determined from Equations 9.35 and 9.36

$$\beta = 8\left(\frac{k_g}{\rho_\ell c_{p,g}}\right)\ln(1+B)$$

$$B = \frac{c_{p_g}(T_\infty - T_b)}{h_{fg}}$$

Assuming the products have the specific heat of nitrogen (Appendix C) and the thermal conductivity of air (Appendix B),

$$B = \left(\frac{35.99 \text{ kJ}}{\text{kgmol} \cdot \text{K}} \cdot \frac{\text{kgmol}}{28 \text{ kg}}\right)\left(\frac{(2000-500) \text{ K}}{1}\right)\left(\frac{\text{kg}}{250 \text{ kJ}}\right) = 7.71$$

$$\beta = 8\left(\frac{0.124 \text{ W}}{\text{m} \cdot \text{K}}\right)\left(\frac{\text{m}^3}{800 \text{ kg}}\right)\left(\frac{\text{kgmol}}{35.99 \text{ kJ}} \cdot \frac{28 \text{ kg}}{\text{kgmol}}\right)\left(\frac{\text{kJ}}{1000 \text{ W} \cdot \text{s}}\right)\frac{1 \times 10^6 \text{ mm}^2}{\text{m}^2}\ln(1+7.71)$$

$$\beta = 2.1 \text{ mm}^2/\text{s}$$

Substituting into Equation 9.29,

$$t_v = \frac{\left(10 \times 10^{-3} \text{ mm}\right)^2}{1} \cdot \frac{\text{s}}{2.1 \text{ mm}^2} = 48 \times 10^{-6} \text{ s}$$

The residence time in the reactor is

$$\tau_{res} = \frac{m_{reactor}}{\dot{m}_{prod}}$$

where $m_{reactor}$ is the reacting mass in the well-stirred reactor and τ_{res} is the residence time. As a first approximation we can set the residence time equal to the evaporation time. Obtaining the total mass in the reactor from the total moles in Example 11.2 and assuming the molecular weight of the products is that of nitrogen, and using the vaporization time as the residence time, it follows that

$$\dot{m}_{prod} = \frac{6.09 \times 10^{-5} \text{ gmol}}{\text{cm}^3} \cdot \frac{28 \text{ g}}{\text{gmol}} \cdot \frac{\text{kg}}{1000 \text{ g}} \cdot \frac{900 \text{ cm}^3}{1} \cdot \frac{1}{48 \times 10^{-6} \text{ s}} = 32.2 \text{ kg/s}$$

Since the mass flow rate into the reactor equals the mass flow rate of products out of the reactor, the fuel flow rate is

$$\dot{m}_{f,in} = (32.2 \text{ kg/s}) \frac{168}{3270} = 1.66 \text{ kg/s}$$

Assuming that the combustion reactions are fast relative to evaporation and noting that the progress of reaction variable, η, has the same value as in Example 11.2, the heat release rate is

$$\eta(\dot{m}_{f,in} \text{LHV}) = (0.905) \frac{1.66 \text{ kg}}{\text{s}} \cdot \frac{42.5 \text{ MJ}}{\text{kg}} = 63.8 \text{ MW}$$

The heat output per unit volume is 71 kW/cm³. This is about half that in Example 11.2; thus it reminds us that when dealing with specific combustors we will need a set of detailed models that account for all the physics and chemistry in the problem. Remember that our vaporization model does not account for radiation heat transfer, we have not included the impact of the high shear rapid mixing environment of the well-stirred reactor, and that we have assumed mono-sized droplets rather than a realistic droplet size distribution. This model is very sensitive to the assumed droplet size. If droplets with a 50 μm diameter are chosen, the heat output drops to 2.5 MW (2.8 kW/cm³). Nonetheless these types of models can provide a starting place for gaining a qualitative understanding of combustion.

For a well-stirred reactor as the fuel flow rate increases, the temperature increases up to the point where blowout occurs. At blowout the temperature is less than the adiabatic flame temperature and η is less than 1. For a given total flow rate less than the blowout value there are two reactor temperatures that satisfy the equations—one with a smaller value of η and one with a larger value of η. In the limit as η approaches 1, the mass flow rate approaches zero and the reactor temperature approaches the adiabatic flame temperature.

Well-stirred reactor calculations are very sensitive to the chemical reaction rate. A single global reaction rate is a large approximation. A more rigorous approach is to use a more detailed set of kinetic reactions such as those presented by Frenklach et al. (1995). Detailed well-stirred reactor calculations allow for a finite vaporization and mixing time as well as multiple kinetic reactions. From the primary zone of the combustor, the products flow to the intermediate and dilution zones to complete the combustion process before entering the turbine.

11.4   LINER HEAT TRANSFER

The heat transfer to the combustion chamber liner comes from both convection and radiation heat transfer. The primary radiation heat transfer component is from the soot particles that are formed in the fuel rich portions of the combustor. These particles are less than 1 mm in size and radiate approximately as black bodies. Because of their small size, their temperature is very close to the local gas temperature. The amount of soot formed by the combustion process increases dramatically with pressure up to about 20 atm. Thus, increasing pressure from 5 to 15 atm increases the radiation heat transfer flux by about four times. Increasing the air-fuel ratio decreases the soot concentration and also lowers the temperature, thus greatly reducing the radiation heat transfer flux. Doubling the air-fuel ratio can decrease the radiation by a factor of three. The radiation heat transfer flux is typically a function of distance from the injector, showing a maximum radiation heat transfer flux at 5–10 cm downstream of the injector. The maximum point of radiation heat transfer flux moves downstream as the air-fuel ratio is decreased. Then the radiation heat transfer flux decreases downstream of this maximum point due to soot oxidation as well as decreasing temperature. The fraction of total heat transfer due to radiation can be as high as 50%. Figure 11.13 shows the total heat flux to the liner wall as a function of liner diameter for several types of fuels. As can be seen, the heat flux is very high, and thus efficient means of cooling the liner are required. Increasing pressure and inlet temperature increases the heat flux and causes a corresponding increase in liner temperature. Increased airflow rate decreases the liner temperature because although both the cooling side and combustion side convection coefficients increase as $\dot{m}_{air}^{0.8}$, the convection component on the combustion side represents only half of the total flux. Thus, a doubling of $\dot{m}_{air}^{0.8}$ will decrease the cooling side thermal resistance by two but decrease the combustion side thermal resistance by only

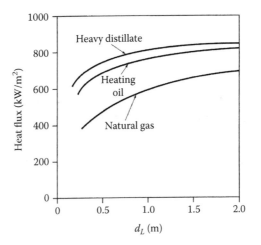

FIGURE 11.13   Typical heat flux to the combustor liner wall versus liner diameter for different fuels. (Lefebvre, A. H., *Gas Turbine Combustion*, 2nd ed., Taylor & Francis, Philadelphia, PA, 1999. Taylor & Francis.)

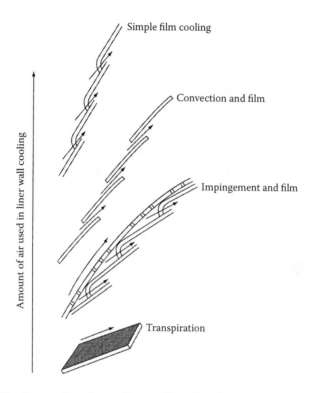

FIGURE 11.14 Types of combustor liner wall cooling designs. (From Lefebvre, A. H., *Gas Turbine Combustion*, 2nd ed., Taylor & Francis, Philadelphia, PA, 1999. Taylor & Francis.)

3/4. Increasing the fraction of the inlet mass flow that goes to the primary zone increases the convection coefficient in the primary zone and thus the liner temperature in that region if the air-fuel ratio is kept constant. The liner temperature reaches a maximum in the primary zone for mixtures 10% richer than stoichiometric, which also corresponds to the maximum flame temperature for a homogeneous mixture.

High performance combustor design depends on film cooling of the liner on the combustion side as well as cooling on the air side of the liner. Film cooling is provided by slots in the liner that introduce a wall jet along the inside of the liner. It is also possible to cool the wall by transpiration cooling; that is, by flowing the cooling air through a porous liner. Use of a liner with many small holes approximates transpiration cooling and is called effusion cooling. Figure 11.14 shows each of these film cooling methods in a conceptual manner.

Because a large portion of the heat transfer is by radiation, slot cooling can only partially reduce the air-side liner coolant load and the gas-side liner temperature. Transpiration or effusion cooling using multi-layered walls with many interconnecting flow passages can be used to provide liner strength while also internally cooling the liner. In all cases, the slots and holes are subject to clogging or partial blockage by soot. An alternative to these methods is to use a refractory liner backed by metal for strength. Refractory liners can operate with surface temperatures up to 1900 K. Such high temperatures prevent carbon build up and reduce wall quenching. The weight and volume of refractory liners restricts their application to industrial turbines. For aircraft, thin ceramic liner coatings, especially in downstream regions, have been attempted with modest success.

11.5   LOW EMISSIONS COMBUSTORS

Over the last 50 years, gas turbine combustors have evolved to achieve high stability and high combustion intensity (high heat release per unit volume) using swirl stabilized diffusion flames. Today, in addition to high stability and intensity, gas turbine combustors must have fewer carbon monoxide, hydrocarbon, nitrogen oxides, and soot emissions. There are four main factors controlling the emissions from gas turbine combustors using distillate liquid fuel or natural gas: (1) the primary zone combustion temperature and equivalence ratio, (2) the degree of mixing in the primary zone, (3) the residence time, and (4) the combustor liner quenching characteristics.

Past practice was to maintain the primary combustion zone at an equivalence ratio of 0.7 to 0.9, which is slightly lean. For leaner mixtures, the CO levels are high because of the slow rate of oxidation and the relatively short residence times. Hydrocarbon (HC) emissions occur due to incomplete vaporization of the fuel spray. Both CO and HC emissions can be reduced by improved fuel atomization, for example by use of well-designed air blast spray nozzles. Air bleed through the liner in the intermediate combustion zone completes combustion of the HC and CO while cooling the wall. Because mixing tends to be non-uniform, cool spots, which quench HC and CO burnout, and hot spots, which tend to generate NO_X, can occur. Reduction of the liner cooling air is effective in reducing HC and CO emissions but at the expense of increased NO_X emissions.

Soot is formed in the fuel rich zones near the fuel droplets. Most of the soot is produced in the primary zone and is consumed in the intermediate and dilution zones in high temperature turbines. Injecting more air into the primary zone reduces soot emissions but at the expense of increased CO and HC emissions. Sooting is more severe at high pressure in part because soot is formed closer to the spray nozzles. Soot emissions can be reduced by improved mixing, water injection, and increased residence time.

The basic idea for reducing NO_X emissions is to reduce the peak temperature during combustion and the time spent at high temperature. Modest reductions in NO_X emissions can be achieved by reducing the residence time at high temperature by improved mixing of the liner air and improved fuel injection. Large reductions in NO_X emissions can be achieved by water injection, premixed lean burn combustion, and staged combustion. Exhaust gas recirculation can also reduce NO_X but requires additional compressor work, hence it is not done.

For utility gas turbines, injecting water at the rate of 0.5–1.5 times the fuel flow rate directly into the primary zone is effective in reducing NO_X emissions (Figure 11.15), while HC and CO emissions increase only slightly as the water-fuel ratio increases to 1. As the water-fuel ratio increases beyond 1, the HC and CO emissions

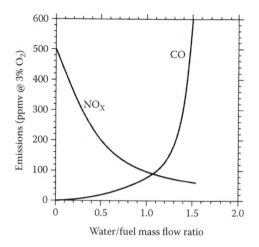

FIGURE 11.15 Effect of water injection on NO_X and CO emissions form large gas turbines. (From Bowman, C. T., "Control of Combustion- Generated Nitrogen Oxide Emissions: Technology Driven by Regulation," *Symp. (Int.) Combust.* 24:859–878, The Combustion Institute, Pittsburgh, PA, 1992, by permission of The Combustion Institute.)

increase rapidly. Injecting water increases the power by as much as 16%; however, the cycle efficiency drops up to 4% due to the lower temperature. In addition to the cost due to lower efficiency, water injection is an added expense because the water needs to be demineralized so as not to foul the turbine. Water injection is sometimes used to meet current federal stationary source emissions standards of about 160 ppm NO_X at 15% oxygen. (The actual emission standard depends on the heat rate and the fuel nitrogen content if any.) Stricter standards, such as the California standard of 25 ppm, are met with premixed lean burn combustion.

With lean combustion the diluent is air rather than water. Reducing the fuel-air ratio to be near the lean combustion limit will reduce the NO_X levels to below 25 ppm at 15% O_2. However, there are two design difficulties to be overcome with lean combustors. First, combustion stability must be assured, and second, turndown capability must be maintained because a gas turbine must ignite, accelerate, and operate over a range of loads. The lowest NO_X levels are achieved with a premixed gaseous fuel.

An example of a premixed low NO_X combustor is shown in Figure 11.16. Extensive air cooling near the combustor walls is used. There are six fuel nozzles, each with its own air swirler. Fuel from the fuel nozzles mixes with the swirl air and flows through a venturi throat. The flame is stabilized by the swirl just downstream of the venturi throat. The venturi throat accelerates the flow and prevents the flame from propagating back into the premixing zone. During startup and part load operation, the six primary fuel nozzles form diffusion flames that burn in the zone formed by the fuel nozzles and venturi throat. In the diffusion flame mode, the NO_X emissions are two to three times higher than the 20 ppm at 15% O_2 achieved when operating in the premixed mode. Carbon monoxide emissions are less than 10 ppm at full load.

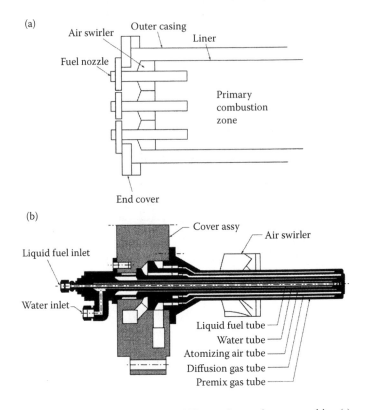

FIGURE 11.16 Schematic drawing of a low NO$_X$ combustor for a gas turbine (a) combustor and (b) fuel nozzle. (Courtesy of General Electric Co., Schenectady, NY.)

With liquid fuels, lean-burn combustors are not able to achieve the reductions in NO$_X$ emissions that are obtained with gaseous fuels. In this case two-stage combustion is used. In this approach, as shown in Figure 11.17, the primary combustion zone is run fuel rich, the combustion products from the primary zone are cooled by heat transfer to the inlet air, and the secondary combustion zone is run lean. The overall stoichiometry is maintained, but heat transfer tends to quench the formation of NO$_X$, and the dilution air tends to reduce hot spots. A 75% reduction in NO$_X$ emissions has been achieved with combustor staging, which is sometimes referred to as the rich-quench-lean approach.

From the above discussion it is apparent that operating performance and emissions from gas turbine combustors are affected by the design details of the front end of the combustor and the subsequent admittance of air into various combustion and dilution zones. For example, minor changes in the liner air distribution can create sizable variations in such parameters as lean blowout and ignition, exit gas temperature profile, and nitric oxide emissions. Improved combustor design requires detailed knowledge of the flow patterns within the combustor and the processes of fuel injection, evaporation, mixing, combustion chemistry, and convective and radiative heat transfer.

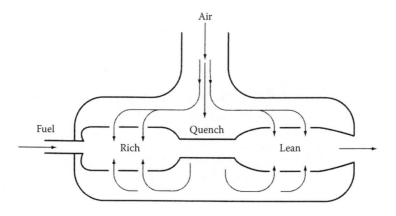

FIGURE 11.17 Two-stage, rich-quench-lean combustor for a low NO_X gas turbine.

11.6 PROBLEMS

1. For the gas turbine cited in Table 11.1 determine the following parameters:
   a. Simple cycle and combined cycle thermal efficiency
   b. Fuel feed rate per nozzle (L/h) assuming No. 2 fuel oil for the simple cycle gas turbine
   c. Combustor inlet temperature assuming no heat loss
   d. Overall excess air
   e. Velocity at the combustor outlet
   f. Water injection rate to meet the specified turbine inlet temperature

   The compressor inlet temperature is 300 K. The combustor chambers are 14 cm in diameter. The stoichiometric air-fuel ratio for No. 2 fuel oil is 14.5. Use the data provided in Table 2.7 for No. 2 fuel oil. Use Appendix B for air properties. Use Appendix C for H_2O properties.

2. Assume an ideal gas turbine with inlet air conditions of 300 K and 1 atm pressure. For combustor temperatures of 1200 and 1800 K, vary the combustor pressure from 1 atm to 35 atm. Determine (a) the ideal gas turbine power per unit mass flow versus combustor pressure, (b) the efficiency, and (c) the heat rate. Assume $c_p = 1.0$ kJ/kg·K and $\gamma = 1.4$.

3. Combustion efficiency is a measure of the completeness of combustion as measured by the energy content of the fuel.

$$\eta_{comb} = \frac{q_{chem}}{\dot{m}_f\left(\text{LHV}\right)}$$

   As discussed in Chapter 4, CO oxidation is the last step in the complex chain of reactions for the combustion of hydrocarbons. Based on this for gaseous and liquid fuels, we can assume that combustion efficiencies less than 100% reflect the energy lost by the incomplete oxidation of CO to CO_2. Consider a gas turbine using kerosene in a lean fuel-air ($F = 0.5$) mixture.

The products are H_2O, CO_2, CO, O_2, and N_2. If the combustion efficiency is 98%, what is the mole fraction (ppm) of CO in the products? Use $C_{12}H_{24}$ for kerosene. The lower heating value of kerosene is 42.5 MJ/kg. Use the heating value of CO given in Table 2.2.

4. The combustion products of the gas turbine combustor of Problem 3 will react further given sufficient time. The mole fractions of CO, O_2, H_2O from Problem 3 are 0.000125, 0.0045, and 0.075, respectively, the pressure is 10 atm and the temperature is 1400 K. Note that the O_2 and H_2O will not change significantly relative to the CO. How long does it take for the CO to decrease to 1 ppm? Assume that there is no back reaction and that

$$\frac{d[CO]}{dt} = -10^{14.60} \exp\left(\frac{-40{,}000}{\hat{R}T}\right)[CO]^{1.0}[O_2]^{0.25}[H_2O]^{0.50}$$

5. A well-stirred reactor with a volume of 900 cm³ operates at 10 atm pressure and 2000 K. Methane at 25°C flows in at 11.2 gmol/s and air at 200°C flows in at 106.6 gmol/s. For a one-step reaction of methane to CO_2 and H_2O, what percentage of the methane reacts before exiting the reactor? Use the lower heating value of methane given in Table 2.2. Assume the products have the enthalpy of N_2 (Appendix C).

6. For the conditions of Example 11.2 vary the reactor temperature from 1600 K to 2300 K in 100 K increments. What reactor temperature provides the maximum heat output, and what is the maximum heat output rate (MW)? Determine the inlet fuel flow rate and the reaction progress variable η at each of these temperatures.

7. For the conditions of Example 11.2 except that the inlet air is preheated to 600 K, vary the reactor temperature from 1600 K to 2300 K in 100 K increments. What reactor temperature provides the maximum heat output, and what is the maximum heat output rate (MW)? Determine the inlet fuel flow rate and the reaction progress variable η at each of these temperatures.

8. For the conditions of Example 11.2 except using 15 atm pressure instead of 10 atm pressure, vary the reactor temperature from 1600 K to 2300 K in 100 K increments. What reactor temperature provides the maximum heat output, and what is the maximum heat output rate (MW)? Determine the inlet fuel flow rate and the reaction progress variable η at each of these temperatures.

9. For the conditions of Example 11.2 determine the average residence time in the reactor.

10. Repeat Example 11.3 using 20 μm droplets and 1600 K. What is (a) the combustion efficiency, (b) the mass flow rate of fuel, and (c) the heat output rate?

REFERENCES

Ballal, D. R. and Lefebvre, A. H., "Ignition and Flame Quenching of Flowing Heterogeneous Fuel-Air Mixtures," *Combust. Flame* 35:155–168, 1979.

Bathie, W. W., *Fundamentals of Gas Turbines*, 2nd ed., Wiley, NY, 1995.

Bowman, C. T., "Control of Combustion-Generated Nitrogen Oxide Emissions: Technology Driven by Regulation," *Symp. (Int.) Combust.* 24:859–878, The Combustion Institute, Pittsburgh, PA, 1992.

Boyce, M. P., *Gas Turbine Engineering Handbook*, 3rd ed., Gulf Professional Publishing, Burlington, MA, 2006.

Brandt, D. E., "Heavy Duty Turbopower: The MS700lF," *Mech. Eng.* 109(7):28–37, 1987.

Claeys, J. P., Elward, K. M., Mick, W. J., and Symonds, R. A., "Combustion System Performance and Field Test Results of the MS7001F Gas Turbine," *J. Eng. Gas Turbines Power* 115(3):537–546, 1993.

Davis, L. B. and Washam, R. M., "Development of a Dry Low NO_x Combustor," ASME paper 89-GT-255, 1989.

El-Wakil, M. M., Chap. 8 in *Powerplant Technology*, McGraw-Hill, New York, 1984.

Frenklach, M., Wang, H., Goldenberg, M., Smith, G. P., Golden, D. M., Bowman, C. T., Hanson, R. K., Gardiner, W. C., and Lissianski, V., "GRI-Mech—An Optimized Detailed Chemical Reaction Mechanism for Methane Combustion," report GRI-95/0058, Gas Research Institute, Chicago, IL, 1995.

Lefebvre, A. H., *Gas Turbine Combustion*, 2nd ed., Taylor & Francis, Philadelphia, PA, 1999.

Lefebvre, A. H. and Ballal, D. P., eds., *Gas Turbine Combustion: Alternative Fuels and Emissions*, 3rd ed., CRC Press, Boca Raton, 2010.

Liu, G.-S. and Niksa, S., "Coal Conversion Submodels for Design Applications at Elevated Pressures, Part II Char Gasification," *Prog. Energy Combust. Sci.* 30(6):679–717, 2004.

Longwell, J. P. and Weiss, M. A., "High Temperature Reaction Rates in Hydrocarbon Combustion," *Ind. Engr. Chem.* 47(8):1634–1643, 1955.

Odgers, J. and Kretschmer, D., *Gas Turbine Fuels and Their Influence on Combustion*, Abacus Press, Tunbridge Wells, Kent, UK, 1986.

Rizk, N. K. and Mongia, H. C., "Three-Dimensional Gas Turbine Combustion Emissions Modeling," *J. Eng. Gas Turbines Power* 115(3):603–611, 1993.

Sawyer, J. W. and Japikse, D., eds., *Sawyer's Gas Turbine Engineering Handbook*, Vols. 1–3, 3rd ed., Turbomachinery International, Norwalk, CT, 1985.

12 Diesel Engine Combustion

The diesel engine is the dominant type of engine for heavy-duty on-road and off-road vehicles, marine transportation, and industrial power sources. While some light-duty vehicles (less than 2700 kg) use diesel engines, issues of weight, noise, odor, and emissions have limited the use of diesel engines for light-duty vehicles, especially in the United States. If light-duty diesel vehicles can be made to meet emissions standards, they offer a means of improving vehicle fuel economy.

Diesel engines follow the same four-stroke cycle as spark ignition (SI) engines. The primary difference is that ignition of the fuel in a diesel engine is by compression, and there is no sparkplug. In the ideal thermodynamic cycle, diesel combustion occurs at constant pressure, while combustion in the Otto cycle occurs at constant volume. In practice, for both types of engines, the pressure and volume change during combustion. Diesel engines, similar to SI engines, have a higher thermal efficiency at an increased compression ratio. The compression ratio in SI engines, which have intake port injection of premixed fuel and air, is limited by knocking combustion. Diesel engines, which have direct injection of liquid fuel into each cylinder, do not experience knock and thus can have a higher compression ratio than SI engines. Whereas diesel engines may have compression ratios between 12 and 18, the compression ratio in SI engines is limited to about 10. In addition, diesel engines control load (power output) by reducing the amount of fuel injected into each cylinder. By contrast, SI engines restrict the air and the fuel at part load by means of a throttle so as to maintain a stoichiometric mixture over all loads. The throttle increases the pressure loss during the intake stroke, reducing the net thermal efficiency of the SI engine.

Before continuing with the discussion of diesel engines, the reader may wish to review the engine terminology given at the end of Chapter 7 for premixed, spark ignition engines.

12.1 INTRODUCTION TO DIESEL ENGINE COMBUSTION

Diesel fuel is injected separately into each cylinder during the compression stroke through a multiple-hole injector. The piston head has a recessed bowl so that droplets formed by the injector spray into the bowl and not onto the piston head. Diesel fuel is less volatile than gasoline but the ignition delay is shorter. Ignition occurs around each drop due to the temperature of compression. Initially, the local fuel-air mixture around each drop is rich, but as vaporization and mixing occur the overall mixture is lean. Because the heat release rate is limited by droplet vaporization, combustion is relatively gradual and knock is not an issue. To complete combustion in the time available, the droplets must be small and thus the injector pressure is high.

There is a tendency to form soot particles, as the temperature during combustion is relatively high, the mixture is rich shortly after injection, and combustion involves local diffusion flames around the droplets. Fortunately, as the overall mixture becomes lean, many of the soot particles burn out in the cylinder; however, some soot particles survive and are emitted. The carbon monoxide and hydrocarbons generally burn completely. However, because of the overall excess oxygen and high temperature, nitrogen oxides are formed and emitted. Because of this, the engineering challenge of diesel engines is to achieve high efficiency in an engine that has a high-energy output per unit weight while maintaining low emissions.

Most diesel engines are turbocharged to obtain more power per unit cylinder volume. A turbocharger is a small centrifugal turbine connected directly to a centrifugal air compressor to recover some of the exhaust pressure and boost the cylinder inlet air pressure and temperature. Heavy-duty diesel engines are sometimes turbocompounded (i.e., part of the turbocharger drives the inlet air compressor, and part of the power is geared to the output shaft), and sometimes an intercooler is also used as well to increase the density of the inlet air.

Diesel engines come in a wide range of sizes, from small industrial engines to enormous ship engines. Although the crank shaft speed ranges from 50 rpm for very large engines to 5000 rpm for very small engines, the average piston speed does not vary nearly as much as the crank shaft speed. Small engines produce less torque than large engines, but by operating at higher revolutions per minute they improve power output per unit of cylinder volume. Table 12.1 shows examples of the wide range of sizes and power outputs produced.

In the following sections of this chapter, combustion chamber geometry and flow patterns, fuel injection, ignition delay, combustion performance, and emissions are discussed. Ways to improve diesel engine combustion, including a concept for dual-fuel partially premixed combustion, are presented.

12.2 COMBUSTION CHAMBER GEOMETRY AND FLOW PATTERNS

The combustion chamber includes the cylinder and piston. Fuel is provided through the fuel injector, air enters the combustion chamber though the intake valve and

TABLE 12.1
Typical Size and Output of Diesel Engines

Bore (mm)	45	80	127	280	400	840
Stroke (mm)	37	80	120	300	460	2900
Displacement (L/cylinder)	0.06	0.40	1.77	18.5	57.8	1607
No. of cylinders	1	4L*	8V+	6–9L*	6–9L*	4–12L*
Output per cylinder (kW)	0.7	10	40	325	550	3380
Rated speed (rpm)	4800	3600	2100	1000	520	55–76
BMEP (atm)	4	7.5	13	22	22	17

V+ designates V-shaped cylinder orientation.
L* designates in-line cylinder orientation.

port, and the combustion products exit through the exhaust valve. The fuel injector is centrally located in the head of the cylinder. The injector sprays the pressurized fuel laterally through four to eight equally spaced holes. The piston head has symmetrically located recessed bowls to delay impingement of the spray on the piston head. One lobe of the cylinder volume near top dead center (TDC) is shown in Figure 12.1. A small crevice volume between the piston and the cylinder wall is also indicated in Figure 12.1.

The flow characteristics in the cylinder depend on the port and valve configuration and the shape of the piston head. The intake valve opens and the piston moves away from TDC drawing air into the cylinder. Swirl around the axis of the cylinder is induced by the curvature of the intake ports. The intake valve closes and as the piston moves upward, the cylinder air is pushed into the bowl of the piston, forcing a radial *squish flow* into the bowl. Conservation of momentum demands that the swirl in the bowl increases as the radius of the swirling gas decreases in the bowl. As the piston moves upward, the air at the piston surface must be at the velocity of the piston. The piston motion imparts a *squash velocity* to the air. The squash velocity is similar in magnitude to the swirl velocity. The *swirl ratio* is defined as the ratio of swirl revolutions per minute to engine revolutions per minute. A swirl ratio of one is considered to be very low, and such chamber designs are often called "quiescent" even though the flows within the chamber are quite turbulent and include complex large-scale flows. Induction swirl ratios of 3–4 are common and can result in piston bowl swirl ratios of 10–15 for small bowl to bore ratios.

Smaller engines tend to have high air swirl in the cylinder, while larger engines tend toward low swirl. High swirl engines have a deep bowl in the piston head, a low number of holes in the fuel injector (typically four), and moderate fuel injection pressures (13–340 atm). Low swirl combustion chamber engines have a shallow bowl in the piston head, a larger number of holes in the fuel injector (typically eight holes), and higher fuel injection pressures (500–1400 atm). There is a somewhat different optimum combination of design parameters for each engine load and speed, and no single design gives the best possible performance over a wide range of speeds and loads.

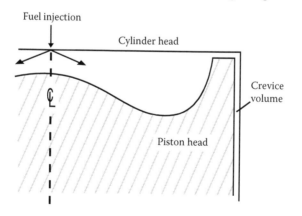

FIGURE 12.1   Schematic drawing of a piston-cylinder showing the cylinder volume near TDC for one lobe of a piston head.

Swirl influences the mixing between the fuel plume and air during the injection period. The swirl also tends to stabilize the flow, thereby reducing cycle-to-cycle flow variations. Chambers with low swirl tend to have tumbling vortices that can shift from one cycle to the next. However, excessive swirl can cause hot, less dense products to move to the center of the chamber, thus reducing mixing and spray penetration, which produces lean mixtures that only partially burn, thereby causing hydrocarbon and CO emissions. Furthermore, excessive swirl causes increased heat transfer. Low swirl engines with shallow, wide bowl combustion chambers and high injection pressures have nozzles with many holes (8–10) to compensate for the lack of tangential mixing that would be caused by swirl. Chambers with low swirl typically lack mixing during the later part of the combustion period.

12.3   FUEL INJECTION

Fuel is injected at a time determined by best torque and emissions considerations. A typical start of fuel injection ranges between 30° before TDC to TDC depending on the fuel cetane number (ignition delay time) and engine speed for a given engine design. The fuel injector in the head of each cylinder sprays intermittent plumes of very small fuel droplets at high pressure across the cylinder head and mixes the droplets with the surrounding air. For example, the droplet size distribution of diesel fuel injected at 1500 atm through a 0.20 mm hole for 2.6 ms was measured 10 mm downstream of the hole by Sakaguchi et al. (2008). Along the centerline the most probable droplet size was 50 microns, and the droplets ranged from 10 to 150 microns in diameter. The maximum fuel-air ratio is determined by the allowable particulate level and is typically about a 0.5 equivalence ratio, and the associated fuel volume per cycle is roughly 100 mm³ per liter of displacement at full load.

Because engine load is controlled by the amount of fuel injected, the injector must handle a volume of fuel that ranges over an order of magnitude. For a fixed amount of fuel injected per cycle and a fixed injection duration, the injection pressure must increase in proportion to the square of the engine speed, or the effective flow area of the injector nozzle must increase in proportion to the engine speed. Alternately, a positive displacement fuel injection system can be used to meter the fuel. Injection flexibility facilitates improved performance, lower emissions, and better fuel tolerance. For example, it is possible to have a pilot injection where a small amount of fuel is injected prior to the injection of the rest of the fuel.

The quality of diesel injection is judged by droplet size distribution, spray penetration history, and spray angle. Other factors that influence combustion rate and emissions include turbulence in the injector nozzle holes, cavitation in the holes, injection rate shape, droplet vaporization, spatial distribution within the spray, spray impingement on the piston bowl surface, and the way the spray influences turbulent mixing rates.

Recall from Chapter 9 that the average droplet size in the diesel spray is proportional to the square root of the pressure drop across the nozzle,

$$\bar{d} \propto \sqrt{\Delta p}$$

and that the spray penetration distance is proportional to one-fourth root of the pressure drop across the nozzle,

$$L \propto (\Delta p)^{1/4}$$

Thus high injection pressure has the advantage of producing small droplets and rapid vaporization, but the high pressure may cause the spray to impinge on the piston bowl, increasing soot emissions. This is of particular concern for smaller engines.

The spray formulas in Chapter 9 were developed for non-vaporizing cases. Spray penetration decreases about 10–20% due to vaporization. Spray penetration is caused by newly formed droplets overtaking and passing previously formed droplets so that the droplets that are least vaporized lead the tip penetration. For very high-pressure injection sprays (2000 atm) with small nozzle holes (0.1 mm), the droplets are very small, and thus the spray vaporizes rapidly, giving a gas-jet-like behavior. Laser sheet diagnostics indicate that at full load the spray becomes vaporized less than 25 mm from the nozzle. The high momentum of the high-pressure sprays greatly increases mixing compared to lower pressure sprays. In general, care must be taken that the spray jet does not impinge on the piston bowl surface.

The effects of droplet size on vaporization rate may be estimated from the theory presented in Chapter 9. There is a narrow core along the spray axis which has an approximate length given by the breakup length (Equation 9.25a). This core has a high momentum and shows some properties of a continuous liquid; it is thus often called the "intact core." Photographs of high-pressure sprays that are vaporizing have shown this core to persist even while all the surrounding spray has vaporized.

12.4   IGNITION DELAY

The parameters that most affect ignition delay are the temperature and pressure of the air during the delay period and the kinetics of the fuel as denoted by the cetane number. Recall from Chapter 2 that the cetane number rating ranks the ignition delay of a given fuel relative to a reference fuel. More specifically, the cetane number is the percentage of cetane (*n*-hexadecane) that must be added to isocetane (heptamethylnonane) to produce a 13 CA° of ignition delay at the same compression ratio that gives a 13 CA° ignition delay with ignition at TDC for the test fuel. Isocetane has a relatively long ignition delay compared to cetane. The cetane number of cetane is arbitrarily set at 100 and the cetane number of isocetane is set at 15. A CFR engine (a standardized single cylinder test engine) is run at 900 rpm with inlet air temperature at 339 K to determine the cetane number.

Long ignition delays are not well tolerated by diesel engines because a large amount of fuel vaporizes and mixes with air prior to ignition if the delay is long. When ignition finally takes place, this vaporized fuel burns rapidly, thus giving a high rate of pressure rise that creates a characteristic sharp noise and high nitrogen oxide emissions. Retarding the injection timing or heating the inlet air may help shorten ignition delay, but this may produce undesirable changes in performance and particulate emissions.

Formulas for predicting ignition delay have been obtained from both engine and constant volume tests. The independent variables are typically average cylinder gas pressure and temperature during the delay, the fuel cetane number, and the overall fuel-air ratio. A typical ignition delay correlation is of the form

$$\Delta t = Cp^a \, F^b \, \exp(E/T) \qquad (12.1)$$

where F is the overall fuel-air equivalence ratio, the constant E depends on the type of fuel, C is a constant, and a and b are negative empirical constants. Consider the following example.

Example 12.1

The following ignition delay formula was obtained for a turbocharged diesel engine running on diesel fuel with a cetane number of 45.

$$t = 0.075\,p^{-1.637}\,F^{-0.445}\,\exp\left(\frac{3812}{T}\right)$$

where Δt is the ignition delay (ms), p is the average pressure (MPa) during the delay, and T is the average temperature (K) during the delay. This diesel engine has a compression ratio of 13.25, and the connecting rod to crank radius ratio is 4.25. At $\theta = 20°$ before TDC for $F = 0.6$ and an engine speed of 1500 rpm, the pressure and temperature of the engine are $p = 3.13$ MPa and $T = 816$ K, respectively.

    a. Compute the ignition delay and the crank angle at start of injection for the average crank angles during the ignition delay period, $\bar{\theta}$, of 0°, 5°, 10°, 15°, and 20°.

    b. Tabulate the results.

Assume that the average pressure and temperature during ignition delay occur at the average crank angle and are given by a polytropic function of the volume ratio (V_θ / V_{20}) with an exponent of 1.35, specifically,

$$\frac{p_\theta}{p_{20}} = \left[\frac{V_{20}}{V_\theta}\right]^{1.35}$$

and

$$\frac{T_\theta}{T_{20}} = \left[\frac{V_{20}}{V_\theta}\right]^{0.35}$$

where V_{20} is the combustion chamber volume at 20° crank angle before TDC ($\theta = 20°$).

Solution

From Section 7.9 Table A,

$$V_\theta = \left[x_\theta (CR-1) + 1 \right] V_c$$

where V_c is the clearance volume and x_θ is the fraction of the stroke. Substituting,

$$\frac{V_{20}}{V_\theta} = \frac{\left[x_{20}(CR-1)+1 \right]}{\left[x_\theta(CR-1)+1 \right]}$$

From Section 7.9 Table A

$$x_\theta = \frac{1}{2}\left[1 - \cos\theta + r_c - r_c\sqrt{1 - \left(\frac{\sin\theta}{r_c}\right)^2} \right]$$

where r_c is the ratio of the connecting rod length to the crank radius. Substituting for a crank angle of 20°

$$x_{20} = \frac{1}{2}\left[1 - \cos 20° + 4.25 - 4.25\sqrt{1 - \left(\frac{\sin 20°}{4.25}\right)^2} \right] = 0.0370$$

Substituting,

$$\frac{V_{20}}{V_\theta} = \frac{\left[0.0370(13.25-1)+1 \right]}{\left[x_\theta(13.25-1)+1 \right]}$$

p_{20} and T_{20} are given in the problem statement. The problem can now be solved by first finding x_θ. Given x_θ, then V_{20}/V_θ, p_θ, and T_θ can be determined. Substituting into the ignition delay formula given in the problem statement provides Δt. At $\theta = 20°$, $\Delta t = 1.555$ ms. At 1500 rpm, $\Delta\theta = 9\Delta t$, since

$$\theta = \frac{1500 \text{ rev}}{\text{min}} \cdot \frac{\text{min}}{60 \text{ s}} \cdot \frac{\text{s}}{1000 \text{ ms}} \cdot \frac{360°}{\text{rev}} \cdot \frac{t \text{ ms}}{1}$$

The start of injection (SOI) is then

$$\text{SOI} = \theta + \frac{\theta}{2}$$

The results are tabulated below.

θ (CA°)	x_θ	V_{20}/V_θ	p (MPa)	T (K)	Δt (ms)	$\Delta\theta$ (CA°)	SOI (CA°)
0	0.0000	0.688	5.19	930	0.383	3.45	1.72
5	0.0023	0.708	4.99	921	0.424	3.82	6.91
10	0.0094	0.767	4.48	895	0.571	5.14	12.6
15	0.0210	0.865	3.81	859	0.893	8.04	19.0
20	0.0370	1.000	3.13	816	1.555	14.0	27.0

Thus for $\bar\theta = 15°$ and $\Delta\theta = 8°$, SOI = 19° and start of combustion (SOC) is at 11° before TDC.

12.5 ONE-ZONE MODEL AND RATE OF COMBUSTION

After a short ignition delay, a small portion of fuel spray vaporizes, mixes with air, autoignites, and burns rapidly. This premixed burning phase is brief because the mixture amount is relatively small and burning occurs near TDC. The main diffusion burning of the spray, which constitutes 85% or more of the heat release, is due to vaporization of the droplets during combustion. Detailed computational modeling of diesel combustion processes is beyond the scope of this book, and instead we follow the lead of Chapter 7 and utilize pressure and volume measurements in a one-zone model to determine the heat release rate. The formulation of the one-zone model is the same as for a gasoline engine, although the conceptual image of the combustion is different. In a spark ignition engine, a flame front sweeps across the combustion chamber, whereas in a diesel engine there are multiple combustion zones around each spray plume and the mixture is not homogeneous.

Initially the cylinder contains only air and then fuel is injected. In the one-zone model the air and fuel are assumed to be perfectly mixed at a rate equal to the burning rate of the fuel. Following the one-zone model in Chapter 7, the heat release rate is determined from cylinder pressure versus volume measurements,

$$q_{chem} = \left(\frac{\gamma}{\gamma-1}\right)p\frac{dV}{dt} + \left(\frac{1}{\gamma-1}\right)V\frac{dp}{dt} + q_{loss} \tag{7.9}$$

where γ is the ratio of the specific heats.

The heat transfer, q_{loss}, includes both radiation and convective heat loss terms. Low sooting engines reduce the radiation component to less than 15% of the total heat transfer to the combustion chamber surfaces. Figure 12.2 shows histories of total and radiation heat flux measured on the head of a fired turbocharged diesel engine. A widely used heat transfer correlation is from Woschni and Anisits (1974). This formula linearizes the radiation term and combines it with the convective term to give

$$q_{loss} = \tilde{h}A_{wall}\left(T - T_{wall}\right) \tag{12.2}$$

where

$$T = T_\theta = \text{mass average gas temperature } (K)$$

$$T_{wall} = \text{temperature of the wall surface area } A_{wall}(K)$$

$$\tilde{h} = 0.82L_b^{-0.2}W^{0.8}p^{0.8}T^{-0.55} \ \left(kW/m^2 \cdot K\right)$$

$$p = p_\theta = \text{cylinder pressure } (MPa)$$

$$L_b = \text{cylinder bore } (m)$$

$$W = 2.28\overline{V}_p + 0.00324\left(p - p_o\right)\left(\frac{V_d T_1}{p_1 V_1}\right)$$

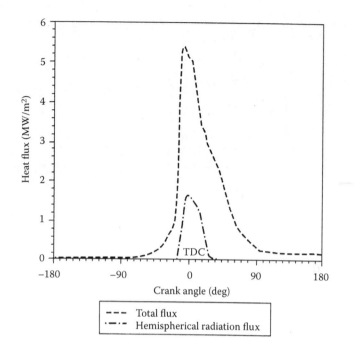

FIGURE 12.2 Total and radiant heat flux measured on the head of an open chamber diesel engine with a 2.33 L displacement at 1500 rpm, part load, $F = 0.4$, and 200 kPa intake pressure. (From McDonald, J., "Construction and Testing of a Facility for Diesel Engines, Heat Transfer and Particulate Research," Master's Thesis, University of Wisconsin–Madison, 1984.)

$p_1, V_1,$ and T_1 are reference values at intake valve closing

$$p_0 = \text{motoring cylinder gas pressure } (\text{MPa})$$

$$\overline{V}_p = \text{mean piston speed } (\text{m/s})$$

$$V_d = \text{displacement volume } (\text{m}^3)$$

Test measurements from a single cylinder diesel engine and calculated results using the one-zone model are shown in Figure 12.3. The cylinder pressure is measured and recorded every ¼ CA° from −360 to +360 CA°, and the displacement volume at every data point is calculated from the engine parameters in Table 12.2. Average cylinder temperature is calculated from

$$\frac{pV}{T} = \text{constant}$$

and heat release per crank angle degree is calculated from Equation 7.9. The mass fraction of fuel burned is calculated from the sum of the heat release at that crank angle divided by the total heat release between intake valve closing (IVC) and exhaust valve opening (EVO). Results from three test runs are plotted in the graphs in Figure 12.3

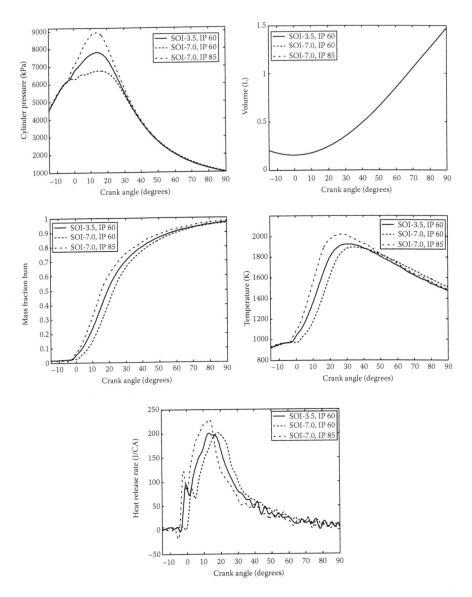

FIGURE 12.3 Three Cummins N14 diesel engine test runs measured cylinder pressure versus crank angle and calculated the mass fraction burned, heat release rate, volume, and gas temperature versus crank angle. The following conditions were tested: (a) start of injection (SOI) at 3.5° before TDC, injector pressure 60 MPa, IMEP = 1.07 MPa; (b) SOI at 7° before TDC, injector pressure 60 MPa, IMEP = 1.13 MPa; and (c) SOI at 7° before TDC, injector pressure 85 MPa, IMEP = 1.2 MPa. See Table 12.2 for engine parameters and test conditions. (Courtesy of J. Ghandhi, University of Wisconsin–Madison, Madison, WI.)

TABLE 12.2

Test Engine Operating Parameters for Cummins N14 Diesel Engine

Parameter	Value
Turbocharger	None
Cylinder bore	139.7 mm
Stroke	152.4 mm
Connecting rod	304.8 mm
Displacement	2.336 L
Compression ratio	16
Fuel type	No. 2 diesel
Fuel HHV	43.8 MJ/kg
Engine speed	1200 rpm
Cycles averaged	200
Exhaust valve closed	−355° after TDC
Intake valve closed	−143° after TDC
Exhaust valve open	130° after TDC
Intake valve open	335° after TDC
Air intake pressure	149 kPa absolute
Air intake Temp.	319 K
Exhaust back pressure	65 kPa gauge
Start of injection (CA° before TDC)	(a) 3.5, (b) 7.0, (c) 7.0
Injector pressure (MPa)	(a) 60, (b) 60, (c) 85
Net IMEP (MPa)	(a) 1.07, (b) 1.13, (c) 1.20

Source: Courtesy of J. Ghandhi, University of Wisconsin–Madison, Madison, WI.

where two different injector pressures and two different start of injection timings (SOI) are used with all other parameters held constant. The higher injection pressure produced a higher peak pressure, a larger premixed heat release spike, a more rapid heat release rate, and an increase in the indicated mean effective pressure (IMEP) from 1.07 to 1.20 MPa (Table 12.2), thus implying higher thermal efficiency. Interestingly, the total burn time did not change, as shown by the mass fraction burned plot. Retarding the SOI from 3.5 to 7.0 CA° BTDC while holding the injection pressure constant reduced the peak pressure, slowed the temperature rise, and shifted the heat release later in the cycle. The engineering challenge is to find the set of operating conditions for a given load and engine speed that gives the best thermal efficiency and lowest emissions.

12.6   ENGINE EMISSIONS

Particulates and nitrogen oxides are the most significant emissions from diesel engines. Because of the overall lean mixture, unburned hydrocarbons and CO emissions are generally insignificant except at light loads. At light loads the fuel is less apt to impinge on surfaces, but because of poor fuel distribution, large amounts of excess air, and a low exhaust temperature, lean fuel-air mixture regions may survive to escape into the exhaust. The white smoke that is sometimes observed at low load conditions is really a fuel particulate fog. The more typical black smoke, which is sometimes observed during periods of rapid load increase or for older engines at higher loads, is primarily carbon particles (soot). At higher loads with higher cylinder temperatures, a high percentage of the carbon particles tend to be oxidized.

Nitrogen oxide production in diesel engines arises from the excess oxygen present during lean combustion and the high temperature of the combustion products favoring thermal NO_X production. Two methods of reducing NO_X emissions are effective. The first method is to retard the start of injection (earlier SOI). This causes some reduction in fuel economy but is effective because it reduces the amount of premixed burning. The second method is to recirculate some of the exhaust gas (EGR) back into the inlet air manifold so that the combustion temperature is lowered. When combined, these two approaches can significantly reduce NO_X emissions, but there is a limit to their use because particulate emissions rise and fuel economy decreases when the temperature is reduced. A plot of particulate mass versus NO_X mass for various injection timings (the "particulate-NO_X trade-off curve") is shown in Figure 12.4 for a heavy-duty single cylinder engine running at 1600 rpm, 75% load, an equivalence ratio of 0.55, and intake air pressure of 185 kPa. The first point on the right for each line represents an injection timing (SOI) of 9° before TDC, and the following points are further retarded by 3° increments.

A large portion of the particulates produced in combustion is oxidized within the cylinder. Only 5–10% of the soot formed actually escapes into the exhaust. The

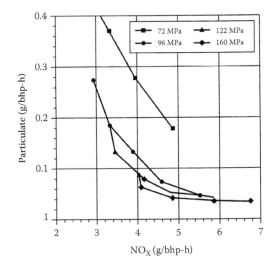

FIGURE 12.4 Effects of fuel injection pressure (72–160 MPa) on the particulate-NO_X trad-eoff. Each curve for a given injection pressure shows data points at 3° intervals of injection retard starting with 9° before TDC on the right side of the curves. Increasing the retard decreases NO_X and increases particulate emissions. Note the units are grams per brake horsepower hour (1 g/bhp · h = 1.36 g/kWh). (From Pierpont, D. A. and Reitz, R. D., "Effects of Injection Pressure and Nozzle Geometry on D.I. Diesel Emissions and Performance," SAE paper 950604, 1995, reprinted with permission from SAE paper 950604. © 1995 SAE International.)

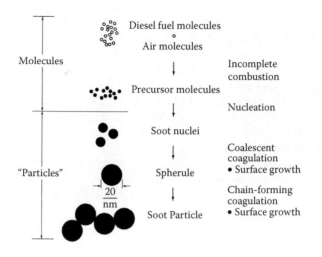

FIGURE 12.5 Soot formation process in a diesel engine starting with gaseous molecules of incomplete combustion that react to produce larger molecules that form soot nuclei which coalesce and grow.

largest peak in soot production is at the start of diffusion burning when the fuel spray is cut off from its air supply and surrounded by very hot products from the premixed burning. Subsequent mixing with air at high temperatures in the cylinder rapidly oxidizes the soot. Soot produced later near the end of burning is less apt to be oxidized because the temperature is lower due to expansion of the products. The tiny soot particles formed initially tend to agglomerate to form larger micron-sized particles before being exhausted to the atmosphere. Figure 12.5 shows the sequence of events in the soot formation process, and the characteristic times for each stage of the process are indicated in Table 12.3.

Formation of soot is further understood by following the spray plume over a period of 10 CA° after the start of injection, as shown in Figure 12.6. First the liquid spray plume begins to form, and then fuel vapor is formed from the smallest droplets.

TABLE 12.3
Time Constants for Various Aspects of Diesel Soot Formation

Process	Approximate duration
Formation of precursors/nucleation	Few μs
Coalescent coagulation	0.5 ms after local nucleation
Spherule identity fixed	After coalescence ceases
Chain-forming coagulation	Few ms after coalescence ceases
Depletion of precursors	0.2 ms after nucleation
Non-sticking collisions	Few ms after nucleation
Oxidation of particles	4 ms
Combustion cycle complete	3–4 ms
Deposition of hydrocarbons on soot	During expansion and exhaust

At 5° after the start of injection, polynuclear aromatic hydrocarbons (PAHs) are formed, but by 6.5° after the start of injection PAHs are not observed. As the diffusion flame grows to the position of 8° after the start of injection, soot concentration is observed. Other laboratory experiments have shown that under certain pressure and temperature conditions, soot is rapidly formed from PAHs. Particulate sampling in engines has shown that by the time the exhaust valve opens, 95% of the soot has been consumed; however, the remaining 5% that is emitted is problematic.

Split injection is a method of reducing both particulate soot and NO_X. In this technique, part of the diesel fuel is injected, there is a short dwell time, and then a second pulse of fuel is injected. Sometimes a small pilot injection is used ahead of the first injection. By optimizing the start of ignition, the separation between the injection pulses, the ratio of the pulses, and the amount of EGR, it is possible to significantly reduce NO_X without increasing soot, but at the expense of reducing thermal efficiency and thus fuel economy. Multiple injection strategies have the thermodynamic disadvantage of injecting and burning the fuel over a longer period.

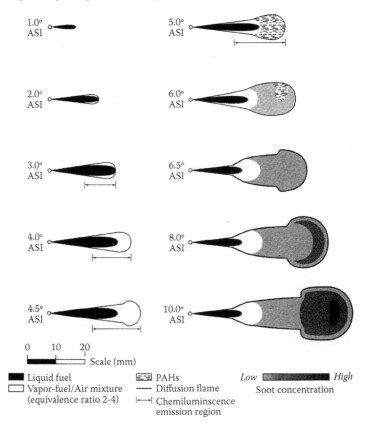

FIGURE 12.6 Soot formation sequence based on diagnostics obtained from an engine with optical access for a period of 10 CA° after the start of injection (ASI). (From Dec, J. E., "A Conceptual Model of DI Diesel Combustion Based on Laser-Sheet Imaging," SAE paper 970873, 1997, reprinted with permission from SAE paper 970873. © 1997, SAE International.)

12.6.1 DIESEL ENGINE EMISSION STANDARDS

All diesel engines must comply with emission standards. Different emission standards apply to light-duty vehicles, heavy-duty trucks, busses, and off-road applications. Different countries have different test procedures and standards. Thus a diesel engine might qualify in one part of the world but not in others. Global standards have been developed for off-road applications, indicating a welcome movement toward global cooperation.

The evolution of United States emission standards from 1973 to the present for heavy-duty trucks is shown in Table 12.4. Regulations have required ever-tightening standards to protect human health and the environment. Particulates are measured in the US EPA procedure by diluting the exhaust and collecting the particulates on a plastic-coated fiberglass filter at a temperature not to exceed 52°C (125°F). The collected mass includes condensed fuel, lubrication oil, and sulfates in addition to carbon particulates. Starting in 1994, diesel engines began to use oxidation catalytic converters to oxidize the soluble organic fraction of the particulate and vapor phase hydrocarbons. In 2004, high-pressure fuel injection and more extensive EGR helped to reduce particulate levels. Very low sulfur diesel fuel (15 ppm by weight, down from 500 ppm) starting in 2007 has facilitated the use of porous ceramic particulate filters. An oxidation catalyst that cuts hydrocarbon emissions precedes the particulate filter. A leading method for NO_x reduction in diesels is increased EGR (up to 40%) coupled with selective catalytic reduction (SCR) with ammonia injection in the exhaust manifold. The ammonia must be carried in a separate tank.

TABLE 12.4
US EPA Heavy-Duty Diesel Truck Emission Standards, g/(kWh)

Year	CO	HC + NO_x		Particulates
1973	–	–		smoke (3)
1974 (1)	53.6	16		(3)
1979 (1)	33.5	10		(3)
		HC	NO_x	
1983 (2)	20.8	1.7	14.3	(3)
1988	20.8	1.7	6.0	0.80
1991	20.8	1.7	6.7	0.33 (4)
1994	20.8	1.7	6.7	0.135
1998	20.8	1.7	5.4	0.135
2010	Std. deleted	0.19	0.27	0.0135 (5)

Notes: (1) hot start, multi-mode emissions testing; (2) cold start, multi-mode emissions testing initiated; (3) particulate standard by opacity rating rather than mass; (4) particulate standard for busses set at 0.10; (5) fuel sulfur reduced to <15 ppm. For the period 2002–2007 the HC and NO_x standard reverted to HC + NO_x = 3.35. HC refers to non-methane hydrocarbons.

12.7   DIESEL ENGINE IMPROVEMENTS

Major energy losses in diesels typically include energy lost during exhaust blowdown (36%), heat transfer to the engine coolant (16%), and pump work and mechanical friction (4%). Net positive work is about 44%. This is clearly better than a conventional SI engine, which is typically about 25%. The main impediment to using diesel engines in automobiles has been meeting the tighter particulate emission standards that apply to automobiles. Increased fuel injection pressure, SOI timing adjustments, split fuel injection, increased EGR, and improved mixing in the cylinder offer incremental steps toward meeting the new emission standards.

Methods to improve net efficiency and reduce emissions are ongoing. The combustion strategy is to avoid locally rich mixtures at high temperatures early in the combustion process (which enhances soot formation) and avoid high temperatures late in the cycle when the combustion is lean (which enhances NO formation) while maintaining good thermal efficiency. To minimize locally rich mixtures, partially premixed combustion is desirable, i.e., some of the fuel should be vaporized and mixed prior to ignition. Retarding the SOI is a way to increase premixing before ignition, but this reduces the IMEP. Another way to achieve partial premixing is to introduce some vaporized fuel with the inlet air. If the fuel was 100% vapor and no spray droplets were used, the rate of pressure rise would be excessive. But if the right mixture of vapor and droplets were selected, the rate and duration of the pressure rise could be optimized. And by using the right amount of EGR, the temperature could be kept low enough to limit NO formation. Since diesel fuel is difficult to vaporize completely outside the cylinder, another method is needed.

TABLE 12.5

Diesel Engine Conditions for a Dual-Fuel, Partially Premixed, Compression Ignition Test

Nominal IMEP (atm)	11
Engine speed (rev/min)	1300
EGR (%)	45
Equivalence ratio	0.77
Intake temperature (°C)	32
Intake pressure (atm)	2.0
Total fuel (mg/cycle)	128
Percent gasoline by mass	78, 82, 85
Diesel injection pressure (atm)	800
Diesel SOI 1st pulse (CA° before DTC)	67
Diesel SOI 2nd pulse (CA° before DTC)	33
Fraction of diesel fuel in 1st pulse	0.65
IVC timing (CA° before DTC)	85

Source: From Hanson, R. M., Kokjohn, S. L., Splitter, D. A., and Reitz, R. D., "An Experimental Investigation of Fuel Reactivity Controlled PCCI Combustion in a Heavy-Duty Engine," SAE paper 2010-01-0864, 2010, reprinted with permission from SAE paper 2010-01-0864. © 2010 SAE International.

Gasoline injection in the inlet air manifold along with direct injection of diesel fuel in the cylinder has been suggested (Hanson and Reitz 2010). This would require separate fuel tanks for the gasoline and diesel fuel, but if successfully implemented the ammonia tank needed for selective catalytic reduction (discussed in Section 12.6.1) could be eliminated. One might question this approach because gasoline mixtures are slow to ignite in compression compared to diesel fuel. The cetane number of diesel fuel is about 50, whereas the cetane number for gasoline is less than 15. However, with dual injection the diesel fuel droplets ignite first and provide many points of ignition for the premixed gasoline. Because of this, combustion in the cylinder is improved significantly.

By running computer simulations and engine tests for a range of gasoline to diesel fuel ratios, diesel fuel injection splits and timings, and amounts of EGR, combustion can be optimized and emissions minimized. For example, tests were run on a single cylinder, four valve Caterpillar engine with a displacement of 2.44 L and an effective

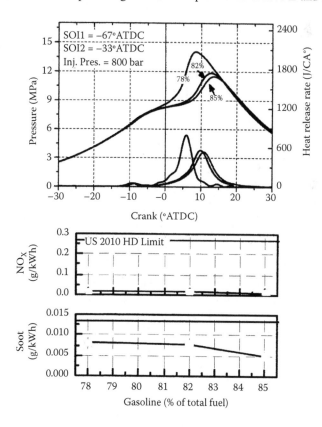

FIGURE 12.7 Dual-fuel, partially premixed compression ignition combustion tests in a Caterpillar single cylinder diesel engine. HD Limit (upper line for both NO_x and Soot plots) is the federal heavy diesel emission limit. See Table 12.5 for engine test conditions. (From Hanson, R. M., Kokjohn, S. L., Splitter, D. A., and Reitz, R. D., "An Experimental Investigation of Fuel Reactivity Controlled PCCI Combustion in a Heavy-Duty Engine," *SAE Int. J. Engines*, 3(1):700–716, 2010. SAE International.)

compression ratio of 10. Table 12.5 shows one set of the test conditions, and Figure 12.7 shows the pressure trace, NO_X, and soot measurements for three gasoline/diesel ratios. The IMEP of 11 atm is about 60% of full load for this engine. The indicated specific fuel consumption was 180 g/kWh, which translates to 53% thermal efficiency, whereas this engine run in the standard diesel mode would have about 45% thermal efficiency. Computations showed that most of the combustion period in the cycle occurred under lean conditions at temperatures below 1900 K. The particulate and NO_X emissions are well below the 2010 U.S. emission standards for heavy duty diesels. For each load condition there is a somewhat different optimum fuel mix. These results are very promising and point the way to significant improvements in diesel engines.

12.8   PROBLEMS

1. Using the data in Table 12.1, calculate the average piston speed for each engine. Plot average piston speed versus displacement.
2. Consider a piston with an on-axis cylindrical bowl with a diameter of 60 mm and a depth of 30 mm. The bore and stroke are each 120 mm. Calculate the volume at bottom dead center (BDC) and the compression ratio. Assume the squish area clearance between piston and head is 1 mm.
3. For the data of Problem 2, the swirl ratio is measured in steady flow bench tests to be 3. (Swirl ratio is the average rotation rate of the gas in rpm divided by the rpm of the engine.) Compute the swirl ratio at TDC by using angular momentum conservation and neglecting friction. Assume solid body rotation of the fluid.
4. Calculate the time averaged squish velocity (the radial inward velocity caused by the flow into the bowl) at the bowl edge and compare this to the average piston velocity and tangential swirl velocity (use the data of Problem 3 for swirl). Assume 2000 rpm.
5. Calculate the discharge velocity and mass flow rate for a 0.2 mm injector hole with a Δp of 55 MPa. Use a flow coefficient of 0.7 and properties of dodecane from Table A.1.
6. A 4-hole injector has a mass flow rate of 25 g/s. Calculate the injection duration (in milliseconds and crank angle degrees) of one cycle for the following parameters: fuel equivalence ratio = 0.8, volumetric efficiency of 95%, and naturally aspirated with an inlet air density of 1.2 kg/m³. The fuel is dodecane and has a stoichiometric fuel-air ratio of 0.067. The cylinder displacement volume is 1500 cm³ and the engine speed is 2000 rpm. To do this first find the trapped mass.
7. Dodecane is sprayed through a 4 hole diesel engine injector with a hole size of 0.20 mm, an injection pressure of 15 MPa, and an injection time of 3.4 ms. Using Equation 9.25b for penetration and the geometry of Problem 2, (a) calculate the distance of spray penetration at the end of injection without regard to the walls of the cylinder. (b) Estimate the time and crank angle when the spray strikes the side of the bowl. Assume the spray starts at 20° before TDC. Assume the time for the linear portion of the penetration is very short and apply Equation 9.25b over the entire time. The cylinder

displacement volume is 1500 cm³, compression ratio = 15, trapped mass of air in the cylinder is 1.7 g.

8. A diesel engine operating at 2000 rpm has a swirl ratio of 12 in the bowl. If the spray plume penetrates 100 mm with no swirl, use Equation 9.26 to calculate the penetration with swirl. The initial velocity of the fuel jet is 270 m/s. Does the spray hit the bowl with the addition of swirl for the geometry of Problem 2 (42 mm from the injector point to the inner corner of the bowl)? (Swirl ratio is the ratio of angular rotation of the gas to rpm of the piston).

9. Using Equation 12.1 with $C = 0.0197$ ms, $a = -1$, $b = -1.75$ and $E = 4500$ K, compare the ignition delay for the following cases. In each case the engine compression ratio is 16:1 and injection starts at 20 CA° before TDC. Assume the volumetric efficiency is 100% for each case and the fuel mass added is the same for each case. The equivalence ratio is $F = 0.8$. Calculate the ignition delay in crank angles for 2500 rpm. Neglect residual fraction effects.

   a. Naturally aspirated: inlet air at 1 atm, 300 K
   b. Turbocharged: inlet air at 2 atm, 380 K
   c. Turbocharged with intercooler: inlet air at 2 atm, 312 K

10. Diesel engines overcome many of the limitations that reduce the engine efficiency of spark ignition engines. Explain qualitatively how each of the following items limit SI engine efficiency, and explain why diesel engines overcome each of these limitations.

   a. Fuel octane number
   b. Flame propagation in lean mixtures
   c. Cycle-to-cycle variations
   d. Fuel trapped in crevice volumes

REFERENCES

Abraham, J. and Magi, V., "Modeling Radiant Heat Loss Characteristics in a Diesel Engine," SAE paper 970888, 1997.

Akihama, K., Takatori, Y., Inagaki, K., Sasaki, S., and Dean, A. M., "Mechanism of the Smokeless Rich Diesel Combustion by Reducing Temperature," SAE paper 2001-01-0655, 2001.

Amann, C. A., Stivender, D. L., Plee, S. L., and MacDonald, J. S., "Some Rudiments of Diesel Particulate Emissions," SAE paper 800251, 1980.

Arcoumanis, C. and Gavaises, M., "Effect of Fuel Injection Processes on the Structure of Diesel Sprays," SAE paper 970799, 1997.

Baumgard, K. J. and Johnson, J. H., "The Effect of Fuel and Engine Design on Diesel Exhaust Particle Size Distributions," SAE paper 960131, 1996.

Benson, R. S. and Whitehouse, N. D., *Internal Combustion Engines: A Detailed Introduction to the Thermodynamics of Spark and Compression Ignition Engines, Their Design and Development*, Pergamon Press, Oxford, UK, 1979.

Bessonette, P. W., Schleyer, C. H., Duffy, K. P., Hardy, W. L., and Liechty, M. P., "Effects of Fuel Property Changes on Heavy-Duty HCCI Combustion," SAE paper 2007-01-0191, 2007.

Borman, G. L. and Brown, W. L., "Pathways to Emissions Reduction in Diesel Engines," Second International Engine Combustion Workshop, C.N.R., Capri, Italy, 1992.

Bosch, R., *Diesel Fuel Injection*, Society of Automotive Engineers, Warrendale, PA, 1974.

Cartellieri, W. P. and Herzog, P. L., "Swirl Supported or Quiescent Combustion for 1990s Heavy-Duty DI Diesel Engines—An Analysis," SAE paper 880342, 1988.

Dec, J. E. and Espey, C., "Ignition and Early Soot Formation in a DI Diesel Engine Using Multiple 2-D Imaging Diagnostics," SAE paper 950456, 1995.

Dec, J. E., "A Conceptual Model of DI Diesel Combustion Based on Laser-Sheet Imaging," SAE paper 970873, 1997.

Ferguson, C. R. and Kirkpatrick, A. T., *Internal Combustion Engines: Applied Thermosciences*, 2nd ed., Wiley, New York, 2001.

Han, Z., Uludogan, A., Hampson, G. J., and Reitz, R. D., "Mechanism of Soot and NO_x Emission Reduction Using Multiple-Injection in a Diesel Engine," SAE paper 960633, 1996.

Hanson, R., Reitz, R. D., Splitter, D., and Kokjohn, S., "An Experimental Investigation of Fuel Reactivity Controlled PCCI Combustion in a Heavy-Duty Engine," SAE paper 2010-01-0864, SP-2279, *SAE Int. J. Engines* 3(1):700–716, 2010.

Heywood, J. B., "Combustion in Compression Ignition Engines," Chap. 10 in *Internal Combustion Engine Fundamentals*, McGraw-Hill, New York, 1988.

Hiroyasu, H., "Diesel Combustion and Its Modeling," COMODIA paper C85 P053, JSME, 1985.

Kalghatgi, G. T., Risberg, P., and Angstrom, H.-E., "Partially Pre-Mixed Auto-Ignition of Gasoline to Attain Low Smoke and Low NO_x at High Load in a Compression Ignition Engine and Comparison with a Diesel Fuel," SAE paper 2007-01-0006, 2007.

Kamimoto, T. and Kobayashi, H., "Combustion Processes in Diesel Engines," *Prog. Energy Combust. Sci.* 17(2):163–189, 1991.

Kong, S.-C. and Reitz, R. D., "Multidimensional Modeling of Diesel Ignition and Combustion Using a Multistep Kinetics Model," *J. Eng. Gas Turbines Power* 115(4):781–789, 1993.

Kong, S.-C. and Reitz, R. D., "Spray Combustion Processes in Internal Combustion Engines" in *Recent Advances in Spray Combustion: Spray Combustion Measurements and Model Simulation*, vol. 2, ed. K. Kuo, 395–424, AIAA, 1996.

McDonald, J., "Construction and Testing of a Facility for Diesel Engines, Heat Transfer and Particulate Research," Master's Thesis, University of Wisconsin–Madison, 1984.

Naber, J. D. and Siebers, D. L., "Effects of Gas Density and Vaporization on Penetration and Dispersion of Diesel Sprays," SAE paper 960034, 1996.

Pierpont, D. A. and Reitz, R. D., "Effects of Injection Pressure and Nozzle Geometry on D.I. Diesel Emissions and Performance," SAE paper 950604, 1995.

Sakaguchi, D., Le Amida, O., Ueki, H., and Ishida, M., "Measurement of Droplet Size Distribution in Core Region of High-Speed Spray by Micro-Probe L2F," *J. Therm. Sci.* 17(1):90–96, 2008.

Shimoda, M., Shigemori, M., and Tsuruoka, S., "Effect of Combustion Chamber Configuration on In-Cylinder Air Motion and Combustion Characteristics of D.I. Diesel Engine," SAE paper 850070, 1985.

Shayler, P. J., Brooks, T. D., Pugh, G. J., and Gambrill, R., "The Influence of Pilot and Split-Main Injection Parameters on Diesel Emissions and Fuel Consumption," SAE paper 2005-01-0375, 2005.

Sihling, K. and Woschni, G., "Experimental Investigation of the Instantaneous Heat Transfer in the Cylinder of a High Speed Diesel Engine," SAE paper 790833, 1979.

Szybist, J. P., Song, J., Alam, M., and Boehman, A. L., "Biodiesel Combustion, Emissions and Emission Control," *Fuel Process. Technol.* 88(7):679–691, 2007.

Van Gerpen, J. H., Huang, C. W., and Borman, G. L., "The Effects of Swirl and Injection Parameters on Diesel Combustion and Heat Transfer," SAE paper 850265, 1985.

Woschni, G. and Anisits, F., "Experimental Investigation and Mathematical Presentation of Rate of Heat Release in Diesel Engines Dependent Upon Engine Operating Conditions," SAE paper 740086, 1974.

Yan, J. and Borman, G. L., "A New Instrument for Radiation Flux Measurement in Diesel Engines," SAE paper 891901, 1989.

Yoshikawa, S., Furusawa, R., Arai, M., and Hiroyasu, H., "Optimizing Spray Behavior to Improve Engine Performance and to Reduce Exhaust Emissions in a Small DI Diesel Engine," SAE paper 890463, 1989.

13 Detonation of Liquid and Gaseous Mixtures

In contrast to gaseous detonation waves, which were discovered over 100 years ago, it was not recognized until the 1960s that a detonation wave can propagate through a two-phase mixture of liquid fuel and gaseous oxidizer. The fuel can be in the form of liquid droplets or a liquid film on the walls of a tube. This chapter will first consider spray detonations and then will discuss liquid layer detonations. Two-phase detonations can occur with nonvolatile fuels as well as with more volatile fuels where the detonation velocity is higher. While these types of detonations are generally to be avoided because of the high transient pressures involved, which can be very destructive, there may yet be some practical applications of two-phase detonations developed.

Phenomena related to a spray detonation may possibly occur in a diesel engine. As the fuel is injected into the cylinder at a very high pressure, the leading droplets of the spray may possibly undergo stripping to form a microspray. The high pressure and temperature and rapid mixing may possibly be conducive to initiation of local pressure waves in the cylinder. Because the distances and times involved are short, this should not be viewed with alarm but rather as another (perhaps speculative) view of what goes on in the cylinder during combustion.

Large liquid fueled rocket motors have occasionally experienced wave-like pressure excursions that have been well documented. This type of combustion instability can be very destructive and can rupture the thin-walled rocket motor. While some combustion instability is acoustic in nature and leads to failure due to high heat transfer, there is another failure mode that is detonation-like with accompanying pressure waves. The cure for acoustic instabilities is appropriately designed liners, while the detonation-like instabilities require suitably placed baffles.

Liquid fuel film detonations have been known to occur in long, partially filled pipelines. In one case in Texas where some cleaning and welding was being done on a pipeline that was partially filled with oil, combustion started at the open end, developed into a detonation, and ruptured the tube in one section, which relieved the pressure and slowed the detonation. Then detonation was reestablished further down the pipe. The pipe was ruptured in various locations for a distance of several miles before the detonation wave hit a valve. Another potential problem is in catapults with oil seals where high pressure air in contact with a layer of oil on the cylinder walls could produce a flame started by frictional heating, and the flame could develop into a detonation.

The rotating detonation combustor, which was mentioned in Chapter 8 as a possible way to utilize the extraordinary combustion intensity of a detonation wave, could perhaps be more easily implemented using liquid fuel rather than gaseous fuel. Because the density of liquid fuels is typically 500 times greater than gaseous fuel, it is easier to feed fresh fuel ahead of the detonation wave using liquid fuel. Additionally, because heterogeneous detonations have an extended reaction zone compared to gaseous detonations, there would be a lesser tendency to form multiple waves in the annulus, which was a factor that plagued the gaseous rotating detonation combustor.

13.1   DETONATION OF LIQUID FUEL SPRAYS

For near stoichiometric mixtures of interest the volume of the liquid droplets is 1/1000th of the volume of the gas. In these dilute fuel droplet sprays shock waves can easily propagate through the mixture. However, because the reaction zone of a gaseous detonation is completed in a few microseconds, one is tempted to assume that fuel from droplets cannot enter into an exothermic chemical reaction rapidly enough to allow a self-sustaining detonation. At least one might think that there is a certain droplet size above which a detonation wave would not be sustained. However, a careful look at the behavior of the fuel droplets behind the leading shock front is required to understand the nature of this type of combustion.

In a manner analogous to a gaseous detonation, a spray detonation may be viewed as a normal shock wave, followed by a dynamic breakup of the droplets and then rapid combustion of the vaporizing microspray, which due to the rapid expansion of product gases, drives the leading shock front. The existence of secondary, transverse shock waves, as in gaseous detonations, should not be ruled out.

13.1.1   DROPLET BREAKUP

Before discussing the nature of spray detonations further, let us consider the breakup of droplets in the convective flow behind a normal shock wave. Imagine a dilute spray of uniform size droplets in a quiescent gas at 1 atm pressure and room temperature. A shock wave passes over the spray subjecting the droplets to a high pressure, high temperature, and high velocity flow. The dynamic pressure ($\rho \underline{V}^2/2$) of the flow tends to distort the droplet by stretching it transverse to the flow. This distortion occurs because the pressure is high on the front of the droplet and lower at 90° from the stagnation point. This flattening effect is opposed by inertia and viscous stresses within the drop. For high speed flow behind a shock wave, surface tension forces are relatively insignificant. The droplet flattens and an internal circulation develops within the drop. The liquid in the droplet flows to the edge of the flattened droplet and is stripped from the surface, thus forming a microspray in the wake of the parent drop, as shown in Figure 13.1. The microspray droplets are of very small size, probably less than 1% of the diameter of the parent drop.

In the case of a gaseous detonation there is little separation between the shock and the combustion, but in a heterogeneous detonation there is a significant time and distance delay before combustion adds heat to the flow. Hence we may treat this flow as a shock wave followed by combustion, but close enough that the combustion drives the shock wave. To say this in another way, a gaseous detonation (Chapter 8) has a three-dimensional structure of shocks with coupled combustion, but the two-phase

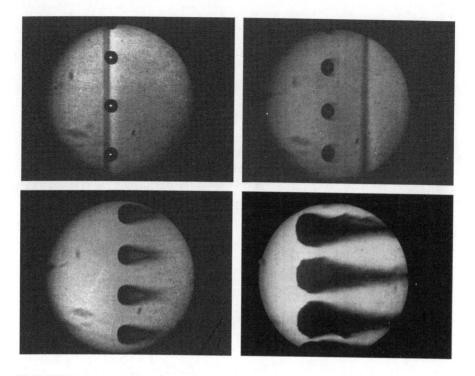

FIGURE 13.1 Shadow photographs of 750 μm water droplets breaking up due to a Ma = 2.7 shock wave in atmospheric air (top left) undisturbed droplets just before shock passes, (top right) 2.6 μs after passage of shock wave, (bottom left) 4.4 μs after shock, and (bottom right) 14.4 μs after shock. (From Dabora, E .K., Ragland, K. W. and Nicholls, J. A., "Drop Size Effects in Spray Detonations," *Symp. (Int.) Combust.* 12(1):19–26, The Combustion Institute, Pittsburgh, PA, 1969, by permission.)

detonation discussed in this chapter has a leading planar shock front with separation between the shock front and the combustion. Because of this when we analyze a two-phase detonation it is helpful to talk about the shock and the combustion as being separate but coupled. However, we cannot make this separation in the case of the gas detonations we discussed in Chapter 8. This carries over to the notation that we will use in the chapter. In Chapter 8 we discussed reactants and products (e.g., $\underline{V}_{react}$ and $\underline{V}_{prod}$) when referring to before and after the shock. In two-phase detonations we have 3 states—before the shock, after the shock and before the combustion, and after the combustion. In the notation of this chapter these states are referred to as State 1, State 2, and State 3; respectively or simply subscript 1, 2, and 3.

The derivation here assumes an initially quiescent air droplet mixture ($\underline{V}_1 = 0$). The droplet breakup time, t_b, for non-burning droplets due to convective flow behind a shock wave is given by

$$t_b = \frac{K_1 d \sqrt{\rho_\ell}}{\sqrt{\rho_2 \underline{V}_2^2 / 2}}$$

(13.1)

where

  d is the initial droplet diameter,
  V_2 is the convective velocity behind the shock front with respect to laboratory coordinates,
  K_1 is a constant that depends weakly on dynamic pressure and may be taken as 7,
  ρ_2 is the gas (air) density behind the shock front, and
  ρ_ℓ is the density of the liquid droplet.

The conditions immediately behind the shock wave front (State 2) are obtained from conservation of mass, momentum, and energy in wave fixed coordinates. It is convenient to formulate these equations in terms of Mach number rather than velocity. Following a similar development as in Chapter 8 (Equations 8.11–8.14) but with $q = 0$ yields

$$\breve{\mathrm{Ma}}_2^2 = \frac{(\gamma-1)\breve{\mathrm{Ma}}_{wave}^2 + 2}{2\gamma\breve{\mathrm{Ma}}_{wave}^2 - (\gamma-1)} \tag{13.2}$$

where $\breve{\mathrm{Ma}}_{wave}$ refers to the Mach number of the propagating wave with respect to State 1.

Equation 13.2 shows that for $\breve{\mathrm{Ma}}_{wave}$ greater than 1, $\breve{\mathrm{Ma}}_2$ is always less than 1. In the same way p_2, T_2, and ρ_2 may be determined,

$$\frac{p_2}{p_1} = \frac{2\gamma\breve{\mathrm{Ma}}_{wave}^2 - (\gamma-1)}{\gamma+1} \tag{13.3}$$

$$\frac{T_2}{T_1} = \frac{\left[2\gamma\breve{\mathrm{Ma}}_{wave}^2 - (\gamma-1)\right]\left[(\gamma-1)\breve{\mathrm{Ma}}_{wave}^2 + 2\right]}{(\gamma+1)^2 \, \breve{\mathrm{Ma}}_{wave}^2} \tag{13.4}$$

$$\frac{\rho_2}{\rho_1} = \frac{(\gamma+1)\breve{\mathrm{Ma}}_{wave}^2}{(\gamma-1)\breve{\mathrm{Ma}}_{wave}^2 + 2} \tag{13.5}$$

From this the droplet time can be evaluated using State 2 conditions and selecting K_1 from experiments to adjust for partial heat release as the droplet microspray is formed and the droplet breaks up. First we need to convert V_2 in Equation 13.1 to wave fixed coordinates.

$$\breve{V}_2 = \breve{\mathrm{Ma}}_2(a_2) = \breve{\mathrm{Ma}}_2\sqrt{\gamma R T_2} \tag{13.6}$$

and converting from wave fixed coordinates to laboratory coordinates

$$V_2 = V_{wave} - \breve{V}_2 \tag{8.5}$$

Also, note that in the lab coordinates, Ma_2 can be either subsonic or supersonic depending on the propagation velocity V_{wave}.

Experiments over a range of drop sizes and shock front speeds have shown that the dynamic pressure during breakup and combustion of the microspray is about one-half that of a normal shock alone. From this and taking K_1 to be 7, Equation 13.1 becomes

$$t_b = \frac{10d}{V_2} \sqrt{\frac{\rho_\ell}{\rho_1}} \sqrt{\frac{\rho_1}{\rho_2}} \tag{13.7}$$

A non-dimensional breakup time is given by

$$\tau_b \equiv t_b \frac{V_{wave}}{d} = 10 \left(\frac{\rho_\ell}{\rho_1} \right)^{1/2} \left(\frac{(\rho_1/\rho_2)^{1/2}}{1-\rho_1/\rho_2} \right) \tag{13.8}$$

Equation 13.8 gives $\tau_b = 162$ at Ma $= 3$, and decreases to $\tau_b = 120$ as Ma $\to \infty$ for decane droplets initially in oxygen at standard conditions. For example, a 3 mm droplet in a Ma $= 3$ detonation in standard oxygen will take 0.5 ms to break up. The speed of sound in standard oxygen is 330 m/s and the breakup distance is

$$3\left(\frac{330 \text{ m}}{\text{s}} \right)(0.5 \text{ ms}) = 0.5 \text{ m}$$

It is possible to have detonations in a fuel spray because of the microspray of fine droplets that are stripped rapidly from the parent drops behind the leading shock front. This occurs even for relatively large drops.

13.1.2 SPRAY DETONATIONS

Experiments have been conducted with polydisperse sprays and sprays with precisely controlled droplet size in which the size was varied from 2 μm–2600 μm diameter. In each case a self-sustaining detonation was achieved, and the propagation velocity was 2%–35% below the detonation velocity of an equivalent gaseous case. These experiments were done with diethylcyclohexane droplets dispersed in standard oxygen. This fuel was chosen because it has an even lower vapor pressure than decane at room temperature and hence precludes initial fuel vaporization. The experiments were conducted in long tubes, and ignition was achieved by a pulse from a small shock tube located at the top of a larger tube. Figure 13.2 shows experimental detonation velocities for three different size droplets and several equivalence ratios compared with the computed ideal gaseous detonation using the Chapman-Jouguet condition. Detonation of sprays of large diameter drops in air requires a more volatile fuel.

Modeling of spray detonations indicates that for non-volatile droplets less than 10 μm in diameter, vaporization alone is sufficiently rapid to allow detonation in oxygen. For droplets between 10 μm and 1000 μm, droplet stripping to form a microspray is sufficiently rapid to support detonation. For droplets greater than 1000 μm, an additional mechanism is needed to create a detonation. Experiments have shown that this additional mechanism is local explosions about the parent drops.

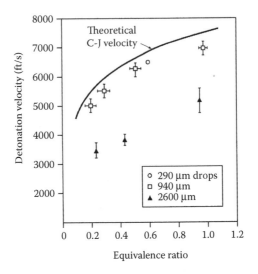

FIGURE 13.2 Comparison of experimental detonation velocity of decane in standard oxygen compared with the computed ideal gaseous detonation velocity for decane using the Chapman-Jouguet condition. (From Ragland, K. W., Dabora, E. K. and Nicholls, J. A., "Observed Structure of Spray Detonations," *Phys. Fluids* 11(11):2377–2388, 1968. © 1968, by permission of American Institute of Physics.)

A schlieren photograph of the shock front and reaction zone of a single stream of 2600 μm diethylcyclohexane droplets in a vertical, 5 cm by 5 cm square tube filled with standard oxygen is shown in Figure 13.3. The schlieren photograph shows density gradients but is not sensitive to the light of combustion. Three droplets are in the field of view. The leading shock front, which is quite planar, is traveling at 1020 m/s or Ma = 3.0 to the left. The first droplet is immediately in front of the shock. Behind

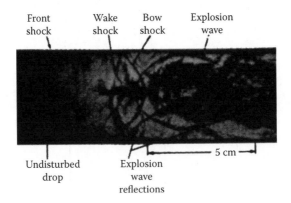

FIGURE 13.3 Spark-source (0.2 μs exposure) photograph of the detonation of a single stream of 2.6 mm fuel droplets showing density gradients in the reaction zone. (From Ragland, K. W., Dabora, E. K. and Nicholls, J. A., "Observed Structure of Spray Detonations," *Phys. Fluids* 11(11):2377–2388. © 1968, by permission of the American Institute of Physics.)

the leading front shock, the second droplet is seen to be forming a microspray in the wake of the drop. Because the convective flow is supersonic with respect to the drop, a bow shock and wake shocks are visible. Self-luminous photographs show that this wake is not yet burning although there is combustion near the stagnation point of the drop. Disintegration of the third droplet has proceeded further and has, in fact, generated a local explosion around the droplet due to the buildup of the vaporized microspray. This local explosion of the sheared droplet propagates outward and further stimulates combustion of other droplets. Details of this process are shown in Figure 13.4. The time for complete disintegration of the 2600 μm diameter droplet at these conditions is about 0.5 ms. Even though the reaction zone extends more than 30 cm behind the leading shock wave front, the combustion is coupled to the shock wave front and drives the entire process at a steady-state velocity. A similar phenomenon occurs with smaller droplets although it is not known if the local explosions occur.

One-dimensional equations of conservation of mass, momentum and energy may be applied to spray detonations as was done for gaseous detonations in Chapter 8, and we will now switch the notation from $\underline{V}_{wave}$ to $\underline{V}_D$ in recognition that the process is a detonation. In addition, because of the extended reaction zone, there is heat

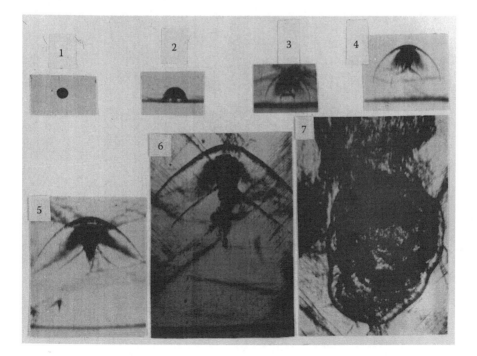

FIGURE 13.4 Schlieren photographs of 2.6 mm fuel droplets within the detonation reaction zone: (1) undistorted, (2) 3 μs after contact with the initial shock front, (3) 6 μs, (4) 11 μs, (5) 23 μs, (6) 40 μs, and (7) 80 μs. (From Ragland, K. W., Dabora, E. K. and Nicholls, J. A., "Observed Structure of Spray Detonations," *Phys. Fluids* 11(11):2377–2388. © 1968, by permission of the American Institute of Physics.)

transfer to the walls and momentum loss due to accelerating the droplets and drag on the walls within the extended reaction zone. Again, the Chapman-Jouguet condition that the velocity at the end of the reaction zone, $\underline{V}_3$, is sonic with respect to the shock front is assumed. Based on the earlier discussion the notation is the following: 1 refers to upstream air, 2 denotes air just after the shock, and 3 refers to combustion products at the end of the reaction zone. The liquid is outside the control volume. The following results have been obtained for the pressure and temperature changes across the detonation wave:

$$\frac{p_3}{p_1} = \frac{\gamma\left(V_D^2/a_1^2\right)(1+f)Z}{1+\gamma} \tag{13.9}$$

$$\frac{T_3}{T_1} = \frac{\gamma^2 M_3 \left(V_D^2/a_1^2\right)Z^2}{\left(1+\gamma\right)^2 M_1} \tag{13.10}$$

where M_3 and M_1 are the molecular weight of the products and the molecular weight of the gas phase in front of the shock wave, respectively. Z is a correction factor that accounts for drag and heat loss to the walls.

$$Z = 1 + \frac{1}{\gamma\left(V_D^2/a_1^2\right)(1+f)} + \frac{C_H A_R V_2^2}{2(1+f)A_s V_D \breve{V}_2}$$

The subscripts are as indicated in Figure 13.5; A_R is the surface area of the reaction zone in contact with the tube walls, and A_s is the frontal area of the leading shock wave. The term $\breve{V}_2$ is the velocity just behind the shock front in a reference system with respect to the shock, which may be calculated or obtained from shock tables

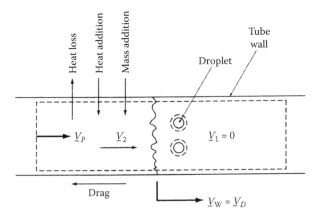

FIGURE 13.5 Control volume for analysis of heterogeneous detonation. The velocities indicated are with respect to stationary laboratory coordinates.

as a function of the Mach number. C_H is the heat transfer coefficient, which has been measured to be about 2.5×10^{-3} for spray detonations, where C_H is dimensionless and is given by

$$C_H = \frac{\int_2^3 q_{wall}\, dx}{\rho_2 \underline{V}_2 \left(h_2 + \underline{V}_2^2/2 - h_{wall} \right)} \tag{13.11}$$

where h_{wall} is the enthalpy of the gas at the temperature of the wall.

The propagation velocity of a spray detonation is less than the equivalent gaseous detonation velocity, $\underline{V}_{D0}$ and is given by

$$\underline{V}_D = \frac{\underline{V}_{D0}}{\left[1 + \dfrac{2\gamma^2 C_H A_R \underline{V}_2^2}{\underline{V}_D \underline{V}_2 (1+f) A_s} \right]^{1/2}} \tag{13.12}$$

The drag coefficient does not appear in Equation 13.12 because the Reynolds analogy that $C_D = 2C_H$ has been used. Since Equation 13.12 is an implicit equation, for predictive purposes start by calculating the equivalent gas phase detonation parameters and then iterating a few times. For droplets less than 1000 μm in size, A_R may be estimated from Equation 13.7 because droplet breakup is the main rate-limiting step. Although the structure of spray detonations is complicated, one-dimensional theory with frictional and heat transfer losses in a reaction zone controlled by the dynamic breakup of the spray can predict the thermodynamic state and propagation velocity with reasonable accuracy.

13.2 DETONATION OF LIQUID FUEL LAYERS

Consider a long tube which contains air or oxygen gas and in which the walls of the tube are coated with a non-volatile fuel. A strong ignition source, such as a pulse from a small shock tube, will cause an accelerating shock followed by combustion of the fuel on the wall. After a transition distance, a steady-state propagation velocity will be reached. The process is similar to that of spray detonations, but there is no spray, only non-volatile fuel on the walls. Behind the shock front the fuel layer on the wall vaporizes and perhaps is stripped off the wall.

Experiments of film detonations show that the fuel burns in a thin region along the walls behind the shock front. For example, in a 5 cm by 5 cm square tube in which two walls were coated with diethylcyclohexane, the propagation velocity reached a steady value of 1370 m/s after 2 m of travel. Two self-luminous photographs of film detonation taken through quartz windows at slightly different times as the detonation moves downward are shown in Figure 13.6. The combustion starts at the wall and spreads inward. Other photographs show that radiating combustion products do not fill the center of the test section until about 0.5 ms or 0.6 m behind the point of ignition.

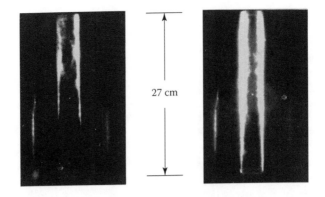

27 cm

FIGURE 13.6 Two self-luminous photographs of film detonation with non-volatile fuel on two walls taken through quartz windows at slightly different times as the detonation moves downward. (From Ragland, K. W. and Nicholls, J. A., "Two-Phase Detonation of a Liquid Layer," *AIAA J.* 7(5):859–863, 1969, by permission of AIAA.)

Spark schlieren photographs in Figure 13.7, taken in the same test setup as Figure 13.6, show the shock structure and the highly turbulent nature of the reaction zone. The detonation is traveling downward. The Schlieren technique uses a collimated light source and is sensitive to density gradients, but not self-light. In this case fuel is on one wall only and the propagation velocity is 1065 m/s. Even though the heat is released mainly near the wall, the leading shock front is surprisingly planar. However, it is evident that the leading shock front is being perturbed by various

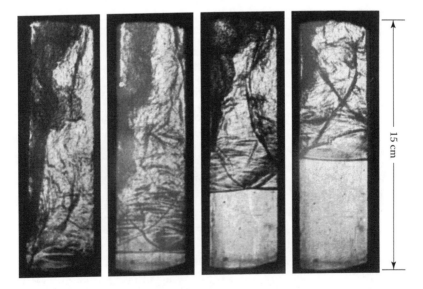

15 cm

FIGURE 13.7 Spark schlieren photographs of film detonation with non-volatile fuel on one wall taken in four consecutive time intervals. The time slices proceed from right to left. The detonation is traveling downward. (From Ragland, K. W. and Nicholls, J. A., "Two-Phase Detonation of a Liquid Layer," *AIAA J.* 7(5):859–863, 1969, by permission of AIAA.)

pressure disturbances as it propagates. The dark zone next to the wall is the burning layer of fuel vapor and associated turbulence. Evidence from pressure transducers and streak schlieren photographs indicates that local explosions occur within the boundary layer behind the shock in a somewhat analogous fashion to large droplet explosions. This is suggested by the extraneous waves in Figure 13.7.

The detonation velocity of decane liquid layers as a function of equivalence ratio in various size circular tubes filled with oxygen is shown in Figure 13.8. Also, a solid layer of cetane was frozen to the walls by chilling the tube walls. Cetane is

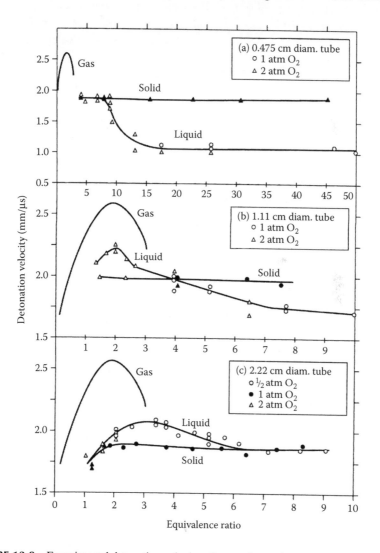

FIGURE 13.8 Experimental detonation velocity of cetane layers in oxygen and comparison with the ideal Chapman-Jouguet detonation velocity for a gas. (From Ragland, K. W. and Garcia, C. F., "Ignition Delay Measurements in Two-Phase Detonations," *Combust. Flame* 18(1):53–58, 1972. © 1972, with permission of Elsevier.)

similar to decane but solidifies at a slightly higher temperature. The gas curve was obtained from computer calculations. The equivalence ratio was changed by varying the thickness of the fuel layer. A length equivalent to an L/d ratio of at least 120 was required to reach a steady-state detonation velocity, after which the detonation velocity was constant to within ±2%. The minimum equivalence ratios examined were not the lean limit but rather the smallest amount of liquid that could be spread uniformly on the tube in the experiment. There is no rich limit for liquid or solid layers. For liquid layers the detonation velocity decreases to a plateau as the equivalence ratio increases. As shown, for solid layers the detonation velocity is independent of the equivalence ratio. It is interesting to note that for rich mixtures in small tubes, the solid velocity curve is higher than the liquid curve. These curves show the different rates at which the fuel enters the reaction for liquid layers of various thicknesses versus various thicknesses of solid layers.

The two-phase detonation velocities shown in Figure 13.8 are all lower than the corresponding gas phase detonation. The equations developed for spray detonations, Equations 13.8–13.10 apply also to the film detonations. Of course, the heat transfer and drag coefficients will be different. In fact, the self-sustaining nature of the film detonation is facilitated because the shear stress at the liquid layer is greatly reduced as the layer is vaporized and burns. Similarly heat transfer to the walls is reduced by the presence of the liquid layer. Turbulent boundary layer heat transfer analysis shows that the vaporization rate is rapid enough to support the observed propagation velocities and reaction zone thickness. In addition the secondary shock waves catch up to the leading shock front and reinforce it.

13.3 PROBLEMS

1. What is the breakup time and breakup distance for a 0.1 mm water droplet subjected to a Ma = 2.5 shock wave propagating through quiescent air-droplet mixture initially at 1 atm and 293 K? The specific heat ratio for air is 1.40.

2. What is the breakup time for a single 50 μm cetane (hexadecane) droplet injected into high pressure air at 50 m/s. The air is at 35 atm and 800 K? These conditions are representative of those in a diesel engine.

3. Estimate the pressure and temperature at the end of the reaction zone for a stoichiometric mixture of cetane droplets in standard oxygen due to a fully developed detonation that is propagating at 2100 m/s. Neglect the effect of heat loss and friction. The specific heat ratio for oxygen is 1.40.

4. For the conditions of Problem 13.3 with 0.5 mm droplets in a 10 cm diameter tube, find the ratio of the spray detonation propagation velocity to the gaseous detonation velocity. Assume that the length of the reaction zone is 33 mm.

5. What is the thickness of a liquid layer of decane that will yield a stoichiometric mixture with (a) standard oxygen and (b) standard air in a 1 cm diameter tube?

REFERENCES

Borisov, A. A., Kogarko, S. M., and Lyubimov, A. V., "Ignition of Fuel Films behind Shock Waves in Air and Oxygen," *Combust. Flame* 12(5):465–468, 1968.

Bowen, J. R., Ragland, K. W., Steffes, F. J., and Loflin, T. G., "Heterogeneous Detonation Supported by Fuel Fogs or Films," *Symp. (Int.) Combust.* 13(1):1131–1139, The Combustion Institute, Pittsburgh, PA, 1971.

Cramer, F. B., "The Onset of Detonation in a Droplet Combustion Field," *Symp. (Int.) Combust.* 9(1):482–487, The Combustion Institute, Pittsburgh, PA, 1963.

Dabora, E .K., Ragland, K. W., and Nicholls, J. A., "Drop Size Effects in Spray Detonations," *Symp. (Int.) Combust.* 12(1):19–26, The Combustion Institute, Pittsburgh, PA, 1969.

Gordeev, V. E., Komov, V. F., and Troshin, Ya. K, "Detonation Burning in Heterogeneous Systems," *Dokl. Akad. Nauk SSSR* 160(4):853–856, 1965. (in Russian)

Komov, V. F. and Troshin, Ya. K., "Characteristics of Detonation in Some Heterogeneous Systems," *Dokl. Akad. Nauk SSSR* 175(1):109–112, 1967. (in Russian)

Lin, Z. C., Nicholls, J. A., Tang, M. J., Kaufmann, C. W., and Sichel, M., "Vapor Pressure and Sensitization Effects in Detonation of a Decane Spray," *Symp. (Int.) Combust.* 20(1):1709–1716, The Combustion Institute, Pittsburgh, PA, 1984.

Pinaev, A. V. and Sobbotin, V. A., "Reaction Zone Structure in Detonation of Gas-Film Type Systems," *Combust. Explos. Shock Waves* 18(5):585–591 (translated from *Fiz. Goreniya Vzryva* 18(5):103–111, 1982), 1982.

Ragland, K. W., Dabora, E. K., and Nicholls, J. A., "Observed Structure of Spray Detonations," *Phys. Fluids* 11(11):2377–2388, 1968.

Ragland, K. W. and Garcia, C. F., "Ignition Delay Measurements in Two-Phase Detonations," *Combust. Flame* 18(1):53–58, 1972.

Ragland, K. W. and Nicholls, J. A., "Two-Phase Detonation of a Liquid Layer," *AIAA J.* 7(5):859–863, 1969.

Webber, W. T., "Spray Combustion in the Presence of a Traveling Wave," *Symp. (Int.) Combust.* 8(1):1129–1140, The Combustion Institute, Pittsburgh, PA, 1961.

Part IV

Combustion of Solid Fuels

The three main types of combustion for firing solid fuels are fixed bed combustion, suspension firing, and fluidized bed combustion. Before considering solid fuel combustion systems, the combustion of a single solid fuel particle in a hot flowing gas stream is discussed. Fixed bed combustion systems range from the three-stone fire to the large spreader stoker boiler for electric power generation and district heating. Solid fuel suspension firing systems are used primarily for large central electric power generating stations. Fluidized bed combustion and gasification systems are also used for heating and electric power generation. The reader may wish to review the material on solid fuels in Chapter 2 before proceeding with Part IV.

14 Solid Fuel Combustion Mechanisms

This chapter examines the combustion of individual solid fuel particles of coal and biomass. The fuel size ranges from finely ground pulverized fuels the size of corn starch; to sawdust; to crushed, chipped, or shredded fuels; to sticks and logs. Investigation of the behavior of individual solid fuel particles provides insight into the design and performance of stoves, furnaces, and boilers, which are discussed in Chapters 15, 16, and 17.

When a solid fuel particle is exposed to a hot flowing gas stream, it undergoes three stages of mass loss—drying, devolatilization (pyrolysis), and char combustion. The relative significance of each of these three processes is indicated by the proximate analysis of the fuel. For example, bituminous coal has less moisture, fewer volatiles, and more fixed carbon (char) than biomass. For very fine particles such as pulverized coal particles drying, devolatilization, and char burn occur sequentially, and the char burn period lasts much longer than the devolatilization and drying stages. For larger particles drying, devolatilization, and char burn occur simultaneously over most of the lifetime of the particle.

14.1 DRYING OF SOLID FUELS

Coal and biomass are porous materials with pore sizes ranging from 0.01–30 μm. Depending on the fuel moisture content, moisture may reside in these pores. Lignite coals contain up to 40% water, while bituminous coals have relatively small pores containing only a few percent water. The as-received moisture content of wood varies from approximately 5% for well dried wood to greater than 50% for green wood. As noted in Chapter 2, within the wood matrix, water exists as water vapor, liquid water chemically bound to the cells of the wood (adsorbed water), and free liquid water within the wood pores. Within the wood, water is first taken up as bound water until all available adsorption sites are occupied, the fiber saturation point, and then becomes free water within the pores. At the fiber saturation point the heat of sorption is near zero. The adsorbed moisture is held with increasing energy as the wood moisture content decreases. This relationship has been expressed algebraically by Stanish, Schajer, and Kayihan (1986) by assuming that the heat of sorption varies quadratically with the bound water content, and at zero moisture content is equal to the heat of vaporization.

$$h_{sorp} = 0.4 h_{fg} \left(1 - \frac{MC_b}{MC_{fsp}}\right)^2 \tag{14.1}$$

where MC_b and MC_{fsp} are the moisture content of the bound water and moisture content at the fiber saturation point, respectively. The fiber saturation point of wood is approximately 30% on a dry basis. Integrating Equation 14.1 from 30% moisture to 0% moisture (dry basis), we see that for a moisture content of 30%, the average heat of sorption is 4% of the heat of vaporization. Based on this, the heat of sorption is generally disregarded for wet wood (moisture content >30%). However, at lower moistures the heat of sorption can be significant and should be considered.

Considering only the heat of vaporization, fuel moisture represents a heat loss of 2400 kJ/kg of water, assuming the heat of vaporization is not recovered in the combustion system. For high moisture fuels it can be advantageous to dry the fuel before feeding it into the combustor. The discussion here is on drying within the combustor, and we consider two cases—small particles where drying occurs before pyrolysis and char burn occur, and larger particles where drying and pyrolysis occur simultaneously.

14.1.1 DRYING OF SMALL PARTICLES

Consider a pulverized coal or small biomass particle that is inserted into a furnace. Upon entry into the gas stream, heat is convected and radiated to the particle surface and conducted into the particle. For a pulverized particle (say 10 μm in size) the water is vaporized and rapidly forced out through the pores of the particle before volatiles are released. For the particle to be considered small in the sense that the temperature gradients within the particle are small, the Biot number (Bi) should be less than 0.2, where

$$\text{Bi} = \frac{\tilde{h}d}{\tilde{k}_p} \tag{14.2}$$

where $\tilde{h}$ is the heat transfer coefficient to the particle, d is the smallest dimension (diameter) of the particle, and $\tilde{k}_p$ is the thermal conductivity of the particle.

The drying time of a small, pulverized particle is the time required to heat the particle to the vaporization point and drive off the water. An energy balance on the small particle may be written as

$$\frac{d}{dt}\left(m_w u_w + m_{dry} u_{dry}\right) = -\dot{m}_w h_{dry} + q \tag{14.3}$$

The first term in Equation 14.3 is the time rate of change of energy within the particle. The term $-\dot{m}_w h_{dry}$ includes the heat rate to vaporize the water in the particle, and q is the net heat transfer rate to the particle by convection and radiation. The subscript w refers to water, and the subscript dry refers to dry fuel.

The rate of heat transfer to the particle, q, depends on the background furnace temperature, T_b, which is assumed to be equal to the surrounding gas temperature. Grey body radiation with emissivity, ε, a view factor of 1, and a convective heat transfer coefficient, $\tilde{h}$, are used.

$$q = \varepsilon\sigma A_p\left(T_b^4 - T_p^4\right) + \tilde{h}^* A_p\left(T_g - T_p\right) \tag{14.4}$$

where $\tilde{h}^*$ is the convective heat transfer coefficient corrected for mass transfer. Integrating Equation 14.3 with Equation 14.4 from the initial temperature to the boiling point of water, using the relation $du = c\,dT$, and noting that the boiling point temperature is much lower than the background furnace temperature, we obtain the particle drying time,

$$t_{dry} \simeq \frac{\left(m_{w,init}\,c_w + m_{dry}\,c_{dry}\right)\left(373 - T_{init}\right) + m_{w,init}\,h_{dry}}{\varepsilon\sigma A_p\left(T_b^4 - T_p^4\right) + \tilde{h}^* A_p\left(T_g - T_p\right)} \tag{14.5}$$

where $m_{w,init}$ is the initial mass of water in the particle and h_{dry} is the energy to dry the particle and includes the heat of sorption and the heat of vaporization.

$$h_{dry} = h_{sorp} + h_{fg} \tag{14.6}$$

For evaluation of the convective heat transfer coefficient, the particle film temperature is used,

$$T_m = \frac{T_p + T_g}{2} \tag{14.7}$$

along with a Nusselt number correlation. Ranz and Marshall (1952) found that for liquid droplets with low rates of vaporization,

$$\mathrm{Nu} = \frac{\tilde{h}d}{k_g} = 2 + 0.6\,\mathrm{Re}_d^{1/2}\,\mathrm{Pr}^{1/3} \tag{14.8}$$

For high rates of vaporization $\tilde{h}$ must be corrected both for the effect of superheating the vapor as it moves away from the surface and for the blowing effect of the vapor motion on the boundary layer. The value of $\tilde{h}$ corrected for the superheating effect is (Bird et al. 2007)

$$\tilde{h}^* = \tilde{h}Z \tag{14.9}$$

where

$$Z = \frac{z}{e^z - 1}$$

and

$$z = -\frac{\dot{m}_w c_{p,vapor}}{\tilde{h}A_p}$$

As should be expected, the value of $\tilde{h}^*$ is a function of the rate of mass transfer, and the mass transfer is a function of the radiative and convective heat transfer to the particle. Additionally, for small vaporization rates, Z, will be less than unity, and $\tilde{h}^* < \tilde{h}$. This is intuitively correct since the streaming vapor leaves the surface and must be heated in the boundary layer until it reaches the ambient temperature at the edge of the boundary layer. At combustion temperatures with very small inert particles, radiation heat transfer and convective heat transfer to the particle are approximately the same. In the case of a wet particle, a quick check of Equation 14.9 will show that the escaping water vapor effectively shields the particle from convection heat transfer. However, radiation heat transfer to the particle is not blocked.

Example 14.1

A 100 μm oak particle with 40% moisture (dry basis) is inserted into a 1500 K furnace. The oak particle was initially at 300 K and had a dry density of 690 kg/m³. Find the drying time.

Solution

Assuming that the slip velocity (relative velocity between the particle and the gas) of the particle is small, from Equation 14.8

$$\mathrm{Nu} = \frac{\tilde{h}d}{\tilde{k}_g} = 2$$

Using a film temperature of 900 K, the thermal conductivity of the air in the furnace is

$$\tilde{k}_g = 0.0625 \ \mathrm{W/m \cdot K} \qquad \text{(Appendix B)}$$

Rearranging and substituting

$$\tilde{h} = 2\frac{\tilde{k}_g}{d} = 2 \cdot \frac{0.0625 \ \mathrm{W}}{\mathrm{m \cdot K}} \cdot \frac{1}{100 \times 10^{-6} \ \mathrm{m}} = \frac{1250 \ \mathrm{W}}{\mathrm{m^2 \cdot K}}$$

Using the average temperature during heating, the specific heat of the wood is

$$c_{\mathrm{dry}} = 0.387(3.36) + 0.103 = \frac{1.4 \ \mathrm{kJ}}{\mathrm{kg \cdot K}} \qquad \text{(Table A.4)}$$

The specific heat of water is

$$c_w = 4.2 \ \mathrm{kJ/kg \cdot K}$$

The heat of vaporization is

$$h_{fg} = 2400 \ \mathrm{kJ/kg}$$

The emissivity of wood is assumed to be

$$\varepsilon = 0.90$$

The mass of dry wood and water are

$$m_{dry} = \rho_{dry}\frac{\pi d^3}{6} = \frac{690\ kg}{m^3}\cdot\frac{\pi\left(100\times10^{-6}\right)^3 m^3}{6} = 3.61\times10^{-10}\ kg$$

$$m_w = 0.40\left(3.61\times10^{-10}\ kg\right) = 1.44\times10^{-10}\ kg$$

The area of the particle is

$$A_p = \pi\left(100\times10^{-6}\right)^2 m^2 = 3.14\times10^{-8}\ m^2$$

Evaluating Equation 14.5 for the drying time,

$$t_{dry} \approx \frac{\left(m_{w,init}C_w + m_{dry}C_{dry}\right)\left(373 - T_{init}\right) + m_{w,init}h_{fg}}{\varepsilon\sigma A_p\left(T_b^4 - T_p^4\right) + \tilde{h}^* A_p\left(T_g - T_p\right)} = \frac{a+b}{c+d}$$

where

$$a = \left[\frac{1.44\times10^{-10}\ kg}{1}\cdot\frac{4.2\ kJ}{kg\cdot K} + \frac{3.61\times10^{-10}\ kg}{1}\cdot\frac{1.4\ kJ}{kg\cdot K}\right]\left(373 - 300\right)K$$

$$a = 8.104\times10^{-8}\ kJ$$

$$b = \left(1.44\times10^{-10}\ kg\right)\left(\frac{2400\ kJ}{kg}\right) = 34.56\times10^{-8}\ kJ$$

Looking at the energies needed for heating the particle given by a and the energy needed for drying the particle given by b, we see that the majority of the energy is used to dry the particle, as expected.

$$c = \left(0.9\right)\left(5.67\times10^{-8}\frac{W}{m^2\cdot K^4}\right)\left(3.14\times10^{-8}\ m^2\right)\left(1500^4 - 336^4\right)K^4$$

$$c = 8.096\times10^{-3}\ W$$

$$d = 1250\frac{W}{m^2\cdot K}\left(3.14\times10^{-8}\ m^2\right)\left(1500 - 336\right)K = 45.71\times10^{-3}\ W$$

Solving for the drying time

$$t_{dry} \simeq \frac{a+b}{c+d} = \frac{\left(8.104 \times 10^{-5} + 34.56 \times 10^{-5}\right) \text{ J}}{\left(8.096 \times 10^{-3} + 45.71 \times 10^{-3}\right) \text{ W}}$$

$$t_{dry} \simeq 7.9 \text{ ms}$$

It should be noted that at this point we have not included self-shielding of the particle from convective heat transfer (Equation 14.9) because we lacked an estimate of the mass transfer rate. Using the currently computed drying time, we can find the average mass transfer rate during drying.

$$\dot{m}_w = \frac{m_w}{t_{dry}} = \frac{1.44 \times 10^{-10} \text{ kg}}{7.9 \times 10^{-3} \text{ s}} = \frac{1.82 \times 10^{-8} \text{ kg}}{\text{s}}$$

From Appendix C the specific heat of the water vapor at the film temperature is

$$c_{p,vapor} = \frac{39.94 \text{ kJ}}{\text{kgmol}_{H_2O} \cdot \text{K}} \cdot \frac{\text{kgmol}_{H_2O}}{18 \text{ kg}_{H_2O}} = \frac{2.22 \text{ kJ}}{\text{kg}_{H_2O} \cdot \text{K}}$$

Solving for z

$$z = -\frac{\dot{m}_w c_{p,vapor}}{\tilde{h} A_p} = \frac{1.82 \times 10^{-8} \text{ kg}_{H_2O}}{\text{s}} \cdot \frac{2.22 \text{ kJ}}{\text{kg}_{H_2O} \cdot \text{K}} \cdot \frac{\text{m}^2 \cdot \text{K}}{1250 \text{ W}} \cdot \frac{1}{3.14 \times 10^{-8} \text{ m}^2} \cdot \frac{1000 \text{ W}}{\text{kJ} \cdot \text{s}}$$

$$z = 1.03$$

Z then becomes

$$Z = \frac{1.03}{e^{1.03} - 1} = 0.572$$

Based on this the heat transfer coefficient is

$$\tilde{h}^* \simeq 0.562(1250) \text{ W/m}^2 \cdot \text{K} = 715 \text{ W/m}^2 \cdot \text{K}$$

and from this

$$t_{dry} \simeq \frac{a+b}{c+0.572d} = \frac{\left(7.841 \times 10^{-5} + 34.56 \times 10^{-5}\right) \text{ J}}{\left(8.096 \times 10^{-3} + 0.572\left(45.71 \times 10^{-3}\right)\right) \text{ W}}$$

$$t_{dry} \simeq 12 \text{ ms}$$

At this point our estimate of the average mass transfer rate has changed, and so we need to repeat the analysis using the new mass transfer rate. If we repeat this process two more times until we have consistent mass transfer value, we find that

$$Z = 0.660$$

and

$$t_{dry} \simeq 11 \text{ ms}$$

While 11 ms may seem like a relatively short time, it can result in long travel distances before ignition. For example, if we were trying to retrofit an oil-fired furnace to burn sawdust, the distance traveled while drying before ignition would be 11 cm if the velocity were 10 m/s. This could create difficulties in trying to stabilize the burner flame, and thus the fuel should be dried before being fed to the burner.

14.1.2 DRYING OF LARGER PARTICLES

For relatively large fuel particles such as stoker coal or woodchips in a convective flow, the assumption of uniform particle temperature and hence uniform drying is not valid, and hence Equation 14.5 is not valid. Because of the temperature gradient within the particle, moisture is evaporated from inside the particle while volatiles are being driven off near the outer shell of the particle. Due to the high pressure in the fuel pores in the drying region during the devolatilization of the outer layer of the particle, some of the moisture is forced towards the center of the particle until the pressure builds up throughout the particle. Hence, drying of large solid fuel particles initially involves inward migration as well as the outward flow of the water vapor. A pyrolysis layer starts at the outer edge of the particle and gradually moves inward while releasing volatiles and forming char. The release of moisture reduces the heat and mass transfer to the particle surface, thus reducing the rate of mass loss of the particle (burning rate). As the rate of moisture loss and devolatilization are reduced, the char surface begins to react.

For example, consider a 10 mm pine cube suspended from an electronic balance into an 1100 K air stream. The mass divided by the initial mass versus time is shown in Figure 14.1 for initial moisture contents of 0%, 15%, and 200% (moisture content on a dry basis). The normalized mass versus time curves were differentiated to yield the normalized burning rates. Gas measurements of H_2O, CO_2, and CO indicate that pyrolysis products are released while the fuel is drying. Drying and pyrolysis are complete at 120 s for the 200% moisture case, at which point some of the char has burned (as indicated by the mass remaining), and the remaining char burns in air generating CO and CO_2.

When a log burns in a cooking fire, campfire, or fireplace; drying, pyrolysis, and char burn occur simultaneously for a significant percent of the burn time. Similarly when a wood chip or coal particle burns on a grate in a furnace or boiler, the center of the particle is not dry until much of the char layer is consumed. Figure 14.2 indicates the three zones for a partially burned particle. The outside char layer is black and porous. The pyrolysis zone is a thin brown layer inside the char layer, and the

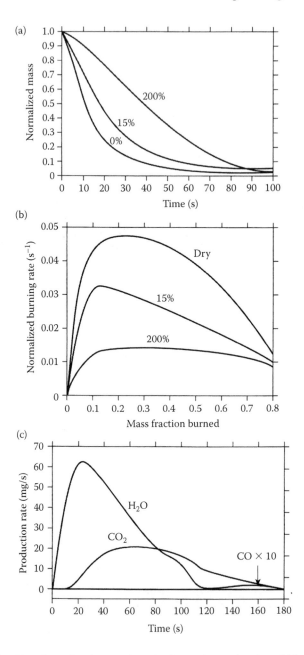

FIGURE 14.1 The effect of moisture (% dry basis) on the combustion of 10 mm pine cubes burned in air at 1100 K and Re = 120 for (a) normalized mass and (b) normalized burning rate (time derivative of mass per initial mass). (c) Combustion rate of products for the combustion of 10 mm pine cubes at 200% moisture (dry basis) burned in air at 1100 K and Re = 120. (From Simmons, W. W. and Ragland, K. W., "Burning Rate of Millimeter Sized Wood Particles in a Furnace," *Combust. Sci. Technol.* 46(1–2):1–15, 1986. © 1986, by permission of Gordon and Breach Publishers.)

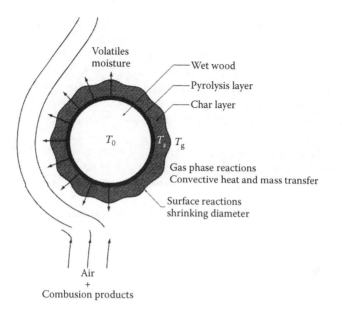

Volatiles
moisture
Wet wood
Pyrolysis layer
Char layer

T_0 T_s T_g

Gas phase reactions
Convective heat and mass transfer

Surface reactions
shrinking diameter

Air
+
Combusion products

FIGURE 14.2 Cross-section of a reacting log showing char, pyrolysis, and undisturbed wood regions.

interior portion of the particle is white. For example, wood begins to turn brown at 250°C and becomes black above 300°C. The moisture and volatiles impede the inward conduction of heat, and some moisture is retained during most of the burn time.

14.2 DEVOLATILIZATION OF SOLID FUELS

When the temperature of a small fuel particle or a zone within a large particle rises to above 250–400°C, solid fuels begin to decompose, releasing volatiles. Since the volatiles flow out of the solid through the pores, external oxygen cannot penetrate into the particle, and hence the devolatilization is referred to as the pyrolysis stage. The rate of devolatilization and the composition of the pyrolysis products depend on the temperature and type of fuel. The pyrolysis products then ignite and form an attached flame around the particle as oxygen diffuses into the products. The flame in turn heats the particle, increasing the rate of devolatilization. While water vapor is flowing out of the pores, the flame temperature will be low and the flame weak. Once all the water vapor is driven from the particle, the flame will be hotter.

Coal pyrolysis begins at 300–400°C and initially CO and CO_2 are the primary volatiles. Ignition of the volatiles occurs at 400–600°C. Carbon monoxide, carbon dioxide, chemically formed water, hydrocarbon vapors, tars, and hydrogen are produced rapidly as the temperature reaches 700–900°C. Above 900°C pyrolysis is essentially complete, and the char (fixed carbon) and ash remain.

The devolatilization of bituminous coal proceeds differently than for lignite or subbituminous coal because bituminous coal contains less oxygen. First the bituminous coal becomes plastic, and some of the bituminous coals swell markedly. Pressure within the particle builds up and tars are squeezed out of the particle. The coal particle may fracture into several pieces due to the internal pressure. Meanwhile, pyrolysis proceeds, releasing carbon monoxide, hydrocarbons, and soot. These pyrolysis products burn as an attached diffusion flame around the particle. The reaction rate of the released volatiles can be calculated approximately using the one- and two-step global hydrocarbon reaction rates in Section 4.3.

For wood the hemicellulose pyrolyzes at 250–325°C, the cellulose at 325–375°C, and the lignin at 300–500°C. Certain extractives such as terpenes, which amount to only a few percent of the wood, escape at less than 225°C. Various hydrocarbon vapors, liquids and tars, and water are formed and quickly break down under combustion conditions; hence the pyrolysis products in a combustion environment may be considered to be short-chained hydrocarbons, carbon dioxide, carbon monoxide, hydrogen, and water vapor. As with coal, the exact composition is a function of the heating rate. The pyrolysis products burn as a diffusion flame around the particle if sufficient oxygen is present.

If the object is to gasify the solid fuel but not to complete combustion, such as for a gasifier to run a boiler or diesel engine, then only enough oxygen is provided to generate an exothermic reaction for drying and pyrolysis of the fuel. A volatile flame is not established. The products are short-chained hydrocarbons, carbon monoxide, hydrogen, carbon dioxide, water vapor, nitrogen, and tar. The tar, char, and ash particles and any vaporized inorganic compounds must be filtered from the hot gaseous pyrolysis products before using them in an engine. This is referred to as hot gas cleanup prior to combustion.

The rate of devolatilization of a solid fuel may be represented approximately as a first order reaction with an Arrhenius rate constant:

$$\frac{dm_v}{dt} = -m_v k_{pyr} \tag{14.10}$$

where

$$k_{pyr} = -k_{0,pyr} \exp\left(-\frac{E_{pyr}}{\hat{R}T_p}\right) \tag{14.11}$$

where *pyr* refers to pyrolysis, and

$$m_v = m_{dry} - m_{char} - m_{ash} \tag{14.12}$$

where m_v, m_{dry}, m_{char}, and m_{ash} are the mass of the volatiles, the mass of the dry particle, the mass of the char, and the mass of the ash, respectively. The pyrolysis rate is independent of the particle size as long as the particle temperature is constant. Generally, the heat up time is short compared to the pyrolysis time for pulverized fuels. For large particles the transient heating of the particle must be considered and the pyrolysis rate of the particle must be summed based on local temperature. More details on this type of analysis are given in Bryden, Ragland, and Rutland (2002).

The activation energy and the pre-exponential factor must be determined experimentally for specific combustion conditions and fuel type. Representative values of the pre-exponential factor and the activation energy are shown in Table 14.1. The mass of the char can be determined from the proximate analysis; however, under high heating rates such as experienced in pulverized fuel flames, the volatile yield is higher and the char yield is lower than suggested by the proximate analysis. Volatile yields of 50% for pulverized coal and 90% for pulverized wood are typical.

If the particle temperature is constant during devolatilization, then k_{pyr} is constant and Equation 14.10 may be integrated to obtain the mass loss as a function of time during pyrolysis,

$$\ln\left[\frac{m_{dry} - m_{char} - m_{ash}}{m_{dry,init} - m_{char} - m_{ash}}\right] = -k_{pyr}t \tag{14.13}$$

Equation 14.13 implies that a single chemical reaction converts a solid fuel to pyrolysis products. In reality solid fuels such as coal and wood are complex compounds that undergo many reactions when they are heated. Some of these reactions are endothermic and some are exothermic, and they proceed at different rates. Nevertheless, the net heat of devolatilization is thought to be near zero, and Equation 14.10 is a useful approximate global pyrolysis rate.

Example 14.2

A pulverized bituminous coal particle has a temperature of 1500 K. Find the time required to devolatilize 90% of the volatile mass.

Solution

From Table 14.1

$$k_{0,pyr} = 700 \text{ s}^{-1}$$

$$E_{pyr} = 11.8 \text{ kcal/gmol}$$

TABLE 14.1
Representative Pyrolysis Parameters for Several Solid Fuels

Fuel	$k_{0,pyr}$ (s^{-1})	E_{pyr} (kcal/gmol)
Lignite	280	11.3
Bituminous coal	700	11.8
Wood	7×10^7	31.0

Thus,

$$k_{pyr} = \left(700\,s^{-1}\right)\exp\left[-\frac{11,800\,cal/gmol}{\left(1.987\,cal/gmol\cdot K\right)\left(1500\ K\right)}\right] = 13.35\,s^{-1}$$

From Equation 14.13,

$$t_{pyr} = -\frac{\ln(0.10)}{13.35\ s^{-1}} = 0.17\,s$$

For larger fuel particles considerable time is required to heat the particles to pyrolysis temperatures after they are inserted into the combustion environment, and the pyrolysis process gradually penetrates into the particle. For example, experimental data obtained by inserting a single coal particle into a hot gas stream are shown in Figures 14.3 and 14.4. The coal particle is attached to a fine quartz rod suspended from an electronic balance. In the case shown by Figure 14.3, a class C bituminous coal particle of 5.3 mm diameter (100 mg) was suspended in an 1100 K air stream. The ignition delay was a few seconds, and then a volatile flame ignited and remained attached to the particle for 30 s. Upon extinction of the volatile flame, 55% of the mass was lost and the rate of mass loss decreased markedly. The proximate analysis

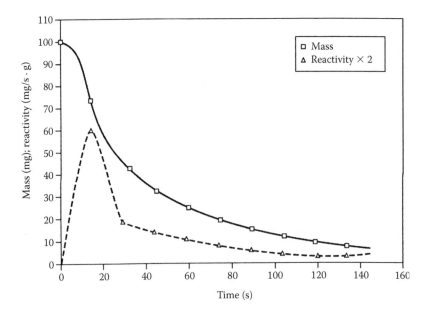

FIGURE 14.3 Typical curves of transient mass and reactivity for 5.3 mm bituminous coal particles in 1100 K air at 2 m/s. (From Ragland, K. W. and Yang, J.-T., "Combustion of Millimeter Sized Coal Particles in Convective Flow" *Combust. Flame* 60:285–297, 1985, by permission of Elsevier Science Inc.)

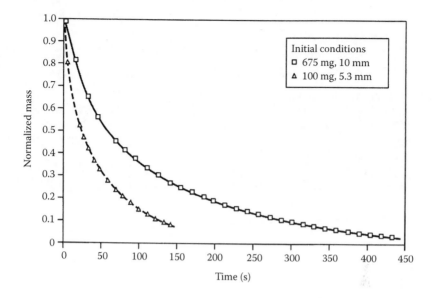

FIGURE 14.4 Effect of particle size on transient mass for bituminous coal particles in 1200 K air flowing at 2 m/s. (From Ragland, K. W. and Yang, J.-T., "Combustion of Millimeter Sized Coal Particles in Convective Flow" *Combust. Flame* 60:285–297, 1985, by permission of Elsevier Science Inc.)

would suggest that only 40% of the mass should be lost at the end of devolatilization. It is typical of solid fuels undergoing high heating rates for the volatile yield to be greater than indicated by the proximate analysis. Figure 14.4 shows that when a 5 mm diameter bituminous coal particle is inserted into a 1200 K stream of air flowing at 2 m/s, the particle requires 22 s to pyrolyze, while a 10 mm particle requires 62 s to pyrolyze. Similarly, a 10 mm cube of dry pine shown in Figure 14.1 requires 25 s to pyrolyze. The 10 mm bituminous coal particle initially weighed 675 mg and pyrolyzed 55% of this weight, while the pine cube weighed 400 mg and pyrolyzed 90% of the initial weight. Hence the average pyrolysis rate was 6.0 mg/s for the 10 mm coal particle and 14.4 mg/s for the 10 mm pine particle. This is in agreement with the general statement that wood is more reactive than coal.

Ignition of solid fuels can occur either by ignition of the fixed carbon (char) on the surface of the fuel, or by ignition of the volatiles in the boundary layer around the particle. Which mechanism actually occurs first depends on the rate of convective and radiative heat transfer to the particle. The ignition will occur first at the surface if the radiative heat transfer is high enough for the surface to quickly heat up to the ignition temperature of the carbon, or if the rate of convective heating is high enough for the surface to rapidly heat, but the volatiles are swept away before a combustible mixture can accumulate. On the other hand, if the surface heating is low, then the volatiles may ignite first since they have a lower ignition temperature than carbon.

Ignition temperatures of various fuels are shown in Table 14.2. Note that char has a much lower ignition temperature than graphite. This implies that char is not pure carbon, and in fact char contains some hydrogen as well as carbon. Pyrolysis under

TABLE 14.2

Typical Ignition Temperature for Selected Solid Fuels

Fuel	Ignition Temperature (°C)
Graphite	820
Bituminous coal char	410
Bituminous coal volatiles	350
Wood char	340
White Pine volatiles	260
Paper	230

the usual combustion conditions does not drive off all of the hydrogen. Note that wood char ignites at a lower temperature than coal char. Similarly, volatiles from wood ignite at a lower temperature than volatiles from coal.

Ignition time delay depends upon particle size and thermal diffusivity as well as heating rate and pyrolysis rate. The ignition time for pulverized fuels is typically a few milliseconds, while for 10 mm particles it can be several seconds under furnace conditions. If the temperature is barely above the ignition temperature, then the ignition delay can be many minutes for large particles. Moisture increases the ignition delay. Ignition delay can be an important consideration in designing burners for pulverized fuels.

14.3   CHAR COMBUSTION

Char combustion is the final step in combustion of pulverized coal and biomass. For larger particles char combustion occurs as the particle is drying and devolatilizing. Char is a highly porous carbon with small amounts of other volatiles and dispersed mineral matter. In this text we assume that the small amount of non-carbon material in the char has a negligible effect on char kinetics. The porosity of wood char is about 0.9 (90% voids), while coal char tends to have a porosity of about 0.7, although this can vary widely. The internal surface area of char is on the order of 100 m^2/g for coal char and 10,000 m^2/g for wood char. If the interior of the particle is dry and pyrolyzed, oxygen can diffuse through the external boundary layer and into the char particles. Otherwise gas phase reactions are limited to the external surface of the particle. The char surface reaction generates primarily CO. The CO then reacts outside the particle to form CO_2. The surface reactions may raise the temperature of the char 100–200°C above the external gas temperature when oxygen is present.

For engineering purposes it is appropriate to use a global reaction rate based on the external char surface area to determine the char burning rate. The alternative is to use intrinsic (elementary) reaction rates; however, this requires detailed calculations in the boundary layer and within the pore structure of the char. The global reaction rate is formulated in terms of the rate of reaction of the char mass per unit of the external surface area and gas concentration outside the particle boundary layer. As with all global reactions, the results should be verified experimentally over the range of operating conditions for a particular application.

The carbon char reacts with oxygen at the surface to form carbon monoxide and carbon dioxide, but generally carbon monoxide is the main product,

$$C + \frac{1}{2}O_2 \rightarrow CO \qquad\qquad (a)$$

The carbon surface also reacts with carbon dioxide and water vapor according to the following reduction reactions:

$$C + CO_2 \rightarrow 2CO \qquad\qquad (b)$$

$$C + H_2O \rightarrow CO + H_2 \qquad\qquad (c)$$

Reduction reactions (b) and (c) are generally much slower than the oxidation reaction (a), and for combustion usually only reaction (a) need be considered. Where the oxygen is depleted, then these reduction reactions are important.

For a global reaction rate of order n with respect to oxygen, the char burning rate is given by

$$\frac{dm_{char}}{dt} = -i\left(\frac{M_C}{M_{O_2}}\right) A_p k_c \left(\rho_{O_2,surf}\right)^n \qquad\qquad (14.14)$$

where i is the stoichiometric ratio of moles of carbon per mole of oxygen (which is 2 for reaction (a)), A_p is the external particle surface area, k_c is the kinetic rate constant, $\rho_{O_2,surf}$ is the oxygen partial density at the surface of the particle, and n is the order of reaction. The oxygen concentration at the particle surface is not known; however, it can be eliminated by equating the oxygen consumed by the char to the diffusion of oxygen across the particle boundary layer. Because a simplification occurs when the reaction order is 1, we consider only this case. Then

$$A_p k_c \rho_{O_2,surf} = A_p \tilde{h}_D \left(\rho_{O_2} - \rho_{O_2,surf}\right) \qquad\qquad (14.15)$$

Solving for the oxygen density,

$$\rho_{O_2,surf} = \frac{\tilde{h}_D}{k_c + \tilde{h}_D} \rho_{O_2,\infty} \qquad\qquad (14.16)$$

Hence Equation 14.14 becomes

$$\frac{dm_{char}}{dt} = -\frac{12}{16} A_p k_e \rho_{O_2,\infty} \qquad\qquad (14.17)$$

where the effective rate constant, k_e, includes both kinetics and diffusion,

$$k_e = \frac{\tilde{h}_D k_c}{\tilde{h}_D + k_c} \tag{14.18}$$

The kinetic rate constant is calculated from the Arrhenius relation

$$k_c = k_{c,0} \exp\left(-\frac{E_c}{\hat{R}T_p}\right) \tag{14.19}$$

Sometimes global char kinetic reaction rate constants are based on oxygen pressure rather than oxygen concentration:

$$k_p = k_{p,0} \exp\left(-\frac{E_c}{\hat{R}T_p}\right) \tag{14.20}$$

Global kinetic parameters for several coal chars are given in Table 14.3. Note that k_p has units of $g_{O_2}/(cm^2 \cdot s \cdot atm_{O_2})$, whereas k_c has units of $g_{O_2}/(cm^2 \cdot s \cdot g_{O_2}/cm^3)$ or cm/s. Values of k_c may be calculated from k_p values from

$$k_c = k_p \frac{T_g \hat{R}}{M_{O_2}} \tag{14.21}$$

The mass transfer coefficient, $\tilde{h}_D$ (cm/s), is obtained from the Sherwood number

$$Sh = \frac{\tilde{h}_D d}{D_{AB}}$$

For very low flow around the particle, $Sh = 2$ by the following reasoning. Consider a spherical char particle surrounded by stagnant air. Diffusion of oxygen to the surface is governed by the diffusion equation, which is analogous to the heat conduction equation. Note that r is the radius here, not the reaction rate.

TABLE 14.3

Representative Global Coal Char Oxidation Rate Constants

Coal Type	$k_{p,0}$ $(g_{O_2}/(cm^2 \cdot s \cdot atm_{O_2}))$	E (cal/gmol)
Anthracite	20.4	19,000
Bituminous (high volatile A)	66	20,360
Bituminous (high volatile C)	60	17,150
Subbituminous (class C)	145	19,970

$$\frac{d}{dr}\left(r^2 \frac{d\rho_{O_2}}{dr}\right) = 0 \qquad (14.22)$$

Integrating twice,

$$\rho_{O_2} = -\frac{a_1}{r} + a_2 \qquad (14.23)$$

At the particle surface it is assumed that all of the oxygen is consumed. Therefore the boundary conditions are

$$\rho_{O_2} = 0 \text{ at } r = d/2$$

and

$$\rho_{O_2} = \rho_{O_2,\infty} \text{ at } r_\infty$$

Hence Equation 14.23 becomes

$$\rho_{O_2} = \rho_{O_2,\infty}\left(1 - \frac{d}{2r}\right) \qquad (14.24)$$

Differentiating Equation 14.24 and setting $r = d/2$, the flux of oxygen to the surface is

$$\left.\frac{d\rho_{O_2}}{dr}\right|_{r=d/2} = \frac{2\rho_{O_2,\infty}}{d} \qquad (14.25)$$

From the definition of the mass transfer coefficient, the oxygen flow to the surface is given by

$$\frac{d\rho_{O_2}}{dr} = \frac{\tilde{h}_D}{D_{AB}}\left(\rho_{O_2,\infty} - \rho_{O_2,surf}\right) \qquad (14.26)$$

Equating Equations 14.25 and 14.26 and assuming zero oxygen at the surface,

$$\frac{\tilde{h}_D d}{D_{AB}} = Sh = 2 \qquad (14.27)$$

When the particle Reynolds number is not much less than 1, then from analogy to heat transfer the Ranz-Marshall equation (Equation 14.8) may be used,

$$\mathrm{Sh} = \left(2 + 0.6\,\mathrm{Re}^{1/2}\,\mathrm{Sc}^{1/3}\right)\phi \tag{14.28}$$

where the Schmidt number, Sc, is typically 0.73 ($\mathrm{Sc} = \upsilon/D_{AB}$). Also a mass transfer screening factor, ϕ, is introduced due to the rapid outward flow of combustion products. The mass transfer screening factor is not well known for char but may vary from 0.6–1.0 depending on the moisture and volatile release rate, and is about 0.9 for coal char.

14.3.1 CHAR BURNOUT

From Equation 14.17 with the ∞ subscript notation dropped for convenience and assuming carbon monoxide is formed at the surface, the global char burnout rate is

$$\frac{dm_{\mathrm{char}}}{dt} = -\frac{12}{16}A_p k_e \rho_{O_2} \tag{14.29}$$

Let us consider the following limiting cases, (1) the char burns out at constant diameter (with decreasing density) or (2) the char burns out at constant density (with decreasing diameter). For constant diameter burnout Equation 14.29 may be integrated directly.

For the constant density case remember that

$$m_{\mathrm{char}} = \rho_{\mathrm{char}}\frac{\pi d^3}{6} \tag{14.30}$$

We can solve Equation 14.30 for d and substitute into Equation 14.29 to obtain

$$\frac{dm_{\mathrm{char}}}{dt} = -\pi\left(\frac{6m_{\mathrm{char}}}{\pi\rho_{\mathrm{char}}}\right)^{2/3}\frac{12}{16}k_e\rho_{O_2} \tag{14.31}$$

With high temperatures and large particles, $k_c \gg \tilde{h}_D$, as a result diffusion is the rate limiting process. In this case, from Equation 14.18 we can quickly see that $k_e = \tilde{h}_D$. Rearranging Equation 14.27

$$\tilde{h}_D = 2\frac{D_{AB}}{d}$$

Substituting into Equation 14.31 and simplifying, the rate of char burnout for a constant density, diffusion limited case is

$$\frac{dm_{\mathrm{char}}}{dt} = -\left(\frac{12}{16}\right)\left[2\pi\left(\frac{6m_{\mathrm{char}}}{\pi\rho_{\mathrm{char}}}\right)^{1/3}\right]D_{AB}\,\rho_{O_2} \tag{14.32}$$

At low temperatures and with small particles, $\tilde{h}_D \gg k_c$, as a result the surface reaction rate is limited by reaction kinetics (kinetic control). Again from Equation 4.18 we can quickly see that $k_e = k_c$. And substituting into Equation 14.31 the rate of char burnout for the constant density, kinetically limited case is

$$\frac{dm_{char}}{dt} = -\pi \left(\frac{6 m_{char}}{\pi \rho_{char}} \right)^{2/3} \frac{12}{16} k_c \rho_{O_2}$$ (14.33)

The burnout time for a pure char particle (containing no volatiles or moisture) can be obtained by integrating Equations 14.29, 14.32, and 14.33 from the initial char mass to zero.

Constant diameter model:

$$t_{char} = \rho_{char, init} \frac{d_{init}}{4.5 k_e \rho_{O_2}}$$ (14.34)

Constant density, diffusion control:

$$t_{char} = \rho_{char} \frac{d_{init}^2}{6 D_{AB} \rho_{O_2}}$$ (14.35)

Constant density, kinetic control:

$$t_{char} = \rho_{char} \frac{\rho_{char} d_{init}}{1.5 k_c \rho_{O_2}}$$ (14.36)

When evaluating the binary diffusion coefficient, we recommend using the molecular diffusion coefficient for oxygen into nitrogen, which is given in Appendix B. Turbulence increases diffusion, and to account for turbulence we recommend increasing the Sherwood number by using an appropriate root-mean-square turbulent velocity in the Reynolds number of Equation 14.28.

A summary of total burn time versus coal particle diameter from 0.1–5 mm in air at 1500 K is given in Figure 14.5. These data cover different types of coal. Ignition delay and total burn time data for several sizes of Douglas-fir bark with 25% excess air and peak temperatures of 1400–1500 K are shown in Table 14.4. By comparing Table 14.4 with Figure 14.5, it can be noted that a particle of bark will burn out roughly twice as fast as a coal particle of the same size. This is because bark has more volatiles and less char than coal reducing the char burnout time.

Solutions of Equation 14.29 for the constant diameter mode and Equation 14.32 for the constant density mode are shown in Figure 14.6, assuming $k_p = 0.1$ $g_{O_2}/$ ($cm^2 \cdot s \cdot atm_{O_2}$) and $k_c \rightarrow \infty$, which implies diffusion control of the burning rate. This figure also implies that the particle Reynolds number is much less than one.

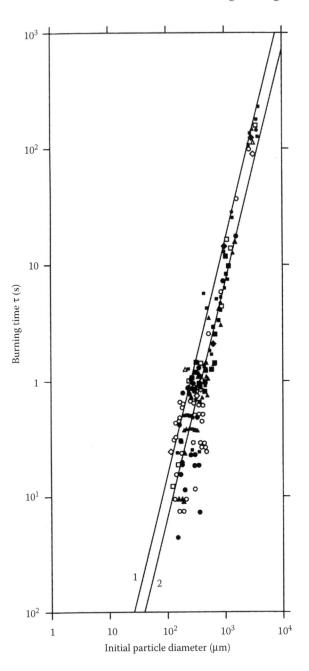

FIGURE 14.5   Single coal particle burn time in air at 1 atm and 1500 K versus particle size. Curves are calculated from Equation 14.36; curve 1, $\rho = 2$ g/cm^3 and curve 2, $\rho = 1$ g/cm^3. (From Essenhigh, R. H., "Fundamentals of Coal Combustion," in *Chemistry of Coal Utilization: Second Supplementary Volume*, ed. M.A. Elliott, 1153–312, Wiley Interscience, New York, 1981, courtesy of the National Academies Press.)

TABLE 14.4
Typical Ignition Delay and Total Burn Times for Douglas-Fir Bark Particles

Particle Size (μm)	Moisture (%)	Ignition Delay (ms)	Burnout Time (ms)
36	10	0	30
300	10	5	540
300	28	30	570
612	10	20	655

Figure 14.6 shows the approximate residence time required to burn out a 100 μm suspended char particle under four different assumptions.

14.3.2 CHAR SURFACE TEMPERATURE

When the surface of a char particle is in an oxidizing environment, the particle surface temperature is typically hotter than the surrounding gas temperature. Heat loss by conduction from the surface into the particle is small and can be neglected. A steady state energy balance equates the heat generation at the surface to the heat loss by convection and radiation;

$$H_{char} \frac{dm_{char}}{dt} = \tilde{h} A_p \left(T_p - T_g \right) + \sigma \varepsilon A_p \left(T_p^4 - T_b^4 \right) \qquad (14.37)$$

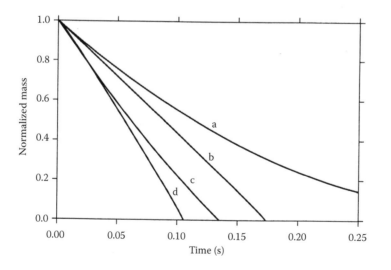

FIGURE 14.6 Burnout of 100 μm diameter coal char particles at 1600 K, 10% oxygen, 1 atm pressure, initial char density 0.8 g/cm³: (a) ρ = constant, $k_p = 0.1$ $g_{O_2}/cm^2 \cdot s \cdot atm_{O_2}$; (b) d = constant, $k_p = 0.1$ $g_{O_2}/cm^2 \cdot s \cdot atm_{O_2}$; (c) ρ = constant, $k_c = h_D$; and (d) d = constant, $k_c = h_D$.

The particle temperature is coupled to the burning rate. Remembering that we have assumed that other than ash the char is entirely carbon, the heat of reaction of the char, H_{char}, is based on the carbon-oxygen reaction,

$$C + \frac{1}{2}O_2 \rightarrow CO$$

For a small char particle the Nusselt number is 2. The particle temperature is determined for a specific case in the following example.

Example 14.3

Determine the char particle temperature due to combustion of C to CO at the char surface for the following conditions: oxygen partial pressure 0.1 atm, particle size 100 μm, gas temperature 1700 K, background temperature 1500 K, particle emissivity 0.9, and a char reaction rate constant of 0.071 $g_{O_2}/(cm^2 \cdot s \cdot atm_{O_2})$.

Solution

The heat of reaction for the C to CO reaction can be determined from the data in Appendix C. This is

$$H_{char} = 9.2 \ kJ/g_C$$

The gas conductivity for air at 1700 K is

$$\tilde{k}_g = 0.105 \ W/m \cdot K \qquad\qquad (Appendix\ B)$$

For a Nusselt number of 2,

$$\tilde{h} = \frac{2\tilde{k}_g}{d} = \frac{2(0.105 \ W/m \cdot K)}{100 \times 10^{-6} \ m} = \frac{2.1 \ kW}{m^2 \cdot K}$$

Substituting Equation 14.29 into Equation 14.37 and dividing by A_p yields

$$H_{char}\left(\frac{12}{16}\right)k_e\, \rho_{O_2} = \tilde{h}\left(T_p - T_g\right) + \sigma\varepsilon\left(T_p^4 - T_b^4\right)$$

Substituting

$$\left(\frac{9.2 \ kJ}{kg_C}\right)\left(\frac{12 \ kg_C}{16 \ kg_{O_2}}\right)\left(\frac{0.071 \ g_{O_2}}{cm^2 \cdot s \cdot atm_{O_2}}\right)\frac{0.1 \ atm_{O_2}}{1}$$

$$= \frac{2.1 \times 10^{-4} \ kW}{cm^2 \cdot K}\left(T_p - 1700\right) K + \frac{5.67 \times 10^{-15} \ kW}{cm^2 \cdot K^4}(0.9)\left(T_p^4 - 1500^4\right) K^4$$

Solving implicitly, $T_p = 1800$ K, which is 100 K above the gas temperature.

14.4 ASH FORMATION

The type and extent of the mineral matter in a solid fuel can influence the reaction rate. Mineral matter in biomass ranges from 1–6%, while in coal the mineral matter can range from a few percent to 50% or more for very low-grade coal. As the char burns, the minerals, which are dispersed as ions and submicron particles in the fuel, are converted to a layer of ash on the char surface. In high temperature pulverized coal combustion the ash tends to form hollow glassy spheres called cenospheres. At lower temperatures the ash tends to remain softer. The ash layer can have a significant effect on heat capacity, radiative heat transfer, and catalytic surface reactions as well as increased diffusive resistance to oxygen, especially late in the char burn stage. In combustion systems the ash that is formed from the mineral matter can slag on radiant heat transfer surfaces and foul convective heat transfer surfaces if the particle temperature is too high. Slagging and fouling will be discussed in Chapter 16. Mineral matter is also a significant factor in the performance of gasification systems. The size and composition of the particulate emissions are influenced by the nature of the mineral matter and time-temperature history.

14.5 PROBLEMS

1. A 100 µm bituminous coal particle initially at 300 K and containing 5% moisture (as-received) is inserted into a furnace at 1500 K. Find the time to reach the ignition temperature assuming that the Nusselt number is 2. Neglect any pyrolysis that may occur. Assume the particle is thermally thin. Neglect radiation heat transfer to the particle. Use coal dry density = 1.3 g/cm^2 and a coal specific heat = 1.3 J/g K.

2. Find the times required for a 100 µm lignite coal particle and a 500 µm lignite coal particle to reach 99% devolatilization assuming that each particle is uniformly at 1200 K and 1500 K. Neglect the particle heatup and drying time. Use data from Table 14.1.

3. Find the times required to release 90% of the volatiles from a woodchip that is abruptly brought to a uniform temperature of 900 K and 1200 K. Use data from Table 14.1.

4. A dry lignite coal particle, which consists of 50% volatiles, 40% char, and 10% ash, is abruptly brought to a uniform temperature of 1500 K. Calculate and plot the particle mass divided by the initial mass versus time during devolatilization. Use data from Table 14.1.

5. A coal char with an apparent activation energy of 17.15 kcal/gmol and a pre-exponential constant ($k_{p,0}$) of 60 g$_{O_2}$/(cm$^2 \cdot$ s $\cdot$ atm$_{O_2}$) is burned in a 1450 K and 1 atm gas stream containing 10% oxygen by volume. Assuming the particle temperature is 1450 K and the slip velocity between the particle and the gas is zero, calculate the char burn time for 10, 100 and 1000 µm particles. Consider the two limiting cases of constant density and constant diameter burning. Make a table of your results. Use a char specific gravity of 0.8.

6. Calculate the char burn times for the 612 and 300 µm Douglas-fir particles given in Table 14.4 assuming that diffusion is the rate limiting step.

Compare the calculated char burn times with the burnout time data given in Table 14.4 and comment on the results. Use a temperature of 1450 K, an oxygen partial pressure of 4.5%, and a char density of 0.20 g/cm³.

7. A 1 mm diameter particle of bituminous (A) coal char burns in a gas stream containing 5% oxygen. The slip velocity between the particle and the gas is 0.15 m/s. What is the char burn time assuming constant char density for gas stream temperatures of 1500 K and at 1700 K? Assume $\rho_{char} = 0.8$ g/cm³ and the particle temperature is the same as the gas temperature.

8. Derive Equation 14.35 for the char burnout time. State all assumptions made.

9. For a 100 µm bituminous (C) coal char particle with an oxygen partial pressure far from the particle of 0.1 atm, find the oxygen partial pressure at the char surface when the gas temperature is 2000 K. Assume that the particle temperature is 100 K greater than the gas temperature and that there is no relative velocity between the particle and the gas. Repeat for a gas temperature of 1500 K.

10. A 50 µm diameter bituminous (A) char particle burns in product gas at 1 atm and 1600 K. As the particle burns to form carbon monoxide, the heat release is 9200 kJ/g$_{char}$. The burning rate constant is 0.05 g$_{O_2}$/(cm²·s·atm$_{O_2}$). Calculate the steady state surface temperature of the particle for oxygen particle pressures of 0.06 atm, 0.13 atm, and 0.21 atm. Account for convection and radiation heat loss from the particle. Assume that the particle emissivity is 0.8, and use a Nusselt number of 2.

11. Consider two subbituminous (C) char particles in quiescent air at 1 atm and 2000 K. One is a 50 µm particle, and the other is a 200 µm particle. Compare the initial burning rates of the two particles.

12. For a lignite char particle in a 1500 K gas stream at 1 atm with a gas velocity of 1 m/s relative to the particle, what is the particle diameter for which the kinetic rate constant equals the diffusion reaction rate constant? Assume $k_c = 24.4$ cm/s and reaction order $n = 1$.

13. A wood char has a total surface area (external + internal) of 1000 m²/g. If this were in the form of a solid sheet with a density of 2 g/cm³, how thick would the sheet be?

REFERENCES

Bird, R. B., Stewart, W. E., and Lightfoot, E. N., *Transport Phenomena*, 2nd ed., Wiley, New York, 2007.

Bryden, K. M. and Ragland, K. W., "Combustion of a Single Wood Log under Furnace Conditions" in *Developments in Thermochemical Biomass Conversion*, vol. 2, eds. A. V. Bridgwater and D. G. B. Boocock, 1331–1345, Blackie Academic and Professional, London, 1997.

Bryden K. M., Ragland, K. W., and Rutland, C. J., "Modeling Thermally Thick Pyrolysis of Wood," *Biomass Bioenergy* 22(1):41–53, 2002.

Bryden, K. M. and Hagge, M. J., "Modeling the Combined Impact of Moisture and Char Shrinkage on the Pyrolysis of a Biomass Particle," *Fuel* 82(13):1633–1644, 2003.

Di Blasi, C., "Modeling and Simulation of Combustion Processes of Charring and Non-Charring Solid Fuels," *Prog. Energy Combust. Sci.* 19(1):71–104, 1993.

Essenhigh, R. H., "Fundamentals of Coal Combustion," in *Chemistry of Coal Utilization: Second Supplementary Volume*, ed. M.A. Elliott, 1153–1312, Wiley Interscience, New York, 1981.

Hagge, M. J. and Bryden, K. M., "Modeling the Impact of Shrinkage on the Pyrolysis of Dry Biomass," *Chem. Eng. Sci.* 57(14):2811–2823, 2002.

Howard, J. B., "Fundamentals of Coal Pyrolysis and Hydropyrolysis," in *Chemistry of Coal Utilization: Second Supplementary Volume*, ed. M. A. Elliott, 665–784, Wiley Interscience, New York, 1981.

Lyczkowski, R. W. and Chao, Y. T., "Comparison of Stefan Model with Two-Phase Model of Coal Drying," *Int. J. Heat Mass Transfer* 27(8):1157–1169, 1984.

Miller, B. and Tillman, D., *Combustion Engineering Issues for Solid Fuel Systems*, Academic Press, Burlington, MA, 2008.

Ragland, K. W. and Yang, J.-T., "Combustion of Millimeter Sized Coal Particles in Convective Flow," *Combust. Flame* 60(3):285–297, 1985.

Ranz W. E. and Marshall, W. R., Jr., "Evaporization from Droplets," *Chem. Eng. Prog.* 48:141–146 and 173–180, 1952.

Simmons, W. W. and Ragland, K. W., "Burning Rate of Millimeter Sized Wood Particles in a Furnace," *Combust. Sci. Technol.* 46(1–2):1–15, 1986.

Smith, I. W., "Combustion Rates of Coal Chars: A Review," *Symp. (Int.) Combust.* 19:1045–1065, The Combustion Institute, Pittsburgh, PA, 1982.

Stanish, M. A., Schajer, G. S., and Kayihan, F., "A Mathematical Model of Drying for Hygroscopic Porous Media," *AIChE J.* 32(8):1301–1311, 1986.

Van Loo, S. and Koppejan, J., eds., *The Handbook of Biomass Combustion and Co-firing*, Earthscan, London, 2008.

Williams, A., Pourkashanian, M., Jones, J. M., and Skorupska, N., *Combustion and Gasification of Coal*, Taylor & Francis, New York, 2000.

15 Fixed Bed Combustion

Fixed bed combustion systems range from simple cookstoves to large stokers that burn solid fuel on a grate for district heating and electric power generation. Efficient fuel utilization, good turndown and control of the heat output, and good combustion intensity (heat output per unit volume) are the desired goals. In addition to the combustion itself, fuel handling and feeding, ash fouling and slagging of internal surfaces, and gaseous and particulate emissions must be considered in any solid fuel combustor design. Fixed bed systems use relatively large pieces of fuel and require the least amount of processing to reduce the size of the fuel compared to pulverized fuel and fluidized bed combustion systems.

In this chapter, several types of fixed bed combustors for solid fuels are considered: biomass cookstoves, biomass space heating stoves, and large grate burning systems for coal and biomass.

15.1 BIOMASS COOKSTOVES

Cooking fires and cookstoves are some of the earliest technologies developed by humankind. The simplest way to boil a pot of water is to lay a small pile of burning sticks on the ground, surround them with three stones, and then rest the pot on the stones. To fry or grill meat, you need only hang the meat on a stick over a pile of burning sticks. The fuel for cooking is often locally available and includes wood, agricultural wastes, dung, and charcoal. Today the household chore of gathering wood and cooking on an open fire is repeated on a daily basis by one-third of the world's people. It is surprising, therefore, that we have not yet developed a low cost, sustainable, locally maintainable cooking system for developing countries. As a result, for one-third of the world's population, the simple act of cooking and heating a home promotes deforestation and increases the risk of disease and injury.

Today, biomass cookstoves are receiving increasing attention, in part due to their impact on the environment. Recent studies have found that black carbon or soot is the second largest cause of global warming, accounting for approximately 18% of global warming compared with 40% for carbon dioxide. Household use of biofuels has been estimated to account for approximately 18% of these black carbon emissions (Bond et al. 2004). In addition, when black carbon settles on the snow and ice of the Arctic, Antarctic, and other snow and glacier filled areas, it reduces the reflectivity of the snow and ice and increases melting in these areas. Recent studies have estimated that black carbon is responsible for 50% of Arctic warming from 1890 to 2007 (Shindell and Faluvegi 2009).

Cooking fires and cookstoves can be thought of as being of three types: three-stone open fires, early improved stove designs that improve fuel efficiency, and more recent stove designs that reduce emissions and indoor air pollution. The three-stone fire, shown in Figure 15.1, has many advantages. It is readily built from local materials, it can burn a wide range of solid fuels, it has good turn down, and it can quickly provide an intense hot flame for boiling water. Despite these advantages, a three-stone fire is inefficient and polluting. The primary cause of the inefficiency is not combustion efficiency, but rather heat transfer to the cooking pot. In a three-stone fire the heat loss to the surroundings by radiation and convection is significant. The combustion products flow unconstrained around the pot, limiting convective heat transfer; in addition, the energy lost to the surroundings from radiation is not recaptured. The emissions of soot and other pollutants from a three-stone fire are high due to poor mixing of fuel and air and incomplete combustion as the combusting gases are quickly quenched by the cool pot and the ambient air.

Cookstove heat transfer efficiency can be improved by enclosing and insulating the combustion chamber, limiting the size of the combustion chamber, and directing the flow of hot gases next to the pot surface by using a pot skirt or other device. In addition, limiting the flow of air improves stove performance by increasing the temperature of the combustion chamber. One example of this type of stove is the rocket stove developed by Dr. Larry Winiarski in the 1980s. As shown in Figure 15.2, the combustion chamber is small, generally 12–20 cm in diameter and 30–50 cm tall. The small chamber limits the amount of fuel being combusted at any given time, ensuring the stove is not overfueled relative to the cooking task. With an open fire there is a temptation to build too large a fire in an attempt to be able to cook more rapidly. But the effective size of the fire is limited by the size of the cooking

FIGURE 15.1   Traditional three-stone fire for cooking.

FIGURE 15.2 Simple rocket stove.

pot. In addition, the combustion chamber is well insulated and limits the heat loss from the cooking fire, and it directs the hot gases under the pot, thereby increasing the heat transfer efficiency. The fuel wood sits on a platform and underfire air is drawn by natural convection to provide complete combustion while limiting the excess air.

Similar principles are used in the design of many types of stoves. Figure 15.3 shows an enclosed cooking stove with a chimney, developed for use in Central America. As shown, the combustion chamber and feed are similar to a rocket stove. The combustion chamber is embedded in a locally available insulative material. The exterior is a reinforced sheet metal box and the top steel plate (sometimes cast iron) is used for cooking. The hot gases flow from the combustion chamber into a thin, flat chamber under the cooking surface, and then exit via the chimney. These

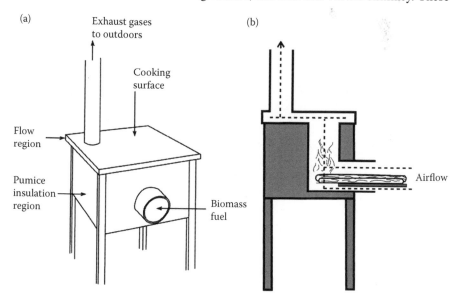

FIGURE 15.3 Enclosed natural draft cookstove using stick wood, (a) isometric view, (b) cross-sectional drawing showing the airflow path.

stoves are relatively efficient and reduce indoor air pollution. However, these stoves are natural draft stoves in which the air flow and the combustion rate are coupled together. Careful design can help limit air flow, but differences in how the stoves are operated make it difficult to limit air sufficiently. The key to lower emissions is to carefully control the air, to ensure good mixing of fuel and air, and to raise the combustion chamber temperatures to promote more complete burnout of combustion products. In almost all cases, better air control requires adding a fan to the stove. The fan may be powered by a battery, a thermoelectric generator, or a thermoacoustic generator. By uncoupling the air flow from the combustion rate, the air flow can be regulated to increase combustion intensity, thus providing better turndown and significant fuel savings. More importantly, a fan allows better mixing of the air and better control of the path the air takes. Increasing the swirl lengthens the path the air takes, thus providing more time for burnout. Better mixing ensures that combustion is not limited by the locally available oxygen.

Example 15.1

A typical cooking cycle in West African villages involves using a three-stone fire to heat 5 L of water from 20°C to 100°C and then maintaining a simmer for 1 h. Assume that the heat transfer efficiency of the three-stone fire is 5% and that the lowest power available with the three-stone fire is 1000 W, which is more than sufficient to simmer the water in the covered pot. The wood has 30% moisture and 0.8% ash on an as-received basis. The bulk density of stacked wood is 45% of the density of the wood. The dry, ash-free composition of the wood is 51.2%, 5.8%, 42.4%, 0.6%, and 0% on a mass basis for C, H, O, N, and S, respectively. The dry density of the wood is 640 kg/m³. The dry, ash-free HHV of the wood is 19.7 MJ/kg.

a. What is the as-received LHV of the wood (kJ/kg)?
b. What is the as-received density of the wood (kg/m³)?
c. What is the as-received bulk density of the wood (kg/m³)?
d. What is the bulk volume (L) and as-received weight (kg) of the wood needed to complete the cooking task?

Solution

Part (a)

First determine the ratio of the as-received mass of wood to the dry, ash-free mass of wood:

$$m_{\text{as-recd}} = m_{\text{daf}} + m_{\text{w}} + m_{\text{ash}}$$

$$\frac{m_{\text{daf}}}{m_{\text{as-recd}}} = \frac{m_{\text{as-recd}}\left(1 - 0.3 - 0.008\right)}{m_{\text{as-recd}}} = 0.692$$

The as-received HHV is

$$HHV_{\text{as-recd}} = HHV_{\text{daf}} \frac{m_{\text{daf}}}{m_{\text{as-recd}}} = \frac{19.7 \text{ MJ}}{\text{kg}_{\text{daf}}} \cdot \frac{0.692 \text{ kg}_{\text{daf}}}{\text{kg}_{\text{as-recd}}} = \frac{13.6 \text{ MJ}}{\text{kg}_{\text{as-recd}}}$$

The mass of water in the products for 100 kg of dry, ash-free wood is

$$m_{H_2O,daf} = \frac{5.8\ kg_H}{1} \cdot \frac{1\ kgmol_H}{1\ kg_H} \cdot \frac{1\ kgmol_{H_2O}}{2\ kgmol_H} \cdot \frac{18\ kg_{H_2O}}{1\ kgmol_{H_2O}}$$

$$m_{H_2O,daf} = 52.2\ kg_{H_2O}$$

The mass of water on an as-received basis is

$$m_{H_2O,as\text{-}recd} = 30\ kg_{H_2O} + 0.692\left(52.2\ kg_{H_2O}\right) = 66.1\ kg_{H_2O}$$

From Equation 2.1

$$LHV_{as\text{-}recd} = HHV_{as\text{-}recd} - \left(\frac{m_{H_2O}}{m_f}\right)h_{fg}$$

$$LHV_{as\text{-}recd} = \frac{13.6\ MJ}{kg_{as\text{-}recd}} - \left(\frac{66.1\ kg_{H_2O}}{100\ kg_{as\text{-}recd}}\right)\frac{2.44\ MJ}{kg_{H_2O}} = 12.0\ MJ/kg$$

Part (b)

In the same way as in Part (a)

$$\rho_{as\text{-}recd} = \frac{640\ kg_{dry}}{m^3} \cdot \frac{kg_{as\text{-}recd}}{0.7\ kg_{dry}} = \frac{914\ kg_{as\text{-}recd}}{m^3}$$

Part (c)

$$\rho_{bulk} = 0.45\ \rho_{as\text{-}recd} = 411\ kg/m^3$$

Part (d)

The amount of energy required is the amount needed to heat the water + the amount needed to simmer the water.

$$Q_{heat} = mc_p\ T = \frac{5\ L}{1} \cdot \frac{1\ kg}{1\ L} \cdot \frac{4.186\ kJ}{kg \cdot K} \cdot \frac{80\ K}{1} = 1.7\ MJ$$

$$Q_{simmer} = \frac{1\ kW(1\ h)}{1} \cdot \frac{1\ kJ}{kW \cdot s} \cdot \frac{3600\ s}{hr} = 3.6\ MJ$$

$$Q_{total} = 5.3\ MJ$$

Considering the efficiency of the stove,

$$Q_{wood} = \frac{Q_{total}}{\eta} = \frac{5.3\ MJ}{0.05} = 106\ MJ$$

The as-received weight of wood is

$$m_{as\text{-}recd} = \frac{106 \text{ MJ}}{1} \cdot \frac{kg_{as\text{-}recd}}{12.0 \text{ MJ}} = 8.8 \text{ kg}$$

The bulk volume of wood needed is

$$V = 8.8 \text{ kg} \cdot \frac{m^3}{411 \text{ kg}} \cdot \frac{1000 \text{ L}}{m^3} = 21 \text{ L}$$

The lower heating value is used because the water vapor in the combustion is not condensed by an open fire or a cookstove.

15.2   SPACE HEATING STOVES USING LOGS

Residential wood stoves are common in many places in the world. In the United States, it is estimated that more than 12 million woodstoves for heating are in use; approximately 75% of these are older, inefficient stoves. The number of stoves nationwide continues to grow as homeowners switch away from more expensive non-renewable fuels to locally available biomass fuels.

The majority of residential woodstoves rely on natural circulation of air; there is no fan. Many space heating stoves burn logs and are hand fed. Some stoves burn pellets of wood or agricultural waste materials that are automatically fed into the stove. Although there are many designs currently available, a generic wood stove design that has several important features will be considered here. A typical residential stove for logs is shown in Figure 15.4. The logs rest on pipes that preheat the air and direct the air upwards from beneath the logs toward the secondary chamber where the volatiles are more completely burned. An internal baffle helps to regulate the flow through the stove and creates turbulence mixing and more complete

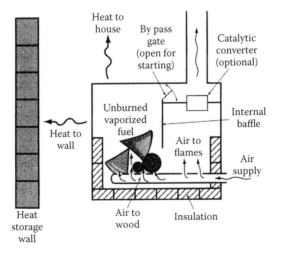

FIGURE 15.4   Schematic diagram of a residential space heating wood stove.

combustion. A bypass gate is used for startup. A catalytic converter, consisting of a ceramic honeycomb impregnated with a thin layer of noble metal is sometimes used to further reduce organic emissions. The lower part of the firebox is insulated to increase the flame temperature, which increases the combustion efficiency.

A large mass of noncombustible material such as stone, brick, or a water tank immediately adjacent to the stove may be used to store heat. In general, the smallest stove that will provide enough heat should be used. A large stove that is burning fuel at the same rate as a small stove will have more heat loss, a lower combustion chamber temperature, and will produce more condensed organic compounds, such as creosote.

The best fuel is hardwood that has been air-dried for a year to reduce the as-received moisture to 25% or less. Air-dried softwoods are slightly harder to burn cleanly due to their higher resin content. Logs that are 10–15 cm in diameter or larger are best. The larger pieces limit the devolatilization rate so that the combustibles can be completely burned by the available air supply within the stove. Wood that has too much moisture, such as freshly cut green wood (50% moisture, as-received basis), will burn with too cool a flame, resulting in unburned organic matter and lower heat output. Wood that is too dry (kiln or oven dried) and wood that is too small in size will burn too fast, resulting in insufficient air and increased organic emissions and soot. Coal, treated wood, and refuse are not recommended in this type of system.

Design improvements for wood stoves include preheated air, a somewhat hemispherical-shaped combustion chamber, a catalytic flame shield, and an automatic thermostatically controlled damper. With these improvements it is possible to average less than 3 g/h of particulate emissions. For example, the state of Oregon has set a standard of 4 g/h for residential catalytic wood stoves and 9 g/h for non-catalytic stoves. This is a 75% reduction in emissions over conventional stoves. These levels can be reached with good design and good operator practice.

The rate of wood burning is related to the air supply. Adding new wood requires opening the dampers. As the wood devolatilizes and mostly char remains, the air dampers may be reduced. For overnight operation when a large charge of wood is desired, charred wood should be accumulated within the stove over a period of several hours and then the damper turned down.

15.3 GRATE BURNING SYSTEMS FOR HEAT AND POWER

Crushed coal and biomass of various types are burned on grates to heat steam for buildings and district heating. When higher pressure steam is used, both heat and electrical power are generated. For a top fuel size of about 50 mm, two types of grates are used to support the fuel: traveling grates and vibrating grates. Both types of grates are either air-cooled or water-cooled, and primary air flows upward through small holes in the grate. When the under-grate air is preheated to achieve a higher heat release rate, a water-cooled grate is required. The feeder for traveling and vibrating grate combustion is either a mass feed type where the fuel falls by gravity onto one end of the grate or a spreader stoker type of feeder where the fuel is flung across the grate. Both types of feeders provide a continuous flow of fuel onto the grate. Grate burning systems are especially effective with biomass such as wood chips and pellets derived from agricultural wastes.

The fuel bed is relatively shallow to ensure that the fuel particles burn out before the ash is dumped. The heat release rate per unit grate area is typically about 1.6 MW/m² for water-cooled mass feed grate systems. For spreader stokers, the heat release rate is about 2.4 MW/m² for coal and 3.5 MW/m² for wood chips or agricultural pellets. Grate systems with steam capacities up to 100 kg/s using wood and 50 kg/s using coal are in operation. Let's consider traveling grate and vibrating grate systems in greater detail.

15.3.1 Traveling Grate Spreader Stokers

A spreader stoker with a traveling grate is shown in Figure 15.5. Fuel is fed by gravity from a hopper to the stoker. The stoker flings the fuel toward the far end of the grate. The larger fuel particles (25–50 mm) tend to go toward the rear, while the mid-size particles drop out in the middle section of the grate. The fine particles (less than 1 mm) are carried upward by the upward motion of the gases and burn in suspension rather than on the grate. For coal-fired systems, the coal falls onto a rotor with paddles that flings the fuel particles across the grate. For biomass fuels typically an air-swept spout projects the fuel particles across the grate.

The endless grate moves forward slowly at speeds up to 12 m/h. The grate speed is adjusted so that the fuel burns out before it reaches the edge of the grate, and the ash

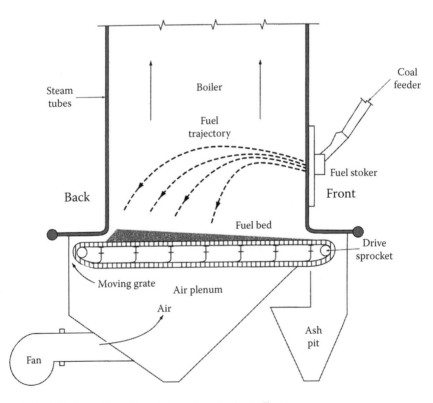

FIGURE 15.5   Spreader stoker with an air-cooled traveling grate.

dumps into the ash pit. The fuel bed is maintained to a depth of 10–20 cm. When the fuel feed rate is adjusted to change the heat load, the airflow rate and grate speed is correspondingly adjusted to control the desired fuel burnout and thickness of the fuel bed. When the fuel size or fuel properties such as moisture or type change, the feed rate and grate speed are adjusted accordingly. The steel grate has holes approximately 6 mm in diameter for the underfire (primary) air. Overfire (secondary) air is supplied through nozzles in the sidewalls to complete the combustion above the fuel bed. For biomass fuels, the ratio of primary to secondary air is typically 40/60, while with coal more primary air and less secondary air is used because coal has fewer volatiles than biomass.

Referring again to Figure 15.5, air flows up through the porous fuel bed as the bed moves horizontally and as fresh fuel falls on the top of the bed. Fine fuel particles and volatile gases burn above the bed. The top of the fuel bed is exposed to radiation from the boiler walls and the combustion above the bed. Heating of the fuel particles in the bed drives out the moisture and volatiles, which consist primarily of hydrocarbon vapors and tars, carbon monoxide, carbon dioxide, and hydrogen. Ignition occurs at the top of the fuel bed and propagates downward in the bed against the air flow. As the reaction front passes, the fuel is heated to a higher temperature, increasing devolatilization and forming char on the surface of the bed particles. The volatiles and char burn in the bed if enough oxygen is available. Depending on the bed depth and air flow rate, at some point in the bed the oxygen is consumed. As the grate moves along, the char layer subsides and burns out, and the ash is dumped into the ash pit.

The primary air flow is set high enough to provide rapid drying, devolatilization, and char burn, but not so high as to create excessively high temperatures in the bed. Ash is mostly retained on the grate, thereby insulating the grate from the highest temperatures. The temperature in the bed is highest in the char oxidation zone. If the ash temperature exceeds the ash softening temperature (about 1300°C for coal and 1200°C for wood ash), clinkers (agglomerated ash) will form. Overfire air jets above the bed are used to complete combustion of the volatiles and small carbon particles. From 40% to 60% of the heat release occurs above the bed. Overfire air jets penetrate into the fireball and promote good mixing for complete combustion.

15.3.2 VIBRATING GRATE SPREADER STOKERS

The vibrating grate spreader stoker, shown in Figure 15.6, is similar to a traveling grate spreader stoker system except that the grate vibrates intermittently instead of moving continuously in a horizontal direction and is sloped at about 6° so that fuel slides along the grate as it burns, and the ash falls into the ash hopper. The vibration cycle is typically 6 cycles/s for 2 s and then off for 2 min (i.e., the grate vibrates less than 2% of the time), and is of low amplitude. The grate is made of metal panels with drilled holes. For high temperature under-grate air (300°C), the grate is water cooled, and this water is included in the boiler water cycle. The advantage of the vibrating grate is that it has fewer moving parts than the traveling grate and hence requires less maintenance. Also the vibrations tend to make

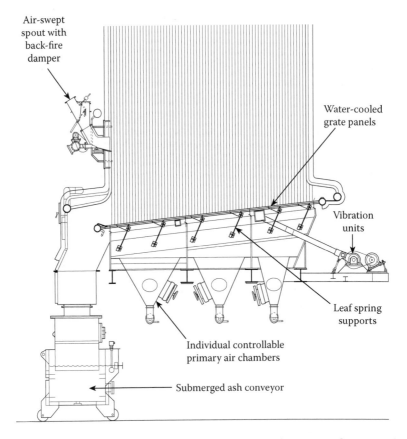

Air-swept
spout with
back-fire
damper

Water-cooled
grate panels

Vibration
units

Leaf spring
supports

Individual controllable
primary air chambers

Submerged ash conveyor

FIGURE 15.6 Spreader stoker with a water-cooled vibrating grate. (Courtesy of The Babcock and Wilcox Company, Barberton, OH.)

the fuel bed more uniform, and thus the air flow is more uniform throughout the fuel bed.

Due to the slope and vibration of the grate, the fuel moves slowly from the back to the front in about 30–60 min. As a generalization it is helpful to visualize the back part of the fuel bed as undergoing drying and devolatilization, and the front part of the fuel bed (near the ash pit) as undergoing char burn. The primary air is zoned to have different flow rates under different parts of the grate.

A full boiler profile including an air-swept spout feeder and vibrating grate is shown in Figure 15.7. Auxiliary wall-fired gas burners are shown. The overfire air (secondary air) jets are located in the furnace arches and the front and back walls. The furnace arches are refractory-lined surfaces that channel the flow and reflect radiation back to the fuel bed. The primary air is preheated and the steam is superheated. The residence time of combustion products and small particles in the furnace zone between the grate and the superheater is typically about 3 s. The boiler for a traveling grate system is similar, and an example of the design parameters for an older boiler is given in Table 15.1.

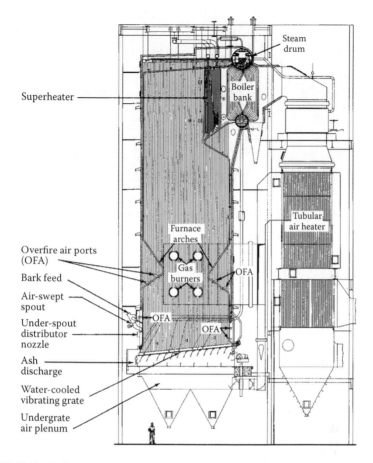

FIGURE 15.7   Boiler with vibrating grate designed for biomass and auxiliary gas burners. (Courtesy of The Babcock and Wilcox Company, Barberton, OH.)

15.4   COMBUSTION EFFICIENCY AND BOILER EFFICIENCY

Combustion efficiency refers to the completeness of combustion. Thus, the combustion efficiency is

$$\eta_{chem} = 1 - \frac{q_{chem,loss}}{\dot{m}_f HHV} \tag{15.1}$$

where $q_{chem,loss}$ is the chemical energy available in the fuel that is lost due to incomplete combustion. In practice, carbon monoxide and unburned hydrocarbon emissions typically represent a negligible heat loss. However, carbon in the bottom ash and fly ash typically represents about a 1% loss. Burnout of carbon fines can be improved by careful design of the overfire air jets. Increasing overall excess air does not necessarily improve carbon burnout and may make it worse if the temperature

TABLE 15.1
Example of Industrial Traveling Grate, Spreader Stoker Boiler

Boiler

Design steaming capacity	56,700 kg/h
Design pressure	2.9 MPa
Final steam temperature	316°C
Boiler heating surface	1288 m²
Waterwall heating surface	240 m²
Economizer heating surface	753 m²
Furnace volume	231 m³

Stoker and Grate

Number of feeders	6
Grate type	Continuous front discharge
Grate length (shaft to shaft)	5.7 m
Grate width	4.1 m
Effective grate area	23 m²
Recommended coal sizing	1.7–32 mm

Overfire Air

Upper rear wall	16 jets, 2.54 cm in diam. located 3.0 m above grate, 30° below horizontal
Lower rear wall	16 jets, 2.54 cm in diam. located 46 cm above grate
Upper front wall	16 jets, 2.54 cm in diam. 3.0 m above grate, 35° below horizontal
Lower front wall	20 jets, 2.2 cm in diam. 30 cm above grate

is too low. Fly ash reinjection into the furnace from cyclone collectors located after the superheater and boiler bank generally improves the carbon burnout but at the expense of increased fine ash emissions.

Boiler efficiency is defined (see also Chapter 6) as the useful heat output divided by the energy input. Referring to Figure 15.8,

$$\eta_b = \frac{q}{\dot{m}_f HHV + \dot{W}_{aux}} \qquad (15.2)$$

where $\dot{W}_{aux}$ is the auxiliary power to run the fans, pumps, and feeder. If the fuel flow rate is known, then the furnace efficiency can be determined directly from Equation 15.2. Frequently the fuel flow rate is not measured, and for this situation the indirect method given by Equation 6.8 may be used. The indirect method is also useful for investigating how to improve the efficiency. In Chapter 6, we noted that the boiler efficiency is increased by reducing the excess air, reducing the extraneous heat loss, and reducing the exhaust (stack) temperature. The excess air should be reduced to the point where the CO, combustibles in the fly ash, and stack opacity just begin to increase. Radiation heat loss from the walls of the boiler is generally less than 1% for large boilers.

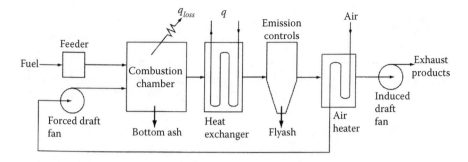

FIGURE 15.8 Diagram of a solid fuel-fired boiler system.

15.5 EMISSIONS FROM GRATE BURNING SYSTEMS

Emissions of particulates, sulfur dioxide, nitrogen oxides, and mercury are of primary concern when burning coal. With biomass, sulfur dioxide and mercury are significantly reduced. Particulate emissions are due to the minerals and unburned carbon in the fly ash. With biomass, the ash in the fuel comes primarily from the dirt and other materials that are entrained with biomass during harvesting, storage, and transport. Particulate emissions depend on furnace design and operating conditions such as load (or grate heat release rate), air preheat, excess air, design and amount of overfire air, and fuel particle size. In spreader stokers, up to 80% of the ash can be retained as bottom ash, as opposed to suspension burning and fluidized beds, which have little bottom ash because the ash is blown out of the combustor. Generally, higher heat release rates per unit area (higher load) result in higher particulate emissions. Particulate emissions are controlled by a multiple cyclone collector (typically located after the economizer) and a fabric filter baghouse or electrostatic precipitator (located after the air preheater).

Emission factors for spreader stokers, derived from field measurements of many spreader stoker units by the U.S. Environmental Protection Agency, are given in Table 15.2. Filterable particulates are particulates collected on a glass fiber filter. Vapors and particles less than 0.3 microns pass through the filter. Fly ash reinjection refers to the process where ash is collected by cyclone collectors downstream of the boiler bank (see Figure 15.7) and is returned to the combustion zone above the bed. In this way the carbon in the fly ash is burned up, thus improving boiler efficiency.

Sulfur in coal is emitted primarily as sulfur dioxide, although approximately 5% of the sulfur combines with the ash to form sulfate salts or sulfur trioxide and gaseous sulfates. Enhancement of sulfur dioxide capture by adding pulverized limestone with the coal has not proved to be an efficient method. Injection of pulverized limestone in the ductwork downstream of the superheater tubes and air heater has resulted in up to 50% conversion of sulfur dioxide to calcium sulfate, which is then removed by particulate control equipment. Flue gas scrubbers with dissolved limestone are used when 90% sulfur dioxide removal is required.

TABLE 15.2

Emission Factors for Uncontrolled Emissions from Spreader Stokers Firing Bituminous Coal and Wood

Pollutant	Coal (g/kg$_{coal}$)	Wood (kg/10^{6}kJ)
Particulates		
without multi-cyclones	33	0.17
with multi-cyclones, without reinjection	8.5	
with multi-cyclones, with reinjection	6	0.13
Nitrogen oxides (as NO$_2$)	5.5	0.21
Sulfur dioxide	19 S	0.011
Carbon monoxide	2.5	0.26

Source: U.S. Environmental Protection Agency, *Compilation of Air Pollutant Emission Factors, Vol. 1: Stationary Point and Area Sources,* 5th ed., EPA-AP-42, 1995, updated 2003.
S is weight percentage of sulfur in the coal (5% sulfur means S = 5).

Nitrogen oxide (NO$_X$) emissions consist of nitric oxide formed from fuel bound nitrogen and nitrogen in the inlet air. Less than 1% of the NO$_X$ emissions are in the form of nitrogen dioxide and nitrate compounds. Note that although the emissions are mostly NO, for regulatory purposes they are reported as NO$_2$ by multiplying the NO emissions by the ratio of the molecular weights (46/30). This is because NO is transformed to NO$_2$ in the atmosphere and NO$_2$ has health and environmental effects, whereas NO does not.

Nitrogen oxide emissions are reduced by reducing excess air up to the point where the carbon monoxide begins to increase. Flue gas recirculation has proved to be a successful way to reduce NO emissions in stoker boilers by reducing the excess oxygen and peak temperature. A fraction of the exhaust gas is ducted back into the underfire air plenum and into the overfire air nozzles. For example, when a certain flow rate of air yielding 10% excess oxygen in the flue gas is replaced by the same amount of flow consisting of 65% air and 35% flue gas, the excess oxygen level is reduced to 4%. This reduction in excess air reduces NO formation and increases the system efficiency. Further reduction of NO emissions requires scrubbing the flue gas with an ammonia solution.

Emission standards are set at the federal level for large sources (greater than 250×10^6 kJ/h) and by state governments for smaller sources. The emission standards, shown in Table 15.3, apply to all types of solid fuel-fired furnaces and boilers, not just stokers. In addition to the above standards, large new furnaces and boilers must meet an opacity standard of 20% based on 6 min averages. Exceedances of 27% opacity for 6 min/h are allowed for soot blowing to clean the boiler tubes. Opacity refers to the light transmissivity of the stack gases and is a measure of visibility degradation. The opacity standard may require more stringent control of particulates than the particulate standard, especially for fine particles.

TABLE 15.3
Federal New Source Emissions Standards for
Large Coal-Fired Boilers[a]

Pollutant	Emission Standard
Particulates[b]	13 g/10^6 kJ
Sulfur dioxide	520 g/10^6 kJ
Nitrogen oxides as NO_2	260 g/10^6 kJ

Source: U.S. Code of Federal Regulations 40, Part 60, Subpart D,
Nov. 15, 2010.

[a] For facilities with greater than 73 MW heat input.

[b] Also requires less than 20% stack opacity.

Example 15.2

A large spreader stoker with multi-cyclone collectors and fly ash reinjection burns 12 T/h of bituminous coal with 9% ash and a heating value of 27 MJ/kg. Assuming that the typical emission factor applies, how much particulate emission control is required?

Solution

The heat input rate is

$$\frac{12 \text{ T}}{\text{h}} \cdot \frac{1000 \text{ kg}}{\text{T}} \cdot \frac{27,000 \text{ kJ}}{\text{kg}} = 324 \times 10^6 \text{ kJ/h} = 90 \text{ MW}$$

Using Table 15.2, the typical particulate emission rate is

$$\frac{8.5 \text{ g}}{\text{kg}} \cdot \frac{12 \text{ T}}{\text{h}} \cdot \frac{1000 \text{ kg}}{\text{T}} = \frac{102,000 \text{ g}}{\text{h}}$$

or

$$\frac{102,000 \text{ g}}{\text{h}} \cdot \frac{\text{h}}{324 \times 10^6 \text{ kJ}} = \frac{315 \text{ g}}{10^6 \text{ kJ}}$$

The percent control needed to meet the particulate standard of 13 g/10^6 kJ from Table 15.3 is

$$\eta_{control} = \left(\frac{315 - 13}{315} \right) 100 = 95.9\%$$

This degree of control will require a fabric filter baghouse or an electrostatic precipitator.

15.6 MODELING COMBUSTION OF SOLID FUELS ON A GRATE

A thin solid fuel bed on a traveling or vibrating grate consists of fuel particles undergoing drying, devolatilization, and char combustion. As the bed is transported laterally, the drying and devolatilization of the surface of the fuel particles propagates from the top of the bed downward towards the grate and also gradually penetrates from the surface to the center of the particles. The char on the particle surface thickens and burns out before the ash is dumped. Note that the fuel particles on a grate are too large to be treated as thermally thin in the sense of Chapter 14. Above the fuel bed the gases consist of CO, CO_2, CH_4, H_2, other hydrocarbons and lesser species, fine carbon particles, and secondary air. Modeling of the reactions and flows above the bed is done with computational fluid dynamics, which is beyond the scope of this text. As a start, let us consider the modeling of a fixed bed of char particles.

15.6.1 MODELING FIXED-BED CHAR COMBUSTION

Consider a fixed bed of stoker size char particles burning on a grate (Figure 15.9). Air flows upward through the grate and the bed of particles. The char flows slowly downward at a velocity of $\underline{V}_s$ as the char burns out. To simplify the situation let us only consider a bed of char so that we can neglect pyrolysis and drying. A one-dimensional model of the fuel bed is used to represent a vertical section out of a large fuel bed.

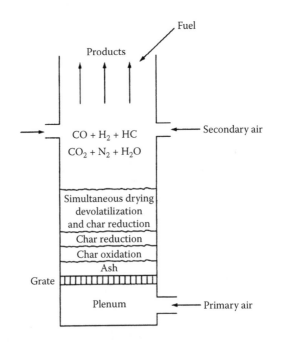

FIGURE 15.9   Diagram showing conceptual layers in combustion of a solid fuel bed.

Air flows through the grate and reacts with the surface of the char to form carbon monoxide:

$$C + \frac{1}{2}O_2 \rightarrow CO$$

Simultaneously the gas phase oxidation of carbon monoxide occurs:

$$CO + \frac{1}{2}O_2 \rightarrow CO_2$$

As the oxygen is consumed in the bed, the carbon dioxide is reduced to carbon monoxide at the surface of the char:

$$C + CO_2 \rightarrow 2\,CO$$

As the velocity of the air through the bed is increased, the burning rate of the char increases, the amount of CO_2 formed is increased, and the position of the reducing zone moves upward. Of course, the air flow must be kept below the point where ash or char particles are blown off the bed. This relationship between O_2, CO, and CO_2 is illustrated in Figure 15.10. Note that the char particles continue to shrink as they react with the CO_2 after the oxygen is consumed.

To obtain the temperature profile and burning rate of the char layer, equations for conservation of mass and energy for the solid and gas phase and conservation of species are needed. Preheated air flows through the grate. The product gas in the bed consists of nitrogen, oxygen, carbon monoxide, and carbon dioxide. The rate constants and heat of reaction for the reactions are given

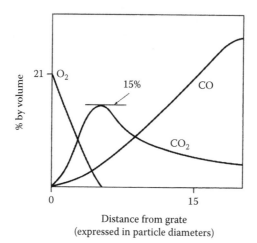

FIGURE 15.10 Typical gas analysis in the char layer.

TABLE 15.4

Typical Rate Constants and Heat of Reaction for Char

Reaction	$k_{co}(g/cm^2 \cdot s \cdot atm)$	E (kcal/gmol)	H (MJ/kg)
(1) $C + 1/2\ O_2 \rightarrow CO$	(see Section 14.3)		9.2 MJ/kg$_C$
(2) $CO + 1/2\ O_2 \rightarrow CO_2$	(see Equation 4.13)	40	10.1 MJ/kg$_{CO}$
(3) $C + CO_2 \rightarrow 2\ CO$	2×10^5	50	−14.4 MJ/kg$_C$
(4) $C + H_2O \rightarrow H_2 + CO$	4×10^5	50	−10.9 MJ/kg$_C$

in Table 15.4. (Water vapor is not present in this instance but is included in the table for completeness.)

Consider a slice Δz thick through the bed of cross-sectional area, A. From conservation of mass for the char, the net rate at which the char particles travel towards the grate equals the rate at which the char is consumed within the slice.

$$\rho_{char}\underline{V}_s A_{char}\Big|_{z+\Delta z} - \rho_{char}\underline{V}_s A_{char}\Big|_z = r_{char}\left(A\Delta z\right) \tag{15.3}$$

where r_{char} is the combustion rate of char per unit volume of the bed, $\underline{V}_s$ is the solid velocity of the char traveling towards the grate, and A_{char} is the cross-sectional area occupied by char. That is,

$$A_{char} = A(1-\varepsilon) \tag{15.4}$$

where ε is the void fraction of the bed (the area not occupied by solid material). Substituting Equation 15.4 into Equation 15.3 yields

$$(1-\varepsilon)\rho_{char}\underline{V}_s A\Big|_{z+\Delta z} - (1-\varepsilon)\rho_{char}\underline{V}_s A\Big|_z = r_{char}\left(A\Delta z\right) \tag{15.5}$$

Assuming that A, ρ, ε, remain constant and dividing by $A\Delta z$ yields

$$\rho_{char}(1-\varepsilon)\frac{d\underline{V}_s}{dz} = r_{char} \tag{15.6}$$

where r_{char} is the combustion rate of char per unit volume of char. From Table 15.4, the two reactions with char in this case are Reaction 1 (char + oxygen) and Reaction 3 (char + carbon dioxide)

$$\rho_{\text{char}}\left(1-\varepsilon\right)\frac{dV_s}{dz} = r_1 + r_3 \qquad (15.7)$$

Introducing the particle area per unit volume, A_v,

$$r_1 = \frac{12}{16} k_{e1} \rho_{O_2} A_v \qquad (15.8)$$

and

$$r_3 = \frac{12}{44} k_{e3} \rho_{CO_2} A_v \qquad (15.9)$$

Recall that the effective rate constants (the k_e's) are a function of the particle surface temperature and include the effects of diffusion (see Section 14.3).

The temperature profile in the bed is obtained from the conservation of energy equation. In order to keep the analysis relatively simple, assume that the gas and particle temperatures are equal, and that energy transport by conduction, radiation, and diffusion is small compared to convective energy transport. Also assume the sensible energy flux of the solids is small compared to that of the gas. The energy equation is

$$\frac{d}{dz}\left(\rho_g \underline{V}_g h_g\right) = r_1 H_1 + r_2 H_2 + r_3 H_3 \qquad (15.10)$$

where $\underline{V}_g$ is the superficial gas velocity (i.e., the velocity that the gas would have if it were flowing through the same cross-sectional area without the char). The temperature is needed to determine the reaction rates and can be determined from h_g. The heat of reaction and reaction rate constants are given in Table 15.4. Note that H_3 is an endothermic reaction and hence has a negative heat of reaction. (To allow for different temperatures between the gas and the solid, an energy equation for both phases must be written and convective heat transfer between the phases would be included using a Nusselt number appropriate to fixed beds. The effect of heat conduction and radiation should be included in a full analysis.)

Species continuity equations for CO_2, O_2, CO, and N_2 with the reaction constants shown in Table 15.4 are used to obtain the density of each of the gas species,

$$\frac{d}{dz}\left(\rho_{CO_2} \underline{V}_g\right) = r_2 \frac{M_{CO_2}}{M_{CO}} - r_3 \frac{M_{CO_2}}{M_C} \qquad (15.11)$$

$$\frac{d}{dz}\left(\rho_{O_2} \underline{V}_g\right) = -r_1 \frac{M_{O_2}}{M_C} - 0.5 r_2 \frac{M_{O_2}}{M_{CO}} \qquad (15.12)$$

$$\frac{d}{dz}\left(\rho_{CO}\underline{V}_g\right) = \left(r_1 + 2r_3\right)\frac{M_{CO}}{M_C} - r_2 \tag{15.13}$$

$$\frac{d}{dz}\left(\rho_{N_2}\underline{V}_g\right) = 0 \tag{15.14}$$

or

$$\left[\rho_{N_2}\underline{V}_g\right]_{z=0} = \rho_{N_2}\underline{V}_g$$

In addition, the gas density at any point is the sum of the partial gas densities,

$$\rho_g = \rho_{O_2} + \rho_{CO} + \rho_{CO_2} + \rho_{N_2} \tag{15.15}$$

and the equation of state,

$$\rho_g = \frac{p}{RT} \tag{15.16}$$

There are 6 primary variables ($\underline{V}_s$, h_g, ρ_{O_2}, ρ_{CO_2}, ρ_{CO}, ρ_{N_2}) that can be found by solving the first-order, ordinary differential conservation equations (Equation 15.7 and Equations 15.10 through 15.14) simultaneously using a stiff ordinary differential equation solver. As noted earlier, the temperature is found from the enthalpy of the gas. The system pressure must be specified and is used in Equation 15.16 to find the gas density. The surface area per unit volume, A_v, is related to the particle diameter for an equivalent sphere by

$$A_v = \frac{6d^2\left(1-\varepsilon\right)}{d^3} = \frac{6\left(1-\varepsilon\right)}{d} \tag{15.17}$$

The boundary conditions need to be specified at the grate ($z = 0$). These include temperature, densities of each gas species, gas velocity, solid velocity, and the particle diameter. The solid velocity is zero at the grate. The particle diameter goes to zero at the grate, but to keep the solution stable, a small but finite diameter must be selected in order to have a finite value for A_v. The char size at the top of the bed is also specified. The solution procedure is to start at the grate and numerically integrate Equation 15.7, and Equations 15.10 through 15.14 until the initial particle diameter is reached indicating the top of the bed.

A solution set for a 10 cm deep bed with 1 cm diameter char particles at the top of the bed is shown in Figure 15.11. The bed height is 10 cm, and the constant density, shrinking sphere model is used. The temperature versus distance from the grate rises

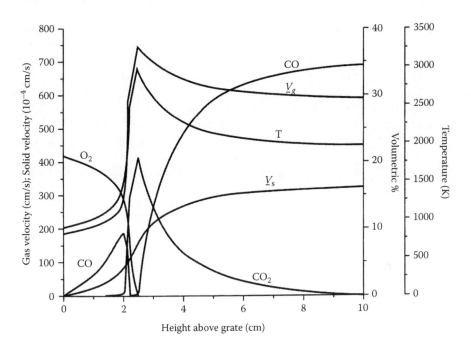

FIGURE 15.11 Fixed bed combustion of char on a grate: solution of Equations 5.7 and 15.10 through 15.17 with d_{init} = 1.0 cm, d_{final} = 0.1 cm, ρ_{char} = 0.8 g/cm³, ε = 0.4, and $\underline{V}_g$ = 2 m/s at 800 K.

slowly at first, peaks rapidly and then decreases at the point where the oxygen is consumed. The CO increases at first, then it is converted to CO_2 as the temperature increases, and then it rises again when the oxygen is depleted. The solid velocity is 0.32 mm/s at the top of the bed and then decreases as the char travels downward. The heat release rate can be determined from the solid velocity at the top of the bed and is 2.5 MW/m² when the overfire air is added. Note that the char-CO_2 reducing reaction adds to the solid velocity and thus the overall heat release rate, but to a lesser extent than the char-O_2 oxidation reaction. Increasing the airflow rate moves the peak temperature away from the grate and increases the heat release rate. If heat conduction and radiation in the solid phase were included, the temperature profile would be more rounded. Additionally, when the inlet air is preheated below the char ignition temperature, it can be shown that increasing the airflow rate too much will blow out the combustion.

15.6.2 Modeling Fixed-Bed Combustion of Biomass

Modeling the combustion of a bed of stoker size coal, woodchips, pellets, or logs is similar to modeling a bed of char, but at least seven gas species must be included in the gas phase of the bed—oxygen, carbon monoxide, carbon dioxide, water vapor, hydrogen, hydrocarbons, and nitrogen—because the particles are devolatilizing and drying as the char reacts on the particle surface. In this case there are

- Seven conservation of species equations
- Conservation of mass for the solid phase
- Conservation of mass overall
- Conservation of energy for the solid phase
- Conservation of energy for the gas phase

for a total of 11 first order, differential equations to be solved simultaneously. Source terms include the rate of drying and devolatilization as a function of solid temperature.

In contrast to suspension burning, particles large enough to be combusted in a packed bed cannot be assumed to be thermally thin. Rather, there are large temperature gradients within the particles, and drying, devolatilization, and char burn occur simultaneously within the particles. Because of this, an intraparticle submodel is needed to model drying and pyrolysis within the particle. These models can be extremely detailed and model the changes within the particle on a differential, time-dependent basis. For examples of these types of detailed models, see Bryden (1998) or Di Blasi (1996).

In many cases, a simpler intraparticle submodel can be used. For example, for logs under typical combustion conditions it has been observed that the rate of pyrolysis can be linked to the rate of char combustion (Bryden and Ragland 1996).

$$r_{pyr} = \delta_{pyr} r_{char} \tag{15.18}$$

where δ_{pyr} is the ratio of wood pyrolyzed and dried to the char consumed, r_{pyr} is the rate of pyrolysis, and r_{char} is the rate of char combustion. In this model the pyrolysis and drying are assumed to occur together in an infinitely thin zone. This is a reasonable assumption. The pyrolysis zone is 1–3 mm thick in logs that are 15–20 cm in diameter. The drying zone occurs adjacent to the pyrolysis zone and is generally very thin. For wood, the ratio of wood pyrolyzed and dried to char consumed is initially high at the top of the bed as the wood is rapidly pyrolyzed and dried and decreases as the diameter shrinks. Experimentally, for logs 12–21 cm in diameter burning at temperatures from 950°C to 1340°C and a variety of oxygen, carbon dioxide, and water vapor concentrations, the thickest char layer observed was 2 cm (Bryden and Ragland 1997). Based on this, when the char thickness reaches 2 cm, it is assumed that the fuel has been completely pyrolyzed. From this and the observation that the wood beneath the pyrolysis-drying layer is essentially undisturbed, the ratio of the wood pyrolyzed and dried to the rate of char consumed can adequately be modeled using

$$\delta_p = C\left(\frac{\Delta_c}{2} - 2\right)^{1/2} \tag{15.19}$$

where Δ_c is the radius of the unpyrolized core and C is a constant chosen such that the char thickness is zero at the top of the bed. It should be noted that this relationship is only approximate and was developed for large 12–20 cm diameter logs. Additionally, this relationship works well because the calculated energy release rate, mass consumption rate, and bed depth are not significantly affected by the form of the relationship. If detailed spatial information about species and combustion within the bed is required, a more detailed two-dimensional or three-dimensional model will be required as well as more detailed submodels.

Consider a bed of logs burning on a fixed grate. This type of system is used for many wood-fired residential furnaces and boilers. A deep bed of logs or whole tree segments on a fixed grate has been proposed as a replacement for wood chips or pellets on a traveling or vibrating grate where large amounts of steam are required (Ostlie 1993). For this large-scale deep bed system, the primary air is maintained high enough to blow the ash upwards off the grate, and logs are fed from above the bed by a ram feeder. The advantage of using logs rather than wood chips or biomass pellets is that fuel preparation is less costly and the heat release per unit plan area of the combustor is greater.

An example of the results of modeling the combustion of a deep bed of logs is shown in Figure 15.12, using the devolatilization products given in Table 15.5 and the submodels described in Equations 15.18 and 15.19. In this example, the initial log diameter is 20 cm, moisture (as-received basis) is 23%, bed depth 3.7 m, bed void fraction 0.65, and inlet air temperature 400°C. The predicted gaseous species profiles in the fuel bed with an inlet air velocity of 3.2 m/s are shown. The oxygen is consumed in the first 35% (1.4 m) of the bed, and the upper 65% is a reducing region in which the char–carbon dioxide and char–water vapor reactions dominate. The hydrocarbons, carbon monoxide, and hydrogen build up in the reducing region and burn out in the overfire air region. The fuel is devolatilizing in the upper 98% of the bed, and pure char exists only in the lowest 2% of the bed. The solid surface temperature just above the grate is high because the high oxygen concentration is reacting with pure char. As the char surface reaction decreases and the volatiles and moisture escape through the surface of the fuel, the surface temperature decreases rapidly, but then rises further above the grate due to heat transfer with the gaseous combustion products. At 1.4 m above the grate, the oxygen is consumed and the char reducing reactions gradually decrease the temperatures in the bed. The predicted burning rate per unit plan area of the bed is 2630 kg/(m² · h), which compares well with measurements in a test rig.

For a fixed fuel bed undergoing combustion on a traveling or vibrating grate, ignition starts at the top of the bed at the back portion of the grate fed by a spreader stoker, and the drying and devolatilization layers propagate downward. For this situation, time-dependent terms need to be added to the conservation equations. In this way the temperature and species profiles and fuel burning rate can be determined as a function of position on the grate, primary airflow, and fuel properties.

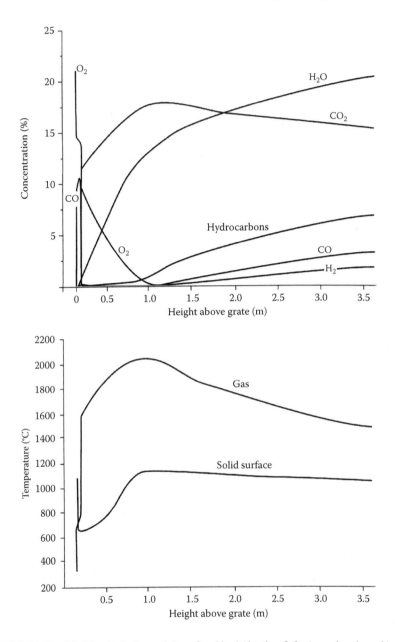

FIGURE 15.12 Model calculations of deep fixed bed (depth = 3.6 m) combustion of hybrid poplar logs on a grate; d_{init} = 20 cm, ρ_{char} = 0.8 g/cm³, ε = 0.65, $\underline{V}_g$ = 3.7 m/s at 400°C underfire air, and initial fuel moisture = 23%. The top graph plots gaseous concentrations, and the bottom graph plots solid surface and gas temperatures. (From Bryden, K. M. and Ragland, K. W., "Numerical Modeling of a Deep, Fixed Bed Combustor," *Energy Fuels* 10(2):269–275, 1996, with permission of the American Chemical Society.)

TABLE 15.5

Mass Fraction Yield of Devolatilization Products from Wood in a Hot Fixed Bed (Dry, Ash Free Basis)

Char	0.200
Water	0.250
Hydrocarbons	0.247
Carbon monoxide	0.183
Carbon dioxide	0.115
Hydrogen	0.005

15.7 PROBLEMS

1. What is the volume of pine woodchips (L) required to heat 1 L of water from 20°C to 100°C using an improved stove? Assume the wood has 20% moisture as-received and assume 30% thermal efficiency based on the lower heating value. Use the information given in Table 2.10 and Appendix A.5.

2. Some cultures use braided grasses for cooking. Assume the grass has 5% moisture and 8.6% ash on an as-received basis. Assume that the dry, ash free composition of the grass is 49.4%, 5.9%, 43.4%, 1.3%, and 0.0% for C, H, O, N, and S, respectively. The bulk density of braided grass is 20% of the density of the grass. The dry density of the grass is 250 kg/m³. The dry, ash-free HHV of the grass is 17.4 MJ/kg. Assume that the heat transfer efficiency of the three-stone fire is 5% and that the lowest power available with the three-stone fire is 1000 W, which is more than sufficient to simmer the covered pot.
   a. What is the as-received LHV of the grass (kJ/kg)?
   b. What is the as-received density of the grass (kg/m³)?
   c. What is the as-received bulk density of the grass (kg/m³)?
   d. What is the volume (liters) and weight (kg) of grass needed to heat 5 L of water from 20°C to 100°C and then simmer the water in a covered pot for 1 h.
   e. Compare your results with the results in Example 15.1.

3. Consider the case of village women gathering wood for cooking. Assume that the village has a population of 1000 people and that 100 women each make one trip to gather 20 kg of wood each day. The forest surrounding the village grows wood at the rate of 3 m³/ha · yr. The density of wood is 600 kg/m³. The energy content of the wood is 20 MJ/kg. Assume that all the wood is used for cooking.
   a. How much land is required to ensure the sustainable harvest of wood?

b.   Assume that the land needed for sustainable energy is in a circle around the village and that on average a woman travels from the center of the village to halfway to the edge of the reserve (a distance of $d/4$) to gather wood at a speed of 3 km/h including gathering time. How much time does each women spend each day gathering wood?

c.   How many tons of carbon dioxide are emitted each year by the cooking fires in the village? If improved stoves are used and the resulting carbon dioxide emissions are reduced by a factor of 3, what would be the value of the carbon dioxide credits to the village? Assume that the carbon dioxide credits can be aggregated and sold for \$25/T. One-half of the value of the carbon dioxide credits will return to the village and one-half of the value of the carbon dioxide credits will be used for program management.

4.   A wood stove has a heat output of 10 kW and an efficiency of 60% based on the HHV. How many kg of wood are needed in a 24 h period, assuming oak with 20% moisture (dry basis)? What is the volume occupied by this charge of small wood logs? Use the data from Table 2.10 and Appendix A.5.

5.   Estimate the required ratio of underfire air to overfire air for top feed, updraft, fixed bed combustion under the assumption that a stoichiometric amount of underfire air is used to burn the char to completion while stoichiometric overfire air is used for the volatiles. Assume the char is pure carbon. Perform the calculation for lignite coal and wood chips with the following analysis:

Proximate Analysis (as-received)	Lignite (wt)	Wood (wt)
Moisture	25%	25%
Volatile matter	35%	59%
Fixed carbon	30%	15%
Ash	10%	1%
Ultimate Analysis (dry, ash-free)	Lignite (wt)	Wood (wt)
C	70%	50%
H	5%	6%
O	25%	44%

6.   For Problem 5, calculate the maximum possible temperature in the bed assuming no radiation or conductive heat loss within the bed or from the bed. Assume air and fuel enter at 25°C. Assume the char is pure carbon. Work the problem assuming no dissociation using the data in Appendix C.

7.   Hybrid poplar with 30% as-received moisture is burned to completion on a traveling grate spreader stoker with 20% excess air. If flue gas recirculation is used at the rate of 25%, (a) what is the O_2% by volume in the flue gas, and (b) what is the volume of flue gas per 100 kg of fuel (as-received) at a stack temperature of 450 K and 1 atm pressure? Use the data given in Table 2.10.

8. It is proposed to harvest and burn kelp in a traveling grate stoker. The kelp will be air-dried to 50% as-received moisture and then burned to completion on a traveling grate spreader stoker with 10% excess air. Use the data given in Table 2.10.
   - How much energy is required to dry the kelp from 50% as-received moisture content in the stoker relative to the energy content of the kelp?
   - What will be the adiabatic flame temperature of the burning kelp? Work the problem assuming no dissociation using the data in Appendix C.
   - What are your comments on the proposed process?

9. Refuse-derived fuel (RDF) is burned to completion in a spreader stoker boiler at the rate of 10,000 kg/h with 30% excess air. Stack gases are emitted at 200°C and 1 atm pressure. For the as-received fuel analysis below find the following:
   a. Volume % of CO_2, H_2O, O_2, and N_2 in stack gas.
   b. Concentration (ppm) of SO_2, HCl, and NO (from fuel N only) in stack gas. Assume all S, N, and Cl go to SO_2, NO, and HCl, respectively.
   c. Volumetric flow rate (m³/h) of stack gases.
   d. Uncontrolled emission rates of SO_2, HCl, and NO (g/s).
   e. Assuming that all the lead is vaporized and then recondenses on the fly ash, that 65% of the ash is fly ash, and that 85% of the fly ash is controlled, what are the emissions of lead (g/s)?

RDF Analysis	% (wt)
Moisture	20.90
Carbon	33.80
Hydrogen	4.50
Nitrogen	0.42
Chlorine	0.31
Sulfur	0.21
Ash	13.63
Oxygen	26.20
Lead	0.03
Sum	100.00

Higher Heating Value	13.44 MJ/kg

10. Show that for a packed bed of spherical particles, the particle surface area per unit volume of the bed is

$$A_v = 6(1 - \varepsilon)/d$$

and show that the Reynold's number is

$$Re = 6\frac{\rho_g V_g}{A_v \mu}$$

REFERENCES

Baldwin, S. F., *Biomass Stoves: Engineering Design, Development, and Dissemination*, Volunteers in Technical Assistance, Arlington, VA, 1987.

Bond, T. C., Streets, D. G., Yarber, K. F., Nelson, S. M., Woo, J.-H., and Klimont, Z., "A Technology-Based Global Inventory of Black and Organic Carbon Emissions from Combustion," *J. Geophys. Res.* 109(D14):D14203/1–D14203/43, doi:10.1029/2003JD003697, 2004.

Bryden, K. M. "Computational Modeling of Wood Combustion," PhD Diss., University of Wisconsin–Madison, 1998.

Bryden, K. M. and Ragland, K. W., "Numerical Modeling of a Deep, Fixed Bed Combustor," *Energy Fuels,* 10(2):269–275, 1996.

Bryden, K. M. and Ragland, K. W., "Combustion of a Single Wood Log under Furnace Conditions" in *Developments in Thermochemical Biomass Conversion*, vol. 2, eds. A. V. Bridgwater and D. G. B. Boocock, 1331–1345, Blackie Academic and Professional, London, 1997.

Bussman, P. J. T., "Woodstoves: Theory and Applications in Developing Countries," PhD Diss., Eindhoven University of Technology, Eindhoven, Netherlands, 1988.

Ceely, F. J. and Daman, E. L., "Combustion Process Technology," in *Chemistry of Coal Utilization, Second Supplementary Volume*, ed. M. A. Elliott, 1313–1387, Wiley-Interscience, New York, 1981.

Di Blasi, C., "Heat, Momentum, and Mass Transport Through a Shrinking Biomass Particle Exposed to Thermal Radiation," *Chem. Eng. Sci.* 51(7):1121–1132, 1996.

Goldman, J., Xieu, D., Oko, A., Milne, R., Essenhigh, R . H., and Bailey, E. G., "A Comparison of Prediction and Experiment in the Gasification of Anthracite in Air and Oxygen-Enriched Steam Mixtures," *Sym. (Int.) Combust.* 20(1):1365–1372, The Combustion Institute, Pittsburgh, PA, 1984.

Johansson, R., Thunman, H. and Leckner, B., "Sensitivity Analysis of a Fixed Bed Combustion Model," *Energy Fuels*, 21(3):1493–1503, 2007.

Kitto, J. B. and Stultz, S. C., eds., *Steam: Its Generation and Use*, 41st ed., The Babcock and Wilcox Co., Barberton, OH, 2005.

Kowalczyk, J. F. and Tombleson, B. J., "Oregon's Woodstove Certification Program," *J. Air Pollut. Control. Assoc.*, 35(6):619–625, 1985.

Langsjoen, P. L., Burlingame, J. O., and Gabrielson, J. E., "Emissions and Efficiency Performance of Industrial Coal Stoker Fired Boilers," EPA-600/S7-81-11a or PB-82-115312 (ECD citations), 1981.

MacCarty, N., Ogle, D., Still, D., Bond, T., and Roden, C., "A Laboratory Comparison of the Global Warming Impact of Five Major Types of Biomass Cooking Stoves," *Energy Sustain. Dev.* 12(2):5–14, 2008.

Ostlie, L. D., Schaller, B. J., Ragland, K. W., Bryden, K. M., and Wiltsee, G. A., "Whole Tree Energy™ Design," Vols. 1–3, EPRI report TR-101564, 1993.

Purnomo, Aerts, D. J. and Ragland, K. W., "Pressurized Downdraft Combustion of Woodchips," *Symp. (Int.) Combust.* 23(1):1025–1032, The Combustion Institute, Pittsburgh, PA, 1991.

Ragland, K. W., Aerts, D. J., and Baker, A. J., "Properties of Wood for Combustion Analysis," *Bioresour. Technol.* 37(2):161–168, 1991.

Reed, T. B. and Das, D., "Handbook of Biomass Downdraft Gasifier Engine Systems," SERI/SP-271-3022, Solar Energy Research Institute, Golden, CO, 1988.

Rönnbäck, M., Axell, M., Gustavsson, L., Thunman, H., and Leckner, B. "Combustion Processes in a Biomass Fuel Bed - Experimental Results," in *Progress in Thermochemical Biomass Conversion*, ed. Bridgewater, A. V., 743–757, Blackwell Science, Oxford, 2001.

Prasad, K. K. and Verhaart, P., eds., *Wood Heat for Cooking*, Indian Academy of Sciences, Bangalore, 1983.

Shelton, J. W., *The Woodburners Encyclopedia*, Vermont Crossroads Press, Waitsfield, VT, 1976.

Shindell, D. and Faluvegi, G., "Climate Response to Regional Radiative Forcing During the Twentieth Century," *Nat. Geosci.* 2:294–300, 2009.

Singer, J. G., ed., *Combustion: Fossil Power System: A Reference Book on Fuel Burning and Steam Generation,* 4th ed., Combustion Engineering Power Systems Group, 1993.

Smoot, L.D. ed., *Fundamentals of Coal Combustion For Clean and Efficient Use*, Elsevier, New York, 1993.

U.S. Environmental Protection Agency, *Compilation of Air Pollutant Emission Factors, Vol. 1: Stationary Point and Area Sources*, 5th ed., EPA-AP-42, 1995.

Wakao, N. and Kaguei, S., *Heat and Mass Transfer in Packed Beds*, Gordon and Breach, New York, 1982.

Yin, C., Rosendahl, L. A., and Kær, S. K., "Grate-Firing of Biomass for Heat and Power Production," *Prog. Energy Combust. Sci.* 34(6):725–754, 2008.

16 Suspension Burning

Suspension burning furnaces and boilers burn pulverized fuel particles that are blown through burner nozzles into a furnace volume that is large enough to allow burnout of the fuel char. Pulverized coal and biomass are burned separately or together (co-firing) in suspension as the small fuel particles flow through the furnace. Coal is relatively easy to pulverize since it is brittle, whereas biomass, being fibrous, is more difficult to pulverize. Much of the electricity in the world today is generated by pulverized coal-fired steam power plants. In this chapter, preparation and combustion of the fuel are considered first and then emissions and emissions control, including carbon dioxide capture and sequestration, are discussed.

The advantages of suspension burning over fixed bed combustion is that the system can be scaled up to very large sizes, and it is more responsive to required changes in load (heat and power output). Large suspension burning systems are more complex than fixed bed systems due to the need to pulverize the fuel and the generally more complex steam cycle.

16.1 PULVERIZED COAL BURNING SYSTEMS

A suspension burning steam power plant is shown schematically in Figure 16.1. As-received coal is fed from coal bunkers or bins into the pulverizers. The pulverizers reduce the size of the coal from centimeters to fine powder with the consistency of face powder. From the pulverizers, the pulverized coal is piped with air to the burners where the fuel is mixed with preheated air from the windbox (air plenum). The burners stabilize the volatile flame, and the char burns out in the radiant section of the furnace. The furnace walls are made of alloy steel tubes with narrow steel strips welded between them to form gas tight walls. The incoming water is pumped up to a high pressure, preheated in the economizer, and sent to the lower steam drum and through downcomers to headers at the bottom of the furnace. The water is heated to a steam-water mixture in the waterwall tubes of the radiant section, and the wet steam rises to the upper drum. In the upper drum the two-phase mixture is separated: the steam flows to the superheater tubes in the convective section, while the water drains to the lower drum and rises as wet steam in the convective boiler tubes. The superheated steam flows to a steam turbine generator. The steam cycle is that of a complex Rankine cycle typically with several steam reheats between sections of the turbine and with multiple steam extractions for the feedwater heaters.

For boilers designed to use *supercritical steam* (steam pressure and temperature above 22.1 MPa and 374°C), steam drums are not necessary because there is no discrete phase change, and the water flows once through the boiler before entering the steam turbine. Ultra-supercritical boilers have pressures and temperatures well above the critical point, e.g., 28 MPa and 730°C. Higher steam pressure and temperature to the turbine inlet results in higher net cycle efficiency. Steam pressure and temperature are limited

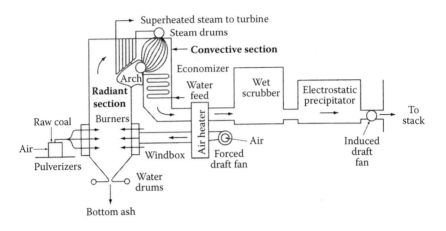

FIGURE 16.1 Utility-scale pulverized coal combustion system.

by the durability of the steam tubes and the temperature and pressure limits of the steam turbines. Steam temperature is designed to be well below the combustion gas temperature, but is gradually being increased as improved turbine materials are developed.

On the combustion gas side, the flue gases flow past the convective waterwall tubes, the economizer tubes, and the air heater to a wet scrubber to control sulfur dioxide emissions and then to an electrostatic precipitator or a fabric filter baghouse to control particulate emissions (fly ash and unburned carbon). The scrubber and fabric filter also help control mercury emissions. Induced draft fans before the stack, forced draft primary inlet air fans, and over-fire air fans balance the system air pressure and maintain the desired burner airflow.

Typical operating parameters for a large supercritical boiler and an ultra-supercritical boiler are shown in Table 16.1. The ultra-supercritical boiler represented by this table is the most current technology of this type and is expected to be in operation in the near future. The higher steam temperature and pressure increases the net power plant efficiency. While the design inlet temperature for gas turbines (operating on distillate fuels) has increased steadily over the years and now exceeds 1200°C, the inlet steam temperature for utility steam turbines (732°C in Table 16.1) has increased more slowly because of the corrosive nature of the combustion gases.

The pulverizer is typically a type of ball mill, the larger of which can handle up to 100 tons of coal per hour. The ball mill grinds the coal down to face powder size to ensure rapid combustion; the typical size requirement is 70% less than 33 μm. Air heated to about 340°C or more is blown through the pulverizer to dry the coal particles and convey them to the burner located in the furnace wall. The coal emerges from the pulverizer at 50°C–100°C. Carbon monoxide is monitored in the pulverizer to prevent an explosion. The air velocity within the conveying lines should be greater than 15 m/s to avoid settling of the pulverized coal, and right angles should be avoided to minimize erosion of the conveying lines.

From the pulverizer, the coal-air mixture is piped to the burners, as shown in Figure 16.2. The conveying air, called primary air, is about 20% of the required combustion air. Typically, the coal and primary air exit the burner nozzle at 75°C and 25 m/s. The

TABLE 16.1

Operating and Performance Parameters for a Large Wall-fired Steam Boiler to Generate 550 MW Net Electric Power

Parameter	Supercritical Steam Boiler	Ultra-supercritical Steam Boiler
Steam pressure[a] (MPa)	24.4	27.9
Steam temperature[a] (°C)	593	732
Steam flow rate (kg/h)	1,618,000	1,388,000
Coal input (kg/h)	185,000	164,000
Primary combustion air (kg/h)	427,000	377,000
Over-fire combustion air (kg/h)	1,390,000	1,228,000
Ammonia injection* (L/h)	480	420
Limestone slurry (kg/h)	61,000	53,000
Gross electrical power generated (MW)	580	577
Net electrical power generated (MW)	550	550
Net power plant efficiency[b] (%)	39.4	44.6

Source: Haslbeck, J. L., Black, J., Kuehn, N., Lewis, E., Rutkowski, M. D., Woods, M., and Vaysman, V., *Pulverized Coal Oxycombustion Power Plants, Volume 1: Bituminous Coal to Electricity Final Report*, DOE/NETL-2007/1291, National Energy Technology Laboratory, U.S. Department of Energy, 2nd rev., 2008.

[a] At the inlet to the high pressure section of the steam turbine.

[b] Based on a HHV of 27.1 MJ/kg.

*For NO_X control.

secondary or main air supply is preheated to 300°C and exits the burner with swirl at about 40 m/s. The flame shape is controlled by the amount of secondary air swirl and the contour of the burner throat. A recirculation pattern extends several throat diameters into the furnace and provides a stabilized zone for ignition and combustion of the volatiles. Air preheat and swirl are the two means of ensuring that ignition occurs in the throat of the burner and that the volatile flame is stable. The pulverized fuels should have a volatile content of at least 20% in order to maintain stability, which includes most solid fuels except anthracite coal. Nozzles with up to 165×10^6 kJ/h heat input per nozzle are used.

The velocity of the primary stream of air and coal in the nozzle must exceed the speed of flame propagation so as to avoid flashback. The flame speed depends on the fuel-air ratio, the amount of volatile matter and ash in the coal, the particle size distribution (fineness of grind), the air preheat, and the tube diameter. The maximum flame speed is comparable to a turbulent premixed gaseous hydrocarbon flame. For low volatile and/or high ash coal, the flame speed is low, and the air from the secondary nozzle is mixed in more slowly to avoid instability.

16.1.1 Location of Fuel and Air Nozzles

Numerous methods of locating and orienting the fuel and air nozzles have been tried, as illustrated in Figure 16.3. In the horizontal firing method, the nozzles

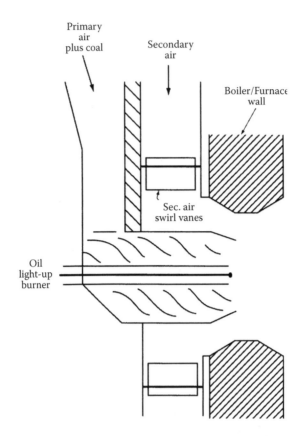

Primary
air
plus coal

Secondary
air

Boiler/Furnace
wall

Sec. air
swirl vanes

Oil
light-up
burner

FIGURE 16.2   Pulverized coal burner nozzle.

are located on the front wall, or on both the front and back walls, which is called opposed horizontal firing. Another approach for firing pulverized coal in large boilers is the tangential firing method, as shown in Figures 16.3(c) and 16.4. The fuel nozzles do not have swirl and are located in the corners of the boiler. The coal-air mixture from the nozzles is directed along a line tangent to a circle in the center of the boiler. There is relatively little mixing in the primary jets, but the interaction of the jets together sets up a large-scale vortex in the center of the boiler that creates intense mixing. The nozzles in the corners consist of a vertical array of coal nozzles and secondary air nozzles. The angle and velocity of the jets can be adjusted to optimize the fireball size and location in terms of heat release and ash slagging on the walls. The opposed inclined firing, and U and double U designs shown in Figures 16.3(d), (e), and (f) are used for harder to burn fuels such as coke, anthracite coal, and high ash coal. Downward firing results in a greater residence time within the furnace.

The firing methods shown in Figure 16.3 are examples of *dry bottom* furnaces, meaning that the temperatures are such that most of the ash does not agglomerate into molten globules and thus does not fall to the bottom of the furnace. The ash

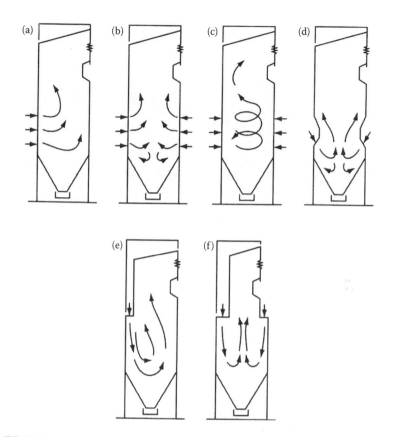

FIGURE 16.3   Dry bottom furnace and burner configurations: (a) horizontal (front and rear), (b) opposed horizontal, (c) tangential, (d) opposed inclined, (e) single U-flame, and (f) double U-flame. (From Elliott, courtesy of National Academy Press, Washington, DC.)

that does fall to the bottom, and this is only 10%–20% of the total ash, remains non-molten because sufficient furnace volume ensures that the peak combustion temperatures are reduced by radiation heat transfer to the walls.

Wet bottom firing is another design approach that has been used. The high temperature necessary for ash melting is produced by locating the nozzles close together near the bottom of the furnace floor and lining nearby walls with refractory. In addition, various types of double chambers or baffled chambers have been built to provide two-stage combustion wherein the first stage is very hot, thereby slagging the ash. Up to 75% ash retention has been achieved in certain wet bottom designs with certain coals and operating conditions. Experience has shown that long-term operation (30 years or more) with a variety of coals and load conditions can best be achieved with dry bottom systems. Wet bottom systems are more prone to slagging and corrosion and have higher NO_x emissions. Dry bottom systems have higher uncontrolled fly ash emissions. In either case, particulate control equipment is required.

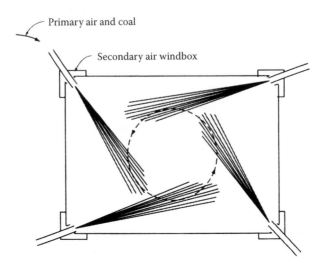

Primary air and coal

Secondary air windbox

FIGURE 16.4 Schematic of a cross section of a corner-fired boiler with tangential injection. Burners are tilting and there are several burner levels.

16.1.2 FURNACE DESIGN

A utility-sized furnace, for which a schematic drawing is shown in Figure 16.5, may be up to 50 m high and 10 m on a side. Combustion begins in the throat of the fuel nozzles and is completed in the radiant section of the boiler before the superheater tubes. Peak temperatures near the nozzles may reach 1650°C, and the temperature drops gradually to 1100°C due to radiant heat transfer to the walls. The radiant section must be large enough to provide sufficient residence time for complete burnout of the fuel particles and must extract enough heat to drop the temperature below the ash softening temperature prior to the convective section of the boiler to avoid fouling of the superheater tubes.

An arch in the upper part of the radiant section is used to provide more uniform flow around the upper corner of the combustion chamber, thereby increasing heat transfer in the convective section. The convection section contains tubes to superheat the steam and provide reheat for multistage turbines. The convective section is designed to extract as much heat in as small a space as possible. Gas velocities in the convective section are restricted to about 20 m/s for coal-fired furnaces to minimize tube erosion due to fly ash. If the ash content of the fuel is high or contains especially hard compounds such as quartz, the gas velocity must be reduced by enlarging the convective section. The flue gas temperature at the inlet to the convective section generally is limited to 1100°C to minimize tube corrosion. Soot blowers, consisting of either steam or compressed air jets, are used regularly on the convective tubes as well as the economizer and air preheater tubes to maintain effective heat transfer and reduce flue gas pressure drop due to deposition of ash on the tube surfaces. Elements such as sodium, potassium and iron in the ash can cause ash deposits to sinter and fuse onto the boiler tubes, thereby reducing the effectiveness of the soot blowers. Sulfate and chloride ions in the deposits can cause corrosion of the tube surfaces.

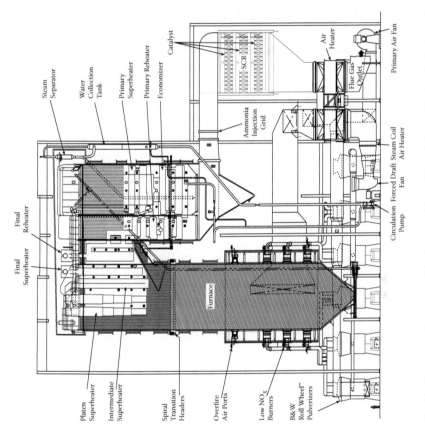

FIGURE 16.5 Pulverized coal-fired boiler generating supercritical steam. (From Kitto, J. B. and Stultz, S. C., *Steam: Its Generation and Use*, 41st ed., Babcock and Wilcox, courtesy of The Babcock and Wilcox Company, Barberton, OH.)

When changing fuels, the effects of slagging, erosion, and corrosion must be considered in relation to the specific boiler design.

16.2   PULVERIZED COAL COMBUSTION

Consider a cloud of coal particles that are blown through a burner into a furnace. Approximately 80% of the particles are between 10 and 100 μm in size. The swirl and backmixing of the burner create a residence time near the burner where radiant heating from the burner quarl (a refractory lined nozzle) ignites the particle cloud. As the particles are transported into the furnace away from the burner, the heat of combustion is radiated to the relatively cool waterwall tubes. As combustion proceeds, the gas temperature drops and the oxygen concentration decreases to a final value of a few percent. Hence, the steps of drying, devolatilization, char combustion, and ash formation proceed under different conditions as the particle moves along its trajectory.

To be more specific, consider an example of a 100 μm coal particle that flows through the burner nozzle and encounters a 1400°C flame zone. In a fraction of a ms the particle reaches 100°C and the remaining particle moisture is driven off. (Some drying also occurs in the pulverizer.) By 1 ms, the particle reaches 400°C and volatile gases and tars start to be forced out of the particle as the particle is pyrolyzed. After 10 ms the particle has reached 1000°C, devolatilization (pyrolysis) is complete, approximately 50% of the particle weight remains, and porosity is established. The initial char particle size is equal to or greater than the original coal particle size at this point. For the first time, oxygen reaches the particle surface and char burning begins.

Between 0.01 and 1 s, char burn continues, and the particle surface temperature is several hundred kelvins hotter than the gas temperature due to the surface reaction with oxygen. After 0.5 s, half of the char is consumed. Fissures are formed in the char particle, the porosity of the particle has increased, and the particle has begun to shrink in size. The mineral matter (which is dispersed throughout the char in nodules mostly less than 2 μm in size) becomes molten, and the trace metals in the mineral matter volatilize. The many molten mineral matter nodules agglomerate into a few nodules, and towards the latter stages of char burnout the char may fragment into several pieces. The mineral matter nodules meanwhile have grown large enough through agglomeration for vapor pressure to overcome surface tension effects. Consequently, they puff up into tiny hollow glass-like spheres called *cenospheres*. These cenospheres are like miniature Christmas tree ornaments that vary in size from 0.1 to 50 μm and are perfectly spherical. After approximately 1 s, char combustion and cenosphere formation are complete, and the combustion products begin to cool as they flow past the convective tubes. In the ambient air, fly ash from pulverized coal-fired boilers (and also residual fuel oil-fired boilers) can be identified by its characteristic cenospheres, as opposed to soil dust for example, which is irregular in shape.

Because drying and pyrolysis occur very rapidly relative to char burnout, the char burning rate is a key factor in determining the size of the boiler. Char consists of porous carbon, mineral matter, and a small amount of organic matter. The gas temperature, particle temperature, and oxygen concentration change as the particles move through the furnace. Char burnout proceeds by the reaction of oxygen, carbon dioxide, and water vapor with the surface of the char; however, the dominant reaction is with oxygen. Combustion of char particles depends on heat transfer and diffusion of reactant gases and products through the external boundary layer of the particle; diffusion and heat transfer within the pores of the char particle; and chemisorption of reactant gases, surface reaction, and desorption of the products. External and internal diffusion occur and chemical kinetics is important. The surface accessibility and surface reactivity of an individual particle change with time.

In order to gain an understanding of the combustion process for suspension burning of solid fuels, let us consider the simplified case of char combustion in a plug flow using a reaction rate based on the external surface area. Isothermal plug flow will be presented first and then followed by non-isothermal plug flow.

16.2.1   ISOTHERMAL PLUG FLOW OF PULVERIZED COAL

Consider an isothermal, uniform flow of uniformly dispersed char particles in a hot gas containing oxygen. The heat release rate due to combustion just equals the heat transfer to the walls of the combustor. Assuming the char particles are small and the diffusion rate is limiting (kinetically controlled combustion) $k_e = k_c$. From Equation 14.29 for a global surface reaction rate constant, and assuming the char goes to CO at the surface, the char particles burn according to

$$\frac{dm_{char}}{dt} = -\frac{12}{16} A_p k_c \, p_{O_2} \tag{16.1}$$

Substituting k_p for k_c (Equation 14.21)

$$\frac{dm_{char}}{dt} = -\frac{12}{16} A_p k_p p_{O_2} \tag{16.2}$$

where p_{O_2} is the oxygen partial pressure in the gas away from the particle surface, and k_p has units of $g_{O_2}/(cm^2 \cdot s \cdot atm_{O_2})$. For a spherical particle the area is related to the mass by

$$A_p = \pi \left(\frac{6 m_{char}}{\rho_{char} \pi} \right)^{2/3} \tag{16.3}$$

The oxygen pressure decreases along the flow path due to combustion of the char,

$$p_{O_2} = p_{O_2,init} \frac{m_{O_2}}{m_{O_2,init}} \tag{16.4}$$

where *init* refers to initial. Introducing excess air, EA (or equivalently, excess oxygen), and letting "*comb*" refer to the mass consumed by combustion and s refer to stoichiometric conditions,

$$p_{O_2} = p_{O_2,init} \left[\frac{m_{O_2(s)}(1+EA) - m_{O_2(comb)}}{m_{O_2(s)}(1+EA)} \right] \tag{16.5}$$

Rearranging,

$$p_{O_2} = p_{O_2,init} \left[1 - \frac{m_{O_2(comb)}}{m_{O_2(s)}} + EA \right] \left(\frac{1}{(1+EA)} \right) \tag{16.6}$$

or remembering that the ratio of the fraction of stoichiometric oxygen consumed is equal to the fraction of the char consumed,

$$p_{O_2} = p_{O_2,init} \left[\frac{m_{char}}{m_{char,init}} + EA \right] \left(\frac{1}{1+EA} \right) \tag{16.7}$$

Substituting Equations 16.3 and 16.7 into Equation 16.2, the char burning rate in isothermal plug flow becomes

$$\frac{dm_{char}}{dt} = -\frac{12}{16} \pi \left(\frac{6m_{char}}{\rho_{char}\pi} \right)^{2/3} (k_p)(p_{O_2,init}) \left(\frac{m_{char}}{m_{char,init}} + EA \right) \left(\frac{1}{1+EA} \right) \tag{16.8}$$

These equations are solved for a reacting cloud of uniformly-sized char particles in the following example. Note that the stoichiometry and the excess air refer to the $C + O_2 \rightarrow CO_2$ reaction, while the surface reaction is for $C + 1/2\ O_2 \rightarrow CO$. If the burnout time is known from Equation 16.8 and the average velocity is known then the average distance for particle burnout can also be determined.

Example 16.1

For a uniform cloud of 50 μm char particles flowing in a stream of hot, isothermal gas containing 20% excess oxygen, determine the particle burnout time based on 99% burnout and plot the particle mass divided by the initial particle mass as a function of time. Use an effective char kinetic rate constant of 0.5 g_{O_2}/ $(cm^2 \cdot s \cdot atm_{O_2})$, a char density of 0.8 g/cm³, and an initial oxygen pressure of 0.1 atm. Repeat for 100 μm particles and compare the burnout times based on 99% burnout.

Solution

Non-dimensionalizing Equation 16.7 by dividing by $m_{char,init}$

$$\frac{d}{dt}\left[\frac{m_{char}}{m_{char,init}}\right] = -\frac{12}{16}\frac{\pi}{m_{char,init}^{1/3}}\left(\frac{6}{\rho_{char}\pi}\frac{m_{char}}{m_{char,init}}\right)^{2/3}$$

$$\cdot\left(k_p\right)\left(p_{O_2,init}\right)\left(\frac{m_{char}}{m_{char,init}}+EA\right)\left(\frac{1}{1+EA}\right)$$

The initial mass of a 50 μm char particle is

$$m_{char,init} = \rho\frac{\pi d^3}{6} = \frac{0.8\text{ g}_{char}}{cm^3}\cdot\frac{\pi}{6}\left[\frac{50\times10^{-4}\text{ cm}}{1}\right]^3 = 5.236\times10^{-8}\text{ g}_{char}$$

Substituting,

$$\frac{d}{dt}\left[\frac{m_{char}}{m_{char,init}}\right] = -\frac{12\text{ g}_{char}}{16\text{ g}_{O_2}}\cdot\frac{\pi}{1}\cdot\frac{1}{\left(5.236\times10^{-8}\text{ g}_{char}\right)^{1/3}}\cdot\left(\frac{cm^3}{0.8\text{ g}_{char}}\cdot\frac{6}{\pi}\frac{m_{char}}{m_{char,init}}\right)^{2/3}$$

$$\cdot\frac{0.5\text{ g}_{O_2}}{cm^2\cdot s\cdot atm_{O_2}}\cdot\frac{0.1\text{ atm}_{O_2}}{1}\left(\frac{m_{char}}{m_{char,init}}+0.2\right)\left(\frac{1}{1+0.2}\right)$$

Simplifying,

$$\frac{d}{dt}\left[\frac{m_{char}}{m_{char,init}}\right] = -46.88\left(\frac{m_{char}}{m_{char,init}}\right)^{2/3}\left(\frac{m_{char}}{m_{char,init}}+0.2\right)$$

This can be readily solved using an equation solver routine. Repeating for a 100 μm particle yields

$$m_{char,init} = \rho\frac{\pi d^3}{6} = \frac{0.8\text{ g}_{char}}{cm^3}\cdot\frac{\pi}{6}\left[\frac{100\times10^{-4}\text{ cm}}{1}\right]^3 = 4.189\times10^{-7}\text{ g}_{char}$$

and

$$\frac{d}{dt}\left[\frac{m_{char}}{m_{char,init}}\right] = -23.44\left(\frac{m_{char}}{m_{char,init}}\right)^{2/3}\left(\frac{m_{char}}{m_{char,init}}+0.2\right)$$

As shown below, the 99% burnout time is 0.126 s for the 50 μm particle and 0.251 s for the 100 μm particle for these conditions.

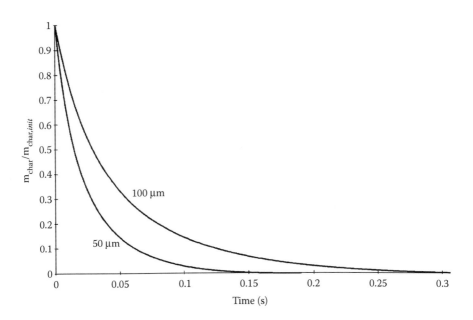

Example 16.2

For the uniform cloud of 50 μm char particles examined in Example 16.1, plot the burnout (particle mass divided by the initial mass versus time) and determine the burnout times for 0%, 5%, 20%, and 40% excess air. Use an effective char kinetic rate constant of 0.5 $g_{O_2}/(cm^2 \cdot s \cdot atm_{O_2})$, a char density of 0.8 g/cm³, and an initial oxygen pressure of 0.1 atm.

Solution

From Example 16.1 the general equation is

$$\frac{d}{dt}\left[\frac{m_{char}}{m_{char,init}}\right] = -\frac{12}{16}\frac{\pi}{m_{char,init}^{1/3}}\left(\frac{6}{\rho_{char}\pi}\frac{m_{char}}{m_{char,init}}\right)^{2/3}(k_p)(p_{O_2,init})\left(\frac{m_{char}}{m_{char,init}}+EA\right)\left(\frac{1}{1+EA}\right)$$

Substituting and simplifying,

$$\frac{d}{dt}\left[\frac{m_{char}}{m_{char,init}}\right] = -\frac{56.26}{(1+EA)}\left(\frac{m_{char}}{m_{char,init}}\right)^{2/3}\left(\frac{m_{char}}{m_{char,init}}+EA\right)$$

Solving this with a numerical integration solver yields the graph below. As shown the burnout times are 0.527, 0.237, 0.126, and 0.092 s for 0%, 5%, 20%, and 40% excess air; respectively. As shown burnout times rise significantly as the amount of excess air decreases. For example, using 5% excess air initially requires 88% more time to reach 99% burnout than does 20% initial excess air initially. Conversely using 40% excess air initially requires 27% less time to reach 99% burnout than does 20% excess air initially. Note that this example has been simplified, k_p is a function of the initial partial pressure of the oxygen and the temperature, both of which can depend on the excess air, and this effect was not included.

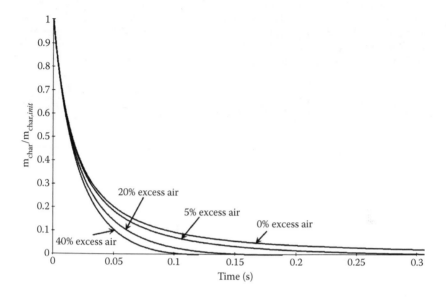

For a distribution of particle sizes, the initial rate of combustion is greater than the combustion rate of a cloud of monosize particles having the same average diameter due to the rapid burnout of the fine particles. In the same way there is a longer tail to the burnout curve due to the time needed to burnout the large particles. The burnout time for the larger particles is slowed because of the size of the particles and because the combustion of the fine particles reduces the oxygen levels available for burnout of the large particles.

The heat input per unit flow area, HR'', depends on the number of char particles per unit volume, n'; the mass of each particle, m_{char}; the flow velocity, $\underline{V}$; and the lower heating value, LHV.

$$HR'' = n'm_{char}\underline{V}(LHV) \tag{16.9}$$

The number of particles per unit volume is the mass of fuel per unit volume divided by the mass per particle,

$$n' = f\frac{\rho_g}{m_{char}} \tag{16.10}$$

where f is the fuel-air ratio. The application of Equations 16.9 and 16.10 for isothermal plug flow of char particles is illustrated in Example 16.3.

Example 16.3

Determine the heat input rate per unit area and the burnout distance for 50 μm char particles in air. The velocity of the particle laden air is 10 m/s. The reaction temperature is 1600 K. Assume the char is composed of carbon and negligible amounts of ash. Assume that the HHV of the char is 31 MJ/kg. Use an effective char kinetic rate constant of 0.5 $g_{O_2}/(cm^2 \cdot s \cdot atm_{O_2})$, a char density of 0.8 g/cm³, and an initial oxygen pressure of 0.21 atm.

Solution

From Example 16.1 the general equation is

$$\frac{d}{dt}\left[\frac{m_{char}}{m_{char,init}}\right] = -\frac{12}{16}\frac{\pi}{m_{char,init}^{1/3}}\left(\frac{6}{\rho_{char}\pi}\frac{m_{char}}{m_{char,init}}\right)^{2/3}$$

$$\cdot\left(k_p\right)\left(p_{O_2,init}\right)\left(\frac{m_{char}}{m_{char,init}} + EA\right)\left(\frac{1}{1+EA}\right)$$

and the mass of a 50 μm char particle is

$$m_{char,init} = \rho_{char}\frac{\pi d^3}{6} = \frac{0.8\ g_{char}}{cm^3}\cdot\frac{\pi}{6}\left[\frac{50\times10^{-4}\ cm}{1}\right]^3 = 5.236\times10^{-8}\ g_{char}$$

Substituting and simplifying,

$$\frac{d}{dt}\left[\frac{m_{char}}{m_{char,init}}\right] = -98.43\left(\frac{m_{char}}{m_{char,init}}\right)^{2/3}\left(\frac{m_{char}}{m_{char,init}} + 0.2\right)$$

Solving using an equation solver, the time for 99% burnout is 0.060 s. For the reaction of one mole of carbon char going to one mole of CO_2 in air with 20% excess air, the balance equation is

$$C + 1.2\left(O_2 + 3.76\ N_2\right) \rightarrow CO_2 + 0.2\ O_2 + 4.51\ N_2$$

From this the fuel-air ratio is

$$f = \frac{1\ mol_C}{5.71\ mol_{air}}\cdot\frac{12\ g_C}{mol_C}\cdot\frac{mol_{air}}{29\ g_{air}}\cdot\frac{1\ g_{char}}{1\ g_C} = 0.0725$$

Since there is no hydrogen or moisture in the char particles, the lower heating value is equal to the higher heating value. The air density is

$$\rho_{air} = 0.000221\ g/cm^3 \qquad\qquad \text{(Appendix B)}$$

The particle mass is

$$m_{char,init} = \rho_{char}\frac{\pi d^3}{6} = \frac{0.8 \text{ g}_{char}}{\text{cm}^3}\cdot\frac{\pi}{6}\left[\frac{50\times10^{-4} \text{ cm}}{1}\right]^3 = \frac{5.236\times10^{-8} \text{ g}_{char}}{\text{particle}}$$

From Equation 16.10,

$$n' = \frac{0.0725 \text{ g}_{char}}{\text{g}_{air}}\cdot\frac{2.21\times10^{-4} \text{ g}_{air}}{\text{cm}^3}\cdot\frac{\text{particle}}{5.236\times10^{-8} \text{ g}_{char}}$$

$$n' = 306 \text{ particles/cm}^3$$

Using Equation 16.9, the heat input rate is

$$HR'' = \frac{306 \text{ particles}}{\text{cm}^3}\cdot\frac{5.236\times10^{-8} \text{ g}_{char}}{\text{particle}}\cdot\frac{1000 \text{ cm}}{\text{s}}\cdot\frac{31 \text{ kJ}}{\text{g}_{char}}$$

$$HR'' = 0.50 \text{ kW/cm}^2 = 5.0 \text{ MW/m}^2$$

As noted earlier, the burnout time for a 50 mm char particle is 0.060 s, so the travel distance, x, is

$$x = \frac{0.060 \text{ s}}{1}\cdot\frac{10 \text{ m}}{\text{s}} = 0.6 \text{ m}$$

From this we can find the average energy density (heat release rate per unit volume)

$$q''' = \frac{HR''}{x} = \frac{5.0 \text{ MW}}{\text{m}^2}\cdot\frac{1}{0.6 \text{ m}} = \frac{8.3 \text{ MW}}{\text{m}^3}$$

This illustrates the relatively high power density of pulverized char combustion.

16.2.2 NON-ISOTHERMAL PLUG FLOW OF PULVERIZED CHAR SUSPENSION

Consider a uniform flow of a homogeneous mixture of char particles suspended in air. There is no relative velocity between the gas and the particles. The particle flow is sufficiently dilute such that the volume occupied by the suspended particles is negligible. The flow is steady and pressure drop is negligible. Heat conduction in the gas phase is negligible compared to convection heat transfer. As the particles flow along the streamlines, the char reacts with the oxygen due to kinetics and diffusion of oxygen from the gas to the solid. This reaction produces CO near the particles' surfaces, which then reacts rapidly in the gas phase to produce CO_2. The reaction rate of CO_2 with char is small relative to the reaction rate of oxygen with char and can be neglected for this discussion. Following the discussion in Section 14.3 the rate of combustion of char per unit volume of gas is

$$r_{char} = \frac{dm'''_{char}}{dt} = \frac{12}{16} k_e \rho_{O_2} A_v \qquad (16.11)$$

where A_v is the fuel external surface area per unit volume of gas and is given by

$$A_v = A_p n' \qquad (16.12)$$

As the char particle cloud moves along, the particle diameter and oxygen density decrease, and the temperature and velocity increase due to combustion.

Since the CO reacts rapidly compared to the char, the CO in the gas is negligible and the gas phase consists of O_2, CO_2, and N_2.

$$r_{CO_2} = -r_C \frac{M_{CO_2}}{M_C} \qquad (16.13)$$

$$r_{O_2} = r_C \frac{M_{O_2}}{M_C} \qquad (16.14)$$

Remembering that we have assumed that char is composed of carbon, $r_{char} = r_C$. Conservation of energy, mass, and species continuity is used to obtain the temperature, velocity, and species profiles, respectively, along the streamlines. Conservation of energy for the gas phase is

$$\frac{d}{dx}(\rho_g h_g \underline{V}) = r_{char} H_{char} - q'''_{rad} \qquad (16.15)$$

where q'''_{rad} is heat loss to the walls of the combustor per unit volume due to radiation. The rate of surface reaction, r_{char}, is governed by the rate of CO formation at the surface. However, compared to the rate of CO formation at the char surface, the CO reacts rapidly to form CO_2. As a result H_{char} is the heat of reaction for

$$C + O_2 \rightarrow CO_2$$

Conservation of mass for the gas phase is

$$\frac{d}{dx}(\rho_g \underline{V}) = r_{char} \qquad (16.16)$$

The species continuity equations are needed to obtain the gas density,

$$\rho_g = \rho_{O_2} + \rho_{CO_2} + \rho_{N_2} \qquad (16.17)$$

$$\frac{d}{dx}\left(\rho_{O_2} \underline{V}\right) = r_{O_2} \tag{16.18}$$

$$\frac{d}{dx}\left(\rho_{CO_2} \underline{V}\right) = r_{CO_2} \tag{16.19}$$

Noting that for every molecule of oxygen consumed one molecule of carbon dioxide is created,

$$\rho_{N_2} = \rho_{N_2,init} \frac{T_{init}}{T} \tag{16.20}$$

For the mixture,

$$h_g = \sum_i h_i y_i \tag{16.21}$$

where

$$y_i = \frac{\rho_i}{\rho_g} \tag{16.22}$$

The solution, which is implicit, proceeds as follows: for a given inlet pressure, p, temperature, T; velocity, $\underline{V}$; particle size, d; and fuel-air ratio, f; the above equations are solved incrementally for a small Δx by obtaining r_{char} from Equation 16.11. The enthalpy of the gas, h_g, is then determined from Equation 16.15 and the temperature of the gas, T_g, is updated from h_g. The gas densities are obtained from Equations 16.17–16.20. Velocity is updated from Equation 16.16. The distance, x, is incremented by Δx and the process is repeated until char burnout is complete. The particle diameter, for use in Equation 16.11 and in obtaining the mass transfer coefficient, $\tilde{h}_D$, (which is part of k_e) is obtained by knowing the time from $x/\underline{V}$ and using Equation 16.11, as is done in section 14.3. The particle temperature, which may be different from the gas temperature, is obtained from the relations in Section 14.3.

Results of applying detailed plug flow relations to bituminous coal suspension burning including drying, particle heat up, and pyrolysis, are shown in Figure 16.6. Particle heat-up requires about 1/3 of the distance in this example. As shown, following heat-up pyrolysis proceeds rapidly. In this example excess oxygen is available at the combustor exit. The time required for burnout of the 85 μm particles exceeds the time available in the combustor length. Of course, in an actual furnace the flow is three-dimensional, but nevertheless these types of calculations provide useful insight into why the furnace height must be so large. Full three-dimensional computational fluid dynamics calculations are used to assist in the design of burners, over-fire air jets, radiant waterwall tubes, and convective boiler tubes.

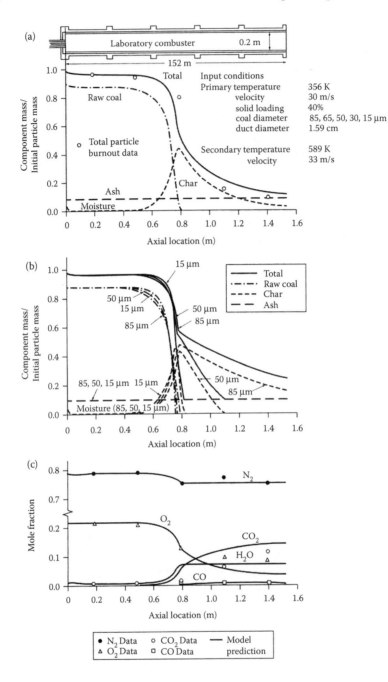

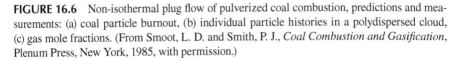

FIGURE 16.6 Non-isothermal plug flow of pulverized coal combustion, predictions and measurements: (a) coal particle burnout, (b) individual particle histories in a polydispersed cloud, (c) gas mole fractions. (From Smoot, L. D. and Smith, P. J., *Coal Combustion and Gasification*, Plenum Press, New York, 1985, with permission.)

16.3 BEHAVIOR OF ASH

Ash is a major factor affecting the performance of boilers. Not only does ash deposit on heat transfer surfaces and in flue gas passages, it also leads to corrosion and erosion of heat transfer tubes. Deposition, erosion, and corrosion can significantly limit the efficiency and lifetime of boiler tubes.

Deposition of ash is called *slagging* and *fouling*. Slagging generally refers to the sticking of molten ash on heat transfer surfaces—usually the radiant tubes. Fouling refers to deposition of non-molten ash on the convective tubes and surfaces by impaction and diffusion. The severity of the slagging and fouling depends on the nature of the mineral matter and the temperatures and velocities involved.

Mineral matter in coal varies widely but typically consists of aluminosilicates (clays), sulfides, carbonates, oxides, and chlorides (Table 16.2). During combustion the mineral compounds are converted to oxides. Typically the ash constituents are reported as separate oxides such as SiO_2, Al_2O_3, Fe_2O_3, CaO, MgO, Na_2O, and K_2O. However, in actuality more complex inorganic compounds are formed. Coal ash may begin to *sinter* at temperatures as low as 650°C. Sintering is chemical bonding of ash particles that occur without melting. At temperatures above 1000°C appreciable melting can occur. Above approximately 1400°C a highly viscous liquid slag is formed. The cone ash fusion temperature test discussed in Section 2.3.5 is a rough indicator of the slagging potential of a particular coal.

Two techniques are used to remove troublesome deposits of slag: (1) high velocity jets of steam, air, or water are used to break the bond between the slag and tube wall, and (2) the firing rate can be reduced to decrease furnace temperatures. Bonding of slag to tube walls is due to alkali compounds such as Na_2SO_4 and K_2SO_4. Sodium and potassium are volatilized in the flame zone, react rapidly with sulfur dioxide, and condense on the external tube surfaces, which are about 500°C, causing the external tube surfaces to become sticky. As the flue gas flows over the tubes, some of the larger fly ash particles are deposited by impaction on the upstream side of the tubes. Particles smaller than about 3 μm can be deposited on the backside of the tubes due to turbulent diffusion.

TABLE 16.2
Common Mineral Matter in Coal

Type	Compound[a]	Formula
Clays	Illite	$KAl_2(AlSi_3O_{10})(OH)_2$
	Kaolinite	$Al_2Si_4O_{10}(OH)_2$
Sulfides	Pyrite	FeS_2
Carbonates	Calcite	$CaCO_3$
	Dolomite	$CaMg(CO_3)_2$
	Siderite	$FeCO_3$
Oxides	Quartz	SiO_2
Chlorides	Halite	$NaCl$
	Sylvite	KCl

[a] Compounds vary widely, only a few common ones are shown.

TABLE 16.3

Ash Fouling Index for Bituminous Coal

Fouling Index[a], R_{foul}	Severity
<0.2	Low
0.2–0.5	Medium
0.5–1.0	High
>1.0	Severe

[a] $R_{foul} = (m_B/m_A)(m_{Na_2O})$, where m_B is a mass of basic compounds such as Fe_2O_3, CaO, MgO, K_2O, and Na_2O; and m_A is a mass of acidic compounds such as SiO_2, Al_2O_3, and TiO_2.

Coals with relatively high sodium and potassium content tend to be more prone to fouling. A fouling index that has been developed for bituminous coals, explained in Table 16.3, provides criteria for rating a coal as having potentially high or low slagging characteristics. Additives, such as alumina powder, are sometimes used to mitigate fouling.

16.4   EMISSIONS FROM PULVERIZED COAL BOILERS

Particulate matter, sulfur dioxide, nitrogen oxides, and carbon dioxide are the most significant pollutant emissions from pulverized coal boilers. Mitigation of carbon dioxide emissions is discussed in the next section. For some coals, mercury emissions are significant. Control equipment for particulates and sulfur dioxide emissions generally captures some of the mercury. Hydrocarbons and carbon monoxide are not

TABLE 16.4

Best Available Emissions Control Technology for Pulverized Coal-Fired Boilers (Emission Output/Heat Input)

Pollutant	Emission Limit	Control Technology
Sulfur dioxide	<0.036 kg/GJ	Wet limestone FGD
Nitrogen oxides	<0.03 kg/GJ	LNB+OFA+SCR
Particulate matter	99.8% or <0.0064 kg/GJ	Fabric filter
Mercury	90% removal	Co-benefit capture[a]

Source: Haslbeck, J. L., Black, J., Kuehn, N., Lewis, E., Rutkowski, M. D., Woods, M., and Vaysman, V., *Pulverized Coal Oxycombustion Power Plants, Volume 1: Bituminous Coal to Electricity Final Report*, DOE/NETL-2007/1291, National Energy Technology Laboratory, U.S. Department of Energy, 2nd rev., 2008.

FGD: flue gas desulfurization.

LNB: low NO_X burners.

OFA: over-fire air.

SCR: selective catalytic reduction with ammonia injection.

[a] Mercury captured in FDG scrubber and fabric filter.

emitted in significant amounts in a properly operating system. New pulverized coal power plants are required to use the best available control technology (Table 16.4).

Ash in the fuel is converted to bottom ash and fly ash. Particulate emissions refer to fly ash. Carbon in the fly ash is very low since carbon burnout is generally good. About 40% of the fly ash tends to be less than 10 μm in size and if emitted tends to remain suspended in the atmosphere for several weeks. Particles less than 2.5 μm can penetrate into human lungs causing adverse health effects. Particulate control equipment in the exhaust stream is required to reduce particulate emissions. Electrostatic precipitators typically collect more than 99% of the particulates in the flue gas, and fabric filters (called baghouses) collect about 99.8% of the particulates.

Sulfur in the fuel is readily converted to sulfur dioxide during combustion. Approximately 4% of the sulfur combines with the calcium and magnesium in the fly ash to form calcium and magnesium sulfate particulate. About 1%–2% of the sulfur forms sulfuric acid that condenses on the submicron particles and contributes to the plume opacity. Wet scrubbers are required to control sulfur dioxide emissions unless the coal has very low sulfur (typically less than 0.6% sulfur). A slurry of pulverized limestone ($CaCO_3$) or hydrated lime ($Ca(OH)_2$) is sprayed into the flue gas, the water is evaporated and the resulting calcium sulfate hydrate ($CaSO_4 \cdot 2H_2O$) particles (gypsum) are separated from the flue gas with particulate control equipment. Generally scrubbers are designed to capture 90% of the sulfur dioxide. Injection of dry limestone directly into the combustion chamber is not effective because the temperature is too high.

Nitrogen oxide emissions from pulverized coal boilers are typically 95% nitric oxide and 5% nitrogen dioxide. The flame temperatures and residence times are such that approximately 75% of the NO_X emissions are formed from fuel bound nitrogen, and 25% is formed thermally from nitrogen in the burner air. Some of the fuel bound nitrogen appears in the volatile flame as HCN and NH_3, and some appears as the char burns out in the radiant section of the boiler. Improved burner design helps to limit the initial formation on NO_X. Two examples of low NO_X burner designs are shown in Figures 16.7 and 16.8. Coal and lean primary air (15%–30% stoichiometric) flow in the center of the burner. Secondary air with carefully controlled inner and outer swirl gradually mix air into the lean core to keep the peak temperature as low as possible while maintaining a stabilized flame in the burner with sufficient carbon burnout.

To achieve sufficiently low NO_X emissions to meet emission standards in new coal-fired boilers even with low NO_X burners, the flue gas is passed through a mesh of catalytic membranes, typically coated with vanadium, that reduces the remaining NO_X to N_2, with 80%–90% efficiency. This is called selective catalytic reduction (SCR). Additionally, urea (($NH_2)_2CO$) is sprayed into the flue gas upstream of the SCR to facilitate the reaction.

16.5   CARBON DIOXIDE CAPTURE AND SEQUESTRATION

Because the volume of flue gas in a utility scale coal-fired power plant is large, elimination of carbon dioxide emissions from a coal-fired power plant to the atmosphere is a significant undertaking. One approach being considered is to add a second wet scrubber to absorb the CO_2 and then desorb the concentrated CO_2 and compress it for transport in a pipeline to a geological site suitable for sequestration by deep

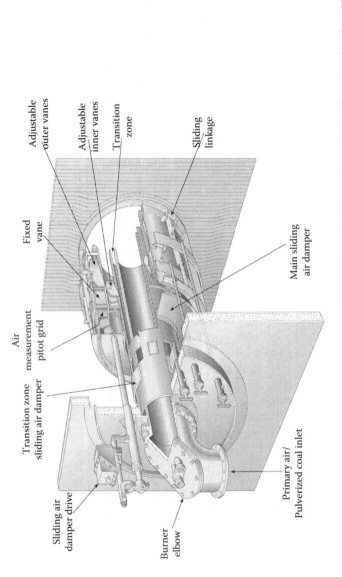

FIGURE 16.7 Low NO$_X$ pulverized coal burner. (From Kitto, J. B. and Stultz, S. C., Steam: *Its Generation and Use*, 41st ed., Babcock and Wilcox, 2005, courtesy of The Babcock and Wilcox Company, Barberton, OH.)

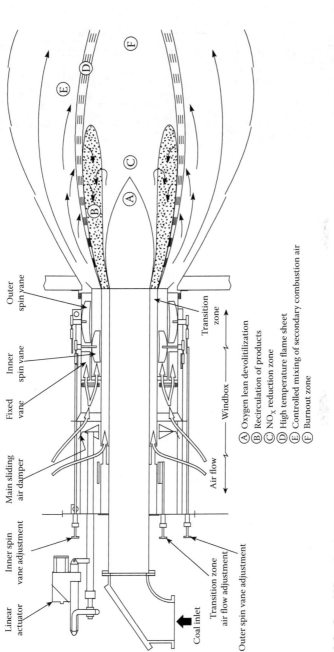

FIGURE 16.8 Low NO$_x$ pulverized coal burner showing combustion zones. (From Kitto, J. B and Stultz, S. C., Steam: Its Generation and Use, 41st ed., Babcock and Wilcox, 2005, courtesy of The Babcock and Wilcox Company, Barberton, OH.)

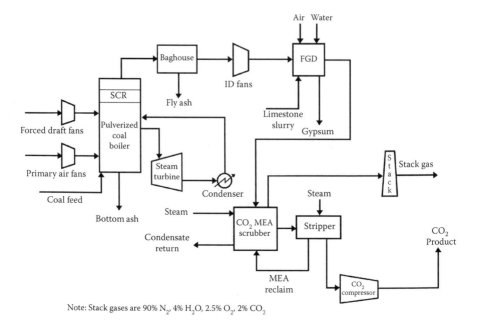

Note: Stack gases are 90% N_2, 4% H_2O, 2.5% O_2, 2% CO_2

FIGURE 16.9   Flow diagram of a pulverized coal and oxygen-fired boiler with NO_X, particulate, SO_X, and CO_2 capture. (From Haslbeck, J. L., Black, J., Kuehn, N., Lewis, E., Rutkowski, M. D., Woods, M., and Vaysman, V., *Pulverized Coal Oxycombustion Power Plants, Volume 1: Bituminous Coal to Electricity Final Report*, DOE/NETL-2007/1291, National Energy Technology Laboratory, U.S. Department of Energy, 2nd rev., 2008.)

injection in geological formations such as old oil wells or deep unusable coal mines where the carbon dioxide is adsorbed onto porous surfaces.

To visualize the carbon capture process, consider the 550 MW ultra-supercritical boiler specified in Table 16.1. The system flow diagram is given in Figure 16.9, and the associated operating parameters are given in Table 16.5. The operating parameters chosen represent advanced technology projected to be available by 2015–2020. Steam conditions at the turbine inlet are 27.9 MPa and 732°C. Without carbon capture the net power plant efficiency based on the higher heating value is 44.6%, but with carbon capture the net power plant efficiency drops to 33.2%. Particulate removal is 98% efficient, NO_X removal 86%, SO_X removal 98%, Hg removal 90%, and CO_2 removal 90% efficient.

Another approach being considered is using oxygen rather than air for combustion of the coal. In this case the volume of flue gas through the carbon dioxide scrubber is reduced, as shown in Table 16.5. Oxygen is obtained by separating nitrogen from air cryogenically or by using ceramic membranes operated at 1.4 MPa and 800°C. However, because of the extra auxiliary power requirements, neither method of oxygen preparation improves the net power plant efficiency, and using oxygen instead of air does not significantly reduce the total power plant cost relative to using air: $1,414/kW using air without CO_2 capture, $2,312/kW using air with CO_2 capture, and $2,238/kW using oxygen with CO_2 capture. Thus carbon capture in this report reduced the power plant efficiency by 25% and increased the capital cost by 58%.

TABLE 16.5

Operating and Performance Parameters for a Large Coal-Fired Steam Boiler to Generate 550 Mw Net Electric Power with CO_2 Capture

Parameter	USC[c] Steam without CO_2 Capture	USC[c] Steam with CO_2 Capture	USC[c] Steam with Oxycombustion and CO_2 Capture
Steam pressure (MPa)	27.9	27.9	27.9
Steam temperature (°C)	732	732	732
Steam flow rate (kg/h)	1,388,000	1,845,000	2,368,000
Coal feed rate (kg/h)	164,000	220,000	221,000
Primary combustion air (kg/h)	377,000	506,000	500,000
Over-fire combustion air (kg/h)	1,228,000	1,648,000	1,628,000[a]
Limestone (kg/h)	18,400	22,400	22,300
Flue gas to baghouse (kg/h)	2,021,000	2,395,000	2,368,000
Flue gas to SO_2 scrubber (kg/h)	2,132,000	2,525,000	2,301,000
Compressed CO_2 product (kg/h)	NA	469,000	598,000
Gross electrical power generated (MW)	577	644	759
Net electrical power generated (MW)	550	550	550
Net power plant efficiency (%)	44.6	33.2	33.0
SO_X emissions[b] (ton/yr)	1,220	348	442
NO_X emissions[b] (ton/yr)	995	1,472	1,484
Particulate emissions[b] (ton/yr)	185	273	276
Hg emissions[b] (ton/yr)	0.016	0.024	0.002
CO_2 emissions[b] (ton/yr)	2,983,000	428,000	267,000

Source: Haslbeck, J. L., Black, J., Kuehn, N., Lewis, E., Rutkowski, M. D., Woods, M., and Vaysman, V., *Pulverized Coal Oxycombustion Power Plants, Volume 1: Bituminous Coal to Electricity Final Report*, DOE/NETL-2007/1291, National Energy Technology Laboratory, U.S. Department of Energy, 2nd rev., 2008.

[a] 26% O_2, 7% N_2, 53% CO_2, 11% H_2O.

[b] Assumes 85% annual capacity factor.

[c] ultra-supercritical

16.6 BIOMASS-FIRED BOILERS

Large electric power plants with suspension burning boilers can co-fire biomass with coal, provided the biomass is ground or pulverized. Bales of switchgrass or other grasses are broken apart and pulverized or ground to fibers with a mass average length of 25 mm or less and with a fiber thickness of about 1 mm. Biomass is harvested relatively close to the power plant to minimize transportation costs if transported by truck, but if barge or rail transport is used distance from the power plant is not as important. Co-firing pulverized agricultural residues or grasses can be done up to 10% on a heat replacement basis without changing burner design. Excess air is maintained at 2%–3% oxygen in the flue gas. Wood can be co-fired by adding woodchips to the coal pulverizers at up to 5% replacement without degrading the coal pulverizers. The moisture content of the woodchips should not exceed 25%.

The design problem is to determine how fine the biomass must be pulverized for proper combustor performance. Since biomass char is more reactive than coal char, the particle size for biomass can be larger than for coal. The particle size distribution must also be taken into account. Both burner stability and particle burnout in the boiler must be considered. For smaller industrial size systems, burners for sander dust and other finely pulverized biomass are commercially available as well.

Biomass gasification is another approach for replacing fossil fuels. In gasification the biomass is gasified in a separate gasifier that produces a syngas that is piped to the burners of a boiler. The syngas may replace part or all of the fossil fuel. Gasification of solid fuels will be discussed in the next chapter.

16.7 PROBLEMS

1. A large pulverized coal-fired dry bottom furnace is used with a Rankine cycle steam turbine to provide 500 MW of electrical power. Subbituminous coal with a higher heating value of 20.1 MJ/kg as-received, 10% ash, and 3% sulfur is used. An electrostatic precipitator removes 99% of the particulate, and a wet limestone scrubber removes 90% of the SO_2 from the exhaust. The overall system efficiency is 35%. Calculate the following:
   a. The coal feed rate (T/h)
   b. The particulate emissions (T/h) and (kg/10^6 kJ input)
   c. The sulfur dioxide emissions (T/h) and (kg/10^6 kJ input)
   d. The annual fuel cost, assuming full power every hour of the year and the coal costs $30/ton
   e. The limestone ($CaCO_3$) feed rate (T/h) if a Ca/S mole ratio of 3 is required

2. Bituminous coal is pulverized and burned in a furnace. The ultimate analysis is 5% H, 82% C, and 13% O on a dry ash free basis. The higher heating value is 32.5 MJ/kg on a dry ash free basis. The ash content is 10% on a dry basis. Assume that the fuel moisture is vaporized in the pulverizer. The fuel and 20% of the air enter the furnace at 298 K. Secondary and tertiary air enter at 600 K. Overall 10% excess air is used. Assume that the coal burns completely near the fuel nozzles. Use the gas property data of Appendix C.
   a. What is the heat transfer per mass of coal (as-received) in MJ/kg to the walls needed to limit the peak flame temperature to 1660 K?
   b. What is the heat transfer per mass of coal (as-received) needed to drop the temperature from 1700 K to 1100 K in the radiant section of the furnace?
   c. What is the heat lost out of the stack per kg of coal (as-received) if the stack gas temperature is 170°C and the ambient temperature is 25°C?

3. A uniform cloud of 50 μm coal char particles flows in a hot gas stream containing 10% excess oxygen. Compute and plot the char particle mass/initial mass versus time. Use an effective char kinetic rate constant of 0.5 g/($cm^2 \cdot s \cdot atm_{O_2}$), a char density of 0.8 g/cm³, and an initial oxygen pressure of 0.1 atm.

4. A uniform cloud of 50 μm wood char particles flow in a hot gas stream containing 10% excess oxygen. Compute and plot the char particle mass/ initial mass versus time. Use an effective char kinetic rate constant of 0.5 $g/(cm^2 \cdot s \cdot atm_{O_2})$, a char density of 0.4 g/cm^3, and an initial oxygen pressure of 0.05 atm (since the volatiles will consume much of the oxygen in the air prior to char combustion).

5. Consider an isothermal plug flow of 40 μm char particles with 20% excess air at 1600 K and 1 atm. What is the number of particles per cm^3? What is the distance between each of the particles?

6. Consider an isothermal cloud of coal particles that contains 50% by weight 50 μm particles and 50% by weight 100 μm particles flowing in a stream of hot, isothermal gas containing 20% excess oxygen. Determine the particle burnout time based on 99% burnout and plot the particle mass divided by the initial particle mass as a function of time. Use an effective char kinetic rate constant of 0.5 $g_{O_2}/(cm^2 \cdot s \cdot atm_{O_2})$, a char density of 0.8 g/cm^3, and an initial oxygen pressure of 0.1 atm. Repeat for 100 μm particles and compare the burnout times based on 99% burnout. Discuss the difference between the burnout of a mono-sized particle cloud of Example 16.1 and the two-size particle cloud.

7. Consider an isothermal plug flow of 50 μm char particles flowing in a stream of hot, isothermal gas containing 20% excess oxygen. Using a char reaction rate constant that is due to diffusional effects only and thus changes as the particle shrinks, determine the particle burnout time based on 99% burnout and plot the particle mass divided by the initial particle mass as a function of time. Use the shrinking sphere constant density model. The gas temperature is 1600 K., a char density of 0.8 g/cm^3, and an initial oxygen pressure of 0.1 atm. Compare the results of your analysis with the results of Example 16.1.

8. For a uniform cloud of 70 μm char particles flowing in a stream of hot, isothermal gas; for 5%, 10%, and 20% excess air determine (a) the particle burnout time based on 99% burnout and (b) plot the particle mass divided by the initial particle mass as a function of time. Use an effective char kinetic rate constant of 0.5 $g_{O_2}/(cm^2 \cdot s \cdot atm_{O_2})$, a char density of 0.8 g/cm^3, and an initial oxygen pressure of 0.1 atm.

9. For the conditions of Example 16.1, estimate the pyrolysis time for bituminous coal particles of 50 μm diameter and of 100 μm diameter using the information given in Chapter 14. Assume the particle temperature is 1000 K and 1500 K. How does this compare to the char burnout times given in Example 16.1?

10. Look up the respective melting points of each of SiO_2, Al_2O_3, Fe_2O_3, CaO, MgO, K_2O, Na_2O, HgO, Hg_2O, and PbO. Comment on the fate of each of these compounds once they leave the convective section and air heater of the boiler.

11. Given a Rosin-Rammler size distribution of char particles,

$$y_i = 1 - \exp\left[-\left(\frac{d_i}{d_0}\right)^q\right]$$

with $d_0 = 50$ μm and $q = 1.2$, find the diameters which divide the size distribution into 10 equally weighted size increments ($y = 0.05, 0.15, 0.25, \ldots,$ 0.95) based on mass.

12. For the 550 MW power plant specified in Table 16.5 and Figure 16.9 without CO_2 scrubbing, assume the average gas temperature in the radiant section of the boiler is 1500 K, the cross section of the radiant boiler is 10 m by 10 m, and the height of the radiant section above the burners is 25 m. Calculate the gas velocity and the particle residence time in the radiant section of the boiler. Is your result reasonable? Explain why the size of the boiler will need to change if a CO_2 scrubber is added and a net power output of 550 MW is specified.

13. For the 550 MW power plant specified in Table 16.5, assume that the coal contains 9.7% ash and 2.51% sulfur. Calculate the degree of emission control (%) of the particulate and sulfur dioxide emissions for the three power plant configurations.

REFERENCES

Aerts, D. J., Bryden, K. M., Hoerning, J. M., Ragland, K. W., and Weiss, C. A., "Co-Firing Switchgrass in a 50 MW Pulverized Coal Boiler," *Proc. Am. Power Conf.* 59(2):1180–1185, 1997.

Backreedy, R. I., Jones, J. M., Pourkashanian, M., and Williams, A., "Burn-out of Pulverized Coal and Biomass Chars," *Fuel* 82(15–17):2097–2105, 2003.

Breen, B. P., "Combustion in Large Boilers: Design and Operating Effects on Efficiency and Emissions," *Symp. (Int.) Combust.* 16:19–35, The Combustion Institute, PA, 1977.

Cooper, C. D. and Alley, F. C., *Air Pollution Control*, 3rd ed., Waveland Press, Long Grove, IL, 2002.

Elliott, M. A. ed., *Chemistry of Coal Utilization: Second Supplementary Volume*, Wiley-Interscience, New York, 1981.

El-Wakil, M. M., *Powerplant Technology*, McGraw-Hill, New York, 1984.

Essenhigh, R. H., "Coal Combustion," in *Coal Conversion Technology*, eds. C. Y. Wen and E. S. Lee, 171–312, Addison-Wesley, Reading, MA, 1979.

Haslbeck, J. L., Black, J., Kuehn, N., Lewis, E., Rutkowski, M. D., Woods, M., and Vaysman, V., *Pulverized Coal Oxycombustion Power Plants, Volume 1: Bituminous Coal to Electricity Final Report*, DOE/NETL-2007/1291, National Energy Technology Laboratory, U.S. Department of Energy, 2nd rev., 2008.

Kitto, J. B. and Stultz, S. C., *Steam: Its Generation and Use*, 41st ed., Babcock and Wilcox, Barberton, OH, 2005.

Lawn, C. J., ed., *Principles of Combustion Engineering for Boilers*, Academic Press, London, 1987.

Rayaprolu, K., *Boilers for Power and Process*, CRC Press, Boca Raton, FL, 2009.

Reid, W. T., "The Relation of Mineral Composition to Slagging, Fouling and Erosion During and After Combustion," *Prog. Energy Combust. Sci.* 10(2):159–175, 1984.

Singer, J. G. and Owens, K. R., *Combustion: Fossil Power Systems: A Reference Book on Fuel Burning and Steam Generation*, 4th ed., Combustion Engineering Power Systems Group, 1993.

Smart, J. P. and Weber, R., "Reduction of Nitrogen Oxides (NO_x) and Optimization of Burnout with an Aerodynamically Air-Staged Burner and an Air-Staged Precombustor Burner," *J. Inst. Energy* 62(453):237–245, 1989.

Smoot, L. D. and Pratt, D. T., *Pulverized Coal Combustion and Gasification: Theory and Applications for Continuous Flow Processes*, Plenum Press, New York, 1989.

Smoot, L. D. and Smith, P. J., *Coal Combustion and Gasification*, Plenum Press, New York, 1985.

Smoot, L. D. ed., *Fundamentals of Coal Combustion For Clean and Efficient Use*, Elsevier, New York, 1993.

Syred, N., Claypole, T. C., and MacGregor, S. A., "Cyclone Combustors," in *Principles of Combustion Engineering for Boilers*, ed. Lawn, C. J., 451–519, Academic Press, London, 1987.

Tomeczek, J., *Coal Combustion*, Krieger, Malabar, FL, 1994.

U.S. Environmental Protection Agency, *Compilation of Air Pollutant Emission Factors, Vol. 1: Stationary Point and Area Sources*, 5th ed., EPA-AP-42, 1995.

Wall, T. F., "The Combustion of Coal as Pulverized Fuel through Swirl Burners," in *Principles of Combustion Engineering for Boilers*, ed. Lawn, C. J., 197–335, Academic Press, London, 1987.

Weber, R., Dugue, J., Sayre, A., and Visser, B. M., "Quarl Zone Flow Field and Chemistry of Swirling Pulverized Coal Flames: Measurement and Computations," *Symp. (Int.) Combust.* 24:1373–1380, The Combustion Institute, Pittsburgh, PA, 1992.

Williams, A., Pourkashanian, M., and Jones, J. M., "The Combustion of Coal and Some Other Solid Fuels," *Proc. Combust. Inst.*, 28(2):2141–2162, 2000.

Wornat, M. J., Hurt, R. H., Yang, N. Y. C., and Headley, T. H., "Structural and Compositional Transformations of Biomass Chars During Combustion," *Combust. Flame*, 100(1–2):131–143, 1995.

17 Fluidized Bed Combustion

Fluidized bed combustion is the third type of approach for utilizing solid fuels. Historically, fixed bed combustors were developed first, followed by suspension burning systems, and more recently by fluidized bed systems. Which of the three types of systems to select depends on the type and size of the solid fuel and the end use. In general fluidized beds have been used for crushed coal where some sulfur capture is required but a flue gas scrubber is not desired, and for biomass fuels that are too big for suspension firing or too variable in size or moisture for grate firing.

A *fluidized bed* is a bed of solid particles that are set into motion by blowing a gas stream upwards through the bed. The velocity of the gas stream must be large enough to locally suspend the particles (fluidize the bed) but small enough to ensure that particles are not blown out of the bed. The bed is similar to a liquid in that it appears to be boiling and exhibits buoyancy and a hydrostatic head. Hence it is called a fluidized bed. The main components of a fluidized bed are the air plenum, the air distributor, the bed, and the freeboard as shown in Figure 17.1. The freeboard is used to disengage the particles that are thrown up above the bed and to complete the combustion of volatiles and fine particles that do not burn completely in the bed. This type of fluidized bed is called a *bubbling fluidized bed*. A second type of fluidized bed—a *circulating fluidized bed*—operates at a higher gas velocity that entrains part of the fuel and bed particles and recirculates them back into the lower part of the bed.

Fluidized beds have been used by the petroleum industry for many years to catalytically crack crude oil to make gasoline, and they are used for many other applications, such as metallurgical ore roasting, limestone calcination, and petrochemical production. While gaseous and liquid fuels may be burned in fluidized beds, the main application of fluidized bed combustion systems is for the combustion of solid fuels such as biomass and coal. Furnaces and boilers utilize fluidized beds operating at atmospheric pressure. Pressurized fluidized beds are being developed for gasification of solid fuels to power gas turbine–steam turbine combined cycle systems. Gasification is like combustion except that a substoichiometric amount of air is used to produce a low heating value gas rich in hydrogen and carbon monoxide.

Fluidized bed boilers overcome several limitations of conventional stoker-fired and pulverized fuel boilers. Fluidized bed boilers have low NO_x emissions because the bed temperature is relatively low. Fluidized beds have good fuel flexibility and can burn a wide range of solid fuels with widely varying ash and moisture contents. Higher moisture fuels can be burned in fluidized beds without danger of combustion instability or loss of ignition. High ash fuels can be burned with less danger of slagging because the combustion temperature is lower. Because the bed is well mixed,

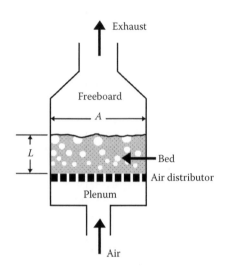

FIGURE 17.1   Schematic diagram of a fluidized bed.

feeding and distributing the fuel is easier. For coal a limestone bed material can be used that captures some of the sulfur dioxide.

This chapter focuses on the fundamentals of fluidization and in-bed combustion processes related to bubbling fluidized beds, circulating fluidized beds, and pressurized fluidized beds. Since fluidized beds are equally well suited to operate as gasifiers or combustors, gasification systems are also discussed.

17.1   FLUIDIZATION FUNDAMENTALS

As noted earlier, a fluidized bed consists of the plenum, distributor, bed, and freeboard. Generally the bed is the primary component in a fluidized bed system. Typically the bed is composed of a large number of solid particles in a variety of sizes and shapes. As a practical matter, the effective diameter of the bed material can be determined from a screen analysis that separates the bed material into a series of size increments. A mean diameter is calculated from the screen analysis based on the mean specific particle surface area (the particle surface area divided by the particle volume), which is appropriate for consideration of pressure drop and surface reaction rates. The mean specific surface area for the particles in the bed is

$$\bar{A}_p = \sum_i \left[y_i \frac{surface\ area_i}{volume_i} \right] = \sum_i \left[y_i \frac{6}{d_i} \right] \tag{17.1}$$

where y_i is the mass fraction of increment i and d_i is the diameter of increment i. The mean diameter, $\bar{d}$, is found by assuming an equivalent sphere with the same surface area to volume ratio as the mean specific particle surface area:

$$\bar{A}_p = \frac{6}{\pi \bar{d}^3} \cdot \frac{\pi \bar{d}^2}{1} = \frac{6}{\bar{d}} \tag{17.2}$$

Combining Equations 17.1 and 17.2,

$$\sum_i \left[y_i \frac{6}{d_i} \right] = \frac{6}{\bar{d}} \qquad (17.3)$$

Simpifying,

$$\bar{d} = \frac{1}{\displaystyle\sum_i y_i / d_i} \qquad (17.4)$$

Note that this is equivalent to the Sauter mean diameter, defined in Section 9.2 (see Problem 17.1), but is obtained from mass fraction rather than number count.

The bed will occupy a certain total volume, and there will be a certain void space containing only gas within that volume. The void fraction, ε, is

$$\varepsilon = \frac{\text{Void Volume}}{\text{Bed Volume}} = \frac{V_{void}}{V_{bed}} \qquad (17.5)$$

The void volume fraction is approximately equal to the fraction of the cross section occupied by voids at any point in the bed. The mean particle surface area per unit volume of bed, $\bar{A}_v$ can be found from

$$V_{bed} = V_{void} + V_p \qquad (17.6)$$

Dividing Equation 17.6 by the bed volume and simplifying with Equation 17.5,

$$\frac{V_p}{V_{bed}} = 1 - \varepsilon \qquad (17.7)$$

Multiplying Equation 17.2 by Equation 17.7,

$$\left(\frac{6}{\bar{d}} \right) \left(\frac{V_p}{V_{bed}} \right) = (1 - \varepsilon) \bar{A}_p$$

The quantity on the left hand side is the mean particle surface area per particle volume multiplied by the particle volume per unit bed volume. Simplifying,

$$\bar{A}_v = (1 - \varepsilon) \bar{A}_p \qquad (17.8)$$

where $\bar{A}_v$ is the mean particle surface area per unit volume of bed. The effective local velocity through the bed is called the *interstitial velocity*, $\underline{V}_I$,

$$\underline{V}_I = \frac{\dot{V}}{A\varepsilon} \tag{17.9}$$

where $\dot{V}$ is the volumetric flow rate through the bed, and A is the cross sectional area of the bed as shown in Figure 17.1. The *superficial velocity*, $\underline{V}_S$, is simply the gas velocity if the bed were not present,

$$\underline{V}_S = \frac{\dot{V}}{A} \tag{17.10}$$

Example 17.1

For the screen analysis below find the mean particle diameter, $\bar{d}$, mean specific surface area for the particles in the bed, $\bar{A}_p$, and the mean particle surface area per unit volume of bed, $\bar{A}_v$. The void fraction is 0.40.

Tyler Mesh No.	Mesh Size (mm)	Weight on Screen (kg)
8	2.36	0
10	1.65	60
14	1.17	80
20	0.83	40
35	0.42	20
48	0.29	0
Total		200

Solution:

First we need to determine the mass fractions in each size range and evaluate Equation 17.4. Note that d_i is the average diameter for a particular mesh. For example, the average diameter for Mesh No. 10 (the first bin of particles) is

$$d_i = \frac{2.36 + 1.65}{2} = 2.00 \text{ mm}$$

Determining the rest of the diameters and compiling the results in a table

d_i (mm)	y_i	y_i/d_i (mm^{-1})
2.00	0.30	0.150
1.41	0.40	0.283
1.00	0.20	0.200
0.62	0.10	0.160
0.35	0.00	0.000
Sum	1.00	0.793

Hence from Equation 17.4 the particle surface mean diameter is

$$\bar{d} = \frac{1}{0.793 \text{ mm}^{-1}} = 1.26 \text{ mm}$$

From Equation 17.2 the mean specific surface area for the particles in the bed is

$$\bar{A}_p = \frac{6}{1.26 \text{ mm}} \cdot \frac{1000 \text{ mm}}{\text{m}} = 4760 \text{ m}^2/\text{m}^3$$

From Equation 17.8 the mean particle surface area per unit bed volume is

$$\bar{A}_v = \bar{A}_p(1-\varepsilon) = 4760(1-0.4) = 2856 \text{ m}^2/\text{m}^3$$

17.1.1 PRESSURE DROP ACROSS THE BED

Consider a bed in which the flow rate is not sufficient to fluidize the bed, but rather it remains as a packed bed. The gas flows up through a tortuous path in the bed. In an analogous manner to pipe flow, the pressure drop across the bed divided by the depth of bed, L, is related to the interstitial velocity, an effective diameter of the tortuous path through the bed, and the density and viscosity of the gas:

$$\frac{\Delta p}{L} = function\left(\underline{V}_I, d_{eff}, \rho_g, \mu\right) \tag{17.11}$$

From dimensional analysis there are two dimensionless groups that may be written as

$$\left(\frac{\Delta p}{L}\right)\left(\frac{d_{eff}}{\rho \underline{V}_I^2}\right) = function\left(\frac{\underline{V}_I d_{eff} \rho_g}{\mu}\right) \tag{17.12}$$

The effective diameter of the tortuous path may be obtained from the hydraulic diameter concept. The hydraulic diameter is four times the volume of fluid divided by the surface area wetted by the fluid.

$$d_{eff} = 4\frac{V_{void}}{V_{bed}}\frac{1}{\bar{A}_v} = \frac{4\varepsilon}{\bar{A}_v} \tag{17.13}$$

Substituting Equations 17.8 and 17.2 yields

$$d_{eff} = \frac{4\varepsilon}{(1-\varepsilon)\bar{A}_p} = \frac{2}{3}\frac{\varepsilon \bar{d}}{(1-\varepsilon)} \tag{17.14}$$

Following the practice with pipe flow, Equation 17.12 is written as

$$f_{pm} = function\left(\text{Re}_{pm}\right) \tag{17.15}$$

where f_{pm} is the porous media friction factor and Re_{pm} is the porous media Reynolds number. The porous media friction factor is

$$f_{pm} = \left(\frac{\Delta p}{L}\right)\left(\frac{d_{eff}}{\rho_g V_1^2}\right) \tag{17.16}$$

Substituting Equations 17.2 and 17.14 into Equation 17.16 and dropping the "2/3" term yields

$$f_{pm} = \left[\frac{\varepsilon^3 \overline{d}}{\rho_g \underline{V}_s^2 (1-\varepsilon)}\right]\left(\frac{\Delta p}{L}\right) \tag{17.17}$$

In the same way the porous media Reynolds number, Re_{pm}, becomes

$$\text{Re}_{pm} = \frac{V_1 d_{eff}\rho_g}{\mu} = \frac{\overline{d}\,\underline{V}_s\,\rho_g}{(1-\varepsilon)\mu} \tag{17.18}$$

Experiments in packed beds have shown that Equation 17.15 holds and has the functional form,

$$f_{pm} = \left(\frac{150}{\text{Re}_{pm}}\right) + 1.75 \tag{17.19}$$

which is the Ergun equation (Kuinii and Levenspiel 1991). The first term represents laminar flow and the second term represents turbulent flow, which because of the extreme roughness, is independent of the Reynolds number. Combining Equations 17.17–17.19, the pressure drop across a packed bed can be written as

$$p = \frac{150\,\underline{V}_s\,\mu\,(1-\varepsilon)^2\,L}{\varepsilon^3\,\overline{d}^2} + \frac{1.75\,\underline{V}_s^2\,(1-\varepsilon)\rho_g\,L}{\varepsilon^3\,\overline{d}} \tag{17.20}$$

17.1.2 Minimum Fluidization Velocity

The onset of fluidization occurs when the drag force on the particles in the bed due to the upward flowing gas just equals the weight of the bed. This is equivalent to stating that the pressure drop across the bed times the area of the bed just equals the weight of the bed at the onset of fluidization:

$$\Delta pA = LA(1-\varepsilon)(\rho_p - \rho_g)g \tag{17.21}$$

Substituting Equation 17.20 into Equation 17.21 and rearranging slightly, a quadratic equation is obtained for the minimum superficial fluidization velocity, V_{mf},

$$\frac{1.75\rho_g}{\bar{d}\varepsilon^3}V_{mf}^2 + \frac{150\mu(1-\varepsilon)}{\bar{d}^2\varepsilon^3}V_{mf} = (\rho_p - \rho_g)g \tag{17.22}$$

The first term is negligible if Re_{pm} is less than 10, and the second term is negligible if Re_{pm} is greater than 1000.

As the flow rate is increased beyond the point of minimum fluidization, the pressure drop remains essentially constant. When the flow rate is increased further, eventually the particles will be blown out of the bed. The limiting case is the single particle terminal velocity, V_t. This behavior is summarized in Figure 17.2. Since V_t is at least ten times greater than V_{mf} for uniform size particles, there is a range of operating flow rates for fluidization. Of course, if the bed contains large particles and small particles, the small particles will be blown out of the bed before the large particles.

17.1.3 SINGLE PARTICLE TERMINAL VELOCITY

The terminal velocity of a single particle is obtained by setting the weight of the particle equal to the drag. The drag is

$$F_D = C_D A_F \frac{\rho_g V_{slip}^2}{2} \tag{17.23}$$

where V_{slip} is the slip velocity between the gas and particle, and A_F is the frontal area of the particle. Thus in equilibrium,

$$\left(\frac{\pi d^3}{6}\right)\rho_p g = C_D \left(\frac{\pi d^2}{4}\right)\left(\frac{\rho_g V_t^2}{2}\right)$$

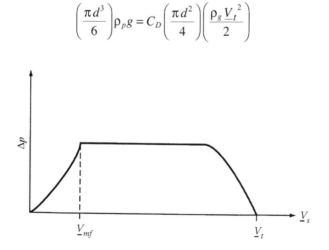

FIGURE 17.2　Pressure drop across a bed versus superficial velocity.

Rearranging,

$$V_t = \left[\frac{4 d \rho_p g}{3 C_D \rho_g} \right]^{1/2}$$

(17.24)

For a spherical particle,

$$C_D = \frac{24}{Re} + \frac{3.6}{Re^{0.313}} \qquad Re < 1000$$

(17.25a)

and

$$C_D = 0.43 \qquad 1000 < Re < 200{,}000$$

(17.25b)

17.1.4 BUBBLING BEDS

At minimum fluidization the bed is homogeneous in the sense that the gas surrounds each particle, and the particles are set into motion. The entire bed behaves as a *dense phase* fluid. As the superficial velocity is increased above minimum fluidization, *bubbles* are formed. These bubbles are essentially void of solids and contain primarily gas. The bubbles are referred to as the *dilute phase*. The initial size of the bubbles depends on the type of air distributor plate used, as indicated in Figure 17.3. A plate with a few large orifice inlets will have larger bubbles near the plate, while a plate

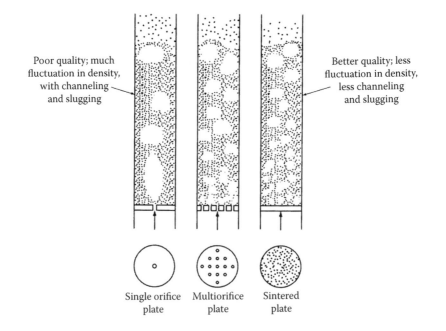

FIGURE 17.3 Quality of fluidization as influenced by the type of air distribution plate.

with many small inlets will have many small bubbles near the plate. However, as the bubbles rise they coalesce, growing in size, and the distribution between dense phase and dilute phase becomes independent of the inlet air distribution design.

The bubbles play a vital role in the behavior of the fluidized bed. In practice bubbles cannot be avoided. As the superficial velocity is increased, the added gas flow goes into the bubbles (dilute phase) rather than into the dense phase. The dense phase essentially remains at minimum fluidization. As the bubbles rise due to buoyancy, they carry along nearby solid particles, thus providing the main mechanism for large scale mixing of the bed. When the bubbles reach the top of the bed, they break through the surface, and the upward momentum flings the particles above the bed. If the superficial velocity above the bed is less than the terminal velocity of a given particle, it will fall back to the bed; otherwise, the particle will be entrained by the exhaust. In vigorously bubbling beds the particles may be flung upwards for distances of several bed depths, and a transport disengagement height or freeboard must be provided to allow the particles to fall back into the bed.

The rise rate of a single bubble of diameter d_B in a fluidized bed is given by

$$V_B = K \sqrt{\frac{g\, d_B}{2}} \qquad (17.26)$$

where K is a constant which depends on the size and shape of the particles and is generally about 0.9. For millimeter size bed particles the velocity of the rising bubbles is typically less than the interstitial velocity, and the bubbles coalesce due to lateral motion rather than being overtaken from below. Heat transfer tubes within the bed tend to break up large bubbles should they form.

With deep beds operating at high superficial velocities without in-bed tubes, the bubbles may grow until they occupy the whole cross-sectional area of the bed. These bubbles carry a group or slug of particles ahead of them until instability occurs and the particles collapse back to the bed. This is referred to as the *slugging* mode of operation, which can cause considerable vibration of the system and is generally undesirable. More fine particles will tend to be elutriated (blown out of the bed) under slugging conditions.

In the case of fluidized bed combustors, the fuel may be of a different size and density than the bed material. If the fuel is lighter and larger than the bed material, the fuel may tend to segregate or float on the top of the bed. In many instances this can be overcome by setting the flow rate so that the bed is bubbling vigorously. Of course, this will also tend to elutriate small particles from the bed.

Due to the bubbles there are pressure fluctuations within the bed, and these pressure fluctuations can feed back across the inlet distributor plate, thus altering the flow rate. Experience has shown that a pressure drop across the inlet air distributor plate of about 12% of the pressure drop across the bed will isolate the inlet plenum sufficiently from the bed pressure fluctuations. Various inlet air distributor plate designs have been utilized to ensure good fluidization. The nozzle standpipe type of distributor, shown in Figure 17.4a, has been found to be well suited for fluidized bed combustion. The air enters the bed from holes at the top of the nozzle, and the static bed forms an insulating layer between the hot fluidizing layer and the base plate. The nozzles are usually arranged on a pitch of 75 to 100 mm over the base plate

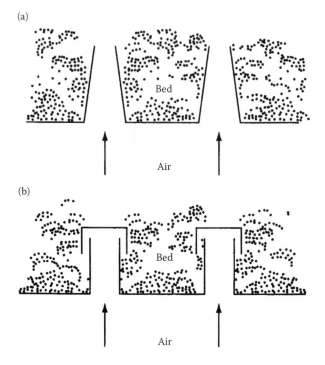

FIGURE 17.4 Air distributor plate for a bubbling fluidized bed: (a) nozzle standpipe type, and (b) bubble cap type.

and have a diameter of 12 to 15 mm and a height of 50 to 100 mm. The nozzle hole size is a compromise between having an excessive number of nozzles and preventing particles from falling through. The bubble cap design, shown in Figure 17.4b, is sometimes used to prevent bed particles from falling through.

17.1.5 HEAT AND MASS TRANSFER IN THE BED

Heat and mass transfer considerations play an important role in the design and operation of fluidized bed combustors. Heat transfer from gas to particles is important in determining the rate of heat up and devolatilization of solid fuel particles. Botterill (1975) reviewed the literature in this area and recommends the following expression for a particle Nusselt number for millimeter sized particles up to 20 atm pressure:

$$\mathrm{Nu}_p = 0.055\,\mathrm{Re}_p^{0.77}\left(\frac{\rho_g}{\rho_{g,0}}\right) \tag{17.27}$$

where $\rho_{g,0}$ is the gas density at atmospheric pressure and ρ_g is the gas density at the working pressure. The increase in heat transfer with increasing pressure is attributed to improvement in the quality of fluidization due to more gas passing through the dense phase.

The heat transfer to tubes immersed in a fluidized bed increases with decreasing bed particle size and with increasing tube size. The heat transfer coefficient is 5–10 times higher than for conventional gas-to-surface heat exchange surfaces, depending on the particle size. For example, in fluidized bed combustors burning coal in beds with millimeter sized bed particles, heat transfer coefficients to water tubes of approximately 200–350 $W/m^2 \cdot K$ are observed.

Mass transfer of oxygen from the gas to the fuel particles sets the burning rate of the char. Part of the inlet air passes through the bubble phase and does not contact the particles. Mass transfer theories to account for the bubbles tend to be complex. A certain fraction of the oxygen, say 20%–40%, does not react with the char surface, but rather reacts with the volatiles in the bed or in the freeboard. Considering diffusion within the dense phase (excluding the bubbles), the oxygen diffusion equation, assuming a relatively quiescent gas, may be written as

$$\frac{d}{dr}\left[\varepsilon r^2 \frac{d}{dr}(\rho_{O_2})\right] = 0 \qquad (17.28)$$

where r is radius not reaction rate. The solution, following the solution method in Section 14.3, is

$$\frac{\tilde{h}_D \bar{d}}{D_{AB}} = Sh = 2\varepsilon \qquad (17.29)$$

where ε, the void fraction, is associated with minimum fluidization, and typically for combustion applications may be taken as 0.4. For significant flow through the dense phase, Equation 17.29 may be modified based on Equation 14.28 to give

$$Sh = \left[2\varepsilon + 0.6 \frac{Re_{mf}^{0.5} Sc^{0.33}}{\varepsilon^{0.5}}\right]\phi \qquad (17.30)$$

Using Equation 17.30, the theory in Chapter 14 may be used for the burning rate of single particles in a fluidized bed with certain additional features discussed in the next section.

17.2 COMBUSTION IN A BUBBLING BED

In fluidized bed combustion the bed temperature is maintained well below the melting point of the ash in the fuel to eliminate slagging of the ash. For biomass combustion the bed is typically made of sand that is about 1 mm in size. When using coal, SO_2 capture is often desired. In this case an active bed such as limestone ($CaCO_3$), typically 0.3–0.5 mm in size, is used. The limestone is reduced to calcium oxide (CaO) in the bed, and the optimum temperature for a CaO reaction with SO_2 to form $CaSO_4$ is 815–870°C. This temperature is low for combustion compared to pre-mixed flames, suspension burning, and grate burning of solid fuels; however, this is a perfectly adequate temperature for most applications. For example, current steam turbine temperatures generally

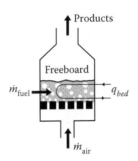

FIGURE 17.5 Schematic diagram for mass and energy balance.

do not exceed 540–700°C. The volatiles burn in the bed and in the freeboard. The superficial gas velocity is set at 2–3 m/s to minimize unburned carbon carryover out of the combustor. In addition to facilitating sulfur dioxide capture and soft ash particles, the low bed temperature results in relatively low NO_x emissions, less volatilization of alkali compounds, and less erosion of in-bed boiler tube surfaces.

To control the bed temperature, boiler tubes are immersed in the bed. To find what fraction of the heat release of the fuel can be removed in the bed without decreasing the bed temperature below a given temperature, let us consider mass and energy balances on a control volume around the bed as shown in Figure 17.5. First we will neglect the effect of the bubbles and assume that combustion is complete within the bed, then we will allow for some combustion above the bed, and finally we will include the effect of the bubbles and combustion above the bed.

17.2.1 Neglect Bubbles and Assume Complete Combustion in the Bed

Conservation of mass across the bed is

$$\dot{m}_{air} + \dot{m}_f = \dot{m}_{prod} \tag{17.31}$$

The flow rate of products depends on density of the gaseous products, the superficial fluidization velocity, and the cross-section area of the bed,

$$\dot{m}_{prod} = \rho_g \underline{V}_s A \tag{17.32}$$

Combining Equations 17.31 and 17.32 and using the fuel-air ratio, f, the fuel feed rate becomes

$$\dot{m}_f = \rho_g \underline{V}_s A \left(\frac{f}{1+f} \right) \tag{17.33}$$

The energy equation across the bed with heat transfer from the bed to in-bed tubes, q_B, and assuming 100% heat release in the bed is

$$\dot{m}_{air} h_{air} + \dot{m}_f h_f = \dot{m}_{prod} h_{prod} + q_{bed} \tag{17.34}$$

Using sensible enthalpies and assuming complete combustion,

$$\dot{m}_{air}h_{air,s} + \dot{m}_f LHV = \dot{m}_{prod}h_{prod,s} + q_{bed} \tag{17.35}$$

Dividing by $\dot{m}_f$ and introducing

$$r_{bed} = \frac{q_{bed}}{\dot{m}_f\, LHV} \tag{17.36}$$

where r_{bed} is the fraction of input heat extracted from the bed, the energy equation becomes

$$h_{air,s} + f(LHV)(1 - r_{bed}) = (1 + f)h_{prod,s} \tag{17.37}$$

Equation 17.37 shows the relationship between the heat extracted and the fuel-air ratio required to maintain a given bed temperature. Of course, the fuel-air ratio should not exceed the stoichiometric fuel-air ratio.

Example 17.2

A fluidized bed burns dry bituminous coal with a lower heating value of 25,000 kJ/kg. The stoichiometric fuel-air ratio for this coal is 0.100 (Table 3.1). The bed temperature is maintained at 877°C, and the inlet air is at 127°C. Neglect the bubbles. Assume that the combustion is complete in the bed and neglect any ash in the fuel. Find the fuel-air ratio for bed heat removal fractions from 0 to 0.7.

Solution

Assuming the products to be air (for simplicity) and using Appendix B with a reference temperature of 300 K for compatibility with the lower heating value,

$$h_{air,s} = (401.3 - 300.4) \; kJ/kg = 100.9 \; kJ/kg$$

$$h_{prod,s} = (1219.4 - 300.4) \; kJ/kg = 919.0 \; kJ/kg$$

Equation 17.37 becomes

$$100.9 + 25,000 f(1 - r_{bed}) = 919.0(1 + f)$$

rearranging,

$$f = \frac{1}{29.44 - 30.56 r_{bed}}$$

Tabulating r_{bed} versus f,

Heat Removal Ratio, r_{bed}	Fuel-air Ratio, f
0	0.0340
0.1	0.0379
0.2	0.0429
0.3	0.0493
0.4	0.0581
0.5	0.0706
0.6	0.0901
0.7	>0.1 (stoich.)

This table gives the relation between r_{bed} and f which is necessary to maintain a bed temperature of 877°C. As shown in the graph below, the relationship between the fraction of heat extracted and the fuel-air ratio is not linear. Rather the fuel required increases significantly as the heat removal ratio increases. Knowing the fuel-air ratio, the fuel feed rate can be obtained from Equation 17.33.

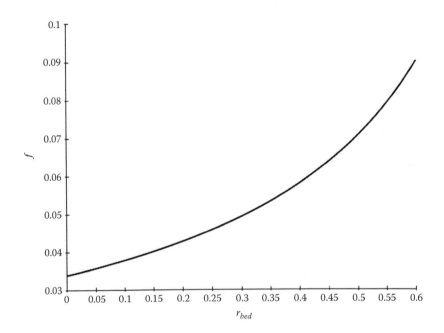

In order to extract the heat, q_{bed}, from the bed with in-bed heat exchanger tubes, the tubes must have sufficient surface area, A_t, such that,

$$q_{bed} = \tilde{h} \, A_t \, \Delta T_{LM}$$

(17.38)

where A_t is the area of the tubes and ΔT_{LM} is the log mean temperature difference across the tubes.

$$T_{LM} = \frac{(T_{bed} - T_{out}) - (T_{bed} - T_{in})}{\ln\left(\dfrac{T_{bed} - T_{out}}{T_{bed} - T_{in}}\right)} \tag{17.39}$$

where T_{in} and T_{out} are the temperatures into and out of the heat exchanger. Assuming cylindrical heat exchanger tubes, the area of the heat exchanger tubes can be related to the volume occupied by the heat exchanger tubes.

$$\frac{A_t}{V_t} = \frac{\pi d L_t}{\pi d^2 L_t / 4} = \frac{4}{d} \tag{17.40}$$

where V_t and L_t are the volume occupied by the heat exchanger tubes. Introducing Φ, the fraction of the bed volume occupied by the tubes, the area of the heat exchanger tubes can be related to the bed depth, the cross-sectional area of the bed, and the tube diameter

$$A_t = \frac{4 \Phi L A}{d_t} \tag{17.41}$$

For a given heat exchanger design, it is apparent from Equation 17.38 and Equation 17.41 that the fluidized bed must have sufficient depth to provide for a given heat removal.

Example 17.3

Find the required bed height for a fluidized bed combustor burning wood waste with the following design parameters:

$\underline{V}_s = 3$ m/s	$d_t = 25.4$ mm
$T_B = 877°C$	$\Phi = 0.17$
$T_{air} = 127°C$	$p = 1$ atm
LHV = 15.0 MJ/kg	$r_{bed} = 0.5$
$T_{in} = 127°C$	$\tilde{h} = 250$ W/(m² · K)
$T_{out} = 827°C$	$f_{air,s} = 0.169$

Solution

Rearranging Equation 17.36,

$$q_{bed} = r_{bed} \dot{m}_f \, (\text{LHV})$$

Substituting Equation 17.33 for $\dot{m}_f$,

$$q_{bed} = r_{bed}\left(\frac{f}{1+f}\right)\rho_g V_{-s} A(\text{LHV})$$

Dividing by the area of the bed and using the ideal gas law,

$$\frac{q_{bed}}{A} = r_{bed}\left(\frac{f}{1+f}\right)\frac{P_g}{RT_{bed}}V_s(\text{LHV})$$

Assuming the molecular weight of the gas is 29.0,

$$\frac{q_{bed}}{A} = 0.5\left(\frac{0.169}{1.169}\right)\left(\frac{101.3 \text{ kPa}}{1} \cdot \frac{29 \text{ kg}}{\text{kgmol}} \cdot \frac{1}{1150 \text{ K}} \cdot \frac{\text{kgmol} \cdot \text{K}}{8.314 \text{ kPa} \cdot \text{m}^3}\right)\frac{3 \text{ m}}{\text{s}} \cdot \frac{15 \text{ MJ}}{\text{kg}}$$

$$\frac{q_{bed}}{A} = \frac{1.00 \text{ MW}}{\text{m}^2}$$

Substituting Equation 17.38 into Equation 17.41 and solving for the bed depth,

$$L = \left(\frac{q_{bed}}{A}\right)\frac{d_t}{4\tilde{h}\Phi\ T}$$

where from Equation 17.39,

$$T = \frac{(877-827)-(877-127)}{\ln\left(\dfrac{877-827}{877-127}\right)} = 258 \text{ K}$$

Substituting

$$L = \left(\frac{1000 \text{ kW}}{\text{m}^2}\right)\frac{0.0254 \text{ m}}{4(0.17)(258 \text{ K})} \cdot \frac{\text{m}^2 \cdot \text{K}}{0.250 \text{ kW}}$$

$$L = 0.58 \text{ m}$$

The minimum bed depth to cover the tubes is 0.58 m, and allowing for clearance at the top and bottom of the bed, a bed height of 1 m is reasonable.

17.2.2 Neglect Bubbles but Include Some Combustion above the Bed

Assume that combustion within the bed consumes completely any pyrolysis gases or char gasification gases rapidly relative to the time to consume the solid fuel. Based on this, the combustion products in the bed are N_2, CO_2, and H_2O and some O_2. Let X equal the fraction of solid fuel burned in the bed and $\dot{m}_{bed,prod}$ equal the mass flow rate of combustion products in the bed. Conservation of mass for the bed is

$$\dot{m}_{air} + X\dot{m}_f = \dot{m}_{bed,prod} \tag{17.42}$$

Conservation of mass in the freeboard above the bed is

$$\dot{m}_{bed,prod} + (1 - X)\dot{m}_f = \dot{m}_{prod} \tag{17.43}$$

Conservation of energy for the bed is

$$\dot{m}_{air}h_{air} + X\,\dot{m}_f h_f = \dot{m}_{bed,prod}h_{bed} + q_{bed} \tag{17.44}$$

Relationships similar to those in Examples 17.2 and 17.3 can be developed for this case also.

17.2.3 INCLUDE THE EFFECT OF BUBBLES AND SOME COMBUSTION ABOVE THE BED

To account for the bubbles, assume that a certain fraction of air, B, flows through the dense phase, and the remaining fraction $(1 - B)$ passes through the bed via the bubbles without reacting with the fuel in the bed. Further assume that there is no heat and mass transfer between the bubbles and the dense phase. Conservation of mass for the dense phase for the bed is

$$(1 - B)\dot{m}_{air} + X\dot{m}_f = \dot{m}_{D,prod} \tag{17.45}$$

Remembering that the mass of gas in the dense phase corresponds to the minimum fluidization velocity with the remainder of the gas in the bubbles,

$$\dot{m}_{D,prod} = \rho_g \underline{V}_{mf} A$$

Conservation of mass for the freeboard is

$$\dot{m}_{D,prod} + B\dot{m}_{air} + (1 - X)\dot{m}_f = \dot{m}_{prod} \tag{17.46}$$

The dense-phase bed energy equation is

$$(1 - B)\dot{m}_{air}h_{air} + X\dot{m}_f h_f = \dot{m}_{D,prod}h_{bed} + q_{bed} \tag{17.47}$$

Relationships similar to Examples 17.2 and 17.3 can be developed for this case. Advanced theories of fluidization account for mass and heat transfer between the dense phase and the bubbles (Howard 1983).

17.2.4   Fuel Hold-Up in the Bed

As indicated in Example 17.2, when the in-bed heat load is increased, the fuel feed rate is increased to maintain the bed temperature. This means that the excess air and the burning rate of the fuel changes. With a fluidized bed the fuel solids circulate in the bed until they are burned up, except for fines that tend to be blown out of the bed. As the burning rate changes, the amount of fuel in the bed at any given instant, called the fuel hold-up, also changes. The airflow must be great enough to fluidize the bed and yet low enough to prevent excessive entrainment of fine particles from the bed.

The amount of fuel in the bed at any given time (the fuel hold-up) may be approximated from the fuel feed rate and the total burnout time of the individual particle,

$$m_f = \dot{m}_f \, t_{char} \tag{17.48}$$

Since the char burn is typically an order of magnitude longer than the volatile burn, the single particle char burn time from Chapter 14 provides a starting point. The shrinking sphere model modified by Equation 17.30 to account for the bed material surrounding the fuel particle may be used to obtain the burn time. However, the oxygen concentration will be lower in the dense phase than in the freeboard due to the bypassing effect of the bubbles. Furthermore, there are two additional complicating factors: coal particles tend to *fracture* into smaller sizes, and the particle surface regression rate is increased by *attrition* due to collisions with the bed material. Hence the apparent burning rate of a single particle in a fluidized bed is more rapid than that of a free burning particle.

Fracturing of coal particles occurs when the coal particles are fed into the high heat transfer environment of the fluidized bed. During devolatilization pressure builds up inside the particle, causing fracture planes and fragmentation of the parent particle. The fracture planes are especially evident in lignite coals, as shown in Figure 17.6.

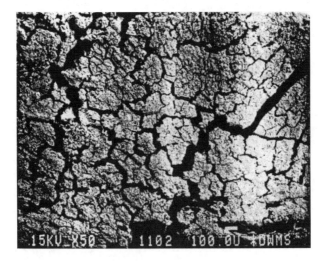

FIGURE 17.6   Microphotograph of a lignite char particle after removal from a fluidized bed combustor. The white mark near the lower right corner is 100 μm.

Fragmentation during devolatilization is called primary fragmentation, and typically one to three primary fragmentations occur per particle. As the char burns, the fracture planes deepen and secondary fragmentation takes place, thus producing an additional 20%–50% increase in the number of char particles. Also, as the char burns, the surface of the char is weakened since the oxygen penetrates into the char, and as bed particles collide with the char surface, small pieces of char are broken off by a process of attrition. In addition, char particles can collide with in-bed tubes causing further fragmentation.

The processes of fragmentation and attrition are shown schematically in Figure 17.7. Fragmentation alters the size distribution, and attrition increases the apparent particle burning rate. Experiments suggest that the attrition rate per unit bed volume can be calculated from an attrition rate constant, k_{attr}, the difference between the superficial velocity and the minimum fluidization velocity, and the surface area per unit volume of the char in the bed,

$$\frac{\dot{m}_{attr}}{V} = k_{attr} \frac{\left(\underline{V}_s - \underline{V}_{mf}\right)\rho_{char}\,\bar{A}_p}{6} \tag{17.49}$$

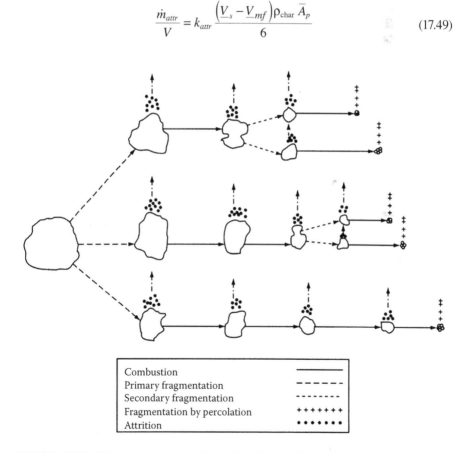

Combustion	———
Primary fragmentation	– – – –
Secondary fragmentation	··········
Fragmentation by percolation	+++++++
Attrition	•••••••

FIGURE 17.7 Schematic representation of coal particle fragmentation and attrition. (From Arena, U., D'Amore, M. and Massimilla, L., "Carbon Attrition During the Fluidized Combustion of a Coal," *AIChE J.* 29(1):40–49, 1983, reproduced with permission of the American Inst. of Chemical Engineers, © 1983, AIChE. All rights reserved.)

The attrition rate constant depends on the bed particle size and the type of char. Figure 17.8 shows that the attrition rate constant is 1×10^{-6} for coal char in 1 mm sand bed material. The effect of attrition is to increase the effective burning rate of char particles in a fluidized bed by a factor of 25%–100% compared to char particles burning in air without attrition. The net result is that the fuel consumption rate, $\dot{m}_f$, due to chemical reaction and attrition is such that the bed consists of roughly 95% inerts and 5% coal.

17.3 ATMOSPHERIC PRESSURE FLUIDIZED BED COMBUSTION SYSTEMS

Fluidized bed combustion systems are attractive because they permit better control over combustion, and thus peak flame temperatures are avoided. Slagging of the ash is minimized, corrosion problems are reduced, nitrogen oxide emissions are reduced, and sulfur control with limestone addition to the bed is possible. Since in-bed heat transfer coefficients are high, fluidized bed boilers are more compact than fixed bed and suspension burning systems. Fuel can be fed from the top or it can be fed under the bed by a ram or pneumatically. A wide range of fuels including high ash and high moisture fuels can be used.

The disadvantages of fluidized bed boilers compared to fixed bed or suspension burning boilers are that the pressure drop across the combustor is greater (thus requiring more fan power), ash carryover (including some bed material) is higher, and in-bed tube erosion can be high. With solid fuels that have a low ash fusion temperature, the bed material tends to agglomerate over long periods of time even when the bed temperature is maintained below 900°C.

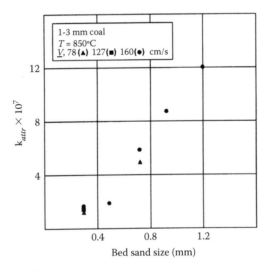

FIGURE 17.8 Attrition rate constant for coal char versus bed sand size. (From Arena, U., D'Amore, M. and Massimilla, L., "Carbon Attrition During the Fluidized Combustion of a Coal," *AIChE J.* 29(1):40–49, 1983, reproduced with permission of the American Inst. of Chemical Engineers, © 1983, AIChE. All rights reserved.)

A fluidized bed boiler, shown schematically in Figure 17.9, contains a bed, in-bed boiler tubes, sufficient freeboard height to avoid excessive elutriation of bed material and unburned carbon, a convective steam raising section, and a superheater section. Coal or biomass are fed from above the bed. The bed temperature is maintained below 900°C to avoid agglomerating the ash and above 800°C to avoid poor combustion efficiency. The gases leave the bed at the bed temperature. Combustion of fines and volatiles continues in the freeboard. In-bed heat transfer allows control of excess air without increasing the bed temperature, and as noted previously, the lower the excess air the higher the boiler efficiency up to the point where combustion is incomplete.

With a low grade, high moisture fuel, the heat release may be only slightly in excess of that required to heat the fuel and air to a bed temperature required for efficient combustion. In this case immersed tubes are not used, and the bed temperature is controlled by supplying excess air to remove heat from the bed for subsequent recovery in convective tube banks. With high-grade fuels combustion in a bed without heat transfer tubes requires 150% excess air or more to maintain a 900°C bed.

For optimum efficiency over the turndown range of a boiler, the excess air level should remain constant as the air supply and fuel feed rate are changed to vary the boiler output. Since the bed temperature can be raised only 100–200°C without increasing emissions of NO_x and SO_x, and since the load turndown should be 2:1 or better and 4:1 is desired, the main alternatives are to provide multiple compartments that can be shut off independently or to provide a means of gradually exposing in-bed tubes. When the fluidizing air is reduced, bed expansion is reduced, which

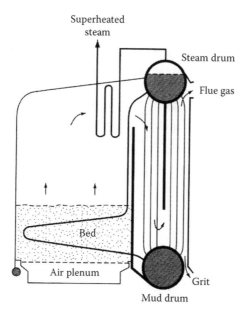

FIGURE 17.9 Schematic of a bubbling fluidized bed boiler. (From Howard, J. R. ed., *Fluidized Beds: Combustion and Application*, Applied Science Publishers, London, 1983.)

can provide a limited method of exposing bed tubes. Bed height can be raised by increasing the bed feed rate, and conversely bed material can be drained off to lower the bed.

17.3.1 EMISSIONS FROM FLUIDIZED BED BOILERS

The emissions of major significance are particulates, sulfur dioxide, and nitric oxide. Particulates include ash, carbon, and bed material. Essentially all of the ash in the fuel is blown out of the bed. About 1%–2% of unburned carbon may also be blown out of the bed. The challenge is to design the system so that the unburned carbon carryover is low. In addition the bed material may elutriate fine particles due to attrition. Particulate carryover must be controlled by a fabric filter baghouse or related device. The emission standards noted in Table 15.3 also apply to fluidized bed combustion of solid fuels.

Sulfur dioxide reacts with the limestone bed to form calcium sulfate. The efficiency of the sulfur dioxide removal depends on the calcium to sulfur molar ratio. Ca/S molar ratios of 2–5 are required for 90% removal of SO_2 in an atmospheric fluidized bed operating at 900°C depending on the reactivity of the limestone. If the bed temperature increases, the SO_2 removal efficiency decreases.

Nitrogen oxide emissions are lower with fluidized bed boilers than with spreader stokers and pulverized combustors since the combustion temperature is lower. Thermally formed NO is negligible. However, the fuel bound organic nitrogen is converted to NO, and as a starting point it can be assumed that half the fuel nitrogen is converted to NO.

17.4 CIRCULATING FLUIDIZED BEDS

As noted above, in a bubbling fluidized bed combustor small, lightweight particles such as char near burnout tend to be blown out of the combustor before combustion is complete. Also, a feed point is needed for approximately every 1 m² for good mixing of the fuel. Circulating fluidized beds have been developed to overcome the carbon carryover of bubbling beds and to facilitate uniform fuel feeding.

Fixed bed, bubbling fluidized bed, and circulating fluidized bed combustors are schematically represented in Figure 17.10. For circulating fluidized beds the velocity is increased beyond the particle entrainment velocity of the smaller particles (3–10 m/s) so that these solids are transported up the full height of the chamber and returned in the downward leg of a cyclone separator. A cyclone collector uses centrifugal force to separate particles from the gases. The pressure drop across the cyclone collector is a function of velocity and particle loading and is not insignificant.

This type of fluidization is characterized by high turbulence, back mixing of the solids, and the absence of a defined bed level. Fuel is fed into the lower part of the combustion chamber, and primary air is introduced through the grid plate. Particles rise with the gas flow in a dilute central core zone (void fraction > 90%). The larger particles successively diffuse toward the wall where they descend back toward the bottom. Secondary air in the upper part of the chamber is used to ensure circulation and combustion of the solids. The lower part of the bed consists of a comparatively

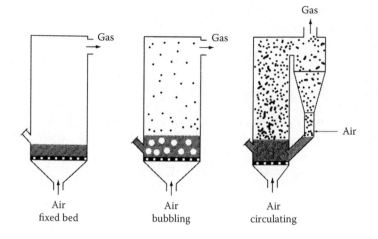

FIGURE 17.10   Fixed bed, bubbling fluidized bed, and circulating fluidized bed combustors.

dense zone (void fraction in the range of 60%–90%) resembling a bubbling fluidized bed. The smaller particles flow through the bed to the downcomer leg of the cyclone separator. A small amount of air is usually introduced near the bottom of the cyclone downcomer to control the return rate of the solids. Mixing of the fuel with the bed material is rapid due to the high turbulence.

In a typical circulating fluidized bed system heat from the cyclone exhaust is absorbed in a boiler bank, a superheater, an economizer, and an air heater as indicated in Figure 17.11. In the design shown, there are no heat transfer tubes in the bed. Velocities of 6 to 10 m/s in the main combustion chamber are typical. The fuel

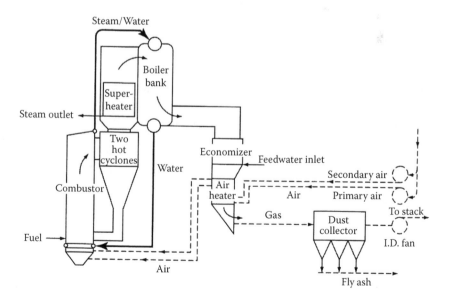

FIGURE 17.11   Circulating fluidized bed combustion system.

size is 25 mm or less (compared to 50 mm for bubbling bed combustors) and the bed particle size is less than 1 mm. The combustion chamber may have water walls; however, the lower portion near the inlet air grid is usually covered with refractory. The cyclone collectors located at the outlet of the combustion chamber are steel vessels lined with hard-faced refractory backed by lightweight insulating refractory. Char and bed particles continue to circulate until they are reduced to 5–10 μm, at which point they escape the cyclone collector. The fuel feed rate and air flow rates are adjusted depending on the steam load so that combustion takes place near 850°C. A turn down ratio of 3:1 can be achieved by reducing the air and fuel flow rates. Load changes of 50% can be made within a few minutes.

In sum, circulating fluidized beds have the ability to burn a wide variety of fuels such as various types of biomass, refuse, coal, and petroleum coke while minimizing unburned carbon carryover, and they are relatively straightforward to scale up in size.

17.5   PRESSURIZED FLUIDIZED BED GASIFICATION OF BIOMASS

Up to this point we have focused on atmospheric pressure systems for heating and steam power plants. As explained in the previous chapter, the efficiency of a steam power plant has been increasing over the years as steam temperature and pressure move into the ultra-supercritical region. For electric power generation a combined cycle power plant (gas turbine combined with a steam turbine) offers another approach to achieving high efficiency. Combined cycles operated on natural gas have net plant efficiencies of 60%. In a combined cycle the combustor is operated at elevated pressure appropriate for a gas turbine (greater than 20 atm). When solid fuels are used, the hot, pressurized combustion products must be cleaned up sufficiently to avoid erosion and corrosion of the turbine blades. The additional pressure drop across the combustor and hot gas cleanup system significantly reduces the net power plant efficiency as well as the reliability of the overall system. Because of this, gasification rather than combustion is considered a better approach for a solid fuel-fired combined cycle since the volume of the gas to be cleaned up is smaller. Pressurized biomass gasification is of particular interest since biomass has less inorganic matter than coal and hence less particulate to filter, and biomass use would not involve carbon dioxide capture and sequestration. Based on this, the last section of this chapter will discuss pressurized biomass gasification. Over the years various types of biomass gasifiers have been tried, and for pressurized combustion applications it turns out that directly heated (i.e., using the fuel itself as the source of heat) pressurized fluidized beds are preferable.

The objective of pressurized fluidized bed gasification is to power a gas turbine using solid fuels. Gasification uses a substoichiometric amount of air to produce as much hydrogen, carbon monoxide, and methane as possible with lesser amounts of carbon dioxide and water. After hot gas clean up the gasification products are burned with air in an integrated gas turbine combustor. A fluidized bed gasifier is like a fluidized bed combustor except that a gasifier is operated fuel rich with just enough air to generate sufficient heat to maintain the bed at a temperature of 900 to 1000°C.

TABLE 17.1
Primary Solid Fuel Gasification Reactions

Exothermic Reactions

Combustion	volatiles + char + $O_2 \rightarrow CO_2$	(a)
Partial Oxidation	volatiles + char + $O_2 \rightarrow CO$	(b)
Methanation	volatiles + char + $H_2 \rightarrow CH_4$	(c)
Water-Gas Shift	$CO + H_2O \rightarrow CO_2 + H_2$	(d)
CO Methanation	$CO + 3\,H_2 \rightarrow CH_4 + H_2O$	(e)

Endothermic Reactions

Steam-Carbon Reaction	char + $H_2O \rightarrow CO + H_2$	(f)
Boudouard Reaction	char + $CO_2 \rightarrow 2\,CO$	(g)

In the gasifier the solid fuel undergoes a multitude of exothermic and endothermic reactions, and the most important are shown in Table 17.1. Pyrolysis requires about 10–15% of the heat of reaction to raise the temperature, devolatilize the fuel, and crack the tars. Gasification at a higher temperature and a higher equivalence ratio results in a lower amount of tars in the product gas; however, this also results in a lower heating value product gas. When air is used as the oxidant, the product gas has a lower heating value of 4–6 MJ/m³ (at standard pressure and temperature). An example of gasification products from wood chips in a circulating fluidized bed gasifier is shown in Table 17.2. This example is from the Varnamo, Sweden

TABLE 17.2
Gasification Products from Wood Chips in a Circulating Fluidized Bed Operating at 950–1000°C and 18 Atm

Species	Dry Composition
Hydrogen	9.5–12 % vol
Carbon monoxide	16–19 % vol
Carbon dioxide	14.4–17.5 % vol
Methane	5.8–7.5 % vol
Nitrogen	48–52 % vol
Benzene[a]	5–9 g/m³
Light tars[a]	2.5–3.7 g/m³
Lower heating value[a]	5. 3–6.3 MJ/m³

Source: Stahl, K., Neergaard, M. and Nieminen, J., "Final Report: Varnamo Demonstration Program," in *Prog. Thermochem. Biomass Convers.* 5, vol. 1, ed. A. V. Bridgewater, 549–563, Blackwell Science Ltd., Oxford, UK, 2001. With permission from Wiley-Blackwell.

[a] At 1 atm, 293 K.

gasification combined cycle project. A simplified schematic of the Varnamo demonstration plant is shown in Figure 17.12, and performance parameters of the system are shown in Table 17.3. In this system wood chips were dried to 5%–20% using a flue gas dryer, then pressurized in a lock hopper and fed to the gasifier. The gasifier was operated at 950–1000°C and 18 atm pressure at an equivalence ratio of 0.27. Output from the gasifier was cooled to 400°C and passed through a bank of fine pore metal candle filters to remove particulates from the product stream. Filters in the shape of candles provide a large surface area and thus a lower velocity and pressure drop. Cooling of the product gas allowed partial condensation of the alkali in the gas, thus extending the life of the filters, but cooling did not result in excessive condensation of tar. Tars were burned up in the gas turbine combustor. Periodically filters were back flushed with a pulse of nitrogen to remove the ash cake that gradually builds up on them. The gas turbine operated stably on the low heat content gas. There was no indication of deposit formation caused by tars or particulate deposition on the piping or turbine blades after many thousands of hours of testing. This was a large research and development project that produced electric power and district heating for the city of Varnamo. Additional design optimization work is needed before this type of system is ready for commercial use.

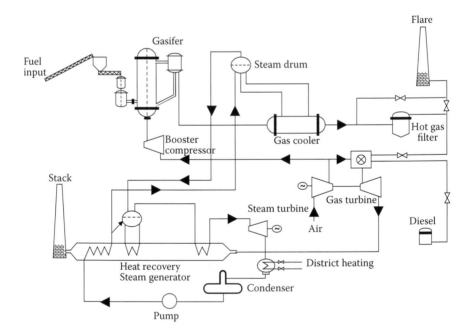

FIGURE 17.12 Process diagram of the Varnamo integrated gasification combined cycle demonstration plant. (From Stahl, K., Neergaard, M. and Nieminen, J., "Final Report: Varnamo Demonstration Program," in *Prog. Thermochem. Biomass Convers.* 5, vol. 1, ed. A. V. Bridgewater, 549–563, Blackwell Science Ltd., Oxford, UK, 2001, with permission from Wiley-Blackwell.)

TABLE 17.3

Performance Data for the Varnamo Integrated Gasification Combined Cycle Demonstration Plant

Heat input (based on LHV of wood chips)	18 MW
Net electrical power to grid	6 MW
Heat to district	9 MW
Net electrical efficiency (based on LHV)	32%
Total net efficiency (based on LHV)	83%
Gasifier temperature/pressure	950°C/18 atm
LHV of gasifier product[a]	5 MJ/m^3
Steam temperature/pressure	455°C/40 atm

Source: Stahl, K., Neergaard, M. and Nieminen, J., "Final Report: Varnamo Demonstration Program," in *Prog. Thermochem. Biomass Convers.* 5, vol. 1, ed. A. V. Bridgewater, 549–563, Blackwell Science Ltd., Oxford, UK, 2001. With permission from Wiley-Blackwell.

[a] At 1 atm, 20°C.

17.6 PROBLEMS

1. Show that Equation 17.3 for the surface mean diameter is equivalent to the Sauter mean diameter (SMD) of Chapter 9.
2. Calculate the minimum fluidization velocity for 0.2 and 2.0 mm diameter limestone particles that have a specific gravity of 2.0. Assume that the void fraction is 0.45 and that the air is at 1 atm and 900°C.
3. Calculate the terminal velocity for limestone particles (specific gravity of 2.0) with diameters of 0.2 mm and 2 mm. The fluid is air at 1 atm and 1200 K. Repeat for char particles with a specific gravity of 0.3.
4. Calculate and plot the pressure drop versus superficial velocity across a 0.5 m deep bed of crushed limestone. Assume the void fraction is 0.4, the bed temperature is 1200 K, and the pressure is 1 atm. The bed particle size is 2 mm. Let the velocity vary from zero to twice the minimum fluidization velocity. Use a limestone specific gravity of 2.0. Repeat for a pressure of 20 atm.
5. Repeat Example 17.2 but assume that as-received LHV is 20,000 kJ/kg and that the coal contains 30% ash. The specific heat of the ash is 0.7 kJ/kg · K. Hint: The ash will need to be included in the conservation of energy equation for the bed.
6. A fluidized bed operates with a superficial velocity of 1.5 m/s at 1 atm and 1200 K. Lignite coal is burned in the bed with 20% excess air. Assume that the fuel burns completely in the bed. The coal and air enter at 298 K. Use air properties to determine the density of the products. Use lignite coal with the following analysis:

Proximate Analysis (as-received)	
Moisture	35% (wt)
Volatile Matter	24%
Fixed Carbon	29%
Ash	12%

Ultimate Analysis (dry, ash-free)	
C	75% (wt)
H	6%
O	19%
HHV (dry, ash-free)	29.2 MJ/kg

Find the following:

a. The coal feed rate per unit area of bed $(kg/(h \cdot m^2))$
b. The heat input rate per unit area of bed (kW/m^2)
c. The heat removal rate from the bed per unit area of bed (kW/m^2)
d. The heat transfer rate from the products above the bed to drop the temperature to 400 K (kW/m^2).

7. For a 0.8 m deep atmospheric fluidized bed with a plan area of 10 m × 10 m using limestone with a bulk density of 1200 kg/m^3, find the pressure drop across the bed and the required blower power. Assume the bed temperature is 1250 K and the superficial velocity is 1.3 m/s.

8. A 250 m^2 atmospheric fluidized bed burns 12,000 kg coal/h. The bed consists of a 1 m depth of crushed limestone with a bulk density of 1400 kg/m^3. If the burn time of a typical coal particle is 180 s, what is the approximate percent by weight of the char in the bed?

9. Repeat Example 17.2 but with air preheat of 500 K, LHV = 25,000 kJ/kg, and bed temperature 1150 K. Find the fuel-air ratio versus the heat removal fraction. Assume combustion is complete in the bed.

10. Repeat Example 17.2 with 10% heat release occurring in the freeboard above the bed. The inlet air is preheated to 500 K, LHV = 25,000 kJ/kg, and the bed temperature is 1150 K. Find the fuel-air ratio versus the heat removal fraction.

11. Make a plot of superficial velocity through a fluidized bed versus heat input (MW) per bed cross-sectional area while holding the excess air constant at 10%. Let the velocity vary from 0.5 to 3.0 m/s. The coal has a higher heating value of 26,000 kJ/kg and a stoichiometric fuel-air ratio of 0.11. Do this for bed pressures of 1 atm and 10 atm, and an inlet air temperature of 300 K.

12. It is desired to build an atmospheric pressure fluidized bed boiler to deliver 100,000 kg/h of saturated steam at 8600 kPa using woodchips with a lower heating value of 17.5 MJ/kg. The desired superficial velocity in the bed is 2.5 m/s. The bed temperature is 1200 K. Estimate the coal feed rate, the cross-sectional area of the bed, and the excess air. Half of the steam is produced by tubes in the bed and half by tubes in the freeboard. Neglect pump and blower power. Assume air bubbles in the bed are minimal. The air enters at 400 K. The overall thermal efficiency is 85% based on the

lower heating value. The stoichiometric fuel-air ratio is 0.10, and all the combustion takes place in the bed. The inlet temperature of the feedwater is 300 K. Steam data provided: enthalpy of water = 88 kJ/kg, enthalpy of the steam leaving the boiler = 2747 kJ/kg. To obtain enthalpy and density of the products treat them as air. Make a sketch of the system. Calculate the fuel feed rate, the bed crosssectional area, and the gas temperature leaving the combustor.

REFERENCES

Arena, U., D'Amore, M., and Massimilla, L., "Carbon Attrition During the Fluidized Combustion of a Coal," *AIChE J.* 29(1):40–49, 1983.

Basu, P., *Combustion and Gasification in Fluidized Beds*, CRC Press, Boca Raton, FL, 2006.

Basu, P., Masayuki, H., and Hasatani, M., eds. *Circulating Fluidized Bed Technology III*, Pergamon Press, New York, 1991.

Botterill, J. S. M., *Fluid-Bed Heat Transfer: Gas-Fluidized Bed Behavous and Its Influence on Bed Thermal Properties*, Academic Press, London, 1975.

Chirone, R., Salatino, P., Scala, F., Sollimene, R., and Urcluolo, M., "Fluidized Bed Combustion of Pelletized Biomass and Waste-Derived Fuels," *Combust. Flame* 155(1–2):21–36, 2008.

Ciferno, J. P. and Morano, J. J., "Benchmarking Biomass Gasification Technology for Fuels, Chemicals and Hydrogen Production," U.S. Department of Energy, National Energy Technology Laboratory, 2002.

Gamble, R. L., "The 100,000-lb/h Fluidized-Bed Steam Generation System for Georgetown University," *Proc. Am. Power Conf.* 41:295–301, 1979.

Hafer, D. R. and Bauer, D. A., "AEP's Program for Enhanced Environmental Performance of PFBC Plants," *Proc. Am. Power Conf.* 55(1):127–132, 1993.

Howard, J. R. ed., *Fluidized Beds: Combustion and Application*, Applied Science Publishers, London, 1983.

Koornneef, J., Junginger, M., and Faaij, A., "Development of Fluidized Bed Combustion—An Overview of Trends, Performance and Cost," *Prog. Energy Combust. Sci.* 33(1):19–55, 2007.

Kunii, D. and Levenspiel, O., *Fluidization Engineering*, 2nd ed., Butterworth-Heinemmann, Newton, MA, 1991.

Maitland, J. E., Skowyra, R. S., and Wilheim, B. W., "Design Considerations for Utility Size CFB Steam Generators," *FACT (Am. Soc. Mech. Eng.)* 19:69–76, 1994.

Makansi, J. and Schwieger, R., "Fluidized-Bed Boilers," *Power* 131(5):Sl–Sl6, 1987.

McClung, J. D., Quandt, M. T., and Froelich, R. E., "Design and Operating Considerations for an Advanced PFBC Facility at Wilsonville, Alabama," *FACT (Am. Soc. Mech. Eng.)* 19:85–92, 1994.

Oka, S., *Fluidized Bed Combustion,* Marcel Dekker, New York, 2004.

Ragland, K. W. and Pecson, F. A., "Coal Fragmentation in a Fluidized Bed Combustor," *Symp. (Int.) Combust.* 22(1):259–265, The Combustion Institute, Pittsburgh, PA, 1989.

Stahl, K., Neergaard, M., and Nieminen, J., "Final Report: Varnamo Demonstration Program," in *Prog. Thermochem. Biomass Convers.* 5, vol. 1, ed. A.V. Bridgewater, 549–563, Blackwell Science Ltd., Oxford, UK, 2001.

Suksankraisorn, K., *Fluidized Bed Combustion and Identification*, VDM Verlag, Saarbrücken, Germany, 2009.

US Department of Energy, *The JEA Large-Scale CFB Combustion Demonstration Project*, Topical technical report 22, 2003.

Appendix A Properties of Fuels

TABLE A.1
Thermodynamic Properties of Alkane Fuels

Formula	Name	M	sg	T_b	p_v	$c_{p,g}$	$c_{p,l}$	T_{fg}	HHV	LHV	h_{fg}	a/f	Octane No. res.	Octane No. mot.	Δh^o	p_c	T_c
CH_4	methane	16.04	0.466	−161		2.21		537	55,536	50,048	510	17.2	120	120	−74.4	45.4	190
C_2H_6	ethane	30.07	0.572	−89		1.75		472	51,902	47,511	489	16.1	115	99	−83.8	48.2	305
C_3H_8	propane	44.10	0.585	−42	12.8	1.62	2.48	470	50,322	46,330	432	15.7	112	97	−104.7	41.9	370
C_4H_{10}	n-butane	58.12	0.579	0	3.51	1.64	2.42	365	49,511	45,725	386	15.5	94	90	−146.6	37.5	425
C_4H_{10}	isobutane[1]	58.12	0.557	−12	4.94	1.62	2.39	460	49,363	45,577	366	15.5	102	98	−153.5	36.0	408
C_5H_{12}	n-pentane	72.15	0.626	36	1.06	1.62	2.32	284	49,003	45,343	357	15.3	62	63	−173.5	33.3	470
C_5H_{12}	isopentane[2]	72.15	0.620	28	1.39	1.60	2.28	420	48,909	45,249	342	15.3	93	90	−178.5	33.5	460
C_6H_{14}	n-hexane	86.18	0.659	69	0.337	1.62	2.27	233	48,674	45,099	335	15.2	25	26	−198.7	29.7	507
C_6H_{14}	isohexane[3]	86.18	0.662	50	0.503	1.58	2.20	421	48,454	44,879	305	15.2	104	94	−207.4	30.9	500
C_7H_{16}	n-heptane	100.20	0.684	99	0.110	1.61	2.24	215	48,438	44,925	317	15.2	0	0	−224.2	27.0	540
C_7H_{16}	triptane[4]	100.20	0.690	81	0.229	1.60	2.13	412	48,270	44,757	289	15.2	112	101		29.2	531
C_8H_{18}	n-octane	114.23	0.703	126	0.036	1.61	2.23	206	48,254	44,786	301	15.1	20	17	−250.1	24.6	569
C_8H_{18}	isooctane[5]	114.23	0.692	114	0.117	1.59	2.09	418	48,119	44,651	283	15.1	100	100	−259.2	25.4	544
C_9H_{20}	n-nonane	128.26	0.718	151	0.012	1.61	2.21		48,119	44,688	288	15.1			−274.7	22.6	595
$C_{10}H_{22}$	n-decane	142.28	0.730	174	0.005	1.61	2.21		48,002	44,599	272	15.1			−300.9	20.9	618
$C_{10}H_{22}$	isodecane[6]	142.28	0.768	161		1.61	2.21					15.1	113	92		24.8	623
$C_{11}H_{24}$	n-undecane	156.31	0.740	196		1.60	2.21		47,903	44,524	265	15.0			−327.2	19.4	639
$C_{12}H_{26}$	n-dodecane	170.33	0.749	216		1.60	2.21		47,838	44,574	256	15.0			−350.9	18.0	658
$C_{13}H_{28}$	n-tridecane	184.35	0.756	236		1.60	2.21				246	15.0				17.0	676
$C_{14}H_{30}$	n-tetradecane	198.38	0.763	253		1.60	2.21				239	15.0				14.2	693
$C_{15}H_{32}$	n-pentadecane	212.45	0.768	271		1.60	2.21				232	15.0				15.0	707
$C_{16}H_{34}$	n-hexadecane[7]	226.43	0.773	287		1.60	2.21		47,611	44,307	225	15.0			−456.1	13.9	722
$C_{17}H_{36}$	n-heptadecane	240.46	0.778	302		1.60	2.21				221	15.0				12.8	733
$C_{18}H_{38}$	n-octadecane	254.50	0.782	317		1.60	2.21			47,542	44,256	214	15.0				11.8	748

Sources: Lide, D. L. and Haynes, W. M., eds., *CRC Handbook of Chemistry and Physics*, 90th ed. CRC Press, Boca Raton, FL, 2009; Bartok, W. and Sarofim, A. F., *Fossil Fuel Combustion: A Source Book*, Wiley, New York, 1991.

Notes: (1) 2-methylpropane; (2) 2,2-methylbutane; (3) 2,3-dimethylbutane; (4) 2,2,3-trimethylbutane; (5) 2,2,4-trimethylpentane; (6) 2,2,3,3-tetramethylhexane; (7) cetane.

Definitions: M, molecular weight; sg, specific gravity (density of substances at 20°C relative to water at 4°C, for gases determined at boiling point of the liquified gas); T_b, boiling point temperature, °C at 1 atm; p_v, vapor pressure at 38°C, atm; $c_{p,g}$, specific heat of the gas at 25°C, kJ/kg °C; $c_{p,l}$, specific heat of the liquid at 25°C, kJ/kg °C; T_{ig}, autoignition temperature, °C; HHV, higher heating value, kJ/kg, first 3 are gas, all others liquid; LHV, lower heating value, kJ/kg, first 3 are gas, all others liquid; h_{fg}, heat of vaporization at 1 atm and boiling point temperature, kJ/kg; a/f, stoichiometric air–fuel mass ratio; Octane-res., research octane no.; Octane-mot., motor octane no.; $\Delta h°$, enthalpy of formation at 25°C, kJ/gmol, first 3 are gas, all others liquid; p_c, critical point pressure, atm; T_c, critical point temperature, K.

TABLE A.2
Thermodynamic Properties of Other Hydrocarbon Fuels

Class	Formula	Name	M	sg	T_b	ρ_v	$c_{p,g}$	$c_{p,\ell}$	T_{ig}	HHV	LHV	h_{fg}	a/f	Octane No. res.	Octane No. mot.	Δh^o	p_c	T_c
Cyclo-alkane	C_5H_{10}	cyclopentane	70.13	0.746	49	0.673	1.135	1.836	385	46,936	43,798	389	14.8	101	85	−76.4	44.5	512
	C_6H_{12}	cyclohexane	84.16	0.779	81	0.224	1.214	1.813	270	46,573	43,435	358	14.8	84	78	−123.2	40.2	553
	C_7H_{14}	cycloheptane	98.19	0.810	119	0.058	1.181	1.826		46,836	43,698	335	14.8	39	41	−118.1	37.6	604
	C_8H_{16}	cyclooctane	112.2	0.835	149	0.021	1.173	1.838		46,943	43,808	309	14.8	71	58	−124.4	35.1	647
Aromatic	C_6H_6	benzene	78.11	0.874	80	0.224	1.005	1.717	592	41,833	40,145	393	13.3		115	82.6	48.3	562
	C_7H_8	toluene[1]	92.14	0.867	111	0.070	1.089	1.683	568	42,439	40,528	362	13.5	120	109	50.4	40.5	592
	C_8H_{10}	ethylbenzene	106.17	0.867	136	0.025	1.173	1.721	460	42,996	40,923	339	13.7	111	98	29.9	35.5	617
	C_8H_{10}	m-xylene[2]	106.17	0.864	139	0.022	1.164	1.692	563	42,873	40,800	342	13.7	118	115	17.3	34.9	617
Olefin	C_3H_6	propylene	42.08	0.519	−47	15.401	1.482	2.450		48,472	45,334	437	14.8	102	85	20.0	45.4	365
	C_4H_8	1-butene	56.11	0.595	−6	4.286	1.487	2.240		48,073	44,937	390	14.8	99	80	−0.1	39.7	420
	C_5H_{10}	1-pentene	70.13	0.641	30	1.293	1.524	2.178	298	47,766	44,528	358	14.8	91	77	−21.3	34.8	465
	C_6H_{12}	1-hexene	84.16	0.673	63	0.408	1.533	2.144	272	47,550	44,312	335	14.8	76	63	−43.5	31.3	504
Diolefin	C_5H_8	isoprene[3]	68.11	0.681	34	1.136	1.495	2.199		46,382	43,798	356	14.2	99	81	75.5	38.0	484
	C_6H_{10}	1,5-hexadiene	82.15	0.688	59	0.483	1.390	2.136		46,796	43,582	312	14.3	71	38	84.1	33.9	507
Cycloolefin	C_5H_8	cyclopentene	68.12	0.772	44		1.064	1.759		45,733	43,149		14.2	93	70	33.9	47.4	506
	C_6H_{10}	cyclohexene	82.15							45,674	42,995					−5.0		
Alcohol	CH_4O	methanol	32.04	0.791	65	0.310	1.370	2.531	385	22,663	19,915	1099	6.5	106	92	−201.5	79.8	513
	C_2H_6O	ethanol	46.07	0.789	78	0.153	1.420	2.438	365	29,668	26,803	836	9.0	107	89	−235.1	60.6	514
	C_3H_8O	1-propanol	60.10	0.803	97	0.061	1.424	2.395		33,632	30,709	690	10.5			−255.1	51.0	537
	$C_4H_{10}O$	1-butanol	74.10	0.805	118	0.022		2.391		36,112	33,142	584	11.1			−274.9	43.6	563
	$C_5H_{12}O$	1-pentanol	88.15	0.814	137			2.361		37,787	34,791	503	11.7			−298.9	38.6	588
	$C_6H_{14}O$	1-hexanol	102.18	0.814	158			2.353		38,994	35,979	436	12.2			−317.8	40.0	611

Sources: Lide, D. L. and Haynes, W. M., eds., *CRC Handbook of Chemistry and Physics*, 90th ed. CRC Press, Boca Raton, FL, 2009; Bartok, W. and Sarofim, A. F., *Fossil Fuel Combustion: A Source Book*, Wiley, New York, 1991.

Notes: 1) methylbenzene, 2) 1,3 dimethllbenzene, 3) 2-methyl-1,3-butadiene.

Definitions: M, molecular weight; sg, specific gravity (density of substances at 20°C relative to water at 4°C, for gases determined at boiling point of the liquified gas); T_b, boiling point temperature, °C at 1 atm; p_v, vapor pressure at 38°C, atm; $c_{p,g}$, specific heat of the gas at 25°C, kJ/kg °C; $c_{p,l}$, specific heat of the liquid at 25°C, kJ/kg °C; T_{ig}, autoignition temperature, °C; HHV, higher heating value of liquid, kJ/kg; LHV, lower heating value of liquid, kJ/kg; h_{fg}, heat of vaporization at 1 atm and boiling point temperature, kJ/kg; a/f, stoichiometric air-fuel mass ratio; Octane res., research octane no.; Octane mot., motor octane no.; Δh°, enthalpy of formation of liquid at 25°C, kJ/gmol; p_c, critical point pressure, atm; T_c, critical point temperature, K.

TABLE A.3
Flame Properties of Selected Fuels in Ambient Air

Fuel	Formula	Flash Point (°C)	Flammability Limits (vol % in air)		Autoignition Temperature (°C)	Stoichiometric Mixture			Laminar Flame Velocity	
			Lean	Rich		Vol % Fuel	Minimum Ignition Energy (10^{-5} J)	Quench Distance (mm)	Vol % Fuel	Max. Flame Velocity (cm/s)
Methane	CH_4	−188	5.0	15.0	537	9.47	33	1.9	9.96	33.8
Ethane	C_2H_4	−130	2.9	10.6	472	5.64	42	2.0	6.28	40.1
Propane	C_3H_8	−104	2.0	9.5	470	4.02	40	2.1	4.54	39.0
n–Butane	C_4H_{10}	−60	1.5	8.5	365	3.12	76	3.0	3.52	37.9
n–Hexane	C_6H_{14}	−20	1.1	7.7	233	2.16	95	3.6	2.51	38.5
Isooctane	C_8H_{18}	22	0.95	6.0	418	1.65	29	2.0	1.90	34.6
Acetylene	C_2H_2	−18	2.5	80	305	3	3	0.8	10.1	141
Carbon monoxide	CO		12.5	74	609	29.5			50	39
Ethanol	C_2H_5OH	12	3.3	19	365	6.52				41.2
Hydrogen	H_2		4	75	400	29.5	2	0.6	50	365

Source: Bartok, W. and Sarofim, A. F., *Fossil Fuel Combustion: A Source Book*, Wiley, New York, 1991.

TABLE A.4
Specific Heats of Selected Fuels

Fuel	c_p (kJ/kg·K)	
Hydrogen	14.3	at 298 K
Methane	$-42.05 + 27.48\theta^{0.25} - 1.55\theta^{0.75} + 20.24\theta^{-0.5}$	300 K < T < 2000 K
Propane	$-0.0918 + 0.692\theta - 0.0357\theta^2 + 0.00072\theta^3$	300 K < T < 1500 K
Petroleum fuel vapor	$0.136 + 0.12\theta (4.0 - sg)$	
Gasoline	2.4	at 298 K
Ethanol	2.5	at 298 K
Coal	1.3	
Wood	$0.387 + 0.103\theta$	280 K < T < 420 K

Notes: $\theta = T/100$ where T is in K; sg is specific gravity of the liquid; for dodecane and octane see Table A.9.

TABLE A.5
Specific Gravity and Bulk Density of Selected Fuels

Fuel Type	Specific Gravity (at 293 K)
Gasoline	0.73
Ethanol	0.79
Diesel fuel (No. 2)	0.85
Biodiesel	0.88
Kerosene	0.80
Fuel oil (No. 2)	0.85
Fuel oil (No. 6)	0.99
Lignite coal[a]	1.2
Bituminous coal[a]	1.4
Anthracite coal[a]	1.7
Wood, pine[a]	0.5
Wood, oak[a]	0.7

Fuel Type	Dry Bulk Density (kg/m³)
Stoker coal (bituminous)	780
Log wood (stacked hardwood)	460
Hardwood chips	300
Softwood chips	200
Straw, miscanthus (block bales)	140
Straw (round bales)	85
Straw (pellets)	500
Refuse-derived fuel (pelletized)	550
Refuse-derived fuel (fluff)	180

[a] dry.

TABLE A.6
Transport Properties of Selected Fuels

Fuel	Viscosity (N·s/m²)	Thermal Conductivity (W/m·K)
Gaseous Fuels		
Hydrogen	8.8×10^{-6}	0.1805 at 300 K
Methane[a]	1.08×10^{-5}	$-1.869 \times 10^{-3} + 8.727 \times 10^{-5}T + 1.179 \times 10^{-7}T^2 - 3.614 \times 10^{-11}T^3$
Propane[a]	–	$1.858 \times 10^{-3} - 4.698 \times 10^{-6}T + 2.177 \times 10^{-7}T^2 - 8.409 \times 10^{-11}T^3$
Ethanol vapor[a]	–	$-7.79 \times 10^{-3} + 4.167 \times 10^{-5}T + 1.214 \times 10^{-7}T^2 - 5.184 \times 10^{-11}T^3$
Liquid Fuels[b]		
Gasoline	5.0×10^{-4}	0.117(1 − 0.00054T)sg
No. 2 diesel	2.4×10^{-3}	0.117(1 − 0.00054T)sg
Ethanol	8.3×10^{-4}	0.169 at 293 K
No. 2 fuel oil	2.3×10^{-3}	0.117(1 − 0.00054T)sg
No. 6 fuel oil	0.36	0.117(1 − 0.00054T)sg
Solid Fuels		
Wood	–	0.238 + sg(0.200 + 0.404MC)
Coal	–	0.25

Notes: sg is specific gravity, T is in K, and MC is moisture content. For dodecane and octane see
        Table A.9.
[a] Viscosity at 288 K, conductivity range 273–1270 K.
[b] Viscosity is at 313 K; thermal conductivity is for the vapour.

TABLE A.7
Surface Tension of Several Fuels and Water (Typical Values)

Substance	Surface Tension (N/m)
Gasoline	0.023
Kerosene	0.025 − 0.03
No. 2 fuel oil	0.029 − 0.032
n-Octane[a]	0.0218
Water	0.0728

[a] 20°C in contact with vapor.

TABLE A.8
Latent Heat of Vaporization of Selected Fuels

Fuel	Latent Heat (kJ/kg)
Ethanol	846
Gasoline	339
Kerosene	291
Light diesel	267
Medium diesel	244
Heavy diesel	232

TABLE A.9
Properties of Dodecane and Octane

n-Dodecane Properties	Equation	Range, K
Density of liquid, kg/m³	$893.09 - 0.31187T - 6.2772 \times 10^{-4}T^2$	311–478
Heat of vaporization, kJ/kg	$269.31 + 0.686T - 1.51 \times 10^{-3}T^2$	366–478
Specific heat of liquid, kJ/kg·K	$0.9741 + 4.07 \times 10^{-3}T$	366–478
Specific heat of vapor, kJ/kg·K	$0.63274 + 3.7145 \times 10^{-3}T$	366–589
Thermal conductivity, vapor, W/m·K	$-1.203 \times 10^{-2} + 7.9963 \times 10^{-5}T$	366–811
Vapor pressure, kPa	$\exp\left(14.0579 - \dfrac{3743.8371}{(T - 93.022)}\right)$	
Viscosity of vapor, N·s/m²	$5.1521 \times 10^{-7} + 1.3052 \times 10^{-8}T$	333–589
Diffusion coefficient[a], fuel-air, cm²/s	$-1.494 \times 10^{-2} + 1.06 \times 10^{-4}T + 3.7703 \times 10^{-7}T^2$	305–466

n-Octane Properties	Equation	Range, K
Density of liquid, kg/m³	$811.8 - 0.31187T - 1.17904 \times 10^{-3}T^2$	322–472
Heat of vaporization, kJ/kg	$372.5 + 0.341T - 1.32 \times 10^{-3}T^2$	255–478
Specific heat of liquid, kJ/kg·K	$0.7141 + 4.95 \times 10^{-3}T$	311–478
Specific heat of vapor, kJ/kg·K	$0.4879 + 4.11 \times 10^{-3}T$	276–500
Thermal conductivity, vapor, W/m·K	$-1.0986 \times 10^{-2} + 8.34 \times 10^{-5}T$	276–528
Vapor pressure, kPa	$\exp\left(13.9271 - \dfrac{3120.2932}{(T - 63.816)}\right)$	
Viscosity of vapor, N·s/m²	$8.04 \times 10^{-7} + 1.823 \times 10^{-8}T$	276–555
Diffusion coefficient[a], fuel-air, cm²/s	$-1.825 \times 10^{-2} + 1.368 \times 10^{-4}T + 5.167 \times 10^{-7}T^2$	305–466

Source: Priem, R. J. "Vaporization of Fuel Drops Including the Heating-up Period," PhD thesis, UW–Madison, 1955.

[a] at 1 atm pressure. T is in units of K.

TABLE A.10
Specifications of Unleaded Gasoline Used in U.S. Emissions Certification

Item	ASTM	Value
Lead (organic), g/L	D3237-06e01	0.013
Distillation Range	D86-07a	
Initial boiling point, °C		24–35[a]
10% point, °C	D86-07a	49–57
50% point, °C		93–110
90% point, °C		149–163
End point, °C, max		213
Total sulfur, mg/kg, max	D1266-07	80
Phosphorous, g/L max.	D3231-07	0.0013
Reid vapor pressure, kPa	D5191-07	60.0–63.4[a,b]
Hydrocarbon composition	D1319-03	
Olefins, m^3/m^3, max.		0.10
Aromatics, m^3/m^3, max.		0.304
Saturates, m^3/m^3		Remainder

Source: Table 1 of CFR 40, Part 1065.710 Gasoline, 30 June 2008.

Note: ASTM is an international society for test standards.

[a] For testing at altitudes above 1219 m the specified volatility range is 52–55.2 kPa and the specified initial boiling point range is 23.9–40.6°C.

[b] For testing unrelated to evaporative emissions control, the specified range is 55.2 – 63.4 kPa.

Appendix B Properties of Air (at 1 atm)

T	u	h	ρ	c_p	$10^5\,\mu$	D_{AB}	$\tilde{k}$	Pr	Sc
300	214.32	300.43	1.177	1.005	1.853	0.21	0.0261	0.711	0.749
400	286.42	401.26	0.882	1.013	2.294	0.34	0.0330	0.703	0.788
500	359.79	503.30	0.706	1.029	2.682	0.50	0.0395	0.699	0.760
600	435.03	607.27	0.588	1.051	3.030	0.68	0.0456	0.698	0.751
700	512.58	713.50	0.505	1.075	3.349	0.89	0.0513	0.702	0.745
800	592.53	822.15	0.441	1.099	3.643	1.11	0.0569	0.703	0.744
900	674.77	933.10	0.392	1.121	3.918	1.35	0.0625	0.703	0.740
1000	759.14	1046.1	0.353	1.141	4.177	1.61	0.0672	0.709	0.735
1100	845.38	1161.1	0.321	1.160	4.44	1.88	0.0732	0.704	0.736
1200	933.28	1277.7	0.294	1.177	4.69	2.17	0.0782	0.704	0.735
1300	1022.7	1395.8	0.271	1.195	4.93	2.47	0.0837	0.704	0.736
1400	1113.3	1515.1	0.252	1.212	5.17	2.79	0.0891	0.703	0.736
1500	1205.1	1635.6	0.235	1.230	5.40	3.13	0.0946	0.702	0.734
1600	1297.9	1757.1	0.221	1.248	5.63	3.49	0.100	0.703	0.730
1700	1391.6	1879.5	0.208	1.266	5.85	3.86	0.105	0.703	0.728
1800	1486.1	2002.7	0.196	1.286	6.07	4.26	0.111	0.703	0.727
1900	1581.3	2126.7	0.186	1.307	6.29	4.67	0.117	0.703	0.724
2000	1677.2	2251.2	0.176	1.331	6.50	5.10	0.124	0.698	0.724
2100	1773.7	2376.4	0.168	1.359	6.72	5.53	0.131	0.696	0.723
2200	1870.7	2502.1	0.160	1.392	6.93	5.97	0.139	0.694	0.725
2300	1968.3	2628.4	0.153	1.434	7.14	6.42	0.149	0.687	0.727
2400	2066.3	2755.2	0.147	1.487	7.35	6.88	0.161	0.679	0.727
2500	2164.8	2882.3	0.141	1.556	7.57	7.36	0.175	0.673	0.729

Source: Data is from Keenan, J., H., Chao, J., and Kaye, J., *Gas Tables*, Wiley, New York, 1983; Kays, W. M. and Crawford, M. E., Table A-1 in *Convective Heat and Mass Transfer*, McGraw-Hill, New York, 1980; Field, M. A., App. Q. in *Combustion of Pulverised Coal*, Cheney & Sons, 1967.

Note: For other pressures multiply ρ by the pressure in atmospheres; divide D_{AB} by the pressure in atmospheres; c_p, μ, $\tilde{k}$ and Pr do not change with pressure. D_{AB} is the binary diffusion coefficient for O_2 into N_2. T(K), u(kJ/kg), h(kJ/kg), ρ(kg/m^3), c_p(kJ/kg·K), μ(kg/m·s), D_{AB} (cm^2/s), $\tilde{k}$(W/m·K).

Appendix C Thermodynamic Properties of Combustion Products

(C (gas), C (graphite), CO_2, CO, H_2, H, OH, CH_4, NO, N_2, N, NO_2, O_2, O, H_2O)

These data are derived from the Stull, D. R. and Prophet, H., *JANAF Thermochemical Tables*, 2nd ed., NSRDS–NBS 37, 1971, National Bureau of Standards, Washington, DC. The absolute enthalpy of a substance is its enthalpy relative to stable elements at the reference state (T_0, 1 atm), which is taken as zero. The absolute entropy of a substance is its entropy relative to 0 K, 1 atm, which is taken as zero. Absolute properties at a pressure of 1 atm are denoted by the superscript °. The reference temperature for these tables is $T_0 = 298$ K. At other temperatures and pressures, absolute enthalpy is given by

$$\hat{h}(T) = \Delta\hat{h}°(T_0) + \hat{h}_s$$

where

$$\hat{h}_s = \int_{T_0}^{T} \hat{c}_p(T)\, dT$$

The absolute entropy is given by

$$\hat{s}(T,p) = \hat{s}°(T) - \hat{R}\, \ln(p/p_0)$$

where

$$\hat{s}°(T) = \int_{T_0}^{T} \frac{\hat{c}_p(T)\, dT}{T}$$

Values of $\hat{g}°/\hat{R}T$ are also tabulated to assist in equilibrium computations.

Carbon (C, gas)

T (K)	$\hat{c}_p$ (kJ/kgmol·K)	$\hat{h}$ (MJ/kgmol)	$\hat{s}°$ (kJ/kgmol·K)	$\hat{g}°/\hat{R}T$
200	20.91	−2.05	149.66	410.753
293	20.84	−0.10	157.66	274.349
298	20.84	0.00	157.99	269.432
400	20.82	2.12	164.11	195.891
500	20.81	4.20	168.76	152.707
600	20.80	6.28	172.55	123.835
700	20.79	8.36	175.75	103.151
800	20.79	10.44	178.53	87.593
900	20.79	12.52	180.98	75.458
1000	20.79	14.60	183.17	65.722
1100	20.79	16.68	185.15	57.733
1200	20.79	18.76	186.96	51.057
1300	20.80	20.84	188.63	45.392
1400	20.80	22.92	190.17	40.523
1500	20.82	25.00	191.60	36.291
1600	20.83	27.08	192.95	32.577
1700	20.85	29.17	194.21	29.291
1800	20.88	31.25	195.41	26.362
1900	20.91	33.34	196.54	23.734
2000	20.95	35.43	197.61	21.362
2100	21.00	37.53	198.63	19.210
2200	21.05	39.64	199.61	17.248
2300	21.11	41.74	200.55	15.452
2400	21.18	43.86	201.45	13.801
2500	21.24	45.98	202.31	12.278

$\Delta\hat{h}°(298 \text{ K}) = 715.00$ MJ/kgmol.

Carbon (C, graphite)

T (K)	$\hat{c}_p$ (kJ/kgmol·K)	$\hat{h}$ (MJ/kgmol)	$\hat{s}°$ (kJ/kgmol·K)	$\hat{g}°/\hat{R}T$
200	5.03	−0.67	3.01	−0.765
293	8.35	−0.04	5.54	−0.684
298	8.53	0.00	5.69	−0.684
400	11.93	1.05	8.68	−0.73
500	14.63	2.38	11.65	−0.828
600	16.89	3.96	14.52	−0.952
700	18.58	5.74	17.26	−1.090
800	19.83	7.66	19.83	−1.233
900	20.79	9.70	22.22	−1.377
1000	21.54	11.82	24.45	−1.520
1100	22.19	14.00	26.53	−1.660
1200	22.72	16.25	28.49	−1.798
1300	23.12	18.54	30.33	−1.932
1400	23.45	20.87	32.05	−2.062
1500	23.72	23.23	33.68	−2.188
1600	23.94	25.61	35.22	−2.310
1700	24.12	28.02	36.67	−2.429
1800	24.28	30.44	38.06	−2.543
1900	24.42	32.87	39.38	−2.655
2000	24.54	35.32	40.63	−2.763
2100	24.65	37.78	41.83	−2.868
2200	24.74	40.25	42.98	−2.969
2300	24.84	42.73	44.08	−3.068
2400	24.92	45.22	45.14	−3.163
2500	25.00	47.71	46.16	−3.256

$\Delta\hat{h}°(298 \text{ K}) = 0.00 \text{ MJ/kgmol}.$

Carbon Dioxide (CO$_2$)

T (K)	$\hat{c}_p$ (kJ/kgmol $\cdot$ K)	$\hat{h}$ (MJ/kgmol)	$\hat{s}°$ (kJ/kgmol $\cdot$ K)	$\hat{g}°/\hat{R}T$
200	32.36	−3.41	199.87	−262.746
293	36.88	−0.19	213.09	−187.161
298	37.13	0.00	213.69	−184.449
400	41.33	4.01	225.22	−144.210
500	44.63	8.31	234.81	−120.904
600	47.32	12.92	243.20	−105.546
700	49.56	17.76	250.66	−94.712
800	51.43	22.82	257.41	−86.693
900	53.00	28.04	263.56	−80.542
1000	54.31	33.41	269.22	−75.693
1100	55.41	38.89	274.45	−71.784
1200	56.34	44.48	279.31	−68.577
1300	57.14	50.16	283.85	−65.907
1400	57.80	55.91	288.11	−63.657
1500	58.38	61.71	292.11	−61.739
1600	58.89	67.58	295.90	−60.091
1700	59.32	73.49	299.48	−58.662
1800	59.70	79.44	302.88	−57.416
1900	60.05	85.43	306.12	−56.322
2000	60.35	91.45	309.21	−55.356
2100	60.62	97.50	312.16	−54.499
2200	60.86	103.57	314.99	−53.737
2300	61.09	109.67	317.70	−53.054
2400	61.29	115.79	320.30	−52.443
2500	61.47	121.93	322.81	−51.892

$\Delta\hat{h}°(298 \text{ K}) = -393.52$ MJ/kgmol.

Carbon Monoxide (CO)

T (K)	$\hat{c}_p$ (kJ/kgmol·K)	$\hat{h}$ (MJ/kgmol)	$\hat{s}°$ (kJ/kgmol·K)	$\hat{g}°/\hat{R}T$
200	29.11	−2.86	185.92	−90.549
293	29.14	−0.15	197.08	−69.111
298	29.14	0.00	197.54	−68.347
400	29.34	2.97	206.12	−57.132
500	29.79	5.93	212.72	−50.746
600	30.44	8.94	218.20	−46.608
700	30.17	12.02	222.95	−43.741
800	31.90	15.18	227.16	−41.658
900	32.58	18.40	230.96	−40.091
1000	33.18	21.69	234.42	−38.881
1100	33.71	25.03	237.61	−37.927
1200	34.17	28.43	240.56	−37.163
1300	34.57	31.87	243.32	−36.543
1400	34.92	35.34	245.89	−36.034
1500	35.22	38.85	248.31	−35.613
1600	35.48	42.38	250.59	−35.262
1700	35.71	45.94	252.75	−34.969
1800	35.91	49.52	254.80	−34.722
1900	36.09	53.12	256.74	−34.514
2000	36.25	56.74	258.60	−34.338
2100	36.39	60.38	260.37	−34.188
2200	36.52	64.02	262.06	−34.062
2300	36.64	67.68	263.69	−33.956
2400	36.74	71.35	265.25	−33.867
2500	36.84	75.02	266.76	−33.792

$\Delta \hat{h}°(298 \text{ K}) = -110.53$ MJ/kgmol.

Hydrogen (H$_2$)

T (K)	$\hat{c}_p$ (kJ/kgmol · K)	$\hat{h}$ (MJ/kgmol)	$\hat{s}^\circ$ (kJ/kgmol · K)	$\hat{g}^\circ/\hat{R}T$
200	27.27	−2.77	119.33	−16.018
293	28.80	−0.14	130.10	−15.707
298	28.84	0.00	130.57	−15.705
400	29.18	2.96	139.11	−15.841
500	29.26	5.88	145.63	−16.100
600	29.33	8.81	150.97	16.391
700	29.44	11.75	155.50	16.684
800	29.65	14.70	159.44	−16.966
900	29.91	17.68	162.95	−17.236
1000	30.20	20.69	166.11	−17.491
1100	30.54	23.72	169.01	−17.734
1200	30.92	26.79	171.68	−17.963
1300	31.34	29.91	174.17	−18.181
1400	31.80	33.06	176.51	−18.389
1500	32.43	36.27	178.72	−18.588
1600	32.73	39.52	180.82	−18.777
1700	33.14	42.81	182.82	−18.959
1800	33.54	46.15	184.72	−19.134
1900	33.92	49.52	186.55	−19.302
2000	34.29	52.93	188.30	−19.464
2100	34.64	56.38	189.98	−19.621
2200	34.97	59.86	191.60	−19.772
2300	35.29	63.37	193.16	−19.918
2400	35.59	66.91	194.67	−20.060
2500	35.88	70.49	196.13	−20.198

$\Delta \hat{h}^\circ(298 \text{ K}) = 0.0$ MJ/kgmol.

Hydrogen, Monotomic (H)

T (K)	$\hat{c}_p$ (kJ/kgmol · K)	$\hat{h}$ (MJ/kgmol)	$\hat{s}°$ (kJ/kgmol · K)	$\hat{g}°/\hat{R}T$
200	20.79	−2.04	106.31	117.077
293	20.79	−0.1	114.28	75.648
298	20.79	0	114.61	74.152
400	20.79	2.12	120.72	51.663
500	20.79	4.2	125.36	38.369
600	20.79	6.28	129.15	29.422
700	20.79	8.35	132.35	22.971
800	20.79	10.43	135.13	18.089
900	20.79	12.51	137.57	14.257
1000	20.79	14.59	139.76	11.163
1100	20.79	16.67	141.75	8.609
1200	20.79	18.75	143.55	6.462
1300	20.79	20.82	145.22	4.628
1400	20.79	22.9	146.76	3.044
1500	20.79	24.98	148.19	1.658
1600	20.79	27.06	149.53	0.436
1700	20.79	29.14	150.8	−0.653
1800	20.79	31.22	151.98	−1.628
1900	20.79	33.3	153.11	−2.508
2000	20.79	35.38	154.17	−3.306
2100	20.79	37.46	155.18	−4.035
2200	20.79	39.53	156.16	−4.703
2300	20.79	41.61	157.08	−5.317
2400	20.79	43.69	157.96	−5.885
2500	20.79	45.77	158.81	−6.412

$\Delta\hat{h}°$(298 K) = 217.99 MJ/kgmol.

Hydroxyl (OH)

T	$\hat{c}_p$	$\hat{h}$	$\hat{s}°$	$\hat{g}°/\hat{R}T$
(K)	(kJ/kgmol·K)	(MJ/kgmol)	(kJ/kgmol·K)	
200	30.78	−2.97	171.48	1.318
293	30.01	−0.15	183.13	−5.896
298	29.99	0.00	183.59	−6.162
400	29.65	3.03	192.36	−10.357
500	29.52	5.99	198.95	−12.995
600	29.53	8.94	204.33	−14.873
700	29.66	11.90	208.89	−16.299
800	29.92	14.88	212.87	−17.433
900	30.26	17.89	216.41	−18.365
1000	30.68	20.93	219.62	−19.151
1100	31.12	24.02	222.57	−19.827
1200	31.59	27.16	225.30	−20.420
1300	32.05	30.34	227.84	−20.945
1400	32.49	33.57	230.23	−21.417
1500	32.92	36.84	232.49	−21.844
1600	33.32	40.15	234.63	−22.235
1700	33.69	43.50	236.66	−22.594
1800	34.05	46.89	238.59	−22.927
1900	34.37	50.31	240.44	−23.236
2000	34.67	53.76	242.22	−23.526
2100	34.95	57.24	243.91	−23.798
2200	35.21	60.75	245.54	−24.054
2300	35.45	64.28	247.12	−24.296
2400	35.67	67.84	248.63	−24.526
2500	35.84	71.42	250.09	−24.745

$\Delta\hat{h}°(298 \text{ K}) = 39.46 \text{ MJ/kgmol}$.

Methane (CH$_4$)

T (K)	$\hat{c}_p$ (kJ/kgmol·K)	$\hat{h}$ (MJ/kgmol)	$\hat{s}°$ (kJ/kgmol·K)	$\hat{g}°/\hat{R}T$
200	33.48	−3.37	172.47	−67.796
293	35.48	−0.18	185.58	−53.113
298	35.64	0.00	186.15	−52.593
400	40.50	3.86	197.25	−45.076
500	46.34	8.20	206.91	−40.924
600	52.23	13.13	215.88	−38.342
700	57.79	18.64	224.35	−36.647
800	62.93	24.67	232.41	−35.501
900	67.60	31.20	240.10	−34.714
1000	71.80	38.18	247.45	−34.175
1100	75.53	45.55	254.47	−33.812
1200	78.83	53.27	261.18	−33.579
1300	81.75	61.30	267.61	−33.442
1400	84.31	69.61	273.76	−33.379
1500	86.56	78.15	279.66	−33.373
1600	88.54	86.91	285.31	−33.411
1700	90.29	95.86	290.73	−33.483
1800	91.83	104.96	295.93	−33.583
1900	93.19	114.21	300.94	−33.705
2000	94.40	123.60	305.75	−33.844
2100	95.48	133.09	310.38	−33.997
2200	96.44	142.69	314.85	−34.161
2300	97.30	152.37	319.15	−34.333
2400	98.08	162.14	323.31	−34.513
2500	98.78	171.99	327.33	−34.697

$\Delta\hat{h}°(298 \text{ K}) = -74.87$ MJ/kgmol.

Nitric Oxide (NO)

T (K)	$\hat{c}_p$ (kJ/kgmol·K)	$\hat{h}$ (MJ/kgmol)	$\hat{s}°$ (kJ/kgmol·K)	$\hat{g}°/\hat{R}T$
200	30.42	−2.95	198.64	28.633
293	29.85	−0.15	210.19	11.703
298	29.84	0.00	210.65	11.087
400	29.94	3.04	219.43	1.672
500	30.49	6.06	226.16	−4.024
600	31.24	9.15	231.78	−7.944
700	32.03	12.31	236.66	−10.835
800	32.77	15.55	240.98	−13.072
900	33.42	18.86	244.88	−14.867
1000	33.99	22.23	248.43	−16.347
1100	34.47	25.65	251.70	−17.596
1200	34.88	29.12	254.71	−18.667
1300	35.23	32.63	257.52	−19.601
1400	35.53	36.17	260.14	−20.424
1500	35.78	39.73	262.60	−21.159
1600	36.00	43.32	264.92	−21.819
1700	36.20	46.93	267.11	−22.418
1800	36.37	50.56	269.18	−20.964
1900	36.51	54.20	271.15	−23.465
2000	36.65	57.86	273.03	−23.929
2100	36.77	61.53	274.82	−24.358
2200	36.87	65.22	276.53	−24.758
2300	36.97	68.91	278.17	−25.132
2400	37.06	72.61	279.75	−25.483
2500	37.14	76.32	281.26	−25.813

$\Delta\hat{h}°(298 \text{ K}) = 90.29$ MJ/kgmol.

Nitrogen (N₂)

T (K)	$\hat{c}_p$ (kJ/kgmol · K)	$\hat{h}$ (MJ/kgmol)	$\hat{s}°$ (kJ/kgmol · K)	$\hat{g}°/\hat{R}T$
200	29.11	−2.86	179.88	−23.353
293	29.12	−0.15	191.04	−23.037
298	29.12	0.00	191.50	−23.033
400	29.25	2.97	200.07	−23.170
500	29.58	5.91	206.63	−23.430
600	30.00	8.89	212.07	−23.724
700	30.75	11.94	216.76	−24.019
800	31.43	15.05	220.91	−24.307
900	32.09	18.22	224.65	−24.584
1000	32.70	21.46	228.06	−24.848
1100	33.24	24.76	231.20	−25.101
1200	33.73	28.11	234.12	−25.341
1300	34.15	31.50	236.83	−25.570
1400	34.53	34.94	239.38	−25.789
1500	34.85	38.40	241.77	−25.999
1600	35.14	41.90	244.03	−26.200
1700	35.39	45.43	246.17	−26.393
1800	35.61	48.98	248.19	−26.579
1900	35.81	52.55	250.13	−26.757
2000	35.99	56.14	251.97	−26.929
2100	36.14	59.75	253.73	−27.095
2200	36.28	63.37	255.41	−27.255
2300	36.41	67.01	257.03	−27.410
2400	36.53	70.65	258.58	−27.560
2500	36.64	74.31	260.07	−27.705

$\Delta\hat{h}°(298\ \text{K}) = 0.00\ \text{MJ/kgmol.}$

Nitrogen Dioxide (NO$_2$)

T (K)	$\hat{c}_p$ (kJ/kgmol · K)	$\hat{h}$ (MJ/kgmol)	$\hat{s}°$ (kJ/kgmol · K)	$\hat{g}°/\hat{R}T$
200	34.38	−3.49	225.74	−9.350
293	36.82	−0.19	239.34	−15.284
298	36.97	0.00	239.92	−15.506
400	40.17	3.93	251.23	−19.084
500	43.21	8.10	260.53	−21.426
600	45.84	12.56	268.65	−23.160
700	47.99	17.25	275.88	−24.531
800	49.71	22.14	282.40	−25.662
900	51.08	27.18	288.34	−26.625
1000	52.17	32.34	293.78	−27.464
1100	53.04	37.61	298.80	−28.207
1200	53.75	42.95	303.44	−28.874
1300	54.33	48.35	307.77	−29.481
1400	54.81	53.81	311.81	−30.037
1500	55.20	59.31	315.61	−30.550
1600	55.53	64.85	319.18	−31.027
1700	55.81	70.42	322.56	−31.472
1800	56.06	76.01	325.75	−31.890
1900	56.26	81.63	328.79	−32.283
2000	56.44	87.26	331.68	−32.655
2100	56.60	92.91	334.44	−33.008
2200	56.74	98.58	337.08	−33.343
2300	56.85	104.26	339.60	−33.662
2400	56.96	109.95	342.02	−33.968
2500	57.05	115.65	344.35	−34.260

$\Delta \hat{h}°(298 \text{ K}) = 217.99$ MJ/kgmol.

Nitrogen, Monotomic (N)

T (K)	$\hat{c}_p$ (kJ/kgmol·K)	$\hat{h}$ (MJ/kgmol)	$\hat{s}°$ (kJ/kgmol·K)	$\hat{g}°/\hat{R}T$
200	20.79	−2.04	144.90	265.669
293	20.79	−0.10	152.86	175.550
298	20.79	0.00	153.19	172.301
400	20.79	2.12	159.30	123.639
500	20.79	4.20	163.94	95.021
600	20.79	6.28	167.73	75.859
700	20.79	8.35	170.94	62.111
800	20.79	10.43	173.71	51.756
900	20.79	12.51	176.16	43.668
1000	20.79	14.59	178.35	37.169
1100	20.79	16.67	180.33	31.829
1200	20.79	18.75	182.14	27.360
1300	20.79	20.82	183.80	23.562
1400	20.79	22.90	185.34	20.293
1500	20.79	24.98	186.78	17.448
1600	20.79	27.06	188.12	14.949
1700	20.79	29.14	189.38	12.734
1800	20.79	31.22	190.57	10.757
1900	20.79	33.30	191.69	8.981
2000	20.79	35.38	192.76	7.376
2100	20.79	37.46	193.77	5.918
2200	20.80	39.53	194.74	4.587
2300	20.80	41.61	195.66	3.366
2400	20.82	43.70	196.55	2.243
2500	20.83	45.78	197.40	1.206

$\Delta\hat{h}°(298 \text{ K}) = 472.79 \text{ MJ/kgmol}$.

Oxygen (O$_2$)

T (K)	$\hat{c}_p$ (kJ/kgmol · K)	$\hat{h}$ (MJ/kgmol)	$\hat{s}°$ (kJ/kgmol · K)	$\hat{g}°/\hat{R}T$
200	29.12	−2.87	193.38	−24.982
293	29.35	−0.15	204.56	−24.664
298	29.37	0.00	205.03	−24.660
400	30.11	3.03	213.76	−24.800
500	31.09	6.09	220.59	−25.067
600	32.09	9.25	226.35	−25.370
700	32.98	12.50	231.36	−25.679
800	33.74	15.84	235.81	−25.981
900	34.36	19.25	239.83	−26.273
1000	34.88	22.71	243.48	−26.553
1100	35.31	26.22	246.82	−26.819
1200	35.68	29.76	249.91	−27.074
1300	36.00	33.35	252.78	−27.317
1400	36.29	36.97	255.45	−27.549
1500	36.56	40.61	257.97	−27.771
1600	36.82	44.28	260.34	−27.983
1700	37.06	47.97	262.58	−28.187
1800	37.30	51.69	264.70	−28.383
1900	37.54	55.43	266.73	−28.571
2000	37.78	59.20	268.65	−28.752
2100	38.01	62.99	270.50	−28.927
2200	38.24	66.80	272.28	−29.096
2300	38.47	70.63	273.98	−29.259
2400	38.69	74.49	275.63	−29.418
2500	38.92	78.37	277.21	−29.570

$\Delta\hat{h}°(298\ \text{K}) = 0.00$ MJ/kgmol.

Oxygen, Monotomic (O)

T (K)	$\hat{c}_p$ (kJ/kgmol·K)	$\hat{h}$ (MJ/kgmol)	$\hat{s}°$ (kJ/kgmol·K)	$\hat{g}°/\hat{R}T$
200	22.74	-2.19	152.05	130.256
293	21.95	-0.11	160.61	82.879
298	21.91	0.00	160.95	81.168
400	21.48	2.21	167.32	55.469
500	21.26	4.34	172.09	40.290
600	21.13	6.46	175.95	30.085
700	21.04	8.57	179.20	22.735
800	20.98	10.67	182.01	17.178
900	20.95	12.77	184.48	12.820
1000	20.92	14.86	186.69	9.306
1100	20.89	16.95	188.68	6.407
1200	20.88	19.04	190.49	3.974
1300	20.87	21.13	192.16	1.897
1400	20.85	23.21	193.71	0.104
1500	20.84	25.30	195.15	-1.462
1600	20.84	27.38	196.49	-2.843
1700	20.83	29.46	197.76	-4.070
1800	20.83	31.55	198.95	-5.170
1900	20.83	33.63	200.07	-6.160
2000	20.83	35.71	201.14	-7.059
2100	20.83	37.80	202.16	-7.877
2200	20.83	39.88	203.13	-8.627
2300	20.84	41.96	204.05	-9.317
2400	20.84	44.04	204.94	-9.954
2500	20.85	46.13	205.79	-10.543

$\Delta\hat{h}°(298\ K) = 249.19$ MJ/kgmol.

Water Vapor (H_2O)

T (K)	$\hat{c}_p$ (kJ/kgmol·K)	$\hat{h}$ (MJ/kgmol)	$\hat{s}°$ (kJ/kgmol·K)	$\hat{g}°/\hat{R}T$
200	33.34	−3.28	175.38	−168.494
293	33.56	−0.17	188.19	−121.920
298	33.58	0.00	188.72	−120.252
400	34.25	3.45	198.67	−95.572
500	35.21	6.92	206.41	−81.333
600	36.30	10.50	212.93	−71.982
700	37.46	14.18	218.61	−65.407
800	38.69	17.99	223.69	−60.557
900	39.94	21.92	228.32	−56.849
1000	41.22	25.98	232.60	−53.937
1100	42.48	30.17	236.58	−51.598
1200	43.70	34.48	240.33	−49.689
1300	44.99	38.90	243.88	−48.107
1400	45.97	43.45	247.24	−46.780
1500	47.00	48.10	250.45	−45.657
1600	47.96	52.84	253.51	−44.697
1700	48.84	57.68	256.45	−43.872
1800	49.66	62.61	259.26	−43.158
1900	50.41	67.61	261.97	−42.536
2000	51.10	72.69	264.57	−41.993
2100	51.74	77.83	267.08	−41.516
2200	52.32	83.04	269.50	−41.095
2300	52.86	88.29	271.84	−40.724
2400	53.36	93.60	274.10	−40.395
2500	53.82	98.96	276.29	−40.103

$\Delta\hat{h}°(298 \text{ K}) = -241.83$ MJ/kgmol.

Appendix D Historical Perspective on Combustion Technology

A brief historical perspective on steam boilers, internal combustion engines, and gas turbines is given in this appendix.

D.1 STEAM BOILERS

Early steam boilers consisted of little more than a kettle of water heated from the bottom. Boilers in the 1700s used the kettle principle but burned the fuel in an enclosed furnace to direct more of the heat to the boiler kettle. To improve efficiency, an integral furnace was developed by 1750 where the fuel was burned in a container enclosed within the water vessel. Flue gas wound through the water vessel to the atmosphere much like a coil in a still. A bellows was used to force air to the combustion zone.

As the demand for power increased, the single flue was replaced by many gas tubes that increased the heating surface. If the flue gas flows inside the tubes, this type of design is called a *fire tube boiler*. The so-called Lancashire boiler designed by William Fairbain in 1845 is an example of a fire tube boiler. However, many disastrous explosions resulted from direct heating of the pressure shell that contained large amounts of saturated water. For example, in 1880 in the United States 259 people were killed and 555 injured by boiler explosions.

The idea that water, instead of flue gas, could flow through the inside of the tubes with heated gas outside, which would boost capacity and lead to safer operation, occurred to a number of early engineers, and the first *water tube boiler* was patented by an American, James Ramsey, in 1788. These early water tube designs were not successful because of construction problems, steam leaks, and internal deposits. It was not until 1856 that a truly successful water tube boiler was designed by Stephen Wilcox. The Wilcox boiler had improved water circulation and increased surface area due to inclined water tubes that connected water spaces at the front and rear and a steam chamber above. His inherently safe design revolutionized the boiler industry. In 1866 Wilcox joined with George Babcock, and their company grew rapidly.

The first commercial steam turbine (4 kW) began operation in 1891 due to the pioneering work of Gustaf DeLaval of Sweden. In the United States, the first steam turbine–electric generator (400 kW) went into operation due to the efforts of George Westinghouse. Steam turbines rapidly gained acceptance over steam engines, and in

GEORGE HERMAN BABCOCK (1832–1893) AND STEPHEN WILCOX (1839–1893); AMERICAN ENGINEERS

George Babcock spent one year at a technical institute in DeRuyter, New York, before going to work as a printer. In 1855, he patented (with his father) a polychromatic printing press, which won a prize in the Crystal Palace Exhibition in London. He also patented a bronzing machine and founded the journal *Literary Echo*. In 1859, Babcock sold his business and became a patent solicitor in Brooklyn. In the early 1860s, he was chief draftsman of the Hope Iron Works in Providence, Rhode Island, and in the evenings he taught mechanical drawing at Cooper Union Academy. In 1866, Babcock became the sixth president of the American Society of Mechanical Engineers, and in 1881, the first President of the Babcock & Wilcox Co., a position he held until his death.

Stephen Wilcox patented a letting-off motion for looms in 1853 and patented his first steam boiler in 1856 at the age of 17. In the 1860s he worked at the Hope Iron Works and became acquainted with Babcock. Wilcox was vice president of Babcock & Wilcox Co. from its incorporation until his death in 1893.

1903 Chicago became the first city to have a central power station designed exclusively for steam turbines (5 MW each). By the early 1920s, steam temperatures were about 300°C and steam pressures were 13–20 atm. By 1929 an 80 atm boiler was built. Today 240 atm steam turbines are common and engineers contemplate higher pressure steam with temperatures of 650°C as a means of increasing the thermodynamic efficiency.

Fuel firing systems underwent a similar evolution. The hand-stoked furnace was replaced by the automatic spreader-stoker in 1822, and the traveling grate stoker was invented in 1833. Pulverized coal firing advanced rapidly in the 1920s, as did stoker firing. Suspension firing became dominant in central power station boilers beginning in the 1930s. With the need to control sulfur dioxide emissions, the first fluidized bed combustor for coal was introduced commercially in 1976.

Because of the ubiquitous nature of furnaces and boilers, smoke control was a major concern in cities by the 1920s, if not before. Cyclone collectors began to be used for particulate control in the 1930s. The electrostatic precipitator was invented in 1910 by F. G. Cottrell and has been used for industrial and utility particulate control starting in the 1950s, as have fabric filters. In the 1970s, control of large sources of particulates, carbon monoxide, nitrogen dioxide, and sulfur dioxide was mandated in the United States by the Clean Air Act.

D.2   SPARK IGNITION ENGINES

The idea of driving the piston by the burning products expansion was first utilized in a production engine of 1/2 hp by Lenoir (1860) who ignited the mixture during the intake stroke by use of an electric spark. The idea of compression before ignition was not considered practical at that time. The Lenoir engine was powered with producer gas made from coal. In 1876, Nikolaus Otto built a 3 hp compression-ignition engine based on the four-stroke concept proposed by Beau de Rochas in 1862, and the basic cycle persists today. In 1878, Dugald Clark developed a two-stroke engine in order to obtain a higher power from the same size engine.

Combustion in the spark-ignited homogeneous charge engine was at first dictated by mechanical considerations of strength, metal temperature, and lubrication constraints. During the era of cheap energy and prior to air pollution regulations, engine design evolved to higher and higher compression ratios using slightly richer than stoichiometric mixtures. Knocking combustion (explosion of the unburned gas giving pressure pulsations) was a serious problem, but was partially overcome by the use of tetraethyl lead (1923) which, when added to gasoline in small quantities, reduces the tendency to knock. Improvements in the quality of gasoline, lubricating oil, and metals allowed compression ratios to rise from the modest 3.6:1 of the 1915 Model T to 8:1 in engines of the early 1960s. The increase in compression ratio might have continued, but new considerations of air pollution emissions, particularly nitric oxide (NO), were to halt this development. Catalysts were developed to reduce the emissions of NO, CO, and unburned hydrocarbons, and these catalysts were poisoned by lead in the gasoline. Thus, considerations of knock again reduced the compression ratio because tetraethyl lead could no longer be used as a fuel additive.

With the growth in urban centers, industrial and automotive sources of air pollution became a serious concern. A most notable problem occurred first in southern California where combustion emissions produced a heavy and health-threatening haze called smog. The term "smog" (smoke plus fog) originated in the early 1900s in Great Britain where burning of high sulfur coal plus natural fog produced deadly sulfuric acid aerosol. The smog in Los Angeles was quite different, however. In 1952, Professor Haagen-Smit from the California Institute of Technology showed that smog could be produced in the laboratory from automobile exhaust plus sunlight. The necessary ingredients for this type of smog are hydrocarbons, nitrogen oxides, air, and strong sunlight. Photochemical and chemical reactions in the atmosphere produce oxidants (ozone, nitrogen dioxide, and peroxyactyl nitrate) and photochemical aerosol, which irritate the eyes and lungs. Due to public demand, federal emission controls for automobiles began with the 1972 models in the United States.

The history of the development of gasoline parallels the history of the Otto (spark ignition) engine and includes some of the events mentioned above. Lead alkyl additive

NIKLAUS AUGUST OTTO; GERMAN ENGINEER; 1832–1891

In spite of a childhood interest in science and engineering, Otto left school at sixteen to become a clerk. However, Otto spent all his spare time and money studying engineering, and had already been inspired by news of an engine built by Lenoir. In 1861, at age 29, Otto built a small four-stroke internal combustion engine. The engine ran extremely roughly because of explosions. Nevertheless he obtained the backing of a wealthy industrialist and formed the Niklaus Otto Company. At the 1867 Paris Exhibition, he won a gold medal in competition with 14 French engines. In 1877, fifteen years after his first attempt to produce a four-stroke engine, Otto produced and patented an 8 hp engine that was a tremendous success. Others sought ways to exploit the new idea, and in 1886 the patent was invalidated. However, over 3000 engines were sold by this time, and the firm had plenty of work modifying the engines to run on liquid fuels. Otto died in 1891, a modest, retiring, and yet truly great engineer who developed the prototype for all modern internal combustion engines.

to prevent knock was introduced in 1923, and its use reached a maximum in 1970. Unleaded gasoline was required in 1974; phase-down of lead began in 1980; a total ban on lead was required by 1996. Although octane quality continues to be a major concern, starting in the 1970s environmental concerns also began to influence gasoline properties. The use of exhaust after-treatment catalysts required the use of non-leaded gasoline, the need to reduce unburned hydrocarbon emissions led to volatility controls, the desire to reduce CO in urban areas led to the use of oxygenates, and the current reformulation efforts are driven by continuing environmental concerns.

D.3 COMPRESSION IGNITION ENGINES

In 1893, a German patent was granted to Rudolf Diesel for the design of a "rational heat engine." Diesel planned a four-stroke engine that would incorporate the constant temperature energy addition of the Carnot cycle. In his first engine, ammonia was injected into the cylinder at the end of the compression stroke to avoid premature ignition. Diesel's second engine (1896) had a water-jacketed cylinder and a pump to supply air to the cylinder to reduce exhaust smoke. In 1898, the first production engine ran on kerosene and gave an amazing 20 hp. By 1901, the external cross-head arrangement used in the original engine design was replaced by a trunk piston design innovated by the American Diesel Company. The M.A.N. Company built a 186 kW (250 hp) two cylinder engine in 1902, and the Sulzers Co. built a three cylinder engine of 225 kW (300 hp) in 1906. Because such engines could not be directly reversed, marine applications used an electric drive for slow speed and maneuvering. A steam engine with oil-fired boilers and a diesel engine were shown side-by-side at the Turin Exhibition of 1911. The steam engine used several times more fuel per horsepower than the diesel engine. This greatly improved fuel economy led to the demise of steam engines.

RUDOLF CHRISTIAN KARL DIESEL;
GERMAN ENGINEER; 1858–1913

Rudolf Diesel was an outstanding student in mechanical engineering in Munich. Professor Linde, one of Diesel's instructors and a worldwide authority on heat engines and refrigeration, helped him get a job at the famous Sulzers factory in Switzerland, where he became a proficient machinist before becoming a factory manager. He began experimenting with a high pressure ammonia engine. His 1893 prototype blew off the cylinder head, but 4 years later a reasonably reliable engine was produced. His new engine was soon accepted throughout the world, and many of his engines were made under license. His wife convinced him to name the engine after himself. However, Rudolf Diesel's enjoyment of fame and fortune was marred by ill health, probably brought on by exhausting legal battles over patent rights and unwise financial speculations. Diesel was a practical genius, noted for his work on engines in the laboratory and for his study of heat engine cycles, including the constant pressure diesel cycle.

By 1910, German and British companies had developed diesel-powered submarines, and the first diesel-powered passenger ship appeared in 1921. Early four-stroke ship engines experienced considerable problems with fouling of the exhaust valves and ports with carbon. This problem led Sulzers to develop the two-stroke diesel. Engines with 1500 kW (2000 hp) per cylinder were in operation before World War I, and by 1939 half of the world's shipping tonnage was diesel powered.

High-speed diesel engines for commercial, farm, and industrial applications developed slowly because advances were needed in the strength of materials and fuel injection systems. Although the principle of airless injection of liquid fuel was pioneered by Herbert Akroyd Stuart as early as 1886, it was not until 1936 that Robert Bosch introduced an ingenious method of metering that did not require a variable-stroke pump. Equally important, the Bosch Company had the ability to produce the high tolerance machining required to fabricate such systems. Diesel engines have displaced spark ignition engines in nearly all applications except automobiles and aircraft.

D.4   GAS TURBINES

Although the use of flue gases or steam to drive a wheel dates to ancient times, the forerunner of the modern gas turbine can be traced to the patent of John Barber in 1791 that utilized a compressor, combustor, and impulse turbine. Early combustors were typically an explosive, intermittent combustion in a closed space that caused a flow through a nozzle to drive an impulse turbine. Although inefficient, this design persisted because development of continuous flow machines was hampered by a lack of knowledge of aerodynamics, resulting in very inefficient compressors.

The first working gas turbine with a constant pressure combustor was that of Aegidius Elling of Norway. He started working on gas turbines in 1882, and 21 years later Elling achieved a net power output of 8 kW (11 hp) with a six-stage centrifugal compressor and an axial impulse turbine with an inlet temperature of 400°C. In 1905, Frenchmen Charles Lemale and Rene Armengaud used a 25-stage Brown Bovari centrifugal compressor (running at 4000 rpm, absorbing 240 kW (325) hp, and giving a 3:1 pressure ratio), a high-temperature combustor, and a two-stage Curtis turbine. The thermal efficiency was 3.5%. By 1939, efficiency had improved dramatically, and a regenerative axial flow compressor and turbine of Hungarian design gave an efficiency of 21%. The first gas turbine in the United States to generate electric utility power was installed in 1949.

The use of gas turbines in aircraft dates to the 1930 patent of Frank Whittle in England. The technical problems that had to be overcome included making a combustor with about twenty times the combustion intensity of stationary gas turbines; improving the compressor and turbine efficiency; and overcoming the mechanical failures that plagued turbines at that time. Meanwhile, Hans von Ohain, with the backing of aircraft manufacturer Ernst Heinkel, independently pioneered the aircraft gas turbine in pre-war Germany. By 1939, both Whittle and von Ohain were well advanced towards a flying prototype. In August 1939, a Heinkel aircraft with von Ohain's engine flew aloft for 7 min. Two years later in May 1941, a Gloster aircraft with a 3.8 kN (850 lbf) thrust Whittle W1 engine flew aloft for 17 min and during the next 12 days logged 10 h of test flights. A General Electric version of the W1 was built and flown soon after that.

The first gas turbine engine in Japan was developed independently by two naval officers, Tokiyasu Tanegashima and Osamu Nagano, starting in 1943. By late 1944, the Germans were sharing technical turbine information with the Japanese. The first Japanese test flight lasted 12 min, and 8 days later the war ended. By the end of World War II in 1945, both the English/American and the German jet aircraft could

---

FRANK WHITTLE; BRITISH ENGINEER; 1907–1996

In 1922 at age 15 Whittle became an apprentice in the British Royal Air Force (RAF). He graduated from the RAF College (Cranwell) in 1928, from the RAF Officers' School of Engineering in 1934, and from Cambridge University in 1936. From an early age he was fascinated by the idea of jet propulsion, and his 1928 thesis, "Future Developments in Aircraft Design," discussed the topic, which led to a patent in 1930. Unfortunately, Whittle's revolutionary invention failed to impress either the British Air Ministry or private companies, who remained skeptical because gas turbines had a long history of failure. During his senior year at Cambridge in 1936 he formed the Power Jets Co. with financial partners. For Whittle, the years 1936–1940 were a period of great technological highs and lows as the W1 engine was gradually made to operate at higher rpm with higher combustion intensity. Always short of cash because the RAF belittled the concept as impractical, Whittle initially received little recognition. Although the design was shared with the United States and progress was more rapid during the war, the British–American gas turbine-powered aircraft did not have an impact on the war. By the end of the war, Whittle's small company and his patents were swept aside by the tides of history, but a new industry had been launched. After the war, he became a technical adviser and consultant. In 1976, he emigrated to the United States. Whittle was awarded numerous honors.

HANS VON OHAIN; GERMAN PHYSICIST; 1911–1998

In 1935, at age 24, von Ohain received his doctoral degree from the University of Goettingen. Struck by the possibilities of jet propulsion while a student and encouraged by a professor who recognized his genius, one of his early engine designs earned him a job at the Heinkel Aircraft Co. Heinkel's backing allowed von Ohain to progress rapidly, and by 1937 (though entirely unaware of Whittle's work, as was Whittle of his) he successfully tested an engine in his workshop. In 1938, the German Air Ministry directed all private aviation companies to begin the development of jet engines, but a year later, as the country went to war, development efforts were shifted back to propeller-driven aircraft in a monumental lack of foresight that lost Germany its lead in jet engine development and possibly the war. After the war, von Ohain came to the United States and continued his work at Wright-Patterson Air Force Base. By 1975, he was responsible for maintaining the quality of all Air Force research and development in turbojet propulsion, and on his retirement in 1979 he joined the University of Dayton Research Institute. He is the holder of 19 United States patents and many honors.

outperform propeller-piston planes on test flights; however, the engines were not durable and did not play a role in the war. The first commercial gas turbine engines entered service in 1953.

As improvements in overall engine design were made, requirements for the combustor increased from the 3:1 pressure ratio of early days to 18:1 in modern aircraft engines. Combustor temperatures have also increased so that cooling the turbine blades and combustor linings has become much more difficult. Fortunately, advances in film cooling of turbine blades, slot and port cooling of combustors, fuel atomization, and flow modeling have allowed the combustor design to keep pace with system demands.

Sir Frank Whittle and Dr Hans von Ohain were the 1991 recipients of the C.S. Draper Prize from the United States National Academy of Engineering "for engineering innovation and individual tenacity in the development and reduction to practice of the turbojet engine, thereby revolutionizing the world's transportation system, improving the world's economy, and transforming the relationship between nations and their peoples."

REFERENCES

Carvill, J., *Famous Names in Engineering*, Butterworths, London, 1981.

Cummins, L., *Internal Fire: The Internal Combustion Engine, 1673–1900*, Society of Automotive Engineers, Warrendale, PA, 1989.

Cummins, C. L. Jr., *Diesel's Engine: From Conception to 1918*, Vol. 1, Carnot Press, Wilsonville, OR, 1993.

Imanari, K., "First Jet Engine in Japan, NE20," *Global Gas Turbine News* 35(March/April):4–6, ASME International Gas Turbine Institute, 1995.

Jones, G., *The Jet Pioneers: The Birth of Jet Powered Flight,* Methuen, London, 1989.

Kitto, J. B. and Stultz, S. C., eds. *Steam: Its Generation and Use*, 41st ed., The Babcock and Wilcox Co., Barberton, OH, 2005.

Rolt, L. T. C., *The Mechanicals*, Heinemann, London, 1967.

Wilson, D. G., Chap. 1 in *The Design of High-Efficiency Turbomachinery and Gas Turbines*, MIT Press, Cambridge, MA, 1984.

Index

An environmentally friendly book printed and bound in England by www.printondemand-worldwide.com

This book is made entirely of sustainable materials; FSC paper for the cover and PEFC paper for the text pages.

#0198 - 010615 - C0 - 234/156/29 [31] - CB - 9781420092509